MW01517692

Soil Biology

Volume 41

Series Editor
Ajit Varma, Amity Institute of Microbial Technology,
Amity University Uttar Pradesh, Noida, UP, India

More information about this series at
http://www.springer.com/series/5138

GEORGIAN COLLEGE LIBRARY

227.30
GEOB-BK

Zakaria M. Solaiman • Lynette K. Abbott •
Ajit Varma

Editors

Mycorrhizal Fungi: Use in Sustainable Agriculture and Land Restoration

Library Commons
Georgian College
One Georgian Drive
Barrie, ON
L4M 3X9

 Springer

Editors
Zakaria M. Solaiman
Lynette K. Abbott
School of Earth and Environment, and
 UWA Institute of Agriculture
The University of Western Australia
Crawley, Western Australia
Australia

Ajit Varma
Amity Institute of Microbial Technology
Amity University Uttar Pradesh
Noida
India

ISSN 1613-3382   ISSN 2196-4831 (electronic)
ISBN 978-3-662-45369-8   ISBN 978-3-662-45370-4 (eBook)
DOI 10.1007/978-3-662-45370-4
Springer Heidelberg New York Dordrecht London

Library of Congress Control Number: 2014958675

© Springer-Verlag Berlin Heidelberg 2014
This work is subject to copyright. All rights are reserved by the Publisher, whether the whole or part
of the material is concerned, specifically the rights of translation, reprinting, reuse of illustrations,
recitation, broadcasting, reproduction on microfilms or in any other physical way, and transmission or
information storage and retrieval, electronic adaptation, computer software, or by similar or dissimilar
methodology now known or hereafter developed. Exempted from this legal reservation are brief excerpts
in connection with reviews or scholarly analysis or material supplied specifically for the purpose of being
entered and executed on a computer system, for exclusive use by the purchaser of the work. Duplication
of this publication or parts thereof is permitted only under the provisions of the Copyright Law of the
Publisher's location, in its current version, and permission for use must always be obtained from
Springer. Permissions for use may be obtained through RightsLink at the Copyright Clearance Center.
Violations are liable to prosecution under the respective Copyright Law.
The use of general descriptive names, registered names, trademarks, service marks, etc. in this
publication does not imply, even in the absence of a specific statement, that such names are exempt
from the relevant protective laws and regulations and therefore free for general use.
While the advice and information in this book are believed to be true and accurate at the date of
publication, neither the authors nor the editors nor the publisher can accept any legal responsibility for
any errors or omissions that may be made. The publisher makes no warranty, express or implied, with
respect to the material contained herein.

Printed on acid-free paper

Springer is part of Springer Science+Business Media (www.springer.com)

Preface

Arbuscular mycorrhizal fungi are ubiquitous soil organisms that form associations with roots of almost all plant species. They facilitate acquisition of nutrients by plants, contribute to processes associated with soil aggregation, and play understated roles in ecosystem function at various scales. They also participate in rhizosphere processes that protect plants against disease and improve access to water during periods of temporary or persistent water deficit. The effective management of mycorrhizal fungi is often an unrecognised component of sustainable agricultural production that contributes to the profitability of farming systems. During the restoration of disturbed lands, arbuscular mycorrhizal fungi contribute with ectomycorrhizal fungi to re-establishing effective nutrient cycling processes and other essential soil biological functions in ecologically significant plant communities.

At a fundamental level, recent advances in the taxonomy and techniques for recognising and assessing the diversity of arbuscular mycorrhizal fungi offer opportunities for reinvigorating research on the management of mycorrhizas in agricultural and natural ecosystems, including evaluation of their economic value. These advances provide the incentive for promoting knowledge of plant–mycorrhizal interactions in debates about soil and land management, fertiliser decision-making, implications for selection of crop rotations, choice of plant cultivars, maintenance of pastures and grasslands for animal production, and environmental impacts of intensive horticultural production. Although it is difficult to quantify the economic benefits of mycorrhizas, ignoring their roles will lead to failure in capturing their benefits. This will be even more important when the challenges of sustaining agricultural production using limited resources with low environmental impacts are highlighted in the coming years.

Appreciation of arbuscular mycorrhizal fungi as dynamic communities in the very contrasting environments of soil and roots is essential to managing their contributions through agronomic practices or inoculation. Competitive interactions among these fungi during colonisation of roots will influence dominance and function of both naturally occurring and introduced fungi as well as the survival from season to season of those which are most effective. Thus, inclusion of arbuscular mycorrhizal fungi in biofertiliser formulations needs to be based on detailed knowledge of biotic and environmental interactions in space and time. A critical evaluation of the selection, technical production, and the use of inoculant arbuscular mycorrhizal fungi—in addition to the marketing of products containing these fungi—needs to be underpinned by sound comprehension of ecological concepts and principles.

Arbuscular mycorrhizas have the potential to mitigate nutrient loss by soil erosion and leaching, as well as increasing nutrient use efficiency. Renewed evaluations of dominant fertiliser inputs of both phosphorus and nitrogen require consideration of mycorrhizal associations, including avoidance of, or compensation for, negative effects of crop management on these associations. This extends to the role of arbuscular mycorrhizas in acquisition of zinc by plants. Furthermore, as arbuscular mycorrhizas can enhance plant survival and growth in extreme environments, research that highlights the potential for acclimation versus adaptation of mycorrhizal fungi will better inform management decisions in disturbed sites or in sites subject to temporary water deficit, salinity, or heavy metal toxicity.

Finally, an understanding of how roots are colonised by communities of these common soil fungi is essential for capturing their benefits. Predictive models that include spatial variability and soil mapping offer the potential for calibrating the impacts of soil properties and land use practices in sustaining the colonising potential of effective communities of mycorrhizal fungi. The role of mycorrhizas in soil carbon sequestration is of increasing interest, as is the potential for moderating their soil and rhizosphere environment by application of ameliorants such as biochar. However, for communities of arbuscular mycorrhizal fungi, their ubiquity and potential are generally hidden from the majority of land managers and thus overlooked. The intensification of agriculture for food production in the coming decades will benefit from the application of knowledge of molecular, physiological, and ecological function of arbuscular mycorrhizas via practical solutions to their use in sustainable agriculture and land restoration.

Crawley, WA, Australia   Zakaria M. Solaiman
Crawley, WA, Australia   Lynette K. Abbott
Noida, Uttar Pradesh, India   Ajit Varma

Contents

Chapter 1
Use of Mycorrhiza in Sustainable Agriculture and Land Restoration

Zakaria M. Solaiman and Bede Mickan

1.1 Introduction

Arbuscular mycorrhizal (AM) fungi form symbiotic relationships with over 80 % of terrestrial plant species (Brundrett 2002). Arbuscular mycorrhizas are ancient and ubiquitous symbioses formed between a relatively small group of soil fungi and higher plant roots which has been traced back 460 million years (Redecker 2002). The symbiosis is primarily characterised by its association with phosphorus (P) uptake by host plants and the enhancement of water uptake through the extraradical fungal hyphal networks. This symbiosis can also trigger physiological and molecular signals at subcellular levels, alter plant community structure and increase plant tolerance to various abiotic and biotic stresses. Mycorrhizal hyphal networks link plants of the same and different species below ground and are able to transfer resources between plants and release signal molecule defence-related proteins, lipochitooligosaccharides and strigolactones. There have been significant recent advances in the understanding of physiological processes and taxonomy of these fungi (Kohout et al. 2014; Saito 2000). They are obligate symbionts belonging to the phylum Glomeromycota (Redecker et al. 2000). Their activity in agricultural ecosystems is well documented (Abbott and Gazey 1994; Bedini et al. 2013; Pellegrino and Bedini 2014) as is their presence during rehabilitation of forest ecosystems (Brundrett and Nanjappa 2013; Solaiman and Abbott 2003, 2008). The distribution of ectomycorrhizal (ECM) fungi is also widespread, but they form associations with only 3 % of terrestrial plant families (Smith and Read 2008). ECM fungi are members of the phyla Ascomycota and Basidiomycota (Hibbett et al. 2000; Siddiqui et al. 2008). Unlike the ECM fungi, AM fungi are dependent

Z.M. Solaiman (✉) • B. Mickan
Soil Biology and Molecular Ecology Group, School of Earth and Environment (M087),
Institute of Agriculture, The University of Western Australia, Crawley, WA 6009, Australia
e-mail: zakaria.solaiman@uwa.edu.au

© Springer-Verlag Berlin Heidelberg 2014   1
Z.M. Solaiman et al. (eds.), *Mycorrhizal Fungi: Use in Sustainable Agriculture and
Land Restoration*, Soil Biology 41, DOI 10.1007/978-3-662-45370-4_1

on plants for their carbon (C) and when a symbiosis is formed, both ECM and AM fungi can demand 20–40 % of photosynthetically fixed plant C (McNear 2013).

1.2 Molecular Identification and New Taxonomic Classification of AM Fungi

Molecular techniques are applied to the identification of AM fungi under both greenhouse and field conditions, but there are many limitations in relation to the detection methods used as these are multinuclear organisms (Kohout et al. 2014; Schüßler et al. 2001). The diversity of AM fungi identified using molecular techniques has been studied at an ecosystem level (Sanders et al. 1995), for particular plant species (Jansa et al. 2003; Kjøller and Rosendah 2001), during seasonal variation in colonisation of roots of arable crops (Daniell et al. 2001), for coexisting species (Gollotte et al. 2004; Vandenkoornhuyse et al. 2003), and in relation to the influence of agricultural practices (Brito et al. 2013).

Despite rapid advances in methodologies, molecular techniques are still considered difficult to use for the identification of AM fungi because DNA extraction from soils typically yields low quantity and quality compared to plant roots (Renker et al. 2003; Olsson et al. 2010). There are also methodological and biological challenges to accurately assess AM fungi community composition which arises from the vast differences in DNA quantity from spores and hyphae in soil used for template DNA (Schüßler et al. 2001). Moreover, the higher quantity of DNA in multi-nucleic AM fungal spores compared to hyphae which contain significantly fewer nuclei which can skew community composition data (Gamper et al. 2008). This leads to potential bias in estimates of relative abundance where AM fungal morphotypes during sporulation will have a greater relative abundance than other morphotypes (Saks et al. 2013). The recent progress of next-generation sequencing techniques will open up a detail examination of fungal communities, and the reduction of cost of analyses will lead to greater accessibility, but there will be challenges.

Kohout et al. (2014) recently used different primers to compare four routinely used AM fungal-specific primer systems for nuclear ribosomal DNA such as (1) the partial small subunit (SSU), (2) the partial large subunit (LSU), (3) the partial SSU and internal transcribed spacer (ITS; "Redecker") and (4) the partial SSU-ITS-partial LSU region ("Krüger"). They also included a new primer combination (5) the ITS2 region in the comparison. They concluded that "Krüger" primers tended to yield the highest AM fungal diversity and higher Shannon diversity index than did SSU primers. They confirmed a strong bias towards the Glomeraceae in the LSU and SSU primer systems and differences in the composition of AM fungal communities based on the "Redecker" primer system.

Willis et al. (2013) considered aspects of AM fungal ecology emphasising past and present importance of the phylum and concluded that it is essential to include the AM symbiosis in studies of higher plants in order to provide a more holistic view of the ecosystems. Concomitant morphological and molecular analyses have led to major breakthroughs in the taxonomic organisation of the phylum Glomeromycota (Oehl et al. 2011). Fungi in this phylum are known to form arbuscular mycorrhizas, and so far the phylum has 3 classes, 5 orders, 14 families and 29 genera described. They present the current classification developed in several recent publications and provide a summary to facilitate the identification of taxa from genus to class level. The history and complexity of the taxonomy and systematics of these obligate biotrophs has been addressed by recognising four periods (Stürmer 2012). The initial discovery period (1845–1974) was categorised mainly by description of sporocarp-forming species and the proposal of a classification for these fungi. The following alpha taxonomy period (1975–1989) established a solid morphological basis for species identification and classification, resulting in report of many new species and the need to standardise the nomenclature of spore subcellular structures. The cladistics period (1990–2000) stated the first classification of AM fungi based only on phenotypic characters. At the end of this period, genetic characters played a role in defining taxa and elucidating evolutionary relationships within the group. The most recent phylogenetic fusion period (2001 to present) started with the proposal of a new classification based on genetic characters using sequences of the multicopy rRNA genes.

Opik et al. (2010) have offered a new database, MaarjAM, that summarises publicly available Glomeromycota DNA sequence and associated metadata. These data have been made accessible in an open-access database (http://maarjam.botany. ut.ee). Two hundred and eighty-two SSU rRNA gene virtual taxa were described based on a comprehensive phylogenetic analysis of all collated Glomeromycota sequences showing limited distribution ranges in most Glomeromycota taxa and a positive relationship between the width of a taxon's geographical range and its host taxonomic range.

Based on morphological and molecular characteristics, 19 genera, viz. *Acaulospora, Ambispora, Archaeospora, Cetraspora, Dentiscutata, Diversispora, Entrophospora, Fuscutata, Geosiphon, Gigaspora, Glomus, Intraspora, Kuklospora, Otospora, Pacispora, Paraglomus, Racocetra, Scutellospora* and *Quatunica*, comprising more than 200 species are documented in AM fungi (Manoharachary et al. 2010). Liu et al. (2009) stated that the AM fungal taxonomy position has moved forward to a phylum with 214 species belonging to 19 genera, 13 families, 4 orders and 1 class now reported. It was suggested that molecular techniques would be the key approaches in the future study of AM fungal species diversity. However, the intraspecific diversity of *Glomus mosseae* in a survey of 186 publications reporting the occurrence of *G. mosseae* from at least 474 different sites from 55 countries resulted in a geographical map of their distribution based on both morphological and molecular techniques (Liu et al. 2009).

While the taxonomy of AM fungi is based on morphological characters of the asexual resting spores, molecular approaches to community ecology have revealed a previously unknown diversity from colonised roots (Rosendahl 2008). The long asexual evolution of the fungi has resulted in considerable genetic diversity within morphologically recognisable species which challenges concepts of individuals and populations. The fossil record and molecular data show that the evolutionary history of AM fungi goes back at least to the Ordovician (460 million years ago), coinciding with the colonisation of the terrestrial environment by the first land plants (Redecker 2002). The fast-growing number of available DNA sequences for molecular identification of the Glomales within roots has been designed and tested. These detection methods have opened up entirely new perspectives for studying the ecology of AM fungi.

The AM fungal phylogeny of the genus *Glomus* of the family Glomaceae has been analysed based on full-length SSU rRNA gene sequences and shows that *Glomus* is not monophyletic (Schwarzott et al. 2001). The separation of *Archaeospora* and *Paraglomus* from *Glomus* was reported. Two ancestral families of AM fungal species such as Archaeosporaceae and Paraglomaceae were discovered from deeply divergent ribosomal DNA sequences in the past (Morton and Redecker 2001). Each family is phylogenetically distant from each other and from other glomalean families despite similarities in mycorrhizal morphology. Morphological characters once considered unique clearly are distributed in phylogenetically distant groups. The spores of some species in *Glomus* have considerable divergence at the molecular level. It is the combination of DNA sequences and mycorrhizal morphology which provides the basis for recognising *Archaeospora* and *Paraglomus*. These investigations reinforce the value of molecular data sets in providing a clearer understanding of phylogeny which in turn can lead to a more robust taxonomy.

1.3 Mechanisms of Nutrient Exchange in Arbuscular Mycorrhizas

In AM associations, nutrient exchange between fungi and host is one of the most important functions. Recent advances in biochemical and molecular biological techniques have revealed a great amount of new information on this topic (Smith and Read 2008; Harrison 1999; Saito 2000; Smith and Smith 2012). This has largely reinforced the ideas concerning symbiotic C and P transfer proposed by classical works in the 1970s and 1980s (Tinker 1975). However, the biochemical mechanisms of nutrient exchange in this symbiosis are still unclear. The following view of phosphorus transfer from fungus to host plant has been widely accepted (Smith and Read 2008). Phosphate in the soil solution is absorbed in the extraradical hyphae via a P transporter (Harrison and van Buuren 1995). The absorbed P is condensed into polyphosphate (poly-P) in the extraradical hyphae and translocated by protoplasmic streaming into the intraradical hyphae (Cox et al. 1975; Cooper and Tinker 1981; Solaiman et al. 1999). At the end, the poly-P may be hydrolysed and released as P across the fungal membrane, probably at the arbuscule. Solaiman and Saito (2001) observed that a substantial proportion of P in mycorrhizal roots was of fungal origin, while the proportion of fungal biomass to root biomass was <2 %. Phosphate efflux and the decrease in poly-P content in the hyphae were both improved by the addition of glucose and 2-deoxyglucose, an analogue of glucose. The rates of enhancement for P efflux and poly-P decrease were analogous, suggesting that P efflux from intraradical hyphae was coupled with poly-P hydrolysis. Based on these findings, the translocation of P from fungus to host was estimated in relation to the distribution of hyphal P in extraradical and intraradical parts of the AM fungus (Solaiman and Saito 2001). By applying the technique of separating intraradical hyphae from host roots, they were able to clarify the translocation and transfer of P in AM symbiosis in a quantitative manner. Further investigation of P efflux using the present experimental system will shed light on the mechanisms of nutrient exchange.

AM fungi rely on the photosynthetic C supplied by their host plants to complete their life cycle. In return, the fungi supply nutrients to the plant especially P, N and Zn. The site of nutrients exchange is primarily the arbuscule (Solaiman and Saito 1997; Harrison 1999; Balestrini and Bonfante 2005). The molecular and structural organisations of arbuscules facilitate and regulate the processes of nutrient exchange. Plant N and P transporter proteins have been reported (Ellerbeck et al. 2013; Javot et al. 2007b). Gene expression studies also showed that specific members of these protein families are expressed in the roots of colonised plants (Harrison et al. 2002; Javot et al. 2007a). Phosphate acquisition via the mycorrhizal pathway begins with the uptake of available P from soil by fungal extraradical hyphae (Bucher 2007). These hyphae extend beyond the host root zone, allowing a greater soil volume to be exploited for P uptake. Uptake at the soil hypha interface is mediated by fungal high-affinity P transporters (Harrison and van Buuren 1995). Following fungal uptake, phosphate is transferred to the fungal vacuole where it is polymerised to form polyphosphate chains and translocated through the vacuolar compartment to the intraradical hyphae (Solaiman et al. 1999). The poly-P is then hydrolysed and phosphate released to the interfacial apoplast from where plant mycorrhizal Pht1 transporters guide the phosphate across the periarbuscular membrane for delivery to other parts of the plant. The extraradical hyphae of AM fungi also absorb ammonium, nitrate and amino acids (Hodge et al. 2001), and the role of mycorrhiza on N delivery is becoming better understood (Chalot et al. 2006; Ellerbeck et al. 2013). The majority of N is thought to be taken up in the form of ammonia via the fungal-encoded AMT1 family transporters such as the protein GintAMT1 characterised from *Glomus intraradices* (Lopez-Pedrosa et al. 2006). There is no evidence for fungal translocation of either ammonium or nitrate and it is thought that nitrogen transport occurs in the form of the amino acid arginine (Govindarajulu et al. 2005), and amino acids may be delivered directly to the interfacial apoplast for plant absorption. However, there is also evidence for an alternative route whereby arginine is broken down by ornithine aminotransferase and urease to release free ammonium. It has been proposed that ammonium is exported by protein-mediated mechanisms and a candidate fungal AMT transporter has been identified that is highly expressed in the internal hyphae (Govindarajulu et al. 2005). Gene expression analyses of medic and rice have identified mycorrhiza-induced transcripts that putatively encode ammonium transporters that are candidates for this function. There is also the possibility for passive ammonia uptake across the periarbuscular membrane, perhaps facilitated by the presence of aquaporin proteins (Uehlein et al. 2007). The use of radiolabeled substrates has demonstrated that AM fungi take up plant carbohydrates in the form of hexose (Solaiman and Saito 1997). The route of this transport is specific to the arbuscule. There is little molecular evidence for the presence of hexose export proteins in the periarbuscular membrane; although a number of mycorrhiza-responsive sugar transporter genes have been identified in medic, they are thought to act as proton-sugar symporters in sugar import rather than export, possibly in support of high metabolic activity in arbuscules (Harrison 1996).

AM fungi deliver both P and N to the root through arbuscules. Previously MtPT4, a *Medicago truncatula* phosphate transporter located in the periarbuscular membrane that is essential for symbiotic P transport and for maintenance, was identified (Javot et al. 2011). In mtpt4 mutants arbuscule degeneration occurs prematurely and symbiosis fails (Javot et al. 2011). The MtPT4 arbuscule phenotype is strongly correlated with shoot N levels. On the other, the transport mechanism of sugars to the apoplast is passive movement. For example, when hexose reached in the apoplast is absorbed by the fungus via specific transport proteins. The characterisation of the GpMST1 hexose transporter from *Geosiphon pyriformis*, a Glomeromycotan fungus, provides a promising direction for further investigation (Schussler et al. 2006). In the intraradical hyphae, much of the C is changed to storage lipids, predominantly as triacylglycerides. Lipids not only act as storage C but also are the main form of C translocated from intra- to extraradical hyphae where they provide the major respiratory substrate.

AM fungi provide benefits to their plant hosts by enhancing mineral nutrition, increasing tolerance to water stress, inducing greater resistance to pathogens and reduce sensitivity to toxic substances present in the soil. However, the cost of colonisation can be ~20 % of the host's fixed C being consumed by the microbial symbiont. Nonetheless, under experimental conditions when nutrients are limiting, mycorrhizal crop plants typically exhibit a better performance over non-mycorrhizal plants in high-input agricultural systems, the relative advantages are reduced while the C costs remain same and the performance of colonised plants can fall below that of non-colonised plants (Janos 2007). Phillips et al. (2013) proposed a framework for considering how tree species and their mycorrhizal symbionts differentially couple C and nutrient cycles in temperate forests. Given that tree species predominantly colonise by a single type of AM fungi or ECM fungi and that the two types of fungi differ in their modes of nutrient acquisition, the abundance of AM and ECM trees in a paddock may provide an integrated index of biogeochemical transformations relevant to C cycling and nutrient retention.

AM fungi can obtain the photosynthates from host plants and can promote N uptake by host plants via the absorption of various N sources by mycorrhizal hyphae leading to improvement in nutrition and stress tolerance of host plants (Li et al. 2013). The symbiont absorbs and transfers N, but the mechanisms behind the N metabolism and translocation from AM fungi to host plants are still in debate.

The roles of AM fungi in N allocation in host plants and the ecological significance at community and ecosystem levels need to be studied in more detail because they vary widely (Hodge and Storer 2014).

1.4   Role of AM Fungi in Sustainable Agriculture

Sustainable agricultural systems use natural processes to achieve satisfactory levels of productivity and food quality while decreasing fertiliser use, dropping input costs and preclude environmental pollution and its impacts (Siddiqui et al. 2008; Harrier and Watson 2004). It should also be ecologically feasible and socially responsible. Several soil factors contribute to sustainable agriculture through control of soil-borne diseases and increased soil microbial activity leading to increased antagonism and parasitism within the rhizosphere level (Jawson et al. 1993; Knudsen et al. 1995). Research approaches are presently focused on the search for suitable alternatives to the use of commercial artificial pesticides. However, progress has also been accomplished in exploring the use of microorganisms for improvement of soil fertility and ultimately increased crop productivity. Greater emphasis is being placed on enhancing exploitation of indigenous soil microbes which will contribute to soil fertility and increase plant growth as well as plant protection.

AM fungi have been difficult to study (Hamel and Strullu 2006), but they are now recognised as key components of soil ecosystems rather than only a plant root component. Recent advances in knowledge conveyed by new techniques for soil microbiology research open the way to AM fungi management in crop nutrition and production. AM fungi can influence crop growth, nutrition and production even in phosphorus-rich soils (Balzergue et al. 2013; Solaiman and Hirata 1997). However, growing crops in soil with lower levels of fertility could enhance the multiple beneficial effects of AM fungi in agroecosystem including decreased nutrient loss to the environment. Inclusion of mycorrhizal bioassays in soil testing protocol (Djuuna et al. 2009) for use in fertilisation recommendations and development of improved inoculants to manipulate AM fungi (Abbott et al. 1987) and screening of crop cultivars with improved symbiotic abilities (Smith et al. 1992) could contribute to agroecosystem stability and sustainability (Hamel and Strullu 2006).

In agriculture, several factors influence plant response and plant benefits from mycorrhizas such as host crop dependency on mycorrhizal colonisation, tillage, fertiliser application and the potential of mycorrhizal fungi inocula. Interest in AM fungal inoculation for sustainable agriculture is based on their roles in the improvement of plant growth through nutrient and water uptake (Augé 2004) and improvements in soil fertility as well as soil aggregate stability (Rillig et al. 2007; Rillig 2004).

1.5 Role of AM and ECM Fungi in Protection of Soil-Borne Diseases

Biological control of soil-borne diseases is currently accepted as a key practice in sustainable agriculture. AM fungi have shown potential in protecting host plants from soil-borne pathogens. While few AM isolates have been tested against these soil-borne pathogens, some appear to be more effective than others (Cameron et al. 2013; Azćon-Aguilar and Barea 1996). The degree of protection varies with the pathogen species involved and can be moderated by soil and other microclimatic conditions of the rhizosphere. Only weak responses to AM fungi colonisation have been found in some activities like lignification, production of phytoalexins and peroxidises and expression of genes encoding for PR proteins, indicating that AM fungi do not elicit typical defence responses (Azćon-Aguilar and Barea 1996). However, these compounds could make roots sensitive to the presence of pathogens and enhance defence mechanisms to subsequent pathogen infection (Benhamou et al. 1994). In this study, responses of AM and non-mycorrhizal transformed carrot roots to infection by *Fusarium oxysporum* f. sp. *Chrysanthemi* were investigated in mycorrhizal roots; the growth of the pathogen was usually restricted to the epidermis and cortical tissues, whereas in non-mycorrhizal roots the infection of pathogen reached at depth even up to the vascular stele. The *Fusarium* hyphae inside mycorrhizal roots exhibited a high level of structural disorganisation, probably induced by a strong reaction of the host cells characterised by the accumulation of phenolic-like compounds and the production of hydrolytic enzymes such as chitinases. This strong reaction was not induced by non-mycorrhizal roots, suggesting that the activation of plant defence responses by mycorrhiza formation provides a protection at a certain level against the pathogen. In contrast to the weak defence response towards AM fungi found in AM hosts, it is noteworthy that in myc⁻ pea mutants, AM fungi trigger a strong resistance reaction. This suggests that the AM fungi are able to elicit a defence response but that symbiosis-specific genes have mechanism to control the expression of the genes related to plant defence during AM establishment (Gianinazzi-Pearson et al. 1996). The expression of several PRs in tobacco plants did not affect the level of colonisation by *Glomus mosseae* which was only reduced in plants constitutively expressing an acidic isoform of tobacco PR-2, a glucanase (Vierheilig et al. 1996).

Ectomycorrhizas (ECM) have a positive effect on the performance of seedlings due to the beneficial relationship between plants and mycorrhizal fungi (Guerin-Laguette et al. 2004; Machón et al. 2006; Minchin et al. 2012). They are also effective against various plant root rot diseases (Duchesne 2000). Many studies have observed the protective role of ECM not only against fungal pathogens (Morin et al. 1999) but also against nematodes (Machón et al. 2006). Pine wilt disease is a globally severe forest disease and demonstrates the importance of ECM relationships (Akema and Futai 2005). In this study, the abundant ECM found in the upper slope enhanced water uptake by the pines, mitigated drought stress and thereby decreased the mortality of pine trees from pine wilt disease.

1.6 Role of AM and ECM Fungi in Restoration of Native Forest Ecosystems

Land clearing of terrestrial ecosystems claims several million hectares annually in Australia (Warren et al. 1996) which causes loss of essential physicochemical and biological soil properties (Skujins and Allen 1986). These properties largely determine soil quality and fertility that supports plant establishment and productivity. Soil degradation limits the potential for restoration of native plants (Agnew and Warren 1996), and erosion and desertification are accelerated. Desertification reduces the inoculum potential of mutualistic symbionts such as AM fungi that are key ecological factors in governing the cycles of major plant nutrients and hence in sustaining vegetative covers in natural habitats.

AM fungi improve the ability of plants to establish and cope in stressful conditions including nutrient deficiency, drought and soil disturbance (Schreiner et al. 1997). The fungal hyphae contribute to the formation of water-stable aggregates necessary for good soil tilth (Jeffries and Barea 2000). Loss of mycorrhizal propagules from degraded ecosystems can overcome either by natural or artificial revegetation where an increase of inoculum may be needed in these ecosystems (Requena et al. 1996). Inoculation of plants with mycorrhizal fungi in revegetation schemes should not only help plant establishment (Herrera et al. 1993) but also improve soil biological, chemical and physical properties thus contributing to soil quality (Carrillo-García et al. 1999). The introduction of a plant species along with a known AM fungus is a successful biotechnological tool to aid the recovery of desertified ecosystems (Azcón-Aguilar et al. 2003).

Many forest trees are dependent on a symbiotic association of their roots with ECM fungi and mobilise minerals from soil and transfer them to the plant. In exchange the trees deliver assimilated C to the fungi. The hyphae of ECM fungi are the source of C to soil microbes and depend on their tree hosts for their energy needs. In return, they take up P, N, S and Zn from soil and translocate them to their host and greatly extend the functional root system of the host plants (Allen 1991). An ECM fungus can connect roots of several trees by fungal hyphae network. Most ECM fungi are basidiomycetes such as *Amanita*, *Cortinarius*, *Lactarius*, *Russula* and *Suillus*, among the best-known ECM genera (Hacskaylo 1972). ECM are widespread particularly in temperate regions where many of the ecologically important tree species involve such as species of *Abies*, *Betula*, *Fagus*, *Picea*, *Pinus*, *Pseudotsuga*, *Quercus* and *Salix*.

Below-ground biodiversity is essential for the maintenance of forest growth and ecosystem functions as well as for reforestation of disturbed lands due to mining activities. ECM fungi are economically symbiotic soil fungi forming a sheath around the root tip and form a special structure called Hartig net (Smith and Read 2008). They gain C from the tree and in return support the tree in taking up nutrients, water and metabolites. The fungus also protects plants from parasites, nematodes and soil pathogens. The importance of ECM in forest plantations has received much attention when it was observed that trees often fail to establish at new sites if the ECM symbiont was absent (Menkis et al. 2012). This effect has been found in exotic pine transplantation in different parts of the world. In Western Australia, *Pinus radiata* and *P. pinaster* failed to establish in nursery beds in the absence of ECM fungi (Lakhanpal 2000). Pine seedlings are known to be tolerant to environmental stresses such as acid mist when colonised with ECM fungi (Asai and Futai 2001). Tropical rainforests harbour the highest known tree diversity on the planet, and many ecological studies have attempted to explain the familiar symbiotic association of so many co-occurring species (Leigh et al. 2004; Valencia et al. 1994).

1.7 Conclusions

Mycorrhizal fungi are well known to have a wide range of benefits to their host plants. They can enhance nutrient uptake especially P, N and Zn. They can also suppress soil pathogens, enhance tolerance to drought stress and reduce sensitivity to toxic substances contaminated to the soil. The suitability of conditions needs to be managed for indigenous fungi to colonise hosts in their natural habitat or to minimise loss of these fungi with high-input farming/disturbance. Highly mycorrhizal host crop cultivars should be selected for use in crop rotations. Conventional plant breeding in soils with high nutrient contents may select against the most efficient fungal communities or even against the mycorrhizal association. Many efforts have been made in recent years to get benefits from mycorrhizas for agriculture, horticulture, forestry, land restoration and contaminated site remediation. The results have generally been consistently positive under controlled conditions, with some difficulties due to complications from diverse variables under field conditions. Mycorrhizal interactions between plants, fungi and the environment are complex and often indivisible. Mycorrhizas are an essential below-ground component in the establishment and sustainability of plant communities, but thorough knowledge is required to achieve maximum benefits from these microorganisms and their associations.

References

Abbott LK, Gazey C (1994) An ecological view of the formation of VA mycorrhizas. Plant Soil 159:69–78
Abbott LK, Robson AD, Gazey C (1987) Selection of inoculant VA mycorrhizal fungi. In: Sylvia DM, Hung LL, Graham JH (eds) Mycorrhizas in the next decade, practical applications and research priorities. Abstract. Proceedings of the 7th North American Conference on Mycorrhizas, pp 10–12. Institute of Food and Agricultural Sciences, University of Florida
Agnew C, Warren A (1996) A framework for tackling drought and land degradation. J Arid Environ 33:309–320

Akema T, Futai K (2005) Ectomycorrhizal development in a *Pinus thunbergii* stand in relation to location on a slope and effect on tree mortality from pine wilt disease. J For Res 10:93–99

Allen MF (1991) The ecology of mycorrhiza. Cambridge University Press, Cambridge, p 184. Chapman and Hall, New York, NY, p 534

Asai E, Futai K (2001) Retardation of pine wilt disease symptom development in Japanese black pine seedlings exposed to simulated acid rain and inoculated with *Bursaphelenchus xylophilus*. J For Res 6:297–302

Augé RM (2004) Arbuscular mycorrhizae and soil/plant water relations. Can J Soil Sci 84:373–381

Azcón-Aguilar C, Barea JM (1996) Arbuscular mycorrhizas and biological control of soil-borne plant pathogens – an overview of the mechanisms involved. Mycorrhiza 6:457–464

Azcón-Aguilar C, Palenzuela J, Roldan A, Bautista S, Vallejo R, Barea JM (2003) Analysis of the mycorrhizal potential in the rhizosphere of representative plant species from desertification-threatened Mediterranean shrublands. Appl Soil Ecol 22:29–37

Balestrini R, Bonfante P (2005) The interface compartment in arbuscular mycorrhizae: a special type of plant cell wall? Plant Biosyst 139:8–15

Balzergue C, Chabaud M, Barker DG, Becard G, Rochange SF (2013) High phosphate reduces host ability to develop arbuscular mycorrhizal symbiosis without affecting root calcium spiking responses to the fungus. Front Plant Sci 4, article no 426. doi: 10.3389/fpls.2013.00426

Bedini S, Avio L, Sbrana C, Turrini A, Migliorini P, Vazzana C, Giovannetti M (2013) Mycorrhizal activity and diversity in a long-term organic Mediterranean agroecosystem. Biol Fertil Soils 49:781–790

Benhamou N, Fortin JA, Hamel C, St-Arnaud M, Shatilla A (1994) Resistance responses of mycorrhizal Ri T-DNA-transformed carrot roots to infection by *Fusarium oxysporum* f. sp. *chrysanthemi*. Phytopathology 84:958–968

Brito I, Carvalho M, Goss MJ (2013) Soil and weed management for enhancing arbuscular mycorrhiza colonization of wheat. Soil Use Manage 29:540–546

Brundrett MC (2002) Coevolution of roots and mycorrhizas of land plants. New Phytol 154:275–304

Brundrett MC, Nanjappa A (2013) Glomeromycotan mycorrhizal fungi from tropical Australia III. Measuring diversity in natural and disturbed habitats. Plant Soil 370:419–433

Bucher M (2007) Functional biology of plant phosphate uptake at root and mycorrhiza interfaces. New Phytol 173:11–26

Cameron DD, Neal AL, van Wees SCM, Ton J (2013) Mycorrhiza-induced resistance: more than the sum of its parts? Trends Plant Sci 18:539–545

Carrillo-García A, Leon de la Luz JL, Bashan Y, Bethlenfalvay GJ (1999) Nurse plants, mycorrhizae, and plant establishment in a disturbed area of the Sonoran Desert. Restor Ecol 7:321–335

Chalot M, Blaudez D, Brun A (2006) Ammonia: a candidate for nitrogen transfer at the mycorrhizal interface. Trend Plant Sci 11:263–266

Cooper KM, Tinker PB (1981) Translocation and transfer of nutrients in vesicular arbuscular mycorrhizas. IV. Effect of environmental variables on movement of phosphorus. New Phytol 88:327–339

Cox G, Sanders FE, Tinker PB, Wild JA (1975) Ultrastructure evidence relating to host/endophyte transfer in a vesicular–arbuscular mycorrhizae. In: Sanders FE, Mosse B, Tinker PB (eds) Endomycorrhizae. Academic Press, London, pp 297–312

Daniell TJ, Husband R, Fitter AH, Young JPW (2001) Molecular diversity of arbuscular mycorrhizal fungi colonising arable crops. FEMS Microbiol Ecol 36:203–209

Djuuna IAF, Abbott LK, Solaiman MZ (2009) Use of mycorrhiza bioassays in ecological studies. In: Varma A, Kharkwal AC (eds) Symbiotic fungi, vol 18, Soil Biology. Springer, Berlin

Duchesne LC (2000) Role of ectomycorrhizal fungi in biocontrol. In: Pfleger FL, Linderman RG (eds) Mycorrhiza and plant health. APS Press, St Paul, MN, pp 27–46

Ellerbeck M, Schüßler A, Brucker D, Dafinger C, Loos F, Brachmann A (2013) Characterization of three ammonium transporters of the Glomeromycotan fungus Geosiphon pyriformis. Eukaryot Cell 12:1554–1562

Gamper HA, Young JPW, Jones DL, Hodge A (2008) Real-time PCR and microscopy: are the two methods measuring the same unit of arbuscular mycorrhizal fungal abundance? Fungal Genet Biol 45:581–596

Gianinazzi-Pearson V, Dumas-Gaudot E, Gollotte A, Tahiri-Alaoui A, Gianinazzi S (1996) Cellular and molecular defence related root responses to invasion by arbuscular mycorrhizal fungi. New Phytol 133:45–57

Gollotte A, van Tuinen D, Atkinson D (2004) Diversity of arbuscular mycorrhizal fungi colonising roots of the grass species Agrostis capillaris and Lolium perenne in a field experiment. Mycorrhiza 14:111–117

Govindarajulu M, Pfeffer P, Jin H, Abubaker J, Douds DD, Allen JW, Bücking H, Lammers PJ, Shachar-Hill Y (2005) Nitrogen transfer in the arbuscular mycorrhizal symbiosis. Nature 435:819–823

Guerin-Laguette A, Shindo K, Matsushita N, Suzuki K, Lapeyrie F (2004) The mycorrhizal fungus Tricoloma matsutake stimulates *Pinus densiflora* seedling growth in vitro. Mycorrhiza 14:397–400

Hacskaylo E (1972) Mycorrhiza: the ultimate in reciprocal parasitism? BioSci 22:577–582

Hamel C, Strullu D-G (2006) Arbuscular mycorrhizal fungi in field crop production: potential and new direction. Can J Plant Sci 86:941–950

Harrier LA, Watson CA (2004) The potential role of arbuscular mycorrhizal (AM) fungi in the bioprotection of plants against soil-borne pathogens in organic and/or other sustainable farming systems. Pest Manag Sci 60:149–157

Harrison MJ (1996) A sugar transporter from Medicago truncatula: altered expression pattern in roots during vesicular-arbuscular (VA) mycorrhizal associations. Plant J 9:491–503

Harrison M (1999) Biotrophic interfaces and nutrient transport in plant/fungal interfaces. J Exp Bot 50:1013–1022

Harrison MJ, van Buuren ML (1995) A phosphate transporter from the mycorrhizal fungus *Glomus versiforme*. Nature 378:626–629

Harrison MJ, Dewbre GR, Liu J (2002) A phosphate transporter from *Medicago truncatula* involved in the acquisition of phosphate released by arbuscular mycorrhizal fungi. Plant Cell 14:2413–2429

Herrera MA, Salamanca CP, Barea JM (1993) Inoculation of woody legumes with selected arbuscular mycorrhizal fungi and rhizobia to recover desertified Mediterranean ecosystems. Appl Environ Microbiol 59:129–133

Hibbett DS, Gilbert LB, Donoghue MJ (2000) Evolutionary instability of ectomycorrhizal symbioses in basidiomycetes. Nature 407:506–508

Hodge A, Storer K (2014) Arbuscular mycorrhiza and nitrogen: implications for individual plants through to ecosystems. Plant Soil. doi:10.1007/s11104-014-2162-1

Hodge A, Campbell CD, Fitter AH (2001) An arbuscular mycorrhizal fungus accelerates decomposition and acquires nitrogen directly from organic material. Nature 413:297–299

Janos DP (2007) Plant responsiveness to mycorrhizas differs from dependence upon mycorrhizas. Mycorrhiza 17:75–91

Jansa J, Mozafar A, Kuhn G, Anken T, Ruh R, Sanders IR, Frossard E (2003) Soil tillage affects the community structure of mycorrhizal fungi in maize roots. Ecol Appl 13:1164–1176

Javot H, Penmetsa RV, Terzaghi N, Cook DR, Harrison MJ (2007a) A *Medicago truncatula* phosphate transporter indispensable for the arbuscular mycorrhizal symbiosis. Proc Natl Acad Sci U S A 104:1720–1725

Javot H, Pumplin N, Harrison MJ (2007b) Phosphate in the arbuscular mycorrhizal symbiosis: transport properties and regulatory roles. Plant Cell Environ 30:310–322

Javot H, Penmetsa RV, Breuillin F, Bhattarai KK, Noar RD, Gomez SK, Zhang Q, Cook DR, Harrison MJ (2011) Medicago truncatula mtpt4 mutants reveal a role for nitrogen in the regulation of arbuscule degeneration in arbuscular mycorrhizal symbiosis. Plant J 68:954–965

Jawson MD, Franzluebbers AJ, Galusha DK, Aiken RM (1993) Soil fumigation within monoculture and rotations—response of corn and mycorrhizae. Agron J 85:1174–1180

Jeffries P, Barea JM (2000) Arbuscular mycorrhiza—a key component of sustainable plant-soil ecosystems. In: Hock B (ed) The mycota. IX. Fungal associations. Springer KG, Berlin, pp 95–113

Kjøller R, Rosendah S (2001) Molecular diversity of glomalean (arbuscular mycorrhizal) fungi determined as distinct *Glomus* specific DNA sequences from roots of field grown peas. Mycol Res 105:127–132

Knudsen IMB, Debosz K, Hockenhull J, Jensen DF, Elmholt S (1995) Suppressiveness of organically and conventionally managed soils towards brown foot rot of barley. Appl Soil Ecol 12:61–72

Kohout P, Sudová R, Janousková M, Ctvrtlíková M, Hejda M, Pánková H, Slavíková R, Stajerová K, Vosátka M, Sýkorová Z (2014) Comparison of commonly used primer sets for evaluating arbuscular mycorrhizal fungal communities: is there a universal solution? Soil Biol Biochem 68:482–493

Lakhanpal TN (2000) Ectomycorrhiza – an overview. In: Mukerji KG, Chamola BP, Singh J (eds) Mycorrhizal biology. Kluwer Academic/Plenum, New York, NY, pp 101–118

Leigh EG, Davidar P, Dick CW, Puyravaud J, Terborgh J, ter Steege H, Wright SJ (2004) Why do some tropical forests have so many species of trees? Biotropica 36:445–473

Li YJ, Liu ZL, He XY, Tian CJ (2013) Nitrogen metabolism and translocation in arbuscular mycorrhizal symbiote and its ecological implications. J Appl Ecol 24:861–868

Liu RJ, Jiao H, Li Y, Li M, Zhu X-C (2009) Research advances in species diversity of arbuscular mycorrhizal fungi. J Appl Ecol 20:2301–2307

Lopez-Pedrosa A, Gonzalez-Guerrero M, Valderas A, Azcon-Aguilar C, Ferrol N (2006) GintAMT1 encodes a functional high-affinity ammonium transporter that is expressed in the extraradical mycelium of *Glomus intraradices*. Fungal Genet Biol 43:102–110

Machón P, Santamaria O, Pajares JA, Alves-Santos FM, Diez JJ (2006) Influence of the mycorrhizal fungus *Laccaria laccaria* on pre-emergence, post-emergence and the late damping-off by *Fusarium moniliforme* and *F. oxysporum* on Scots pine seedlings. Symbiosis 42:153–160

Manoharachary C, Kunwar IK, Tilak KVBR, Adholeya A (2010) Arbuscular mycorrhizal fungi – taxonomy, diversity, conservation and multiplication. Proc Natl Acad Sci Ind Sect B Biol Sci 80:1–13

McNear DH Jr (2013) The rhizosphere – roots, soil and everything in between. Nat Educ Knowl 4 (3):1

Menkis A, Lygis V, Burokiene D, Vasaitis R (2012) Establishment of ecto-mycorrhiza inoculated Pinus sylvestris seedlings on coastal dunes following a forest fire. Baltic Forestry 18:33–40

Minchin RF, Ridgway HJ, Condron L, Jones EE (2012) Influence of inoculation with a Trichoderma bio-inoculant on ectomycorrhizal colonisation of *Pinus radiata* seedlings. Ann Appl Biol 161:57–67

Morin C, Samson J, Dessureault M (1999) Protection of black spruce seedlings against Cylindrocladium root rot with ectomycorrhizal fungi. Can J Bot 77:169–174

Morton JB, Redecker D (2001) Two new families of Glomales, Archaeosporaceae and Paraglomaceae, with two new genera Archaeospora and Paraglomus, based on concordant molecular and morphological characters. Mycologia 93:181–195

Oehl F, Sieverding E, Javier P, Kurt I, da Silva GA (2011) Advances in *Glomeromycota* taxonomy and classification. IMA Fungus 2:191–199

Olsson PA, Rahm J, Aliasgharzad N (2010) Carbon dynamics in mycorrhizal symbioses is linked to carbon costs and phosphorus benefits. FEMS Microbiol Ecol 72:125–131

Opik M, Vanatoa A, Vanatoa E, Moora M, Davison J, Kalwij JM, Reier U, Zobel M (2010) The online database MaarjAM reveals global and ecosystemic distribution patterns in arbuscular mycorrhizal fungi (Glomeromycota). New Phytol 188:223–241

Pellegrino E, Bedini S (2014) Enhancing ecosystem services in sustainable agriculture: biofertilization and biofortification of chickpea (*Cicer arietinum* L.) by arbuscular mycorrhizal fungi. Soil Biol Biochem 68:429–439

Phillips RP, Brzostek E, Midgley M (2013) The mycorrhizal-associated nutrient economy: a new framework for predicting carbon-nutrient couplings in temperate forests. New Phytol 199:41–51

Redecker D (2002) Molecular identification and phylogeny of arbuscular mycorrhizal fungi. Plant Soil 244:67–73

Redecker D, Morton JB, Bruns TD (2000) Ancestral lineages of arbuscular mycorrhizal fungi (*Glomales*). Mol Phylogen Evol 14:276–284

Renker C, Heinrichs J, Kaldorf M, Buscot F (2003) Combining nested PCR and restriction digest of the internal transcribed spacer region to characterize arbuscular mycorrhizal fungi on roots from the field. Mycorrhiza 13:191–198

Requena N, Jeffries P, Barea JM (1996) Assessment of natural mycorrhizal potential in a desertified semiarid ecosystem. Appl Environ Microbiol 62:842–847

Rillig MC (2004) Arbuscular mycorrhizae, glomalin, and soil aggregation. Can J Soil Sci 84:355–363

Rillig MC, Caldwell BA, Wosten HAB, Sollins P (2007) Role of protein in soil carbon and nitrogen storage: controls on persistence. Biogeochemistry 85:25–44

Rosendahl S (2008) Communities, populations and individuals of arbuscular mycorrhizal fungi. New Phytol 178:253–266

Saito M (2000) Symbiotic exchange of nutrients in arbuscular mycorrhizas: transport and transfer of phosphorus. In: Douds DD, Kapunik Y (eds) Arbuscular mycorrhizas: physiology and function. Kluwer, Dordrecht, The Netherlands, pp 85–106

Saks Ü, Davison J, Öpik M, Vasar M, Moora M, Zobel M (2013) Root-colonizing and soil-borne communities of arbuscular mycorrhizal fungi in a temperate forest understorey. Botany 92:277–285

Sanders IR, Alt M, Groppe K, Boller T, Wiemken A (1995) Identification of ribosomal DNA polymorphisms among and within spores of the Glomales – application to studies on the genetic diversity of arbuscular mycorrhizal fungal communities. New Phytol 130:419–427

Schreiner RP, Milhara KL, McDaniel H, Bethlenfalvay GJ (1997) Mycorrhizal fungi influence plant and soil functions and interactions. Plant Soil 188:199–209

Schüßler A, Schwarzott D, Walker C (2001) A new fungal phylum, the Glomeromycota: phylogeny and evolution. Mycol Res 105:1413–1421

Schussler A, Martin H, Cohen D, Fitz M, Wipf D (2006) Characterization of a carbohydrate transporter from symbiotic glomeromycotan fungi. Nature 444:933–936

Schwarzott D, Walker C, Schussler A (2001) Glomus, the largest genus of the arbuscular mycorrhizal fungi (Glomales), is nonmonophyletic. Mol Phylogenet Evol 21:190–197

Siddiqui ZA, Akhter MS, Futai K (eds) (2008) Mycorrhizae: sustainable agriculture and forestry. Springer Science + Business Media B.V., 365 pp

Skujins J, Allen MF (1986) Use of mycorrhizae for land rehabilitation. MIRCEN J 2:161–176

Smith SE, Read DJ (2008) Mycorrhizal symbiosis. Academic Press, London, 800 pp

Smith SE, Smith FA (2012) Fresh perspectives on the roles of arbuscular mycorrhizal fungi in plant nutrition and growth. Mycologia 104:1–13

Smith SE, Robson AD, Abbott LK (1992) The involvement of mycorrhizas of genetically-dependent efficiency of nutrient uptake and use. Plant Soil 146:169–179

Solaiman ZM, Abbott LK (2003) Phosphorus uptake by a community of arbuscular mycorrhizal fungi in jarrah forest. Plant Soil 248:313–320

Solaiman ZM, Abbott LK (2008) Influence of arbuscular mycorrhizal fungi, inoculum level and phosphorus placement on growth and phosphorus uptake of *Phyllanthus calycinus* in jarrah forest soil. Biol Fertil Soils 44:815–821

Solaiman ZM, Hirata H (1997) Effect of arbuscular mycorrhizal fungi inoculation of rice seedlings at the nursery stage upon performance in the paddy field and greenhouse. Plant Soil 191:1–12

Solaiman ZM, Saito M (2001) Phosphate efflux from intraradical hyphae of Gigaspora margarita in vitro and its implication for phosphorus. New Phytol 151:525–533

Solaiman ZM, Saito M (1997) Use of sugars by intraradical hyphae of arbuscular mycorrhizal fungi revealed by radiorespirometry. New Phytol 136:533–538

Solaiman ZM, Ezawa T, Kojima T, Saito M (1999) Polyphosphates in intraradical and extraradical hyphae of arbuscular mycorrhizal fungi. Appl Environ Microbiol 65:5604–5606

Stürmer SL (2012) A history of the taxonomy and systematics of arbuscular mycorrhizal fungi belonging to the phylum Glomeromycota. Mycorrhiza 22:247–258

Tinker PB (1975) Effects of vesicular–arbuscular mycorrhizas on higher plants. Symp Soc Exp Biol 29:325–350

Uehlein N, Fileschi K, Eckert M, Bienert GP, Bertl A, Kaldenhoff R (2007) Arbuscular mycorrhizal symbiosis and plant aquaporin expression. Phytochemistry 68:122–129

Valencia RH, Balslev H, Paz H, Mino CG (1994) High tree alpha-diversity in Amazonian Ecuador. Biodiv Conserv 3:21–28

Vandenkoornhuyse P, Ridgway KP, Watson IJ, Fitter AH, Young JPW (2003) Co-existing grass species have distinctive arbuscular mycorrhizal communities. Mol Ecol 12:3085–3095

Vierheilig H, Alt M, Gut-Rella M, Lange J, Boller T, Wiemken A (1996) Colonization of tobacco constitutively expressing pathogenesis-related proteins by arbuscular mycorrhizal fungi. In: Azcón-Aguilar C, Barea JM (eds) Mycorrhizas in integrated systems: from genes to plant development. European Commission, EUR 16728, Luxembourg, pp 270–273

Warren A, Sud YC, Rozanov B (1996) The future of deserts. J Arid Environ 32:75–89

Willis A, Rodrigues BF, Harris PJC (2013) The ecology of arbuscular mycorrhizal fungi. Crit Rev Plant Sci 32:1–20

Chapter 2
Assessing Economic Benefits of Arbuscular Mycorrhizal Fungi as a Potential Indicator of Soil Health

L.K. Abbott and S. Lumley

2.1 Introduction to Soil Health Indicators

Arbuscular mycorrhizal (AM) fungi have the potential to influence the economic benefits of agricultural systems through both direct and indirect processes related to plant nutrition (e.g. Smith and Smith 2012), access to moisture in water-limiting situations (e.g. Manoharan et al. 2010), building soil structure (e.g. Rillig and Mummey 2006), protection of soil carbon in aggregates (e.g. Jastrow et al. 1998) and strengthening plant resilience to disease (e.g. Azćon-Aguilar and Barea 1996). In some situations, AM fungi may have negative influences, particularly in relation to carbon transfer (Smith and Smith 2012). However, despite the demonstrated potential for AM fungi to contribute to soil physical, chemical and biological processes under controlled conditions, their contributions can be overridden in farming systems by management decisions that do not take them into account.

Although contributions of mycorrhizas are well documented (Smith and Read 2008), it is generally difficult to quantify their economic benefits (Miller et al. 1994). This is because there has been little work done either to identify systematically all such benefits or to identify how variables that influence mycorrhizal function might interact with each other to influence overall benefits. To complicate matters further, it is possible that the nature and magnitude of such benefits might be site specific, requiring all possible mycorrhizal impacts for defined rotations in a specific location to be considered. The emphasis in this overview is to determine the relevance of AM fungi in '*normal agricultural field*

L.K. Abbott (✉) • S. Lumley
School of Earth and Environment, The University of Western Australia, Crawley 6009, Australia

Institute of Agriculture, The University of Western Australia, Crawley 6009, Australia
e-mail: Lynette.Abbott@uwa.edu.au

© Springer-Verlag Berlin Heidelberg 2014

Z.M. Solaiman et al. (eds.), *Mycorrhizal Fungi: Use in Sustainable Agriculture and Land Restoration*, Soil Biology 41, DOI 10.1007/978-3-662-45370-4_2

conditions', including field inoculation with AM fungi where this is demonstrated to be commercially practical.

Factors known to affect the magnitude of mycorrhizal influence under field conditions (both positive and negative) include the availability of soil phosphorus in relation to the requirements of the plant, the diversity and abundance of the AM fungi present, the plant host species growing in the farming system (either within rotations or in continuous planting of one crop or communities of pasture species), the size of the plant and its stage of development, and the levels of soil carbon and nitrogen. Other issues likely to influence mycorrhizal colonisation include plant stressors such as disease (Azćon-Aguilar and Barea 1996), soil constraints such as salinity or acidity (Evelyn et al. 2009; Juniper and Abbott 1993; Sano et al. 2002), heat and water limitation (Manoharan et al. 2010), and the presence of other soil organisms which interact directly with mycorrhizal hyphae such as soil mesofauna (Endlweber and Scheu 2007). In addition, the various factors that influence AM fungi may interact with one another, leading to negative, synergistic or additive effects (Pearson et al. 1993, 1994; Thonar et al. 2014).

2.2 Introduction to Economic Evaluation of Environmental Contributions

Arbuscular mycorrhizas are but one element of soil biodiversity, which strongly influences soil health. As well as being influenced by the presence of other soil biota, such as saprophytic fungi (Albrechtova et al. 2012), the abundance and role of AM fungi are in turn influenced by soil treatments such as tillage and soil amendments (Brito et al. 2012; Lehmann and Joseph 2009). While their complete range of impacts on agricultural and natural ecosystems is yet to be fully appreciated, their potential for beneficial effects in all types of ecosystems has been acknowledged (Chaurasia 2004). However, the need for inoculation is controversial (Schwartz et al. 2006) and cannot be determined without clear understanding of the benefits of AM fungi present in the soil and the suitability of inoculants (Abbott et al. 1992).

A problem with valuing any aspect of biodiversity is that it is generally held to be an economic intangible, that is, it has no market price (Baker and Ruting 2014; Bishop 2013; Pearce 1995). In common with many other environmental goods and services that have vast overall intangible benefits to society, biodiversity itself cannot be bought and sold, making its value very difficult to quantify (Martinez-Alier 1987). This is unlike goods and services for which a market, and therefore a price, exists (Baker and Ruting 2014). Environmental and ecological economists have long attempted to develop methodologies for valuing intangibles because without some measure of their economic benefit, these valuable resources tend to be ignored or neglected in a world where the market and its attendant prices are treated with an almost religious reverence (Dobell 1995; Loy 1997; Pearce 2002;

Lumley 2013). Thus, it is difficult to make financial comparisons of their worth in comparison with resources like minerals and timber which have tangible values. This reverence strongly influences policy globally, and decision makers have come to rely on comparative financial values to prioritise budget allocations and other important determinations (Bishop 2013; Lumley 2013).

Various attempts have been made to quantify biodiversity value because its benefits are known amongst biologists and other scientists to be far-reaching and because biodiversity loss can have long-term, sometimes catastrophic, consequences for the human economy. Pimental et al. (1997) conducted an economic analysis of the benefits of biodiversity in which they concluded that they were worth $300 billion annually to the US economy alone. In their article, the authors disaggregated various biodiversity services into 21 activities while trying to place a financial value on each activity. One of the activities they identified was 'soil formation' about which they stated (Pimental et al. 1997, p. 748): 'Diverse soil biota facilitate soil formation and improve it for crop production'. They estimated the biodiversity value of soil formation to be worth $5 billion to the US economy and $25 billion to the world economy annually. Given that this estimate was in 1997 US dollars, it will now be substantially higher. Arbuscular mycorrhizas constitute a significant subset of soil fungi, and while it is not possible to extrapolate the value of mycorrhizas alone from the figures for soil biota estimated by Pimental et al., it is likely that their economic benefits are globally significant. Schulz (2001, p. 111) while investigating the effect of arbuscular mycorrhizas on the development of micropropagated oil palms noted that: 'While the economic benefits of arbuscular mycorrhizas per se have not been calculated, it has long been recognised that they do indeed have substantial worth and overall significance to soil health. In recent years the interest in mycorrhizas has increased, partly due to economic benefits because most of the economically important plants in agriculture, horticulture and forestry have been found to be mycorrhizal'. Delian et al. (2011) claimed that the presence of mycorrhizas in soil can increase economic profitability and it is widely recorded that mycorrhizas influence crop productivity (e.g. Smith and Read 2008; Gazey et al. 2004), although Ryan and Kirkegaard (2012) question their benefits. In a modelling study of the apparent diversity of mycorrhizal effects, Veresoglou and Malley (2012) claimed that potentially beneficial versus damaging relationships between plants and mycorrhizal fungi depended upon the number and nature of mycorrhizal species that colonised the plant. In response to a suggestion that mycorrhizal colonisation might be damaging in some Australian cropping situations, Smith and Smith (2011, p. 73) state 'We know of no convincing evidence for deleterious effects in the field that can confidently be ascribed to AM symbiosis'.

The potential breadth of contributions of AM fungi to important aspects of plant health and soil quality, underlying the notion that they might be used as an indicator of soil health, 'have received less emphasis than increases in production, probably because the economic benefits are less easily quantified or appreciated' (Smith and Read 1996, p. 454). Smith and Read (1996, ibid) also state that 'The possible economic benefits of managing mycorrhizal populations in agriculture and horticulture need to be critically assessed in the context of the ecology of the systems, not simply in the growth of the crops'. Acknowledging the difficulties inherent in such an analysis, we propose a framework as a means of assessing the economic benefits of arbuscular mycorrhizas in the context of agro-ecosystems (e.g. Smith and Smith 2011) while recognising their broader ecological and global context (e.g. Chaurasia 2004). Furthermore, the same roles that are exhibited in agricultural soils extend into diverse natural ecosystems, and some of these environmental resources indirectly benefit agricultural ecosystems (Ryan and Kirkegaard 2012). Indeed, as Ryan and Kirkegaard (2012, p. 50) state, 'the role of AMF in restoration of native plants and ecosystems on agricultural lands in Australia may merit investigation. Plants in Australian native ecosystems are colonised by AMF; although there may be a significant nonmycorrhizal component in some instances'.

In order to estimate economic values of mycorrhizas at either paddock or farm level, factors affecting the life cycle of AM fungi, especially the colonisation of roots by communities of these fungi, need to be quantified. However, there are risks to making such estimates if they are based on (1) inaccurate measurement of mycorrhizal hyphae in soil and in roots including discrepancies associated with measurement of root density and/or root architecture (see Gutjahr and Paszkowski 2013); (2) misunderstanding of the behaviour and measurement of colonisation of roots by AM fungi according to the method of identification of species, 'strain' or morphotype (see Shi et al. 2012); (3) inaccurate measurement of mycorrhizal function, including estimation of variation in contributions of different AM fungi throughout stages of the plant growth cycle (see Mickan et al. submitted); (4) inaccurate assessment of benefits and disbenefits due to failure to account for mycorrhizal variation within crop rotations (see Koide and Peoples 2012); (5) not recognising the discreet effects of C and N cycles on mycorrhizas and their interactions with P cycles through plant uptake and use (see Johnson 2010); (6) lack of recognition of effects of other soil organisms which may be both under- and overstated (see Lewandowski et al. 2013; Steinaker and Wilson 2008); (7) lack of recognition of effects of plant disease and other stressors leading to distorted quantification of mycorrhizal contributions (Hilou et al. 2014; Singh et al. 2013); and (8) inaccurate assessment associated with independent and inter-related climate or environmental attributes.

Risk minimisation strategies can be taken into account to deal with some or all of the factors that impede realistic economic valuation of mycorrhizas. Some of the risks apply widely, but others are more farm or paddock specific. Without even rudimentary local knowledge of AM fungi in agricultural ecosystems, there is potential that management practices will be used that fail to consider and consequently fail to capture potential benefits. Within the rhizosphere, AM fungi occur at the interface of soil biophysical and biochemical processes, and this central position warrants clarification of their role as an indicator of soil health.

AM fungi occur ubiquitously in agricultural systems and have a close affiliation with roots of most agricultural plants (Smith and Read 2008). Therefore, factors which influence their distribution, abundance, diversity, infectivity and longevity in roots and soil have the potential to be incorporated into an integrated indicator of soil health.

2.3   Arbuscular Mycorrhizal Measurement

Most demonstrations of benefits of AM fungi have been made in terms of increased early plant vigour associated with mycorrhizal function under controlled conditions, including inoculation in the field. In parallel, detrimental impacts have been widely reported, particularly during early stages of plant growth (Graham and Abbott 2000; Johnson and Graham 2013). It is more difficult to demonstrate mycorrhizal function under field conditions (Ryan and Angus 2003; Watts-Williams and Cavagnaro 2012). Gazey et al. (2004) demonstrated mycorrhizal benefits in terms of P uptake and growth of subterranean clover under field conditions in southwestern Australia using a phosphorus response curve approach that included an inoculation control. Ryan and Kirkegaard (2012) concluded there was little evidence of benefits of mycorrhizas in agricultural production systems commonly used in Australia, and indeed they found that some of these agronomic practices may reduce colonisation of roots by AM fungi. Given that the practices involved are based on considerable research to identify the best agronomic practices for sustaining production, there is an opportunity to explore whether this level of production uses practices that do not capture some components of soil biological fertility (Abbott and Murphy 2003) and that further investigation of the basis of 'sustainable production' that does not maximise contributions of soil flora and fauna is required. Generally, claims of mycorrhizal benefits in agricultural soils that relate to improving profitability rather than maximising productivity, as well as their possible role in the decontamination of soils polluted by residual organophosphates and their contribution to sustainability of crop production (Smith and Read 2008; Gazey et al. 2004; Delian et al. 2011; Albrechtova et al. 2012; Brito et al. 2012), are all in need of investigation within a framework that highlights intangible economic benefits.

Overall, while it is relatively easy to demonstrate mycorrhizal benefits under controlled conditions, including controlled field experiments, it is not easy to extend this to assessment of their benefits under '*normal agricultural field conditions*' because the fungi are ubiquitous. Even though different fungi have been shown to differ in their effectiveness (e.g. Smith et al. 2000; Graham and Abbott 2000), the extent to which this is translated into field soils where competition between fungi leads to differences in relative abundance in roots and in infectivity (based on relative inoculum potential) is difficult to measure. However, despite their ubiquity, the contributions of different AM fungi during plant growth stages under 'normal' agricultural field conditions are not well established. While it is known that different AM fungi have different capabilities to scavenge for P under P-limited conditions for plant growth (Schweiger et al. 2007; Thonar et al. 2011), the extent to which this plays out during stages of plant development is not clarified in 'normal' agricultural field conditions. Diversity in the life cycles of AM fungi in association with plants leads to changes in their relative abundance in root systems and in soil over time. For example, Pearson and Schweiger (1993, 1994) showed how understanding the life cycles of AM fungi in both roots and soil helped identification of the mechanism of competition between two fungi that occur

Table 2.1 Risks in assessing economic value of arbuscular mycorrhizas and potential remedies for overcoming such risks

Risks in assessing economic value of mycorrhizas	Remedy for overcoming risks in assessing economic value of mycorrhizas
Inaccurate measurement of mycorrhizal hyphae in soil and in roots associated with variation in root density and/or root architecture	Understand the relationship between root growth and mycorrhizal colonisation throughout the life cycle of plants in agricultural rotations
Inaccurate measurement of mycorrhizal function, including varying contributions of different fungi throughout the stages of the plant growth cycle	Understand the extent to which different mycorrhizal fungi colonise roots during the plant growth cycle and how this affects mycorrhizal contributions at different stages
Misunderstanding of the behaviour and measurement of colonisation of roots by species, strains and/or morphotypes of mycorrhizal fungi	Understand how communities of mycorrhizal fungi interact with one another in roots and whether this affects their ability to access P and water, and the ramification of hyphae in soil
Inaccurate assessment of benefits and disbenefits due to failure to account for mycorrhizal variation according to crop rotation	Understand how mycorrhizas contribute in sequences of crop and pasture species so that benefits can be accounted for seasonally
Not recognising the discreet effects of C and N cycles on mycorrhizas and their interactions with P	Understand interrelationships between mycorrhizas and C and N cycles in soil to calculate P and N fertiliser requirements that do not override potential mycorrhizal contributions
Lack of recognition of effects of other soil organisms which may be under or overstated	Understand how other soil organisms interact with mycorrhizal fungi
Lack of recognition of effects of plant disease and other stressors leading to distorted quantification of mycorrhizal contributions	Understand how mycorrhizal fungi interact with plant pathogens either to alleviate plant disease or to influence quantification of their abundance
Inaccurate assessment associated with independent and interrelated environmental and/or climate attributes	Understand how soil conditions such as salinity, acidity, compaction and waterlogging influence the life cycles of mycorrhizal fungi

commonly within roots of agricultural plants in southwestern Australia. Factors of significance were the dynamics of colonisation of roots associated with changes in sporulation and soluble carbohydrates. Given this degree of complexity, measurement of mycorrhizal fungi as '% root colonised' at one point in time may be of little relevance to estimation of the potential contribution of mycorrhizal fungi over an entire plant production cycle. Examples of the limitations in measurement of mycorrhizas and their function in 'normal' agricultural field conditions are illustrated below.

If mycorrhizas are not accurately measured, there will be risks in assessment of their potential contributions (Table 2.1). The measurement most commonly used is the proportion of root length colonised. However, there can be large variation in the density of root colonised and fungal structures within roots (McGonigle et al. 1990) and in the diameter of roots, all of which influence the total biomass of fungi present both inside the root and in the surrounding soil (Abbott and Robson 1984, 1985). Furthermore, these differences are not usually recorded (Gazey et al. 1992) and change with time (McGonigle 2001).

2.4 Mycorrhizal Benefits and Costs

Field studies of benefits of AM fungi are fewer than are glasshouse studies primarily because of the difficulties in establishing and monitoring experiments (McGonigle 1988). However, another factor in assessing the benefit of mycorrhizas in agricultural systems is that their contribution may be diffusely distributed amongst a number of areas, none of which reaches a threshold level, but when considered together, there is a benefit. Most studies focus on one aspect, and quantification relevant to assessing a wider suite of contributions can be prohibitive in terms of time and cost (Schnepf et al. 2008).

Where AM fungi contribute to P uptake, the benefit can be measured in terms of savings in fertiliser (e.g. Schweiger et al. 2007). There has been little consideration of potential savings in nitrogen fertiliser, but the close links between P and N cycles (Johnson 2010) mean that such attention is warranted. Evaluation of phosphorus-use efficiency of plants in crop rotations, continuous cropping or pasture production could include estimations of contributions of AM fungi. If this were done, there will be a clearer estimation of nitrogen fertiliser needs in agricultural systems. While there has been in-depth analysis of P and N fertiliser requirements for agricultural production (according to crop or pasture species for particular rotations and tillage practices), little attention has been paid to the potential roles that effective communities of AM fungi might contribute to these calculations. Where such contributions are not considered, there is a greater chance for potentially useful contributions of AM fungi to be overlooked. A logical stepwise process for N and P fertiliser recommendations could include first an estimate of P requirements that takes into account the potential benefit of AM fungi that are present. This would form a baseline for estimation of N fertiliser requirements. Where AM fungi were demonstrated to be likely to provide a benefit (because the 'right' fungi were present in the 'right' amounts for the crop/pasture sequence), then this could be taken into account. Where AM fungi were demonstrated to be unlikely to provide a benefit (because the 'wrong' fungi were present in the 'wrong' amounts for the crop/pasture sequence), then this could also be taken into account in terms of remediation required through agricultural management to restore mycorrhizal communities to a state where they can make close to their potential contribution (i.e. a state of equilibrium). Thus, understanding the state of the existing community of AM fungi underpins decisions about N and P fertiliser use for a given agricultural sequence. Clearly, AM fungi will have less to contribute under some circumstances than others, but the emphasis needs to be on the extent to which they are achieving their potential in a given situation.

Other benefits of AM fungi such as (1) facilitating plant access to moisture under drying soil conditions, (2) increasing retention of soil carbon by protecting it from microbial degradation via enhanced aggregation of soil particles and (3) creating a soil and rhizosphere environment that is more resilient to development of plant disease may be co-benefits of more effective supply of nutrients to plants, but they can also stand alone in situations where the AM fungi have no particular role in nutrient-use efficiency. This could occur in soils that are already well supplied with P and N for plant growth.

2.5   Is There a Link Between Mycorrhiza Measurement and Benefit?

The only way to obtain an idea of the economic benefits of arbuscular mycorrhizas is to ascertain the link between their presence and function and the impacts that they have on agricultural ecosystems or, more particularly in this instance, on productivity and/or profitability. In some cases, there may be a negative impact, or disbenefit, on plant growth, although Smith and Smith (2011) disputed this, and Veresoglou and Malley (2012) suggested that any potential disbenefits depended on the number and type of colonising mycorrhizal fungi. This is necessarily a complex process because of the number of variables involved.

Table 2.2 Variables, impacts and risks of assessing the economic benefits of mycorrhizas: fungal factors

Variable	Potential effect	Impact	Risk	Risk minimisation strategy
Growth rate of mycorrhizal fungi in roots and in soil	Mycorrhizal fungi promote soil aggregation and plant growth	Positive	May use an inaccurate measure of mycorrhizal growth and function	Use both proportion of root colonised and absolute amount and quantify mycorrhizal biomass
Type of mycorrhizal fungi present	Different species or subspecies might grow at different rates and have differing benefits to plant and soil	Positive	Misunderstanding behaviour of individual species or subspecies could cause inaccurate assessment of their benefits	Identify growth attributes and behaviour of species and subspecies present and their interactions
Number of mycorrhizal species of subspecies present	There may be several species or subspecies present in varying amounts and they might interact competitively or synergistically	Positive or negative	Ignorance of how mycorrhizal species or subspecies interact could result in ignorance of competition or synergism	Identify the way mycorrhizal species or subspecies interact and give value for synergistic or competing effects

Table 2.3 Variables, impacts and risks of assessing the economic benefits of mycorrhizas: soil and plant variables

Variable	Potential effect	Impact	Risk	Risk minimisation strategy
Level of soil phosphorus	Promotes plant and mycorrhizal growth but needs to be balanced	Positive or negative	P see-saw effect. Both too much and too little P inhibit mycorrhizal growth	Assume ~40 ppm is optimal level of soil P for mycorrhizal growth and soil quality
Plant characteristics (e.g. size and growth stage)	Plant attributes such as size and growth stage affect mycorrhizal colonisation and function	Positive or negative	The role of plant size and growth stage might lead to inaccurate assessment of number and size of hyphae	Identify impact of plant attributes such as size and growth stage on measure of hyphae
Crop cycle characteristics	Attributes of plant type and rotation type could affect mycorrhiza activity	Positive or negative	May lead to inaccurate assessment of benefits and disbenefits due to failure to account for mycorrhizal variation according to crops in cycle	Account for mycorrhizal attributes and association for each plant in a rotation
Soil carbon and nitrogen levels	Levels of soil carbon and N affect soil quality and may interact with P	Positive or negative	Not recognising the discreet effects of C and N cycles on mycorrhizas and interaction with P	Account for carbon and nitrogen cycles and interaction with phosphorus

Table 2.4 Variables, impacts and risks of assessing the economic benefits of mycorrhizas: other environmental or climatic factors

Variable	Potential effect	Impact	Risk	Risk minimisation strategy
Presence of other key soil organisms	Other soil organisms may have a positive or negative effect on mycorrhizal function	Positive or negative	If possible effects of other soil organisms are not recognised, the effects of mycorrhizas might be under or overstated	Identify any organisms that affect soil quality, plant growth and mycorrhizal function and quantify impact if possible
Presence of plant diseases and disease vectors	The presence of plant diseases and their spread by vectors will inhibit plant growth and may affect mycorrhizal function	Negative	If presence of plant diseases and other stressors is not recognised, their impact on plant growth and/or mycorrhizal function may distort mycorrhizal benefit assessment	Identify the impacts of plant diseases on plant growth and mycorrhizal function

(continued)

Table 2.4 (continued)

Variable	Potential effect	Impact	Risk	Risk minimisation strategy
Climate attributes	Variation in temperature, sunlight and rainfall might influence plant growth and mycorrhizal function	Positive or negative	Independent and interrelated climate attributes might lead to inaccurate assessment of mycorrhizal benefits	Identify independent and interrelated climate impacts on plant growth and mycorrhizal function
Interaction of variables	Identified variables might have a linear or exponential effect on mycorrhizal function	Positive or negative	Lack of recognition of interaction of variables might lead to inaccurate assessment of mycorrhizal benefits	Identify the extent and nature of all possible interactions between variables

In order to estimate economic values of mycorrhizas at paddock or farm level, various factors affecting mycorrhizal influences on plants and soil need to be assessed, characterised and quantified (Tables 2.2, 2.3 and 2.4).

2.6   Risk Minimisation Strategies

A simplistic way to obtain an estimate of the economic benefits of mycorrhizas is to estimate the value of crop production with and without mycorrhizas present, although this is difficult to do under field conditions (see Gazey et al. 2004). Given the wide range of variables influencing either production outcomes or profitability, as well the difficulties associated with accurate measurement of the mycorrhizas themselves, it is important to employ risk minimisation strategies and to monitor and control, as far as possible, the conditions under which such an estimate is made.

Risk minimisation strategies can be taken into account to deal with some or all of the factors that impede realistic economic valuation of mycorrhizas. Some of the risks apply widely, but others are more farm specific or even paddock specific. Clearly the range of crops, soil, disease and climate conditions is almost limitless although we have attempted to identify the risks and variables inherent in this type of assessment. In the first instance, case studies should be implemented on a farm-by-farm basis whereby the independent variables associated with cropping regime, climate, soil conditions, disease organisms and vectors can be held reasonably constant with the presence and nature of mycorrhizas being characterised. While it may not be possible to cultivate a plot devoid of mycorrhizas if, within the same vicinity, a plot with a significantly different mycorrhizal profile can be identified, then any difference in productivity can be attributed to the difference in mycorrhizal profile (see Gazey et al. 2004). A dollar value can then be calculated for the mycorrhizas present, at least in terms of production.

If case studies for multiple farms that accommodate identified risks and conditions can be designed for a range of cropping regimes, their benefits for different production systems and environments can be estimated and the magnitude of their influence on soil health can be inferred. In this way, evidence of the overall economic benefits of mycorrhizas in agricultural and horticultural ecosystems can be painstakingly constructed (Table 2.4). Because different crops have different responses to and aptitude for mycorrhizal colonisation, it is very important to ensure that the case studies cover a wide range of crops. As Smith and Read (1996, p. 454) have observed: 'Both cultivation and monoculture appear to change the species composition of the fungal populations and reduce their diversity, but the impact of these changes on crop production has not been adequately evaluated'. It is thus likely that mycorrhizas not only respond differently to different regimes but that their benefits might vary significantly between agricultural and natural ecosystems: they not only constitute an important element of biodiversity but they also respond to ecosystem biodiversity.

2.7 Conclusion

Although some of the contributions of mycorrhizas are well documented for reasons mentioned earlier, it is difficult to quantify their economic benefits in agricultural ecosystems. This is because there has been little work done, either to identify systematically all such benefits or to identify how variables that influence mycorrhizal function might interact with each other to influence overall benefits. To complicate matters further, it is possible that the nature and magnitude of such benefits might be site specific, so that all possible mycorrhizal impacts for specific rotations in specific paddocks during a particular season might need to be considered. Numerous studies have claimed explicit benefits for soil health and agricultural production from mycorrhizal colonisation. For example, Chaurasia (2004) viewed AM as having universal benefits for agriculture as well as for forests and other ecosystems, Smith and Read (1996, 2008) and Gazey et al. (2004) discussed their potential for improving crop productivity, while Delian et al. (2011) specifically referred to their role in increasing profitability. Albrechtova et al. (2012) mentioned their 'numerous benefits for sustainable crop production' as well as their possible role in the decontamination of soils polluted by residual organophosphates. Brito et al. (2012) also saw arbuscular mycorrhizas as having an important role in sustainable crop production, while other authors (e.g. Smith and Read 1996; Schulz 2001) explicitly mentioned economic benefits. It is important to note that all of the benefits mentioned above are, in fact, economic benefits. While most people tend to think of economics as particularly relating to commerce or finance, anything through which benefits accrue to humanity is deemed to be economic ('economics' means 'humanity's household', while 'ecology' means 'nature's household'). This is one reason that the importance of nonmarket (intangible) values has been stressed here, especially as it relates to soil biodiversity. In its briefing paper, 'Valuing Nature', UNEP (2014, p. 1) observed that 'Part of the challenge is that the sheer range of benefits from ecosystems is often poorly understood. The term "ecosystem services"—the benefits derived from nature—is a useful concept for making the value of nature more explicit and relevant to human well being'. As mycorrhizas are part of soil biodiversity, and that they are part of an agricultural ecosystem, the 'sheer range' of benefits even from a relatively small-scale ecosystem is difficult to reflect accurately. While it is possible that unidentified elements and unknown benefits of mycorrhizas might be omitted from agricultural studies, thus reducing perceptions of their economic worth, it is also probable that their presence in agricultural ecosystems will have wider, undervalued, benefits to natural ecosystems, and vice versa.

References

Abbott LK, Murphy DV (eds) (2003) Soil biological fertility: a key to sustainable land use in agriculture. Kluwer, Dordrecht, The Netherlands

Abbott LK, Robson AD, Gazey C (1992) Selection of inoculant VAM fungi. In: Norris JR, Read DJ, Varma AK (eds) Methods in microbiology: experiments with mycorrhizas. Academic Press, London, pp 1–21

Abbott LK, Robson AD (1984) The effect of root density, inoculum placement and infectivity of inoculum on the development of vesicular-arbuscular mycorrhizas. New Phytol 97:285–299

Abbott LK, Robson AD (1985) Formation of external hyphae in soil by four species of vesicular-arbuscular mycorrhizal fungi. New Phytol 99:245–255

Albrechtova J, Latr A, Nederost L, Pokluda R, Posta K, Vosatka M (2012) Dual inoculation with mycorrhizal and saprophytic fungi applicable in sustainable cultivation improves the yield and nutritive value of onion. Sci World J, Article ID 374091, 8 pp

Azćon-Aguilar C, Barea JM (1996) Arbuscular mycorrhizas and biological control of soil-borne plant pathogens – an overview of the mechanisms involved. Mycorrhiza 6:457–464

Baker R, Ruting B (2014) Environmental policy analysis: a guide to non-market valuation. Staff Working Paper, Productivity Commission, Australian Government. Accessed online 1 April 2014; www.pc.gov.au/research/staff-working/non-market-valuation

Bishop J (ed) (2013) The economics of ecosystems and biodiversity in business and enterprise. Routledge, Abingdon, Oxon, p 296

Brito I, Goss MJ, De Carvalho M (2012) Effect of tillage and crop on arbuscular mycorrhiza colonisation of winter wheat and triticale under Mediterranean conditions. Soil Use Manage 28:202–208

Chaurasia B (2004) Vesicular arbuscular mycorrhiza: a potential biofertiliser. ENVIS Newsl: Himal Ecol 1:1–2

Delian E, Chira A, Chira L, Savulescu E (2011) Arbuscular mycorrhizae: an overview. SW J Horticult Biol Environ 2:167–192

Dobell AR (1995) Environmental degradation and the religion of the market. In: Coward H (ed) Population, consumption and the environment. State University of New York Press, Albany, pp 229–250

Endlweber K, Scheu S (2007) Interactions between mycorrhizal fungi and Collembola: effects on root structure of competing plant species. Biol Fertil Soils 43:741–749

Evelyn H, Kapoor R, Giri B (2009) Arbuscular mycorrhizal fungi in alleviation of salt stress: a review. Ann Bot 104:1263–1280

Gazey C, Abbott LK, Robson AD (1992) The rate of development of mycorrhizas affects the onset of sporulation and production of external hyphae by two species of *Acaulospora*. Mycol Res 96:643–650

Gazey C, Abbott LK, Robson AD (2004) Indigenous and introduced arbuscular mycorrhizal fungi contribute to plant growth in two agricultural soils from south-western Australia. Mycorrhiza 14:355–362

Graham JH, Abbott LK (2000) Wheat responses to aggressive and non-aggressive arbuscular mycorrhizal fungi. Plant Soil 220:207–218

Gutjahr C, Paszkowski U (2013) Multiple control levels of root system remodelling in arbuscular mycorrhizal symbiosis. Front Plant Sci 4: Article 204. doi:10.3389/fpls.2013.00294

Hilou A, Zhang H, Franken P, Hause B (2014) Do jasmonates play a role in arbuscular mycorrhiza-induced local bioprotection of *Medicago truncatula* against root rot disease caused by *Aphanomyces euteiches*? Mycorrhiza 24:45–54

Jastrow JD, Miller RM, Lussenhop J (1998) Contributions of interacting biological mechanisms to soil aggregate stabilization in restored prairie. Soil Biol Biochem 30:905–916

Johnson NC (2010) Resource stoichiometry elucidates the structure and function of arbuscular mycorrhizas across scales. New Phytol 185:631–647

Johnson NC, Graham JH (2013) The continuum concept remains a useful framework for studying mycorrhizal functioning. Plant Soil 363:411–419

Juniper S, Abbott LK (1993) Vesicular-arbuscular mycorrhizas and soil salinity. Mycorrhiza 4:45–57

Koide RT, Peoples MS (2012) On the nature of temporary yield loss in maize following canola. Plant Soil 360:259–269

Lehmann J, Joseph S (eds) (2009) Biochar for environmental management, science and technology. Earthscan, London

Lewandowski TJ, Dunfield KE, Antunes PM (2013) Isolate identify determines plant tolerance to pathogen attack in assembled mycorrhizal communities. PLoS One 8:e61329. doi:10.1371/journal.pone.0061329

Loy DR (1997) The religion of the market. J Am Acad Relig 65(2):275–290

Lumley S (2013) Sordid Boon? The context of sustainability in historical and contemporary global economics. Academica Press, Palo Alto

Manoharan PT, Shanmugaiah V, Balasubramanian N, Gomathinayagam S, Sharma MP, Muthuchelian K (2010) Influence of AM fungi on the growth and physiological status of *Erythrina variegata* Linn. grown under different water stress conditions. Eur J Soil Biol 46:151–156

Martinez-Alier J (1987) Ecological economics. Energy, environment and society. Basil Blackwell, Oxford

McGonigle TP, Miller MH, Evans DG, Fairchile GL, Swan JA (1990) A new method which gives an objective measure of colonization of roots by arbuscular-mycorrhizal fungi. New Phytol 115:495–501

McGonigle TP (1988) A numerical analysis of published field trials with vesicular-arbuscular mycorrhizal fungi. Funct Ecol 2:473–478

McGonigle TP (2001) On the use of non-linear regression with the logistic equation for changes with time of percentage root length colonized by arbuscular mycorrhizal fungi. Mycorrhiza 10:249–254

Mickan B, Abbott LK, Stephanova K, Solaiman ZM (submitted) Demonstrated mechanisms for interactions between biochar and mycorrhizal fungi in water-deficient agricultural soil

Miller M, McGonigle T, Addy H (1994) An economic approach to evaluate the role of mycorrhizas in managed ecosystems. Plant and Soil 159:27–35

Pearce D (1995) Blueprint 4. Capturing global environmental value. Earthscan, London

Pearce D (2002) An intellectual history of environmental economics. Annu Rev Energy Environ 27:57–81

Pearson JN, Abbott LK, Jasper DJ (1993) Mediation of competition between two colonizing VA mycorrhizal fungi by the host plant. New Phytol 123:93–98

Pearson JN, Abbott LK, Jasper AD (1994) Phosphorus, soluble carbohydrates and the competition between two arbuscular mycorrhizal fungi colonizing subterranean clover. New Phytol 127:101–106

Pearson JM, Schweiger P (1993) *Scutellospora calospora* (Nicol and Gerd) associated with subterranean clover – dynamics of colonization, sporulation and soluble carbohydrates. New Phytol 127:697–701

Pearson JM, Schweiger P (1994) *Scuttelosposa calospora* (Nicol and Gerd) Walker and Sanders associated with subterranean clover produces non-infective hyphae during sporulation. New Phytol 124:215–219

Pimental D, Wilson C, McCallum C, Huang R, Dwen P, Flack J, Tran Q, Saltman T, Cliff B (1997) Economic and environmental benefits of biodiversity. Bioscience 47:747–757

Rillig MC, Mummey DL (2006) Tansley review – mycorrhizas and soil structure. New Phytol 171:41–53

Ryan MH, Angus JK (2003) Arbuscular mycorrhizae in wheat and field pea crops on a low P soil: increased Zn uptake but no increase in P-uptake or yield. Plant Soil 250:225–239

Ryan MH, Kirkegaard JA (2012) The agronomic relevance of arbuscular mycorrhizas in the fertility of Australian extensive cropping systems. Agricult Ecocsyst Environ 163:37–53

Sano SM, Abbott LK, Solaiman Z, Robson AD (2002) Influence of liming, inoculum level and inoculum placement on root colonization of subterranean clover. Mycorrhiza 12:285–290

Schnepf A, Roose T, Schweiger P (2008) Growth model for arbuscular mycorrhizal fungi. J R Soc Interface 5:773–784

Schulz C (2001) Effect of (vesicular-) arbuscular mycorrhiza on survival and post-vitro development of micropropagated oil palms (*Elaeis guineensis* Jacq.). Doctoral Thesis, University of Göttingen

Schwartz MW, Hoeksema JD, Gehring CA, Johnson NC, Klironomos JN, Abbott LK, Pringle A (2006) The promise and the potential consequences of the global transport of mycorrhizal fungal inoculum. Ecol Lett 9:501–515

Schweiger PF, Robson AD, Barrow NJ, Abbott LK (2007) Arbuscular mycorrhizal fungi from three general induce two-phase plant growth responses on a high P-fixing soil. Plant Soil 292:181–192

Shi P, Abbott LK, Banning NC, Zhao B (2012) Comparison of morphological and molecular genetic quantification of relative abundance of arbuscular mycorrhizal fungi within roots. Mycorrhiza 22:501–513

Singh R, Sunit KS, Alok K (2013) Synergy between *Glomus fasciculatum* and a beneficial *Pseudomonas* in reducing root diseases and improving yield and forskolin content in *Coleus forskohlii* Briq. under organic field conditions. Mycorrhiza 23:35–44

Smith SE, DJ Read (1996) *Mycorrhizal symbiosis*, 2nd edn. Academic, London

Smith SE, Read DJ (2008) Mycorrhizal symbiosis. Academic Press, London, p 800

Smith FA, Smith SE (2011) What is the significance of the arbuscular mycorrhizal colonisation of many economically important crop plants? Plant Soil 348:63–79

Smith SE, Smith FA (2012) Fresh perspectives on the roles of arbuscular mycorrhizal fungi in plant nutrition and growth. Mycologia 104:1–13

Smith FA, Jakobsen I, Smith SE (2000) Spatial differences in acquisition of soil phosphate between two arbuscular mycorrhizal fungi in symbiosis with *Medicago truncatula*. New Phytologist 147:357–366

Steinaker DF, Wilson SD (2008) Scale and density dependent relationships among roots, mycorrhizal fungi and collembola in grassland and forest. Oikos 117:703–710

Thonar C, Frossard E, Smilauer P, Jansa J (2014) Competition and facilitation in synthetic communities of arbuscular mycorrhizal fungi. Mol Ecol 23:733–746

Thonar C, Schnepf A, Frossard E, Roose T, Jansa J (2011) Traits related to differences in function among three arbuscular mycorrhizal fungi. Plant Soil 339:231–245

UNEP (2014) 'Valuing nature', green economy briefing paper. United Nations Environment Program. www.unep.org/greeneconomy/ResearchProducts/GEBriefingPapers/. Accessed on 3 April 2014

Veresoglou SD, Malley JM (2012) A model that explains diversity patterns of arbuscular mycorrhizas. Ecol Model 231:146–152

Watts-Williams S, Cavagnaro TR (2012) Arbuscular mycorrhizas modify tomato responses to soil zinc and phosphorus addition. Biol Fertil Soils 48:285–294

Chapter 3
Contribution of Dynamics of Root Colonisation by Arbuscular Mycorrhizal Communities to Ecosystem Function

Sutarman Gafur

3.1 Introduction

Arbuscular mycorrhizas (AM) form potentially symbiotic associations between species of Glomeromycota fungi and the roots of the majority of vascular plant species (Smith and Read 2008). Roots from natural ecosystems contain various mycorrhizal taxa (Brundrett and Abbott 1991, 1994; Clapp et al. 1995; Merryweather and Fitter 1998b; Moyersoen and Fitter 1999; Helgason et al. 1999). Most soils contain communities of AM fungi (Cuenca et al. 1998; Smilauer 2001). Investigation of root colonisation by communities of AM fungi has been hampered by the difficulty in distinguishing among fungi inside roots, but some morphological characteristics (Abbott 1982; Merryweather and Fitter 1998a) and molecular techniques (Turnau et al. 2001; Kohout et al. 2013) are overcoming these limitations.

The dynamics of root colonisation by communities of AM fungi is associated with their capacity to form propagules, their tolerance of environmental conditions and their competitive ability (Abbott and Gazey 1994). It is also influenced by the host plant through the availability of carbon substrates needed for fungal growth (Pearson et al. 1994) and root architecture (Hetrick 1991). The form and infectivity of propagules of AM fungi within soil are also important (Brundrett and Abbott 1991), and this in turn is associated with prior colonisation of roots, and relative susceptibility of plant species present to colonisation by different fungi present in the community.

Mycorrhizal communities are complex (Read 1991; Holland et al. 2014) and the fungi can interact with each other within these communities (Pearson et al. 1994). As a result of these interactions, the relative abundance of infective spores and

S. Gafur (✉)
Department of Soil Science, The University of Tanjungpura, West Kalimantan 78124, Indonesia
e-mail: sutarmangafur@gmail.com

© Springer-Verlag Berlin Heidelberg 2014
Z.M. Solaiman et al. (eds.), *Mycorrhizal Fungi: Use in Sustainable Agriculture and Land Restoration*, Soil Biology 41, DOI 10.1007/978-3-662-45370-4_3

hyphae in soil varies in association with different life cycle strategies of AM fungi. An understanding of the dynamics of colonisation of roots by communities of AM fungi is required for predicting the likely success of inoculation with introduced fungi (Dodd et al. 2002; Bell et al. 2003) and for selecting soil management procedures (Abbott and Gazey 1994) for maximising their benefits.

3.2   Colonisation of Roots by AM Fungi

The emergence of molecular tools for identifying AM fungi in roots has highlighted the complexity of root colonisation by these ubiquitous soil organisms (Holland et al. 2014) although not all are extremely diverse (Shi et al. 2012). As root colonisation by communities of AM fungi is dependent on the characteristics of the host plant, the soil as a medium for the plant to grow in and the characteristics of the fungi, the relative abundance of fungi present can change over time (Sanchez-Castro et al. 2012).

The infectivity of fungi can be associated both with differences in inoculum level and differences in their ability to colonise roots (Srivastava et al. 1996; Wilson and Tommerup 1992). Abbott et al. (1992) also explained that competitiveness among AM fungi depends on their relative infectivity but that the quantity of inoculum can interfere with this relationship. The types of propagules include hyphal fragments, living and dead/dried roots and spores. The relative abundance of forms of inocula can contribute to the infectivity of the fungi to different extents. Although dried roots of subterranean clover contained an effective form of inoculum for several AM fungi, they were ineffective as inocula for other AM fungi (Tommerup and Abbott 1981). Differences among AM fungi include physiological processes such as in bidirectional nutrient transport in fungal hyphae or in carbon metabolism as was indicated by the decline in infectivity following sporulation by *S. calospora* (Pearson and Schweiger 1994) and *A. laevis* (Jasper et al. 1993). In addition to the propagule characteristics of AM fungi, their capacity to colonise roots also depends upon how they react to chemical, physical and biological properties of soil. Soil organic matter, including plant root residues, influences soil structure, soil pH, nutrient and soil water-holding capacity, all of which, alone or in combination, can influence mycorrhizal colonisation (Gaur et al. 1998; Hamel et al. 1997; Nadian et al. 1998).

Mycorrhizal fungi have the potential to interact with a wide range of other soil organisms in the root, in the rhizosphere and in the bulk soil (Linderman 1992; Andrade et al. 1998). These interactions have a range of effects; some are competitive and others may be mutually beneficial. Effects can be seen at all stages of the AM fungal life cycle, from spore population dynamics through to root colonisation and external hyphal growth (Fitter and Garbaye 1994). Increased branching and orientation of the hyphae towards the root may enhance the subsequent process of colonisation (Tamasloukht et al. 2003), and AM fungi may produce substances that are antagonistic to other rhizosphere organisms (Anderson 1992) which in turn could influence colonisation of roots.

Root anatomy can influence mycorrhizal colonisation, but AM fungi can also influence root structure. Plant species that have few root hairs are strongly mycorrhiza-dependent in phosphorus-deficient soils (Crush 1974; Schweiger et al. 1995). Secondary thickening of cell walls can affect colonisation, with invading fungi tending to penetrate and colonise cells that have little or no suberin deposition (Srivastava et al. 1996). AM fungal hyphae can grow easily on and inside young roots without secondary thickening; certain areas of a root may also be more susceptible to colonisation. Intercellular hyphae are formed more frequently as the fungi spread into the inner cortex and colonisation can be intense in the deeper layers of root cortex where fungi form complex intracellular arbuscules (Bonfante-Fasolo and Vian 1989). Differences in colonisation due to root diameter were highlighted by Fitter and Merryweather (1992) supporting evidence that plants with coarser roots and few root hairs were more typically mycorrhizal (Baylis 1970), but the extent of colonisation of fine-rooted plants varies with fungal isolate (Schenck and Smith 1982).

Root density, distribution and growth rate are each relevant to the formation of a primary entry point by AM fungi because the fungus must first intercept a root (Abbott et al. 1984). The intercepting hyphae may grow from fungal propagules in soil or from ramification of hyphae associated with mycorrhizal roots. Root density differs among plant species and over time and has the potential to influence colonisation by AM fungi. Therefore, consideration of the structure and distribution of roots (Hetrick 1991) are important factors to be taken into account in preparing a strategy to improve colonisation of roots by AM fungi.

Root exudates can influence root colonisation by AM fungi through influences on hyphal growth in soil and spore germination both positively and negatively (Gianinazzi-Pearson et al. 1990; Graham 1982). Elias and Safir (1987) observed that hyphal elongation of *G. fasciculatus* was enhanced by exudates from *Trifolium repens* but only when the plants were grown under phosphate deficiency. The function of the exudates as a promoter of fungal growth decreased when phosphate deficiency was overcome. In contrast, root exudates from non-mycorrhizal and mycorrhizal peas inhibited hyphal growth of *Gigaspora margarita* (Balaji et al. 1995). Mycorrhizal *Pisum sativum* and its non-mycorrhizal isogenic mutant were not different with respect to effects of their root exudates on *G. mosseae* (Giovanetti et al. 1993). The amount, type and complexity of the volatile compounds exuded from roots have the potential to influence fungal reactions before the fungus meets the root as well as after the hyphae contact the root. In addition, later stages of mycorrhiza development, such as the formation of appressoria and hyphal penetration of the root surface, may involve exudate molecules in recognition processes (Koske and Gemma 1992).

3.3   Relative Abundance of AM Fungi Within Roots

The relative abundance of AM fungi within roots is influenced by changes in environmental conditions and interactions with host plants (Sanchez-Castro et al. 2012). As this depends on interactions between fungi, environmental changes and presence of host plants, it is difficult to formulate a simple mechanism for the process of root colonisation by communities of AM fungi. Knowledge of phenomena related to both the quantity and types of propagules within the soil may not be sufficient for predicting the dynamics of root colonisation (Bowen 1987). Furthermore, one of the characteristics of communities of AM fungi is strong unevenness in the relative abundance of fungi (Allen et al. 1995; Brundrett and Abbott 1995). Thus, investigations of characteristics of individual AM fungi merely provide a starting point for understanding processes involved during simultaneous colonisation of roots by genetically diverse fungi.

Before microscopic observation of colonisation of AM fungi in roots, root sample is first cleared and stained using classic stains such as trypan blue (Phillips and Hayman 1970), acid fuchsin (Saito et al. 1993) and chlorazol black E (Brundrett et al. 1984). Other visual techniques involving staining include enzyme analysis (Rosendahl and Sen 1994; Tisserant et al. 1998) and fluorescent antibody techniques (Wilson et al. 1983). The emergence of molecular tools has greatly expanded the ability to detect communities of AM fungi in roots (Kohout et al. 2013), and they are occasionally used in conjunction with morphological methods (Shi et al. 2012). Morphological observations enable structural features of AM fungi in roots (Dickson 2004). There are discrepancies between quantifications of relative abundance of AM fungi in roots using both morphological (Abbott 1982; Merryweather and Fitter 1991) and molecular approaches (Robinson-Boyer et al. 2009) which are further highlighted when the same roots are assessed using different methodologies (Shi et al. 2012).

There have been many studies of the diversity of AM fungi in field soils based on spore type and abundance (e.g. Brundrett and Abbott 1994; Cuenca et al. 1998; Franke-Snyder et al. 2001). However, there is little relationship between the presence and abundance of spores of particular species of AM fungi in soil and the extent to which they are present within roots growing in the soil (Merryweather and Fitter 1998a; Scheltema et al. 1987). This may be partly due to preferential colonisation of roots (Helgason et al. 2002) and to host-dependent patterns of colonisation (Bever et al. 1996).

The abundance, distribution, effectiveness and aggressiveness of each AM fungus species within a community are determining factors of competitive success (Graham and Abbott 2000; Wilson and Tommerup 1992). High density of fungal propagules and localised distribution increased the rate of colonisation of roots compared to that of low density and dispersion of fungal propagules in soil (Wilson and Trinick 1983). Critical density levels for propagules may vary with species of fungus. In field soil, propagule densities can be highly variable even between adjacent soil cores (Brundrett and Abbott 1995). Furthermore, AM fungi differ in their biological characteristics such as spore dormancy period and type of propagules (Abbott et al. 1992; Tommerup 1983; Tommerup and Abbott 1981) which combined with heterogeneity in distribution of infective propagules in soil will lead to difference in the proportion of each AM fungus within roots from time to time.

As AM fungi differ in their tolerance of conditions such as soil pH, nutrient content, water-holding capacity, soil organic matter, soil disturbance and other soil organisms, it is expected that these characteristics may result in the dominance of certain AM fungi in soil. Consequently, the dominance of fungi within the roots might change, but not necessarily in direct proportion to the abundance of spores (Merryweather and Fitter 1998a; Scheltema et al. 1987).

While most AM fungi can associate with a wide range of hosts, their performance relative to each other depends on the host characteristics. A comparison of inoculation with AM fungi on different host plants showed that *Polianthes* was highly colonised compared to *Capsicum* (Gaur et al. 1998). Studies of host dependence and species diversity of AM fungi in grassland (Bever et al. 1996) demonstrated that co-occurring plant species supported very different rates of sporulation by AM fungi. These differences were not affected by the time of sampling, suggesting that they reflect host-dependent differences in fungal growth rates, rather than host-dependent timing of sporulation. It was hypothesised that the host dependence of the relative growth rates of fungal populations may play an important role in the maintenance of AM fungal species diversity (Bever et al. 1996). Thus, the relationship between sporulation and the extent of colonisation by AM fungi is complex (Douds and Schenck 1990; Gazey et al. 1992) further compounding the dynamics of mycorrhiza formation by individuals within communities of AM fungi.

There is a high degree of variability in mycorrhizal dependency among host plants (Hoeksema et al. 2010). Plants range from highly dependent, whereby the plants are unable to survive without mycorrhizas in highly phosphate-deficient soils, to low dependency, where plants can survive under some conditions without mycorrhizas when phosphate is highly deficient (Janos 1980). This could be related to the fact that species of AM fungi differ in their ability to take up phosphorus and transfer it to the host plant (Solaiman and Abbott 2008) as well as to preferential colonisation of roots of some plants by some AM fungi (Helgason et al. 2002). Further investigation of the relevance of diversity within communities of AM fungi and mycorrhizal dependency is recommended.

Soil disturbance, including soil removal during mining operations (Jasper et al. 1989a, b, c, 1992), soil erosion (Powell 1980), mechanical disruption (Cuenca and Lovera 1991) and tillage (Kabir et al. 1997; Jansa et al. 2003) can reduce the abundance of AM fungi in roots. Species diversity in a community of AM fungi may also be reduced (Cuenca and Lovera 1991). Soil disturbance may have an indirect effect on the presence and abundance of AM fungi within roots by reducing the inoculum potential of members of the community of AM fungi present. However, the percentage of root length colonised by AM fungi in soil from an annual pasture was not decreased after disturbance, whereas colonisation of plants grown in disturbed soil from forest or heathland was only half that of the undisturbed soil

(Jasper et al. 1991). These differences were correlated with the number of infective propagules that survived the disturbance treatment. Another study on propagules of AM fungi in a disturbed habitat in the Kakadu region of tropical Australia revealed that propagules of AM and ectomycorrhizal (ECM) fungi occurred in all sites (Brundrett et al. 1996) and were sporadically distributed in highly disturbed areas. Both the relative abundance and frequency of occurrence of inoculum of AM and ECM fungi increased with vegetation cover in the older disturbed sites examined in this study.

Soil disturbance may change the effect that these communities of AM fungi have on host plants depending on how the disturbance influences each fungus and how it affects the way the fungi influence each other. In highly disturbed environments, low inoculum densities can lead to slow or low levels of colonisation of roots (Bellgard 1993; Jasper et al. 1989b). In contrast, seedlings in undisturbed vegetation may become rapidly colonised when their roots contact existing mycorrhizal hyphae in shallow layers of soil, reflecting a greater concentration of propagules in surface soils (Bellgard 1993).

Mining activities can reduce the abundance of propagules of AM fungi (Jasper et al. 1992) and destroy their infectivity in relocated topsoil (Jasper et al. 1989b). Severe soil disturbance has the pronounced effects of separating much of the external hyphae from the host root and of breaking up the soil hyphal network (-Jasper et al. 1989c; Miller and Jastrow 1990). Therefore, AM fungi which depend on intact hyphae as propagules may be less competitive in forming mycorrhizas than those which rely on more robust propagules such as spores when soil is disturbed.

3.4   Function of Communities of AM Fungi

As field soils contain communities of AM fungi, there can be no single mycorrhizal effect on plant growth. However, due to the difficulties in studying AM fungi as communities, most investigations of the function of AM fungi focus at the level of an individual species or isolates of AM fungus under controlled conditions. Nevertheless, the function of AM fungi varies depending on the environmental conditions, plant and AM fungus species, and observations of function under field conditions are the result of combined effects of dynamic communities.

The hyphae of AM fungi can extend up to several centimetres from the root, effectively extending the zone of nutrient depletion around roots to absorb immobile elements from the bulk soil (Jakobsen et al. 1992). A high diversity of AM fungi may be important for buffering an ecosystem against disturbance (Vogt et al. 1997). The number of AM fungal spores and abundance of mycorrhizal roots fluctuate with season (Brundrett and Kendrick 1988) associated with differences in life cycles and spore dormancy periods, but the effect of this instability could be minimised when the fungi are present in communities if they have complementary effects (Koide 2000). Consequently, the fluctuation in root colonisation levels is expected to be less when roots are colonised by several species of fungi than if a single fungal species is present. The extent to which this influences mycorrhizal function is less clear.

3.5 Conclusion

The dynamics of root colonisation by communities of AM fungi occur in parallel with the changes in abundance of individual species of AM fungi. Characteristics of the fungi play a significant role in the fluctuation of root colonisation and demonstrate that it is not easy to predict root colonisation dynamics by communities of AM fungi based on the knowledge of the individual characteristics of AM fungi within the community. Differences in physiological characteristics of members of the AM fungal community are likely to influence the dynamics of root colonisation. Soil disturbance can also affect different AM fungi in different ways and this will alter the relative abundance of AM fungi in soil and in roots. This differential sensitivity and tolerance of particular fungi to soil characteristics and soil disturbance will also be dependent on the propagule potential in soil and emphasises that short-term studies overlook the possibility of identifying long-term contributions of communities of AM fungi in field soils.

References

Abbott LK (1982) Comparative anatomy of vesicular-arbuscular mycorrhizas formed on subterranean clover. Aust J Bot 30:485–499

Abbott LK, Gazey C (1994) An ecological view of the formation of VA mycorrhizas. Plant Soil 159:69–78

Abbott LK, Robson AD, De Boer G (1984) The effect of phosphorus on the formation of hyphae in soil by the vesicular arbuscular mycorrhizal fungus, *Glomus fasciculatum*. New Phytol 97:437–446

Abbott LK, Robson AD, Gazey C (1992) Selection of inoculant vesicular-arbuscular mycorrhizal fungi. In: Norris JR, Read DJ, Varma AK (eds) Techniques for mycorrhizal research. Academic Press, London, pp 1–21

Allen EB, Allen MF, Helm DJ, Trappe JM, Molina R, Rincon E (1995) Patterns and regulation of mycorrhizal plant and fungal diversity. Plant Soil 170:47–62

Anderson AJ (1992) The influence of the plant root on mycorrhizal formation. In: Allen MF (ed) Mycorrhizal functioning. Chapman and Hall, New York, NY, pp 37–64

Andrade G, Linderman RG, Bethlenfalvay GJ (1998) Bacterial associations with the mycorrhizosphere and hyphosphere of the arbuscular mycorrhizal fungus Glomus mosseae. Plant Soil 202:79–87

Balaji B, Poulin MJ, Vierheilig H, Pichié Y (1995) Responses of an arbuscular mycorrhizal fungus, *Gigaspora margarita*, to exudates and volatiles from the Ri T-DNA transformed roots of nonmycorrhizal and mycorrhizal mutants of *Pisum sativum* L Sparkle. Exp Ecol 19:275–283

Baylis GTS (1970) Root hairs and phycomycetous mycorrhizas in phosphorus deficient soil. Plant Soil 33:713–716

Bell J, Wells S, Jasper DA, Abbott LK (2003) Field inoculation with arbuscular mycorrhizal fungi in rehabilitation of mine sites with native vegetation, including *Acacia* spp. Aust J Syst Bot 16:131–138

Bellgard SE (1993) The topsoil as the major store of the propagules of vesicular-arbuscular mycorrhizal fungi in southeast Australian sandstone soils. Mycorrhiza 3:19–24

Bever JD, Morton JB, Antonovics J, Schultz PA (1996) Host-dependent sporulation and species diversity of arbuscular mycorrhizal fungi in a mown grassland. J Ecol 84:71–82

Bonfante-Fasolo P, Vian B (1989) Cell wall architecture in mycorrhizal roots of *Allium porrum* L. Annales des Sciences Naturelles-Botanique et Biologie Vegetale 10:97–109

Bowen G (1987) The biology and physiology of infection and its development. In: Safir GR (ed) Ecophysiology of VA mycorrhizal plants. CRC, Boca Raton, FL, pp 27–57

Brundrett MC, Abbott LK (1991) Roots of jarrah forest plants. I. Mycorrhizal associations of shrubs and herbaceous plants. Aust J Bot 39:445–457

Brundrett MC, Abbott LK (1994) Mycorrhizal fungus propagules in the jarrah forest. I. Seasonal study of inoculum levels. New Phytol 127:539–546

Brundrett MC, Abbott LK (1995) Mycorrhizal fungus propagules in the jarrah forest. II. Spatial variability in inoculum levels. New Phytol 131:461–469

Brundrett MC, Kendrick B (1988) The mycorrhizal status, root anatomy and phenology of plants in a sugar maple forest. Can J Bot 66:1153–1173

Brundrett MC, Piche Y, Peterson RL (1984) A new method for observing the morphology of vesicular-arbuscular mycorrhizae. Can J Bot 62:2128–2134

Brundrett MC, Bougher N, Dell B, Grove T, Malajczuk N (1996) Working with mycorrhizas in forestry and agriculture. ACIAR Monograph 32

Clapp JP, Young JPW, Merryweather JW, Fitter FH (1995) Diversity of fungal symbionts in arbuscular mycorrhizas from a natural community. New Phytol 130:259–265

Crush JR (1974) Plant growth response to vesicular arbuscular mycorrhiza. VII. Growth and nodulation of some herbage legumes. New Phytol 73:743–749

Cuenca G, Lovera M (1991) Vesicular arbuscular-mycorrhizae in disturbed and revegetated sites from La Gran Sabana, Venezuela. Can J Bot 70:73–79

Cuenca G, De Andrade Z, Escalante G (1998) Diversity of glomalean spores from natural, disturbed and revegetated communities growing on nutrient-poor tropical soils. Soil Biol Biochem 30:711–719

Dickson S (2004) The Arum-Paris continuum of mycorrhizal symbioses. New Phytol 163:187–200

Dodd JC, Dougall TA, Clapp JP, Jeffries P (2002) The role of arbuscular mycorrhizal fungi in plant community establishment at Samphire Hoe, Kent, UK – the reclamation platform created during the building of the Channel tunnel between France and UK. Biodivers Conserv 11:39–58

Douds DD, Schenck NC (1990) Relationship of colonisation and sporulation by VA mycorrhizal fungi to plant nutrient and carbohydrate contents. New Phytol 116:621–627

Elias KS, Safir GR (1987) Hyphal elongation of Glomus fasciculatus in response to root exudates. Appl Environ Microbiol 53:1928–1933

Fitter AH, Garbaye J (1994) Interactions between mycorrhizal fungi and other soil organisms. Plant Soil 159:123–132

Fitter AH, Merryweather JW (1992) Why are some plants more mycorrhizal than others? An ecological enquiry. In: Read DJ, Lewis DH, Fitter AH, Alexander IJ (eds) Mycorrhizas in ecosystems. C.A.B. International/University Press, Cambridge, pp 26–36

Franke-Snyder M, Douds DD, Galvez L, Phillips JG, Wagoner P, Drinkwater L, Morton JB (2001) Diversity of communities of arbuscular mycorrhizal (AM) fungi present in conventional versus low-input agricultural sites in eastern Pennsylvania, USA. Appl Soil Ecol 16:35–48

Gaur A, Adholeya A, Mukerji KG (1998) A comparison of AM inoculants using Capsicum and Polianthes in marginal soil amended with organic matter. Mycorrhiza 7:307–312

Gazey C, Abbott LK, Robson AD (1992) The rate of development of mycorrhizas affects the onset of sporulation and production of external hyphae by two species of *Acaulospora*. Mycol Res 96:643–650

Gianinazzi-Pearson V, Branzati B, Gianinazzi S (1990) In vitro enhancement of spore germination and early hyphal growth of a vesicular-arbuscular mycorrhizal fungus by host root exudates and plant flavonoids. Symbiosis 7:243–255

Giovanetti M, Avio L, Sbrana C, Citernesi AS (1993) Factors affecting appressorium development in the vesicular arbuscular mycorrhizal fungus *Glomus mosseae* (Nicol. & Gerd.) Gerd & Trappe. New Phytol 123:114–122

Graham JH (1982) Effects of citrus root exudates on germination of chlamydospores of the vesicular-arbuscular fungus *Glomus epigaeum*. Mycologia 74:831–835

Graham JH, Abbott LK (2000) Wheat responses to aggressive and non-aggressive arbuscular mycorrhizal fungi. Plant Soil 220:207–218

Hamel C, Dalpe Y, Furlan V, Parent S (1997) Indigenous population of arbuscular mycorrhizal fungi and soil aggregate stability are major determinants of leek (*Allium porrum* L.) response to inoculation with *Glomus intraradices* Schenk and Smith or *Glomus versiforme* (Karsten) Berch. Mycorrhiza 74:187–196

Helgason T, Fitter AH, Young JPW (1999) Molecular diversity of arbuscular mycorrhizal fungi colonising Hyacinthoides non-scripta (bluebell) in a semi-natural woodland. Mol Ecol 8:659–666

Helgason T, Merryweather JW, Denison J, Wilson P, Young JPW, Fitter AH (2002) Selectivity and functional diversity in arbuscular mycorrhizas of co-occurring fungi and plants from a temperate deciduous woodland. J Ecol 90:371–384

Hetrick BAD (1991) Mycorrhizas and root architecture. Experientia 47:355–361

Hoeksema JD, Chaudhary VB, Gehring CA, Johnson NC, Karst J, Koide RT, Pringle A, Zabinski C, Bever JD, Moore JC, Wilson GWT, Klironomos JN, Umbanhowar J (2010) A meta-analysis of context-dependency in plant response to inoculation with mycorrhizal fungi. Ecol Lett 13:394–407

Holland TC, Bowen P, Bogdanoff C, Hart MM (2014) How distinct are arbuscular mycorrhizal fungal communities associating with grapevines? Biol Fertil Soils 50:667–674

Jakobsen I, Abbott LK, Robson AD (1992) External hyphae of vesicular arbuscular mycorrhizal fungi associated with *Trifolium subterraneum* L. 1. Spread of hyphae and phosphorus inflow into roots. New Phytol 120:371–380

Janos DP (1980) Vesicular-arbuscular mycorrhizae affect lowland tropical rain forest plant growth. Ecology 61:151–162

Jansa J, Mozafar A, Kuhn G, Anken T, Ruh R, Sanders IR, Frossard E (2003) Soil tillage affects the community structure of mycorrhizal fungi in maize roots. Ecol Appl 13:1164–1176

Jasper DA, Abbott LK, Robson AD (1989a) Acacias respond to additions of phosphorus and to inoculation with VA mycorrhizal fungi in soils stockpiled during mineral sand mining. Plant Soil 115:99–108

Jasper DA, Abbott LK, Robson AD (1989b) The loss of VA mycorrhizal infectivity during bauxite mining may limit the growth of *Acacia pulchella* R.Br. Aust J Bot 37:33–42

Jasper DA, Robson AD, Abbott LK (1989c) Soil disturbance reduces the infectivity of external hyphae of vesicular-arbuscular mycorrhizal fungi. New Phytol 112:93–99

Jasper DA, Abbott LK, Robson AD (1991) The effect of soil disturbance on vesicular-arbuscular fungi in soils from different vegetation types. New Phytol 118:471–476

Jasper DA, Abbott LK, Robson AD (1992) Soil disturbance in native ecosystems – the decline and recovery of infectivity of VA mycorrhizal fungi. In: Read DJ, Lewis DH, Fitter AH, Alexander IJ (eds) Mycorrhizas in ecosystems. CAB International, Wallingford

Jasper DA, Abbott LK, Robson AD (1993) The survival of infective hyphae of vesicular-arbuscular mycorrhizal fungi in dry soil: an interaction with sporulation. New Phytol 124:473–479

Kabir Z, O'Halloran IP, Fyles JW, Hamel C (1997) Seasonal changes of arbuscular mycorrhizal fungi as affected by tillage practices and fertilization: Hyphal density and mycorrhizal root colonization. Plant Soil 192:282–293

Kohout P, Sudová R, Janoušková M, Čtvrtlíková M, Hejda M, Pánková H, Slavíková R, Štajerová K, Vosátka M, Sýkorová Z (2013) Comparison of commonly used primer sets for evaluating arbuscular mycorrhizal fungal communities: is there a universal solution? Soil Biol Biochem 68:482–493

Koide RT (2000) Functional complementarity in the arbuscular mycorrhizal symbiosis. New Phytol 147:233–235

Koske RE, Gemma JN (1992) Fungal reactions to plant prior to mycorrhizal formation. In: Allen MF (ed) Mycorrhizal functioning. Chapman and Hall, London, pp 3–36

Linderman RG (1992) Vesicular-arbuscular mycorrhizae and soil microbial interactions. In: Bethlenfalvay GJ, Linderman RG (eds) Mycorrhizae in sustainable agriculture. ASA Special Publication no 54, pp 29–70

Merryweather J, Fitter A (1991) A modified method for elucidating the structure of the fungal partner in a vesicular-arbuscular mycorrhiza. Mycol Res 95:1435–1437

Merryweather J, Fitter AH (1998a) The arbuscular mycorrhizal fungi of *Hyacinthoides* non scripta. I. Diversity of fungal taxa. New Phytol 138:117–129

Merryweather J, Fitter AH (1998b) The arbuscular mycorrhizal fungi of *Hyacinthoides* non scripta. II. Seasonal and spatial patterns of fungal populations. New Phytol 138:131–142

Miller RM, Jastrow JD (1990) Hierarchy of root and mycorrhizal fungal interactions with soil aggregation. Soil Biol Biochem 22:579–584

Moyersoen B, Fitter AH (1999) Presence of arbuscular mycorrhizas in typically ectomycorrhizal host species from Cameroon and New Zealand. Mycorrhiza 8:247–253

Nadian H, Smith SE, Alston AM, Murray RS, Siebert BD (1998) Effects of soil compaction on phosphorus uptake and growth of *Trifolium subterraneum* colonized by four species of vesicular-arbuscular mycorrhizal fungi. New Phytol 139:155–165

Pearson JN, Schweiger P (1994) *Scutellospora calospora* (Nicol. & Gerd.) Walker and Sanders associated with subterranean clover produces non-infective hyphae during sporulation. New Phytol 127:697–701

Pearson JN, Abbott LK, Jasper DA (1994) Phosphorus, soluble carbohydrates and the competition between two arbuscular mycorrhizal fungi colonizing subterranean clover. New Phytol 127:101–106

Phillips JM, Hayman DS (1970) Improved procedure for clearing roots and staining parasitic and vesicular-arbuscular mycorrhizal fungi for rapid assessment of infection. Trans Br Mycol Soc 55:158–161

Powell CL (1980) Effect of phosphate fertilizers on the production of mycorrhizal inoculum in soil. NZ J Agric Res 23:216–223

Read DJ (1991) Mycorrhizas in ecosystems. Experientia 474:376–390

Robinson-Boyer L, Grzyb I, Jeffries P (2009) Shifting the balance from qualitative to quantitative analysis of arbuscular mycorrhizal communities in field soils. Fungal Ecol 2:1–9

Rosendahl S, Sen R (1994) Isozyme analysis of mycorrhizal fungi and their mycorrhiza. In: Norris JR, Read DJ, Varma AK (eds) Techniques for mycorrhizal research. Academic Press, London, pp 629–654

Saito M, Stribley DP, Hepper CM (1993) Succinate dehydrogenase activity of external and internal hyphae of a vesicular-arbuscular mycorrhizal fungus, *Glomus mosseae* (Nicol. & Gerd.) Gerdemann and Trappe during mycorrhizal colonization of roots of leek (*Allium porrum* L.), as revealed by in situ histochemical staining. Mycorrhiza 4:59–62

Sanchez-Castro I, Ferrol N, Cornejo P, Barea JM (2012) Temporal dynamics of arbuscular mycorrhizal fungi colonizing roots of representative shrub species in a semi-arid Mediterranean ecosystem. Mycorrhiza 22:449–460

Scheltema MA, Abbott LK, Robson AD (1987) Seasonal variation in the infectivity of VA mycorrhizal fungi in annual pastures in a Mediterranean environment. Aust J Agric Res 38:707–715

Schenck NC, Smith GS (1982) Responses of six species of vesicular-arbuscular mycorrhizal fungi and their effects on soybean at four soil temperatures. New Phytol 92:193–201

Schweiger PF, Robson AD, Barrow NJ (1995) Root hair length determines beneficial effect of a *Glomus* species on shoot growth of some pasture species. New Phytol 131:247–715

Shi P, Abbott LK, Banning NC, Zhao B (2012) Comparison of morphological and molecular genetic quantification of relative abundance of arbuscular mycorrhizal fungi within roots. Mycorrhiza 22:501–513

Smilauer P (2001) Communities of arbuscular mycorrhizal fungi in grassland: seasonal variability and effects of environment and host plants. Folia Geobotanica 36:243–263

Smith SE, Read DJ (2008) Mycorrhizal symbiosis. Academic Press, Cambridge

Solaiman ZM, Abbott LK (2008) Influence of arbuscular mycorrhizal fungi, inoculum level and phosphorus placement on growth and phosphorus uptake of *Phyllanthus calycinus* in jarrah forest soil. Biol Fertil Soils 44:815–821

Srivastava D, Kapoor R, Srivastava SK, Mukerji KG (1996) Vesicular-arbuscular mycorrhiza – an overview. In: Mukerji KG (ed) Handbook of vegetation science: concepts in mycorrhizae research, vol 19/2. Kluwer, Academic Publishers, Netherland, pp 1–39

Tamasloukht M, Sejalon-Delmas N, Kluever A, Jauneau A, Roux C, Becard G, Franken P (2003) Root factors induce mitochondrial-related gene expression and fungal respiration during the developmental switch from asymbiosis to presymbiosis in the arbuscular mycorrhizal fungus *Gigaspora rosea*. Plant Physiol 131:1468–1478

Tisserant B, Brenac V, Requena N, Jeffries P, Dodd JC (1998) The detection of *Glomus* spp (arbuscular mycorrhizal fungi) forming mycorrhizas in three plants, at different stages of seedling development, using mycorrhiza-specific isoenzymes. New Phytol 138:225–239

Tommerup IC (1983) Spore dormancy in vesicular arbuscular mycorrhizal fungi. Trans Br Mycol Soc 81:37–45

Tommerup IC, Abbott LK (1981) Prolonged survival and viability of VA mycorrhizal hyphae after root death. Soil Biol Biochem 13:431–434

Turnau K, Ryszka P, Gianinazzi-Pearson V, van Tuinen D (2001) Identification of arbuscular mycorrhizal fungi in soils and roots of plants colonizing zinc wastes in southern Poland. Mycorrhiza 10:169–174

Vogt K, Asbjornsen H, Ercelawn A, Montagnini F, Valdes M (1997) Roots and mycorrhizas in plantation ecosystems. In: Nambiar EKS, Brown AG (eds) Management of soil, nutrients and water in tropical plantation forests. ACIAR, Canberra, Australia, pp 247–296

Wilson JM, Tommerup IC (1992) Interactions between fungal symbionts: VA mycorrhizae. In: Allen MF (ed) Mycorrhizal functioning. Chapman and Hall, New York, NY, pp 199–248

Wilson JM, Trinick MJ (1983) Infection development and interactions between vesicular-arbuscular mycorrhizal fungi. New Phytol 93:543–553

Wilson JM, Trinick MJ, Parker CA (1983) The identification of vesicular-arbuscular mycorrhizal fungi using immunofluorescence. Soil Biol Biochem 15:439–445

Chapter 4
Biofertilizers with Arbuscular Mycorrhizal Fungi in Agriculture

Olmar B. Weber

4.1 Introduction

An increase in crop production and land productivity in agriculture is necessary to meet the demand for food. If agricultural systems are to be sustainable in maintaining soil fertility and soil structure over a long period of time, they need management strategies that are economically viable, environmentally safe, and socially fair with soils managed to safeguard food security (Killham 2011). One strategy involves fertilizer (NPK) replacement by compounds that are less polluting, less persistent in soil, and less energy consuming (Lichtfouse et al. 2009). Considering that fluctuating crop and fertilizer prices occur and farmers rarely reduce expenditure on fertilizers, a knowledgeable farmer will use inputs to reach economical profitability, environmental safety, and social fairness farming systems. Another strategy involves applying ecological concepts and principles to the design, development, and management of sustainable agricultural systems (Lichtfouse et al. 2009).

Microbial inoculants, including arbuscular mycorrhizal (AM) fungi and plant growth-promoting rhizobacteria, are potential components of sustainable management systems (Adesemoye and Kloepper 2009). AM fungal inoculants are marketed as an important biological component to commercial horticulture and agriculture, but for successful application of AM fungi with economically profitable results, the soil environment must be suitable for the development of the AM fungal symbiosis (Baar 2008).

Companies have taken different market approaches for microbial inoculants, ranging from products with AM fungi alone to mixed products (Baar 2008). In order to exploit beneficial effects of AM fungi in sustainable agriculture,

O.B. Weber (✉)
Brazilian Agricultural Research Corporation, Embrapa Tropical Agroindustry, Sara Mesquita 2270, 60511-110 Fortaleza, Ceara, Brazil
e-mail: olmar.weber@embrapa.br

© Springer-Verlag Berlin Heidelberg 2014   45
Z.M. Solaiman et al. (eds.), *Mycorrhizal Fungi: Use in Sustainable Agriculture and Land Restoration*, Soil Biology 41, DOI 10.1007/978-3-662-45370-4_4

appropriate management practices have to be applied (Vosátka and Albrechtova 2009). By better understanding mycorrhizal symbioses for optimization of plant–soil systems, the need for biofertilizers that include inoculants of AM fungi in agricultural production will be clarified.

4.2   Fertilizer Use in Agriculture

The increase in food production and land productivity in agriculture has been due to the use of fertilizers, especially NPK, and significant increases in their use in the future are predicted (FAO 2010). As a result of extensive use of fertilizers (Vance 2001), the agrochemicals and agronomic practices commonly adopted in intensive production systems have generated environmental problems including deterioration of soil quality, surface water, and groundwater, and reduced biodiversity and function of ecosystems. An alternative way to increase soil fertility and supply nutrients based on the efficient use of mineral and manufactured fertilizers is to intensify the use of biofertilizers. The ability of the root systems to establish symbiotic relationships with soil microorganisms of the rhizosphere represents one strategy that land plants have developed to survive the abiotic and biotic stresses (Allen 1996). This approach is economically viable, environmentally safe, and socially fair (Lichtfouse et al. 2009). For instance, sustainable crop productivity depends largely on soil biological fertility, defined by Abbott and Murphy (2007) as the ability of soil organisms to contribute to the nutritional requirements of plants and foraging animals for productivity, reproduction, and quality while maintaining the biological processes that contribute positively to the physical and chemical states of the soil.

The use of biofertilizers as a source of plant nutrients for more sustainable agricultural practices (Gentili and Jumpponen 2006) involves a range of soil microorganisms including AM fungi which have potential to enhance productivity in combination with a reduction in application of mineral phosphate fertilizer (Marin 2006). AM fungi form symbiotic associations with the vast majority (>80 %) of families of land plants, but some members of the families *Brassicaceae*, *Chenopodiaceae*, *Caryophyllaceae*, *Cyperaceae*, *Juncaceae*, and *Amaranthaceae* do not form mycorrhizal associations (Cardoso and Kuyper 2006). Plant growth response to mycorrhizas can be positive, neutral, or negative (Smith and Smith 2011). In contrast to nitrogen-fixing bacteria, AM fungi do not add mineral nutrients to the soil and are therefore not strictly biofertilizers (Cardoso et al. 2010). There are various types of microbial cultures and inoculants available on the market today, and results are often difficult to extrapolate to field conditions.

4.3 Analysis of Biofertilizers with AM Fungal Used in Agriculture

AM fungal inoculants (Table 4.1) in commercial products (or biofertilizers) usually contain one or a few species of *Glomus*, particularly *G. intraradices* (Akhtar and Siddiqui 2008; Baar 2008; Öpik et al. 2008). Specially designed agricultural machines can be used to inoculate the seed or distribute AM fungal inocula with a relatively high speed over a large area (Baar 2008).

Murphy et al. (2007) reported that the performance of microbial inoculants in the field can be inconsistent. Measurement of mycorrhizal colonization of roots is usually used in bioassays with commercial products, but calibrations that take into account an understanding of the relationships between mycorrhizal colonization and soil conditions are necessary (Djuuna et al. 2009). Consistency of plant responsiveness to inoculation with selected strains of AM fungi should be a prerequisite to adoption of AM inoculation practices.

AM fungal inocula are commonly produced in association with host plants in a greenhouse or in growth chambers using "pot cultures" containing expanded clay and vermiculite (Gentili and Jumpponen 2006; Vosátka et al. 2008; IJdo et al. 2011). Other production techniques as well as substrate-free culture techniques (hydroponic and aeroponic) and in vitro cultivation methods have been attempted (IJdo et al. 2011), and higher quality of commercial products can be expected in the future.

The producers of AM fungal inoculants should specify the quantity and quality of the mycorrhizal fungi present in the marketed products. Accordingly to Wiseman et al. (2009), if manufacturers of commercial AM fungal inoculants desire large-scale acceptance of mycorrhizal technology, they must better demonstrate that their products are compatible with current retail distribution methods and can promote mycorrhizal colonization under the conditions of their intended use. Vosátka et al. (2008), when analyzing the market development for mycorrhizal technology, noted that while industry has developed inoculum using AM fungal strains, quite often they are not well characterized in terms of ecological requirements and stability. The success of an inoculum formulation depends on whether it is (1) economically viable to produce, (2) unaltered in viability and function, and (3) easy to carry and disperse during application.

A basic step is the AM fungi selection that depends on environmental conditions of the locality, soil, and host plant. AM fungi must be able to colonize roots rapidly after inoculation, absorb phosphate from the soil effectively and transfer it to the plant, persist in soil and reestablish mycorrhizal symbiosis during the following seasons, and form propagules that remain viable during and after inoculum production (Abbott et al. 1992; Tanu Prakash and Adholeya 2006). Selection of AM fungi may also consider tolerance to abiotic stress and resistance to soil pathogens. There are studies and comprehensive reviews exploring possibilities for AM fungi to protect endangered plants and habitats (Bothe et al. 2010), alleviation of salt stress (Evelin et al. 2009; Kaya et al. 2009; Mardukhi et al. 2011), bioprotection

Table 4.1 Examples of the use of biofertilizers containing AM fungal inocula

Biofertilizer or microbial inoculant	Experimental conditions	Main results	Selected references
Commercial MicoFert® (Cuba) containing a mixture of AM fungi-colonized soil and AM fungi-colonized root fragments was used in high-input coffee plants grown in different coffee plantation soils	Study based on the results of 62 inoculation trials using soils from Cuban plains with a history of high-input agriculture and 68 trials conducted with pristine or seminatural soils, which were under low-input under-story coffee production	There was no significant relationship between plant response to AM fungal strains and soil properties in the high-input agriculture soil data set, may be due to variation induced by the use of different host plant species and to modification of soil properties by a history of intensive production. Findings indicate that AM strains must not only be highly effective, they must be able to function in the soil environment where they are introduced	Herrera-Peraza et al. (2011)
Commercial Mycorise (Canada) AM-1207 containing *G. intraradices* combined with one exotic *Glomus intraradices* (isolate AM-1004 from Canada) and mixed AM fungi (AM-1209, native from India) containing *Glomus, Gigaspora,* and *Scutellospora* spp.	Field trial with carrot (*Daucus carota* L.) to compare three AM inocula on the formation of infectious propagules in different fractions of inocula in a marginally sandy loam Alfisol amended with farmyard manure	The exotic isolate AM-1004 and mixed AM fungal (*Glomus, Gigaspora,* and *Scutellospora* spp.) inocula used either as roots, soil, or a mixture of both have a greater potential in producing more prop-agules in the shortest span of time. *G. intraradices* (AM-1004), root based, and the indigenous mix-ture (AM-1209), as crude inoculum, can be used as starter cultures for on-farm production of these AM fungal inocula	Sharma and Adholeya (2011)
Commercial Aegis Argilla (Italy) containing clays as granular carriers, leek root pieces, and *Glomus intraradices* spores	Greenhouse experiment with two cucumber (*Cucumis sativus* L.) genotypes (hybrid "Ekron" and variety "Marketmore") were inoculated with AM fungi grown in pots containing quartziferous sand mixed with slow-release fertilizer, pH values 6.0 and 8.1	The inoculated plants under alkaline conditions had higher total, market-able yield, and total bio-mass than noninoculated plant Mycorrhizal cucumber plants grown under alka-line conditions had a higher macronutrient concentration in leaf tis-sue compared to noninoculated plants	Rouphael et al. (2010)

(continued)

Table 4.1 (continued)

Biofertilizer or microbial inoculant	Experimental conditions	Main results	Selected references
Commercial TerraVital Hortimix (UK) containing *Glomus mosseae*, *G. intraradices*, *G. claroideum*, and *G. microaggregatum*, >50 infective units ml^{-1}) combined without and 135 mg P kg^{-1}	Greenhouse trial with cowpea (*Vigna unguiculata* L. Walp cv. IT18) on washed sand and rock phosphate treatments supplemented with nutrient solution and mycorrhizal inoculum	AM fungi are able to increase plant availability of RP when NO^{-3} is the major form of N in the soil and the substrate pH is in a neutral range Increased supply of NH^{-4} relative to NO^{-3} improved plant P availability from rock phosphate but also had a negative effect on the extent of AM fungal root colonization, irrespective of the plant P-nutritional status	Ngwene et al. (2010)
Commercial Endol® Biorize Sarl (France) combined with the inoculants of *Glomus mosseae* (isolates BEG 12, BEG 167), *G. intraradices* (BEG 141), and *G. etunicatum* (isolates BEG 168 and HB-Bd45)	Sweet potatoes (*Ipomoea batatas* L.) were inoculated with various combinations of AM fungi and grown under traditional condition in China	Inoculation with *G. intraradices* BEG 141 or *G. etunicatum* (HB-Bd45-Gsp4, BEG 167, and BEG 168) increased yields (10.2 and 14.0 %) AM fungi varied in their ability to establish after inoculation and in their effect on yield and quality of sweet potato tubers	Farmer et al. (2007)
Commercial products: TerraVital Hortimix (UK) containing *Glomus mosseae*, *G. intraradices*, *G. claroideum*, and *G. microaggregatum* (>50 infective units per ml inoculum); Endorize-Mix (France) with the *G. mosseae*, *G. intraradices*, and *Glomus* sp. (infective units not specified); and AMYkor (Germany) with *G. mosseae*, *G. intraradices*, and *G. etunicatum* (50 infective units per ml inoculum)	Pot experiment with horticultural pelargonium (*Pelargonium peltatum* L'Her.) on compost substrates and commercial AM inocula	The inoculation of three different commercial AM inocula resulted in colonization rates of up to 36 % of the total root length. Addition of compost in combination with mycorrhizal inoculation can improve nutrient status and flower development of plants grown on peat-based substrates	Perner et al. (2007)

(continued)

Table 4.1 (continued)

Biofertilizer or microbial inoculant	Experimental conditions	Main results	Selected references
Two commercial biofertilizers: BioLife (USA), containing a combination of 13 bacterial strains, and Media Mix (USA), containing spores of four species mycorrhizal fungi as well as N_2-fixing and P-solubilizing bacteria combined with mineral fertilizers	Greenhouse trial with tomato plants (*Solanum lycopersicum* L. var. Belle) applying commercial inoculants and conventional fertilization (115 kg N ha^{-1}, 69 kg P ha^{-1}, 366 kg K ha^{-1} and 100 kg Mg ha^{-1})	The tomato yield can be increased by 32.0 % when using bacterial fertilizer BioLife and by 22.9 % when using mycorrhizal inoculum Media Mix, compared with conventional fertilization. The total energy input required in unheated greenhouses increases by 1.88 % with the application of BioLife and by 1.38 % with the application of Media Mix; however, they decrease by 22.31 % and 16.92 %, respectively, per ton of tomato production	Mihov and Tringovska (2010)
Commercial MYKE® PRO SG2 (Canada) containing a *G. intraradices*, covered by 1.3 kg of pasteurized soil, MYKE® PRO covered by unpasteurized soil, and control with pasteurized soil	Glasshouse experiment with maize (*Zea mays* L. hybrid IC 192) and AM fungal inoculant	The inoculant significantly improved the P content of the host in the presence of the resident AM fungal community. In contrast to inoculation, soil disturbance had a significant negative impact on species richness of AM fungi and influenced the AM fungal community composition as well as its functioning	Antunes et al. (2009)
Nine commercial AM fungal inoculants (seven granular and two root dipped) purchased through typical consumer channels	Greenhouse experiments using corn (*Zea mays*), sorghum (*Sorghum bicolor*), trident maple (*Acer buergerianum*), and sweet bay magnolia (*Magnolia virginiana*) as host plants and AM fungal inoculants	Commercial AM fungal inoculants had little effect on corn root mycorrhizal colonization when obtained anonymously through typical consumer channels and applied at the manufacturers' recommended rates In corn and sorghum, colonization rarely exceeded 5 % when plants were treated with commercial inoculants.	Wiseman et al. (2009)

(continued)

Table 4.1 (continued)

Biofertilizer or microbial inoculant	Experimental conditions	Main results	Selected references
		Despite the near absence of colonization, commercial inoculants generally improved shoot growth and increased soil nutrient concentrations in a dose-dependent manner. Commercial inoculants tested had no effect on mycorrhizal colonization or shoot growth of trident maple or sweet bay magnolia liners	
Commercial Biorize Sarl (France) containing *Glomus mosseae* or *Glomus intraradices* was combined with a mixture of *Azotobacter chroococcum*, HKN-5, *Bacillus megaterium*, HKP-2, *Bacillus mucilaginosus*, HKK-2, organic fertilizer, and two levels (50 and 100 %) of NPK (300 mg N, 92.3 mg P, and 184.6 mg K per kilogram of dry weight of soil)	Greenhouse trial with maize (*Zea mays* L.) on peat moss-based bacterial inoculum and/or 15.0 g sand-based mycorrhizal inoculum	The use of biofertilizer (*G. mosseae* and three bacterial species) resulted in the highest biomass and seedling height Microbial inoculum increased the nutritional assimilation of plant (N, P, and K) and improved soil properties, such as organic matter content and total N in soil The presence of mycorrhizal fungi had different influence on the population of rhizobacteria. *G. intraradices* was able to stimulate the introduced beneficial bacterial growth in the rhizosphere soil. However, the high mycorrhizal infection with *G. mosseae* showed a strong inhibition of P and K solubilizers	Wu et al. (2005)
Commercial Mycogold (Malaysia) containing *Glomus* spp. (infective units not specified) combined with two native communities of AM fungi (Brazil):	Greenhouse experiment with dwarf cashew (*Anacardium occidentale* L. clone CCP76) without and with 87 mg de P l^{-1} soil and AM fungal inocula	The cashew seedlings presented a low response to the phosphorus treatment. The symbiotic plant association with *Glomus etunicatum*, *G. glomerulatum*,	Weber et al. (2004)

Table 4.1 (continued)

Biofertilizer or microbial inoculant	Experimental conditions	Main results	Selected references
(1) *Glomus etunicatum*, *G. glomerulatum*, *Scutellospora* sp., and *Acaulospora foveata* and (2) *G. etunicatum*, *Entrophospora* sp., and *Scutellospora* sp.		*Scutellospora* sp., and *Acaulospora foveata* (native community) and the commercial product allowed a better plant growth 4 months after cashew nut sown	
Seven commercially mycorrhizal inoculants obtained from different companies (USA)	Experiments with strawberry (*Fragaria* × *ananassa* Duch.) cultivars were evaluated side by side in organic strawberry production fields in central California	None of the seven commercially prepared mycorrhizal inoculants tested resulted in an increased marketable fruit yield in organic or non-fumigated fields. In one of six experiments, a commercial inoculant increased total yield over the nontreated control but did not influence marketable fruit yield	Bull et al. (2005)
Eight commercial AM fungal inoculants obtained anonymously (USA)	Greenhouse trial with maize (Z. mays L. var. Golden Cross) on sand/peat medium inoculated with AM fungi	Only three of the commercial inocula formed mycorrhizas when used at the recommended rate, and the extent of colonization ranged from 0.4 to 8 %. The failure of five of the eight commercial inocula to colonize roots when applied at the recommended rate suggests that preliminary trials should be made before commercial AM fungal inocula are used in important landings	Tarbell and Koske (2007)
Ten commercial mycorrhizal inoculants (six contains *G. intraradices*, three with one or more fungal species, and one endo-/ectomycorrhizal inoculum) were used in nursery conditions	Experiments with sweet corn (*Zea mays* L. hybrid Silver Queen) grown in a soil-based medium and in two different soilless substrates, a potting mix prepared with redwood bark, pine sawdust, calcined clay and sand, and the commercial Sunshine mix, mainly composed of sphagnum peat moss	Only the plants inoculated with the products that did not promote mycorrhizal colonization increased their growth relative to the noninoculated controls, suggesting the presence of other growth promoters in the inoculum products. Based on these results, nurseries should	Corkidi et al. (2004)

(continued)

Table 4.1 (continued)

Biofertilizer or microbial inoculant	Experimental conditions	Main results	Selected references
		conduct preliminary tests to determine which inoculants will perform in their potting mixes to assure the best fit of inoculum with their particular conditions	
Different AM fungal species *Glomus mosseae*, *G. etunicatum*, and *G. intraradices* isolated from saline soils (Iran) were tested under saline conditions	Greenhouse and field trials with three wheat cultivars (Roshan, local; Kavir, genetically modified; and one mutated line, Tabasi) combined with and without AM fungal species	In both experiments, AM fungi significantly enhanced the concentrations of macro and micro-elements. The results indicated that the selected combination of AM species and wheat cultivar can result in the highest rate of nutrient uptake by wheat plants under saline conditions	Mardukhi et al. (2011)
Different AM fungal inoculants from soils of high-input continuous maize fields, low-input continuous maize fields, and undisturbed native vegetation were used	Greenhouse trials with maize (*Zea mays* L. cv. Landrace and hybrid H5), to assess the mutualistic functioning of AM fungi and the mycorrhizal responsiveness of maize genotypes	When maize was grown in field soil brought into the greenhouse, AM fungi and communities of other soil organisms did not benefit plant growth in high-fertility soil, but they did improve maize growth in low-fertility soil Landrace maize was more responsive to mycorrhizas than hybrid maize, and novel soil inoculum was more beneficial than inoculum from sites where the crop and organisms have long coexisted	Martinez and Johnson (2010)
Different AM fungal inoculants were used in greenhouse: *Glomus mosseae* (isolate IMA1 from the UK and isolate AZ225C from the USA) and *Glomus intraradices* (isolate IMA5 from Italy and isolate IMA6 from	Greenhouse and field trials to assess the effect of native and exotic selected AM fungal inocula on plant growth and nutrient uptake in a low-input, 2-year forage legume (*Trifolium alexandrinum*) cv. Tigri	The use of AM fungal inocula may be very effective in improving crop productivity and quality in low-input agricultural systems and their effects are persistent at least until 2 years after inoculation.	Pellegrino et al. (2011)

(continued)

Table 4.1 (continued)

Biofertilizer or microbial inoculant	Experimental conditions	Main results	Selected references
France). In field experiment were used the four *Glomus* isolates single and mixed and a native AM fungal inoculum (mixed fungal population)	and maize (*Zea mays*) crop rotation	Differences in isolate performances indicate that the choice of the best AM fungal inoculum for field utilization is pivotal for the success of inoculation practices The use of native AM fungi, produced on farm with mycotrophic plants species, might represent a convenient alternative to commercial AM fungal inocula and offer economically and ecologically important advantages in sustainable or organic cropping systems	
Communities of fungi obtained from soils of conventional and low-input cropping systems. The conventional cropping system with a nonleguminous 6-year crop rotation (barley–barley–rye–oat–potato–oat) received different fertilizer rates	Pot experiment with leek (*Allium porrum* cv. Titan), barley (*Hordeum vulgare* cv. Arra), flax (*Linum usitatissimum* cv. Linetta), alfalfa (*Medicago sativa*, unknown cultivar), red clover (*Trifolium pratense* cv. Björn), white clover (*Trifolium repens* cv. Huia), and subclover (*Trifolium subterraneum*, unknown cultivar)	*Glomus claroideum* was the most commonly identified single species in the experimental area. A bioassay using roots as inoculum for isolation and culture of dominating AM fungi was successfully developed and yielded only *G. claroideum*. Such dominating AM fungi seem to compete successfully against other indigenous AM fungal species and would therefore ensure a predictable impact of AM fungal inoculation in field conditions	Vestberg et al. (2011)
Mycorrhizal inoculants with species *Glomus mosseae* and *Entrophospora colombiana* (=*Kuklospora colombiana*)	Plants of papaya (*Carica papaya* L cv. Maradol) inoculated with AM fungi under controlled conditions and transplanted into an experimental orchard fertilized with NPK (235–42–222 kg ha^{-1})	Inoculation with *G. mosseae* and *E. colombiana* increased papaya yield by improving setting and fruit weight *G. mosseae* had better influence in various aspects, possibly due to greater association with	Vázquez-Hernández et al. (2011)

(continued)

Table 4.1 (continued)

Biofertilizer or microbial inoculant	Experimental conditions	Main results	Selected references
		papaya plants and better adaptability to edaphic and environmental conditions	
Mycorrhizal inoculant with *Glomus mosseae* was combined with P levels (18.3, 48, 79.4, and 100 mg kg^{-1})	Pot experiment with papaya (*Carica papaya* L. cv. Sunrise), and pineapple [*Ananas comosus* L. Merr.] were inoculated with AM fungal and cultured for 5 (papaya) and 7 months (pineapple)	Mycorrhizal papaya plants exhibited higher biomass and macroelement contents in shoots than plants without mycorrhizas at any P level Mycorrhizal effects on pineapple at the lowest P level were significant in terms of plant development and P shoot contents. Differential benefits derived from mycorrhization seem to be correlated to each crop's internal P requirements	Rodriguez-Romero et al. (2011)
Mycorrhizal inoculant with *Glomus mosseae* combined without, 80, and 160 kg N ha^{-1} and tilled and no-tilled soils	Pot experiment with wheat (*Triticum aestivum* L.) plants combined with AM fungal inoculant and conventional tilled and no-tilled soils	In no-tillage, the plant colonization was greater than in conventional tillage, but it was reduced by the N fertilization. In conventional tillage, the inoculation with *G. mosseae* increased colonization. Both conventional tillage and N fertilization promoted wheat root growth. Inoculation did not affect root growth but enhanced N concentration in roots when fertilizer was not applied	Shalamuk et al. (2011)
Different AM fungal inoculants: *Acaulospora foveata* (HR0602), *A. appendicula* (HR0201), *A. denticulata* (RA2106), *Glomus dimorphicum* (WH0101), *G. tenerum*	Greenhouse experiment with chili (*Capsicum frutescens* L.) inoculated with AM fungi and cultivated on fumigated soil	*Acaulospora appendicula* HR0201, *A. denticulata* RA2106, and *G. clarum* RA0305 were found to be efficient chili growth promoters, with *G. clarum* RA0305 being the best. These findings suggest	Boonlue et al. (2012)

(continued)

Table 4.1 (continued)

Biofertilizer or microbial inoculant	Experimental conditions	Main results	Selected references
(WH0102), *G. clarum* (A0305), *A. denticulata* (HR0406), *G. globiferum* (PY0109), *G. globiferum* (PY0103), and *G. globiferum* (PY0107)		the potential of *G. clarum* RA0305 for use as an AM fungal inoculum for the production of organic chili in Thailand	
Twenty-three different AM fungal inoculants: *Acaulospora delicata*, *A. rugosa*, *Gigaspora candida*, *Glomus aggregatum*, *G. albidum*, *G. aurantium*, *G. claroideum*, *G. clarum*, *G. coronatum*, *G. etunicatum*, *G. fasciculatum*, *G. geosporum*, *G. glomerulatum*, *G. hoi*, *G. intraradices*, *G. macrocarpum*, *G. mosseae*, *G. occultum*, *G. versiforme*, *G. xanthium*, *Glomus* sp. 2, *Glomus* sp. 4, and *Glomus* sp. 5	Greenhouse experiment with long pepper (*Piper longum* L.) and indigenous AM fungi from India	Considering the shoot length, total biomass, nutrient content, chlorophyll content, and root infection, pre-inoculation with 6 AM fungal species (*Glomus fasciculatum*, *G. versiforme*, *G. clarum*, *Glomus* sp. 2, *G. mosseae*, and *G. etunicatum*) appeared to be promising AM fungi for inoculating this medicinal plant	Gogoi and Singh (2011)
Mycorrhizal inoculant with *Glomus mosseae* combined with N_2-fixing *Bradyrhizobium* sp. (strain BXYD3)	Field and greenhouse trials with soybean (*Glycine max* L. Merr.) genotypes HN89 (P-efficient) and HN112 (P-inefficient) on soil with low available N and P content	Co-inoculation with rhizobia and AM fungi increased soybean growth under low P and/or low N conditions as indicated by increased shoot dry weight, along with plant N and P content. A synergistic relationship dependent on N and P status exists between rhizobia and AM fungi on soybean growth. The deep root genotype, HN112, benefited more from co-inoculation than the shallow root genotype, HN89	Wang et al. (2011)

(continued)

Table 4.1 (continued)

Biofertilizer or microbial inoculant	Experimental conditions	Main results	Selected references
Seven AM fungal consortia isolated from coffee plantations with different agricultural inputs (low, intermediate, and high) were used	Greenhouse and field trials with coffee plants (*Coffea arabica* L.) inoculated with different AM fungal consortia under greenhouse and transplanted onto a coffee field in Mexico	The most effective AM fungal consortia on plant growth promotion and survival under field conditions were collected from intermediate-input agricultural plantations, which also had the greatest number of AM fungal species	Trejo et al. (2011)
Different AM fungal inoculants (mix of *Glomus clarum* and *Gigaspora margarita* and five isolates of *Glomus etunicatum*) were combined with P rates (0, 20, 40, 80, and 160 g P2O5 plant^{-1})	Field trial with coffee (*Coffea* sp.) inoculated under glasshouse conditions and transplanted onto a farm with Oxisol in Minas Gerais (Brazil)	Coffee bean yield measured for 5 consecutive years, showing consistent effects of P application. Based on the total yield of five harvests, maximal productivity was achieved with a mix of *Glomus clarum* and *Gigaspora margarita* at 20 g P2O5 plant^{-1} and with the same mixed inocula and *G. etunicatum* (isolate Var) at the highest P rate. Pre-colonization of coffee outplants with selected AM fungi and application of low to moderate P rates at planting is advantageous for coffee production in low-fertility soils	Siqueira et al. (1998)
Different inoculants of N-fixing bacteria (*Azospirillum brasilense*, *Azotobacter chroococcum*) combined with AM fungi (*Glomus fasciculatum* and *Glomus mosseae*)	Field conditions and nursery with pomegranate (*Punica granatum* L.) and microbial inoculant	In both conditions tested, the combined treatment of *Azotobacter chroococcum* and *Glomus mosseae* was found to be the most effective. A significant improvement in the plant height, plant canopy, pruned material, and fruit yield was evident in 5-year-old pomegranate plants in field conditions	Aseri et al. (2008)

(continued)

Table 4.1 (continued)

Biofertilizer or microbial inoculant	Experimental conditions	Main results	Selected references
Biofertilizer (mycorrhizal, nitrogen-fixing bacteria, phosphorous-solubilizing bacteria) combined with no fertilizer and chemical (135 kg ha^{-1} urea + 185 kg ha^{-1} triple superphosphate)	Field experiments with annual medic (*Medicago scutellata* L. cv. Robinson) under dry farming conditions in Iran	The biological fertilizers can modify the adverse effects of moisture stress conditions. The highest pod yield was obtained after applying nitrogen-fixing bacteria + mycorrhiza	Shabani et al. (2011)

against plant pathogens (Akhtar and Siddiqui 2008; Saldajeno et al. 2008), and interactions with rhizobacteria (Adesemoye and Kloepper 2009) and other soil microorganisms (Das and Varma 2009; Javaid 2010; Reis et al. 2010; Miransari 2011).

In studies of the performance of AM fungal inoculants (Table 4.1), species of AM fungi are commonly obtained from work collections, where a small number of species are maintained (e.g., Stürmer and Saggin 2010). This may be one reason for using a limited number of AM fungal species in glasshouse or field experiments. In Brazil, for example, there are collections containing a variable number of isolates (3–50) and species of AM fungi (2–28) which are held in university laboratories and research organizations. Greater exploration of AM fungal diversity can be expected by assessing the substantial international collections of AM fungi including INVAM (http://invam.caf.wvu.edu), IBG (http://www.kent.ac.uk/bio/beg), and GINCO (http://emma.agro.ucl.ac.be/ginco-bel/ordering.php) which guarantee delivery of well-identified species and traceability.

The efficiency of AM fungi in promoting plant growth is commonly evaluated under controlled conditions, including micro-plots (Table 4.1). Such control allows study of certain effects of AM fungi on plants under specific environmental conditions. However, this does not mean that the same performance of plants inoculated with selected AM fungi will occur under field conditions. The overall contribution of AM to plant nutrition and crop productivity is determined by the interactions among the host plant, the AM fungal partner, soil properties, and other environmental factors (Kahiluoto et al. 2012). Environmental conditions (Baar 2008; Kahiluoto et al. 2012), production systems (Sieverding 1991), and agricultural practices (Brito et al. 2008) can all interfere in plant responsiveness to AM fungal inocula. Furthermore, Antunes et al. (2009) showed that soil disturbance may under certain conditions have greater consequences for AM fungal effects on plant productivity and the structure of indigenous AM fungal communities than certain AM fungal introduction through commercial inoculation.

In agroecosystems where production is dependent on indigenous AM fungi and there is low potential inoculum of these fungi in the soil, a response in crop yield can be expected after application of selected isolates of AM fungal inocula (Sieverding 1991). Similarly, in areas where plant production has been limited by stresses, especially low Pi available in soil (Sieverding 1991; Cuenca et al. 2008; Osorio and Habte 2009), salt stress (Evelin et al. 2009; Kaya et al. 2009; Mardukhi et al. 2011), and stresses caused by soil bacteria (Miransari 2011) and plant pathogens (Akhtar and Siddiqui 2008; Saldajeno et al. 2008), responses to inoculation may be expected. Higher plant performance was observed after pre-inoculation of coffee plants (Siqueira et al. 1998), papaya (Vázquez-Hernández et al. 2011; Rodriguez-Romero et al. 2011), banana, and some vegetables (Cuenca et al. 2008) on soils with low and medium fertility, after application of suboptimal dosages of phosphate.

Kahiluoto et al. (2009) suggested that the low-input cropping system with recycled organic matter composted before incorporation favors AM contribution to crop performance in the long term compared with a conventional cropping system. Advantages have been observed for (1) co-inoculation of AM fungi and N_2-fixing bacteria (species not specified) under dry conditions for annual medic production (Shabani et al. 2011), (2) AM fungi and phosphate-solubilizing bacteria (species not specified) with reduction of phosphate fertilization under dry conditions for corn production (Zarabi et al. 2011), (3) AM fungi and rhizobia under low P and/or low N conditions for soybean growth (Wang et al. 2011), (4) AM fungi (*Glomus fasciculatum* and *G. mosseae*) combined with the N_2-fixing bacteria *Azospirillum brasilense* and *Azotobacter chroococcum* (Aseri et al. 2008), (5) - co-inoculation of AM fungi and phosphate-solubilizing bacteria under drought stress condition for corn grain production (Zarabi et al. 2011), and (6) inclusion of AM fungi in biofertilizer in organic systems (Tanu Prakash and Adholeya 2006).

In cropping systems, on-farm production of AM fungal inocula may offer an alternative to a commercial AM fungal inocula (Douds et al. 2010; Sharma and Adholeya 2011; Pellegrino et al. 2011; Vestberg et al. 2011). Of importance for exploiting mycorrhizal technology in agriculture is the function of AM fungi in the soil and cropping systems where they are managed (Vosátka and Albrechtova 2009; Antunes et al. 2009; Barea et al. 2011; Herrera-Peraza et al. 2011).

4.4 Principles of AM Fungal Effectiveness in Cropping Systems

The majority of crop plants naturally form arbuscular mycorrhizas. AM fungi live in two environments, the root from which they receive C and to which they deliver nutrients and the soil from which they absorb those nutrients. Factors related to both the soil conditions and internal plant environment are involved in the function of AM symbiosis. Soil management and agronomic practices controlling the effectiveness of AM fungal strains need to be understood for a reliable use of AM inoculation in agriculture.

Smith and Smith (2011) identified factors that may influence mycorrhizal growth responses of plants to colonization by AM fungi as hyphae, interfaces, and the root and soil environments (Herrera-Peraza et al. 2011). Evaluated the performance of coffee in different cropping systems after inoculation with AM fungi and observed that under seminatural soils, typically from mountainous areas of Cuba, their performance was not the same in all soils. The diversity in fungal efficiency can be related to ecological factors in plant–soil systems (Solaiman and Abbott 2004; Yang et al. 2010; Barea et al. 2011). The plant genotype, phosphate and its equilibrium with other nutrients, and the need for alleviation of abiotic and biotic stresses all need to be considered in AM fungi selection programs (Abbott et al. 1992).

AM fungal effectiveness and plant responsiveness involve the interaction of independent plant and fungus genomes (Janos 2007). The mycelia (extra and intraradical) produced by different AM fungi have quite varied characteristics in terms of hyphal diameter, the extent of growth away from the root, and the ability to absorb nutrients from soil away from the root (Jansa et al. 2003). These authors observed acquisition and transport of substantial amounts of phosphorus and zinc located 15 cm away from the roots by *Glomus intraradices* in symbiosis with maize. Thonar et al. (2011) observed for mycorrhizal *Medicago truncatula* that *G. intraradices*, *G. claroideum*, and *Gigaspora margarita* were able to take up and deliver Pi to plants from distances of 10, 6, and 1 cm from the roots, respectively. The differences among *Glomus* species related to C requirements. *G. margarita* provided low P benefits to plants that formed dense mycelium networks close to the roots where P was probably transiently immobilized. Tracer studies also provide insights into the role of the fungal symbionts in determining diversity in plant responsiveness (Grace et al. 2009).

Diversity in responsiveness to AM fungi may reflect the diversity in function of the symbiosis. Phosphate uptake by mycorrhizal plants involved two pathways: (1) the soil–root system which involves uptake from the rhizosphere by root epidermis and root hairs and (2) the AM fungal pathway which involves uptake by extraradical mycelium, rapid translocation over many centimeters, delivery to the symbiotic interfaces, and transfer to the plants (Javot et al. 2007; Smith and Smith 2011). These pathways involve different cell types and nutrient transporters, providing capacity for independent and coordinated regulation (Smith and Smith 2011). In nonresponsive associations, the mycorrhizal pathway can also be functional (Grace et al. 2009) and may reflect an estimation of mycorrhizal effectiveness when P is measured. Similarly, the pathway for N uptake has been demonstrated in soil-grown plants using $^{15}NH_4$ and $^{15}NO_3$, but it is not known what proportion of total plant N requirement is delivered via this route (Smith and Smith 2011).

The effectiveness of mycorrhizal inoculation can involve a mixture of native AM fungi or exotic AM fungi and compatible hosts. Antunes et al. (2011) compared the growth responses of bluegrass (*Poa pratensis*) and Bermuda grass (*Cynodon dactylon*) inoculated with AM fungi assemblages originating from distant areas with contrasting climates and observed that AM fungal isolates originating from contrasting climates consistently and differentially altered plant growth, suggesting that AM fungi from contrasting climates have altered symbiotic function. AM fungi may adapt to different climatic conditions but of importance is how they compete with indigenous AM fungal communities in cropping systems. Nevertheless, the use of native AM fungi produced on-farm with mycorrhizal plants species (Douds et al. 2010; Sharma and Adholeya 2011; Vestberg et al. 2011) may represent a convenient alternative to commercial inoculum production. Pellegrino et al. (2011), evaluating native and exotic AM fungal inocula on plant growth and nutrient uptake in a low-input *Trifolium alexandrinum* and *Zea mays* in crop rotation, observed that the native fungal inoculum was as effective as highly efficient single exotic fungi.

Soil disturbance and cropping systems can affect the abundance and diversity of AM fungi and the function of the symbiosis. Soil disturbance associated with tillage systems can disrupt extraradical mycelium (Kabir et al. 1997). However, extraradical mycelia can survive the summer dry conditions in Mediterranean climate, allowing the next crop to benefit from the mycelium developed by the previous crop in the rotation (Brito et al. 2011). Duan et al. (2011) tested effects of soil disturbance and residue retention on the function of the symbiosis between medic and two AM fungi (*Glomus intraradices* and *Gigaspora margarita*) in an experiment simulating a crop rotation of wheat followed by medic and observed that the AM fungi responded differently to disturbance. *G. intraradices*, which was insensitive to disturbance, compensated for lack of contribution by the sensitive *G. margarita* when they were inoculated together. With respect to intercropping systems, Bainard et al. (2011) showed that intercropping systems supported a more abundant and diverse AM fungal community compared to conventionally managed systems. A positive influence on diversity of AM fungi can be related to more compatible host species, and the common mycorrhizal networks allow different plants to communicate with each other (Song et al. 2010).

In the field, the AM fungal interactions with other microorganisms (Das and Varma 2009; Javaid 2010; Reis et al. 2010) may affect mycorrhizal function. Wang et al. (2011) evaluated the co-inoculation of *G. mosseae* and *Bradyrhizobium* sp. on soybean genotypes HN89 (P-efficient) and HN112 (P-inefficient) and observed a synergistic relationship on plant growth, especially under low P and low N conditions for HN112, but there were no effects of inoculation under adequate N and P conditions. The N and P status influenced the effectiveness of inoculation, but AM fungal colonization reduced total root length, root surface area, and root volume in soybean. Furthermore, the genotype root architecture may affect the symbiosis. Yao et al. (2009) inoculated trifoliate orange (*Poncirus trifoliata*) seedlings with *G. intraradices*, *G. caledonium*, *G. margarita*, and *G. versiforme* and observed that AM colonization affected the distribution of root classes, increasing the proportion of fine roots (0–0.4 mm) and decreasing the proportion of coarse roots (0.6–1.2 mm), and reducing the total root length and root volume.

4.5 Conclusions

Agricultural practices and soil nutrient management practices can affect the mycorrhizal symbiosis and plant performance. As AM fungi inoculation becomes more
popular, inoculation techniques need to be improved, always ensuring that inoculants are applied as close to seeds and roots as possible. Training programs for
farmers may be necessary for successful use of mycorrhizal technology to avoid
ineffective use of costly biofertilizers that contain AM fungi. Studies with AM
fungal inoculants have demonstrated that some products can improve plant uptake
of nutrients and thereby increase the use efficiency of applied artificial fertilizers.

Acknowledgments This manuscript was prepared in part with support from project AUX-PE-
PGGI 267/2010 from the Coordination of Improvement of Higher Education Personnel (Brazil).

References

Abbott LK, Murphy DV (2007) What is biological fertility? In: Abbott LK, Murphy DV (eds) Soil
   biological fertility: a key to sustainable land use in agriculture. Springer, Dordrecht, pp 1–15
Abbott LK, Robson AD, Gazey C (1992) Selection of inoculant vesicular-arbuscular mycorrhizal
   fungi. In: Norris JD, Read DJ, Varma AK (eds) Methods in microbiology, vol 24. Academic
   Press, London, pp 1–21
Adesemoye O, Kloepper JW (2009) Plant–microbes interactions in enhanced fertilizer-use efficiency. Appl Microbiol Biotechnol 85:1–12
Akhtar SM, Siddiqui ZA (2008) Arbuscular mycorrhizal fungi as potential bioprotectants against
   plant pathogens. In: Siddiqui ZA, Akhtar MS, Futai K (eds) Mycorrhizae: sustainable agricul-
   ture and forestry. Springer, Dordrecht, pp 61–97
Allen MF (1996) The ecology of arbuscular mycorrhizas: a look back into the 20th century and a
   peek into 21st. Mycol Res 100:769–782
Antunes PM, Koch AM, Dunfield KE, Hart MM, Downing A, Rillig MC, Klironomos JN (2009)
   Influence of commercial inoculation with Glomus intraradices on the structure and functioning
   of an AM fungal community from an agricultural site. Plant Soil 317:257–266
Antunes PM, Koch AM, Morton JB, Rillig MC, Klironomos JN (2011) Evidence for functional
   divergence in arbuscular mycorrhizal fungi from contrasting climatic origins. New Phytol
   189:507–514
Aseri GK, Jain N, Panwar J, Rao VA, Meghwal PR (2008) Biofertilizers improve plant growth,
   fruit yield, nutrition, metabolism and rhizosphere enzyme activities of Pomegranate (*Punica
   granatum* L.) in Indian Thar Desert. Sci Hortic 117:130–135
Baar J (2008) From production to application of arbuscular mycorrhizal fungi in agricultural
   systems: requirements and needs. In: Varma A (ed) Mycorrhiza. Springer, Heidelberg,
   pp 361–372
Bainard LD, Klironomos JN, Gordon AM (2011) Arbuscular mycorrhizal fungi in tree-based
   intercropping systems: a review of their abundance and diversity. Pedobiologia 54:57–61
Barea JM, Palenzuela J, Cornejo P, Sánchez-Castro I, Navarro-Fernández C, Lopéz-García A,
   Estrada B, Azcón R, Ferrol N, Azcón-Aguilar C (2011) Ecological and functional roles of
   mycorrhizas in semi-arid ecosystems of Southeast Spain. J Arid Environ 75:1292–1301.
   doi:10.1016/j.jaridenv.2011.06.001

Boonlue S, Surapat W, Pukahuta C, Suwanarit P, Suwanatit A, Morinaga T (2012) Diversity and efficiency of arbuscular mycorrhizal fungi in soils from organic chili (*Capsicum frutescens*) farms. Mycoscience 53:10–16

Bothe H, Turnau K, Regvar M (2010) The potential role of arbuscular mycorrhizal fungi in protecting endangered plants and habitats. Mycorrhiza 20:445–457

Brito I, Goss MJ, Carvalho M, van Tuinen Y, Antunes PM (2008) Agronomic management of indigenous mycorrhizas. In: Varma A (ed) Mycorrhiza. Springer, Heidelberg, pp 375–402

Brito I, Carvalho MD, Goss MJ (2011) Summer survival of arbuscular mycorrhiza extraradical mycelium and the potential for its management through tillage options in Mediterranean cropping systems. Soil Use Manag 27:350–356

Bull CT, Muramoto J, Koike ST, Leap J, Shennan C, Goldman P (2005) Strawberry cultivars and mycorrhizal inoculants evaluated in California organic production fields. Crop Manag 4(1). doi: 10.1094/CM-2005-0527-02-RS

Cardoso IM, Kuyper TW (2006) Mycorrhizas and tropical soil fertility. Agric Ecosyst Environ 116:72–84

Cardoso EJBN, Cardoso IM, Nogueira MA, Baretta CRDM, Paula AM (2010) Micorrizas arbusculares na aquisição de nutrientes pelas plantas. In: Siqueira JO, de Souza FA, Cardoso EJBN, Tsai SM (eds) Micorrizas: 30 anos de experiência no Brasil. Universidade Federal de Lavras (UFLA), Lavras, pp 153–214

Corkidi L, Allen EB, Merhaut D, Allen MF, Downer J, Bohn J, Evans M (2004) Assessing the infectivity of commercial mycorrhizal inoculants in plant nursery conditions. J Environ Hortic 22:149–154

Cuenca G, Cáceres A, González MG (2008) AM inoculation in tropical agriculture: field results. In: Varma A (ed) Mycorrhiza. Springer, Heidelberg, pp 403–417

Das A, Varma A (2009) Symbiosis: the art of living. In: Varma A, Kharkwal AC (eds) Symbiotic fungi, soil biology, vol 18. Springer, Heidelbereg, pp 1–28

Djuuna IAF, Abbott LK, Solaiman ZM (2009) Use of mycorrhiza bioassays in ecological studies. In: Varma A, Kharkwal AC (eds) Symbiotic fungi, soil biology, vol 18. Springer, Heidelberg, pp 41–50

Douds DD Jr, Nagahashi G, Hepperly PR (2010) On-farm production of inoculum of indigenous arbuscular mycorrhizal fungi and assessment of diluents of compost for inoculum production. Bioresour Technol 101:2326–2330

Duan T, Facelli E, Smith SE, Smith FA, Nan Z (2011) Differential effects of soil disturbance and plant residue retention on function of arbuscular mycorrhizal (AM) symbiosis are not reflected in colonization of roots or hyphal development in soil. Soil Biol Biochem 43:571–578

Evelin H, Kapoor R, Giri B (2009) Arbuscular mycorrhizal fungi in alleviation of salt stress: a review. Ann Bot 104:1263–1280

FAO (Food and Agriculture Organization) of the United Nations (2010) Current world fertilizer trends and outlook to 2014. Available at: ftp://ftp.fao.org/ag/agp/docs/cwfto14.pdf. Accessed 22 July 2011

Farmer MA, Li X, Feng G, Zhao B, Chatagnier O, Gianinazzi S, Gianinazzi-Pearson V, van Tuinen D (2007) Molecular monitoring of field-inoculated AMF to evaluate persistence in sweet potato crops in China. Appl Soil Ecol 35:599–609

Gentili F, Jumpponen A (2006) Potential and possible uses of bacterial and fungal biofertilizers. In: Rai MK (ed) Handbook of microbial biofertilizers. Food Products Press, New York, pp 1–28

Gogoi P, Singh RP (2011) Differential effect of some arbuscular mycorrhizal fungi on growth of *Piper longum* L. (Piperaceae). Indian J Sci Technol 4:119–125

Grace EJ, Smith FA, Smith SE (2009) Deciphering the arbuscular mycorrhizal pathway of P uptake in non-responsive plant species. In: Azcón-Aguilar C, Barea JM, Gianinazzi S, Gianinazzi-Pearson V (eds) Mycorrhizas – functional processes and ecological impact. Springer, Heidelberg, pp 89–106

Herrera-Peraza RA, Hamel C, Fernández F, Ferrer RL, Furrazola E (2011) Soil-strain compatibility: the key to effective use of arbuscular mycorrhizal inoculants? Mycorrhiza 21:183–193

IJdo M, Cranenbrouck S, Declerck S (2011) Methods for large-scale production of AM fungi: past, present, and future. Mycorrhiza 21:1–16

Janos DP (2007) Plant responsiveness to mycorrhizas differs from dependence upon mycorrhizas. Mycorrhiza 17:75–91

Jansa J, Mozafar A, Frossard E (2003) Long-distance transport of P and Zn through the hyphae of an arbuscular mycorrhizal fungus in symbiosis with maize. Agronomie 23:481–488

Javaid A (2010) Beneficial microorganisms for sustainable agriculture. In: Lichtfouse L (ed) Genetic engineering, biofertilisation, soil quality and organic farming, sustainable agriculture. Springer, New York, pp 347–369

Javot H, Pumplin N, Harrison MJ (2007) Phosphate in the arbuscular mycorrhizal symbiosis: transport properties and regulatory roles. Plant Cell Environ 30:310–322

Kabir Z, O'Halloran IP, Fyles JW, Hamel C (1997) Seasonal changes of arbuscular mycorrhizal fungi as affected by tillage practices and fertilization: hyphal density and mycorrhizal root colonization. Plant Soil 192:282–293

Kahiluoto H, Ketoja E, Vestberg M (2009) Contribution of arbuscular mycorrhiza to soil quality in contrasting cropping systems. Agric Ecosyst Environ 134:36–45

Kahiluoto H, Ketoja E, Vestberg M (2012) Plant-available P supply is not the main factor determining the benefit from arbuscular mycorrhiza to crop P nutrition and growth in contrasting cropping systems. Plant Soil 350:85–98

Kaya KC, Ashraf M, Sonmez O, Aydemir S, Tuna AL, Cullu MA (2009) The influence of arbuscular mycorrhizal colonisation on key growth parameters and fruit yield of pepper plants grown at high salinity. Sci Hortic 121:1–6

Killham K (2011) Integrated soil management—moving towards globally sustainable agriculture. J Agric Sci 149:29–36

Lichtfouse E, Navarrete M, Debaeke P, Souchère V, Alberola C, Ménassieu J (2009) Agronomy for sustainable agriculture: a review. Agron Sustain Dev 29:1–6

Mardukhi B, Rejali F, Daei G, Ardakani MR, Malakouti MJ, Miransari M (2011) Arbuscular mycorrhizas enhance nutrient uptake in different wheat genotypes at high salinity levels under field and greenhouse conditions. C R Biol 334:564–571

Marin M (2006) Arbuscular mycorrhizal inoculation in nursery practice. In: Rai MK (ed) Handbook of microbial biofertilizers. Food Products Press, New York, pp 289–324

Martinez TN, Johnson NC (2010) Agricultural management influences propagule densities and functioning of arbuscular mycorrhizas in low- and high-input agroecosystems in arid environments. Appl Soil Ecol 46:300–306

Mihov M, Tringovska I (2010) Energy efficiency improvement of greenhouse tomato production by applying new biofertilizers. Bulg J Agric Sci 16:454–458

Miransari M (2011) Interactions between arbuscular mycorrhizal fungi and soil bacteria. Appl Microbiol Biotechnol 89:917–930

Murphy DV, Stockdale EA, Brookes PC, Gouling KWT (2007) Impact of microorganisms on chemical transformations in soil. In: Abbott LK, Murphy DV (eds) Soil biological fertility: a key to sustainable land use in agriculture. Springer, Dordrecht, pp 37–59

Ngwene B, George E, Claussen W, Neumann E (2010) Phosphorus uptake by cowpea plants from sparingly available or soluble sources as affected by nitrogen form and arbuscular-mycorrhiza-fungalinoculation. J Plant Nutr Soil Sci 173:353–359

Öpik M, Saks Ü, Kennedy J, Daniell T (2008) Global diversity patterns of arbuscular mycorrhizal fungi–community composition and links with functionality. In: Varma A (ed) Mycorrhiza. Springer, Heidelberg, pp 89–111

Osorio NW, Habte M (2009) Strategies for utilizing arbuscular mycorrhizal fungi and phosphate-solubilizing microorganisms for enhanced phosphate uptake and growth of plants in the soils of the tropics. In: Khan MS, Zaidi A, Musarrat J (eds) Microbial strategies for crop improvement. Springer, Heidelberg, pp 325–351

Pellegrino E, Bedini S, Avio L, Bonari E, Giovannetti M (2011) Field inoculation effectiveness of native and exotic arbuscular mycorrhizal fungi in a Mediterranean agricultural soil. Soil Biol Biochem 43:367–376

Perner H, Schwarz D, Bruns C, Mäder P, George E (2007) Effect of arbuscular mycorrhizal colonization and two levels of compost supply on nutrient uptake and flowering of pelargonium plants. Mycorrhiza 17:469–474

Reis VM, Andrade G, Faria SM, Silveira APD (2010) Interações de fungos micorrízicos arbusculares com outros microrganismos do solo. In: Siqueira JO, de Souza FA, Cardoso EJBN, Tsai SM (eds) Micorrizas: 30 anos de experiência no Brasil. Universidade Federal de Lavras (UFLA), Lavras, pp 361–413

Rodriguez-Romero AS, Azcón R, Jaizme-Vega MDC (2011) Early mycorrhization of two tropical crops, papaya (*Carica papaya* L.) and pineapple [*Ananas comosus* (L.) Merr.], reduces the necessity of P fertilization during the nursery stage. Fruits 66:3–10

Rouphael Y, Cardarelli M, Mattia ED, Tullio M, Rea R, Colla G (2010) Enhancement of alkalinity tolerance in two cucumber genotypes inoculated with an arbuscular mycorrhizal biofertilizer containing *Glomus intraradices*. Biol Fertil Soils 46:499–509

Saldajeno MGB, Chandanie WA, Kubota M, Hyakumachi AM (2008) Effects of interactions of arbuscular mycorrhizal fungi and beneficial mycoflora on plant growth and disease protection. In: Siddiqui ZA, Akhtar MS, Futai K (eds) Mycorrhizae: sustainable agriculture and forestry. Springer, Dordrecht, pp 211–226

Shabani G, Ardakani MR, Chaichi MR, Friedel JK, Khavazi K, Eshghizaderh HR (2011) Effect of different fertilizing systems on seed yield and phosphorus uptake in annual medics under dryland farming conditions. Not Bot Hort Agrobot Cluj 39:191–197

Shalamuk S, Cabello MH, Chidichimo H, Golik S (2011) Effects of inoculation with *Glomus mosseae* in conventionally tilled and nontilled soils with different levels of nitrogen fertilization on wheat growth, arbuscular mycorrhizal colonization, and nitrogen nutrition. Commun Soil Sci Plant Anal 42:586–598

Sharma MP, Adholeya A (2011) Developing prediction equations and optimizing production of three AM fungal inocula under on-farm conditions. Exp Agric 47:529–537

Sieverding E (1991) Vesicular-arbuscular mycorrhiza management in tropical agrosystems, vol 224. Deutsche Gesellschaft für Technische Zusammenarbeit (GTZ), Eschborn. ISBN 3-88085-462-9

Siqueira JO, Saggin-Júnior OS, Flores-Aylas WF, Guimarães PTG (1998) Arbuscular mycorrhizal inoculation and superphosphate application influence plant development and yield of coffee in Brazil. Mycorrhiza 7:293–300

Smith SE, Smith FA (2011) Roles of Arbuscular mycorrhizas in plant nutrition and growth: new paradigms from cellular to ecosystem scales. Annu Rev Plant Biol 62:227–250

Solaiman MZ, Abbott LK (2004) Functional diversity of arbuscular mycorrhizal fungi on root surfaces. In: Varma A, Abbott LK, Werner D, Hampp R (eds) Plant Surface Microbiology. Springer, Heidelberg, pp 331–349

Song YY, Zen RS, Xu JF, Li J, Shen X, Yihdego WG (2010) Interplant communication of tomato plants through underground common mycorrhizal networks. PLoS One 5(10):e13324

Stürmer SL, Saggin O Jr (2010) Bancos de germoplasma de Glomeromycota no Brasil. In: Siqueira JO, de Souza FA, Cardoso EJBN, Tsai SM (eds) Micorrizas: 30 anos de experiência no Brasil. Universidade Federal de Lavras (UFLA), Lavras, pp 525–550

Tanu Prakash A, Adholeya A (2006) Potential of arbuscular mycorrhizae in organic farming system. In: Rai MK (ed) Handbook of microbial biofertilizers. Food Products Press, New York, pp 223–239

Tarbell TJ, Koske RE (2007) Evaluation of commercial arbuscular mycorrhizal inocula in a sand/peat medium. Mycorrhiza 18:51–56

Thonar C, Shnepf A, Frossard E, Roose T, Jansa J (2011) Traits related to differences in function among three arbuscular mycorrhizal fungi. Plant Soil 339:231–245

Trejo D, Ferrera-Cerrato R, Garcia R, Varela L, Lara L, Alarcon A (2011) Efectividad de siete consorcios nativos de hongos micorrízicos arbusculares en plantas de café en condiciones de invernadero y campo. Rev Chil Hist Nat 84:23–31

Vance CP (2001) Symbiotic nitrogen fixation and phosphorus acquisition. Plant nutrition in a world of declining renewable resources. Plant Physiol 127:390–397

Vázquez-Hernández MV, Arévalo-Galarza L, Jaen-Contreras D, Escamilla-García JD, Mora-Aguilera A, Hernández-Castro E, Cibrián-Tovar J, Téliz-Ortiz D (2011) Effect of *Glomus mosseae* and *Entrophospora colombiana* on plant growth, production, and fruit quality of 'Maradol' papaya (*Carica papaya* L.). Sci Hortic 128:255–260

Vestberg M, Kahiluoto H, Wallius E (2011) Arbuscular mycorrhizal fungal diversity and species dominance in a temperate soil with long-term conventional and low-input cropping systems. Mycorrhiza 21:351–361

Vosátka M, Albrechtová J (2008) The international market development for mycorrhizal technology. In: Varma A (ed) Mycorrhiza: state of the art, genetics and molecular biology, eco-function, biotechnology, eco-physiology, structure and systematics. Springer, Heidelberg, pp 419–438

Vosátka M, Albrechtova J (2009) Benefits of arbuscular mycorrhizal fungi to sustainable crop production. In: Khan MS, Zaidi A, Musarrat J (eds) Microbial strategies for crop improvement. Springer, Heidelberg, pp 205–225

Wang X, Pan Q, Chen F, Yan X, Liao H (2011) Effects of co-inoculation with arbuscular mycorrhizal fungi and rhizobia on soybean growth as related to root architecture and availability of N and P. Mycorrhiza 21:173–181

Weber OB, Souza CCM, Gondin DMF, Oliveira FNS, Crisóstomo LA, Caproni AL, Saggin O Jr (2004) Inoculação de fungos micorrízicos arbusculares e adubação fosfatada em mudas de cajueiro-anão-precoce. Pesq agropec bras 39(5):477–483

Wiseman PE, Colvin KH, Wells CE (2009) Performance of mycorrhizal products marketed for woody landscape plants. J Environ Hortic 27:41–50

Wu SC, Cao ZH, Li ZG, Cheung KC, Wong MH (2005) Effects of biofertilizer containing N-fixer, P and K solubilizers and AM fungi on maize growth: a greenhouse trial. Geoderma 125:155–166

Yang C, Hamel C, Schellenberg MP, Perez JC, Berbara RL (2010) Diversity and functionality of arbuscular mycorrhizal fungi in three plant communities in semiarid grasslands national park, Canada. Microb Ecol 59:724–733

Yao Q, Wanga LR, Zhu HH, Chen JZ (2009) Effect of arbuscular mycorrhizal fungal inoculation on root system architecture of trifoliate orange (*Poncirus trifoliata* L. Raf.) seedlings. Sci Hortic 121:458–461

Zarabi M, Alahdadi I, Akbari GA, Akbari GA (2011) A study on the effects of different biofertilizer combinations on yield, its components and growth indices of corn (*Zea mays* L.) under drought stress condition. Afr J Res 6:681–685

Chapter 5
Mycorrhizal Inoculum Production

Shivom Singh, Kajal Srivastava, Suvigya Sharma, and A.K. Sharma

5.1 Introduction

The reliance on chemical fertilizers and pesticides in current agriculture and horticulture can cause pollution of natural environments as crop production rises rapidly. Therefore, development and management of sustainable agriculture is a primary issue in the agriculture sector worldwide. To enhance this practice, one feasible way is to generalize the recyclable organic agricultural cultivation methods such as utilizing naturally occurring plant growth-promoting microbes (PGPM).

Arbuscular mycorrhizal (AM) fungi can form symbiosis with virtually 80 % of all cultivated plants. They infect the roots and colonize invasively inside root cells. AM fungi are mainly characterized by arbuscules which are formed by fine, bifurcate branching hyphae in cortical cells. Hyphae of AM fungi extend outwards from the root surface, expanding the accessibility of the root system for nutrient uptake. AM fungi can therefore contribute as a "biofertilizer" by facilitating access to nutrients (Sylvia 1990; Leyval et al. 2002; Srivastava and Sharma 2011; Turnau and Haselwandter 2002).

Commercial application of AM fungal inoculum is increasing. In 2001, Sylvia listed 21 companies in North America, eight in Europe, two in South America, and two in Asia with involvement in production of inocula of AM fungi, but there are many more established companies which aim to produce and use inocula in various

S. Singh
Department of Environmental Science, ITM University, Gwalior 475001, MP, India

K. Srivastava • S. Sharma • A.K. Sharma (✉)
Department of Biological Sciences, CBSH, G. B. Pant University of Agriculture and Technology, Pantnagar 263145, UK, India
e-mail: anilksharma_99@yahoo.com

© Springer-Verlag Berlin Heidelberg 2014   67
Z.M. Solaiman et al. (eds.), *Mycorrhizal Fungi: Use in Sustainable Agriculture and Land Restoration*, Soil Biology 41, DOI 10.1007/978-3-662-45370-4_5

sectors of plant production (Gianinazzi and Vosátka 2004). Reasons for development of this agricultural biotechnology industry include the fact that AM fungi are
increasingly being considered as a "natural plant health insurance" (Gianinazzi and
Gianinazzi-Pearson 1988). Their positive impacts on plant development and health,
land reclamation, and phyto-remediation are well recognized (Leyval et al. 2002;
Turnau and Haselwandter 2002), and there is higher awareness of biodiversity
issues, including those concerning soil microbial communities, and acceptance of
these natural technologies as alternatives to agrochemicals (Barea 2000; Gryndler
2000). Furthermore, society is demanding more sustainable means of production,
with a consequent feedback to farmers and land conservationists.

Producing microbial inoculum is a complex procedure that involves the biotechnological expertise and the ability to respond to associated legal, ethical, educa-
tional, and commercial requirements. This is particularly the case for obligate
endosymbiotic microorganisms such as AM fungi because satisfying the aforemen-
tioned requirements is closely associated with the particular method of inoculum
production.

5.2 Techniques Employed to Cultivate AM Fungi
Propagules

"Pot culture" (Wood 1985) is the main technique employed for the large-scale
production of inocula of AM fungi. It is a traditional and widely practiced method
which employs trap plants (Chellappan et al. 2002). Potty (1985) reported that
Glomus mosseae multiplied on cassava (*Manihot esculenta*) tuber peel yielding 3–
4/cm^2 and suggested that this peel could be used for mass multiplication of AM
fungi. Subsequently, Ganesan and Mahadevan (1998) claimed that hyphae,
arbuscules, and vesicles of *G. aggregatum* developed on the surface of cassava
tuber could be used as inoculum. Selvaraj and Kim (2004) used sucrose-agar
globule with root exudates (SAGE) as a source of inoculum to increase the
production of spores of AM fungi. The exudates led to higher percentage of root
colonization (by about 10 % more) and increases in the number of spores (by about
26 %) and dry matter content (by more than 13 %) for the inoculum of AM fungal
spores compared with their soil inoculum. A range of techniques such as the
nutrient film technique (Mathew and Johri 1988), aeroponic culture systems
(Hung and Sylvia 1987, 1988), and root-organ cultures (Declerk et al. 1996) have
been employed for the production of AM fungal inocula in near-sterile environment
(Raja and Mahadevan 1991). For all methods, including pot culture, only a few
spores are usually used to initiate the production of the inocula.

The sources of AM fungal inocula are defined by the biology of the fungi. All
infective structures of the fungi including fungal spores and/or the mycelium
produced inside or outside the host root can be used as inoculum. However,
arbuscules and the auxiliary cells formed by some fungi are not known to be a

source of inoculum. Infected roots and substrates (carriers which contain infected root and/or mycelium and spores) are commonly used. Spores are an important source for the establishment of clean cultures of AM fungi on host plants in a previously sterilized substrate. This is because (1) a small number of spores can be isolated relatively easily from a soil substrate, (2) spores can be morphologically distinguished for the identification of the fungi, and (3) spore surfaces can be satisfactorily disinfected (with the object of producing inoculum free from contaminants).

It is well known that a single spore can initiate the mycorrhizal symbiosis (Sieverding 1991) and that spores are suitable sources of inoculum for experiments (Nopamornbodi et al. 1987) and other special cases of plant growth (e.g., in nurseries or in conditions where aseptic inoculum is required). Spores of AM fungi can be produced on artificial media at low cost or established as by-products in the process of manufacturing other forms of AM inoculum such as those embedded in expanded clay as the carrier. Although it is technically possible to inoculate crops with spores, the use of spore inoculum is questionable for other reasons. Some spores require several days or longer to germinate and spore dormancy has been reported (Tommerup 1987). Due to the slow initial development and slow spread in the root system, it cannot be expected that a spore inoculum can compete with indigenous AM fungal propagules in the presence of other soil microorganisms. A rapid and high level of colonization of roots by the inoculated fungus is a prerequisite for the desired inoculant fungus to be beneficial to the host plant (Douds et al. 2005).

Widespread application of AM fungal inocula has been limited by difficulties in obtaining large quantities of pure inoculum. Mass production of AM fungal inocula became technically feasible with the introduction of the pot culture technique (Wood 1985) by culturing with plant hosts on substrates such as sand, peat, expanded clay, perlite, vermiculite, soilrite (Mallesha et al. 1992), rock wool (Heinzemann and Weritz 1990), and glass beads (Redecker et al. 1995).

Soil inoculum contains all AM fungal structures and can be highly infective as an inoculum for some fungi. A soil inoculum is generally chopped up before being applied in quantities depending on the inoculation technique. Many plants including maize, sorghum, Bahia grass, and Sudan grass have been shown to be suitable hosts for inoculum production. Soils and climates vary regionally, and locally available materials for inoculum production need to be tested. An important consideration for the selection of the host is that it should not have pathogens in common with those crops which are to be inoculated.

The time required for cultivation of AM fungal inocula in soil-based culture systems is relatively long, and the product quality can be inconsistent due to the possibility of introducing contaminants. Spores produced in pot cultures are generally more conducive to identification than those collected from the field which usually consist of mixtures of species. Cultures may be produced by various methods, including soil trap cultures, pot substrate cultures, plant trap cultures, and soil culture (Brundrett et al. 1999). Soil trap cultures involve growing plants in field soil for up to 6 months and then separating spores of the different AM fungi into separate pot cultures. These cultures can be used as base for further purification. For pot substrate culturing, a small quantity of substrate from an existing pot culture can be mixed thoroughly with disinfected substrate, added to a mycorrhiza-free plant or placed under seeds in a sterilized soil. These cultures usually establish quickly, and spores were normally produced within a month or two of the subculturing attempt depending on the fungal species. Plant trap cultures involve removing plants from an area of interest, washing the roots thoroughly to remove all traces of soil and external mycelium, and planting them in a suitable sterile substrate. Mixed culture produced by this method can be used for subsequent further purification. Soil culture involves adding a layer of pot culture or other inoculum soil in sterilized media over several cycles to enhance mycorrhiza formation by the fungi present.

5.3 On-Farm Production of Arbuscular Mycorrhizal Fungus Inoculum

On-farm production of AM fungus inocula is an alternative to commercially produced inocula as the quantities of inoculum necessary for large-scale agriculture may be costly. Producing the inoculum on-site saves processing and shipping costs. These factors are the primary reason why most on-farm methods have been utilized in developing nations. Another benefit of on-farm production of inoculum is that locally adapted isolates can be used which may be more effective than introduced ones in certain situations (Sreenivasa 1992) and a taxonomically diverse inoculum can be produced.

Significant advances in on-farm production of AM fungal inocula have been made in developing countries in the tropics. For example, Sieverding (1987, 1991) produced inoculum of an effective strain of the AM fungus *Glomus manihotis* in Columbia using 25-m^2 fumigated field plots. Dodd et al. (1990) used a similar method in Columbia to produce an inoculum containing three AM fungi. Gaur (1997) and Douds et al. (2000) used raised beds of fumigated soil to produce inocula. Douds et al. (2005) also developed a modified raised bed method for on-farm production of AM fungus inoculum in temperate climates. The raised bed enclosures, 0.75 m × 3.25 m × 0.3 m, were constructed with silt fence walls, weed barrier cloth floors, and plastic sheeting dividing walls between 0.75-m square sections. The enclosures were 20-cm deep with mixtures of compost and vermiculite. In all cases, the choice of fumigants needs to comply with government regulations.

The compost utilization trial at the Rodale Institute Experimental Farm, Kutztown (Reider et al. 2000), used three treatments for comparing bulked AM fungal inocula which were (1) a commercially available AM fungus inoculum, (2) an on-farm inoculum, and (3) a control treatment. The commercially available AM fungus inoculum (MYKE® Garden, Premier Tech Biotechnologies, Rivièredu-

Loup, Quebec) contained 30 propagules of the AM fungus *Glomus intraradices*, as determined by the manufacturer, in a peat vermiculite mixture. The on-farm inoculum (compost–vermiculite = 1:9 vol/vol) contained 3,225 propagules of *G. mosseae*, *G. etunicatum*, *G. claroideum*, and AM fungi indigenous to the small amount of soil mixed into the compost. The two soil fertility management regimes in this experiment were a conventional chemical fertilizer and dairy cow manure + leaf compost. Final yields of tubers showed that the mycorrhizal treatments outproduced the control by 33-45 % and that the on-farm inoculum performed as well as the commercial mix (Douds et al. 2005).

Colonized roots contain internal fungal mycelium as well as external mycelium (and sometimes AM fungal spores) and can be used as an inoculum. Before use, roots are often chopped into smaller pieces. The infectiveness of colonized roots can be higher than that of spores, with new root colonization occurring within 1–2 days of inoculation with infected roots (Sieverding 1991). In addition to using colonized roots in greenhouse experiments, this form of inoculum has been used to inoculate plants in nursery experiments (Janos 1980). The quantities of roots used in inocula vary from up to 20 g fresh weight per plant, but even 1 g per plant was sufficient to obtain growth responses in nurseries and greenhouse experiments (Howeler 1985) in sterilized soil. When the colonized roots contain AM fungal spores, they can be dried and stored at ambient temperature or in a cold room for up to a month without any loss of their infectiveness, but when stored at 4 °C in water or moist vermiculite, infectivity was maintained for 2 months (Hung and Sylvia 1987).

5.4 Aeroponic and Hydroponic Inoculum Sources

Aeroponic culture of AM fungi is a complete culture system that starts with relatively few spores of the selected fungus to inoculate culture plants (Singh et al. 2012). These plants were then transferred into the aeroponic environment for more extensive root growth, colonization, and sporulation of the fungus. The resulting colonized root and spores may be used in a variety of ways (research, horticulture, floriculture, vegetable crops, forestry, scrublands, landscaping). Large-scale production of colonized roots with single AM fungi is practicable in aeroponic culture (Hung and Sylvia 1987; Sylvia and Hubbel 1986). When horticulture crops including tomato, sweet potato, and strawberry were produced with this system, their roots were by-products which can be used as inocula. Hence, the cost of this inoculum source may be fairly low in regions where these systems are used. Aeroponic culture of AM fungi is a biotechnology that allows both efficient production of AM fungal inoculum and soil-free investigation of mycorrhizas.

Aeroponic culture was first explored for legume rhizobia interaction by Zobel et al. (1976) and then for AM fungi by Sylvia and Hubbel (1986). It is a more aerated system than hydroponics and has proven to be an efficient system for growing AM fungal inoculum without a physical substrate (Hung and Sylvia

1988) as used in a nutrient flow system (Mosse and Thompson 1984). Inoculum production of AM fungi in aeroponic culture allows easy extraction of spores, hyphae, and roots. In addition, the roots may be sheared to produce inocula with a high propagule density (Sylvia and Jarstfer 1992a, b). In aeroponic culture, plants are grown in a closed or semi-closed environment by spraying the roots with a nutrient-rich solution where the environment is kept free from pests and diseases so that the plants may grow healthier and quicker than plants grown in a medium. However, if aeroponic environments are not completely sealed, then pests may pose a problem.

Problems are encountered during inoculum production at an industrial scale, including maintaining non-contaminated conditions. Martin-Laurent et al. (1997) demonstrated that soilless culture method such as aeroponic culture was a promising way to produce pure inoculum. Saplings inoculated with AM fungi in pots and subsequently grown in aeroponic conditions showed significantly higher rates of mycorrhizal colonization and P content than saplings grown in soil (Martin-Laurent et al. 1997). The potential of AM inoculum production in aeroponic culture for industrial applications was similarly demonstrated on *Paspalum notatum* and *Ipomoea batatas*. These plants were inoculated with *Glomus etunicatum* using a water-soluble polymer as the sticking agent for AM inoculum (Hung et al. 1991). However, the conventional spray nozzle and ultrasonic fog to produce fine mist of nutrient solution result in rapid loss of nutrient solution through evaporation and diminishing the possibility of rapid absorption of nutrients by the aeroponically cultured roots (Carruthers 1992). Another disadvantage of spraying roots with a nutrient solution with larger droplets is stifled root growth (Carruthers 1992). These problems limit rapid growth and AM fungal colonization of roots in the system. To overcome these limitations, piezo-ceramic element technology was used by Carruthers (1992), employing high-frequency sound that blasted the nutrient solution and nebulized it into microdroplets size of 1 mm in diameter.

It has been demonstrated that colonized roots sporulate rapidly in aeroponic culture (Martin-Laurent et al. 1997; Sylvia and Jarstfer 1992a, b). It was also demonstrated that both colonized roots and spores produced in aeroponic chambers can serve as infective AM fungal inocula which can be mixed directly and thoroughly with growing media if plants are to be immediately planted or transplanted. However, inoculum viability declines with storage time (Sylvia and Jarstfer 1992a).

Hydroponic culture of mycorrhizal plants provides a controlled nutrient environment and allows the harvest of mycorrhizal plants free from soil. This type of culture is infrequently used for growing and studying mycorrhizal plants. Hydroponic culture of mycorrhizal fungi was reported first by Peuss (1958) for *Glomus mosseae* with *Nicotiana tabacum*. Mycorrhizal plants were also grown in nutrient solution culture by Cress et al. (1979, 1986) and Karunaratne et al. (1986), and Dugassa et al. (1995) reported a culture of *Glomus intraradices* with *Linum usitatissimum*. To avoid microbial contaminants, frequent refreshment of the nutrient liquid is needed. The soaked state of the host plants and AMF therein caused by aquatic environment used in this method is not the natural growth condition for AM fungi and may limit sporulation.

Nutrient film culture is a technique developed for commercial production that entails continuous recycling of a large volume of nutrient liquid in a film which flows over the roots of the plant. MacDonald (1981) used a compact autoclave hydroponic culture system for the production of axenic mycorrhizas between *Trifolium parviflorum* and *Glomus caledonium* and others (Elmes et al. 1984; Elmes and Mosse 1984; Howeler et al. 1982; Mathew and Johri 1988; Mosse and Thompson 1984).

Besides the variation of techniques used in aseptic inoculation of AM fungi to the host plant, the major concern in the nutrient film technique system is the concentration of nutrients. The preferred values for the various nutrient elements vary from one particular mycorrhizal system to another depending particularly on the size and other features of the plant (Sharma et al. 2000). Another important factor is the compromise between plant growth and mycorrhizal colonization as waterlogged conditions affect mycorrhizal growth adversely (Tarafdar 1995).

As AM fungi are integrated components of most terrestrial plants, the nutrient exchange and other benefits due to AM fungi are sufficient for research conditions without contaminations (Diop 2003). The development of the root-organ culture technology system has opened new avenues for studying the symbiosis (Elsen et al. 2001).

5.5   Root-Organ Cultures

The root-organ culture technique was developed by White (1943) and others (Mosse 1962; Butcher and Street 1964; Butcher 1980) using excised roots on synthetic mineral media supplemented with vitamins and carbohydrates. The formation of lower-order roots is essential for rapid increase in root biomass and the establishment of continuous cultures. Both axenic and monoxenic approaches using different sources of propagules of AM fungi aimed to acquire root-organ cultures of AM fungi (Bécard and Piché 1992; Chabot et al. 1992; Diop 1990, 1995; Declerk et al. 1996).

Initiation of roots requires pre-germination of seeds after surface sterilized with classical disinfectants (sodium hypochlorite, hydrogen peroxide) and then thoroughly washed in sterile distilled water. Following germination of seeds on water agar or moistened filter papers, the tips (2 cm) of emerged roots can be transferred to a nutrient-rich media such as modified White medium (Bécard and Fortin 1988) or Strullu and Romand medium (Strullu and Romand 1987). With pH of the medium adjusted to 5.5, fast-growing roots can be cloned by repeated subcultures. This method of artificial culture is a valuable tool for the study and inoculum production of arbuscular mycorrhizal fungi as it avoids the interaction of other inhabitants of the rhizosphere.

The vegetative development of AM fungi in monoxenic cultures has been followed using either transformed or non-transformed roots (Bécard and Piché 1992; Fortin et al. 2002). The long-term behavior of *G. margarita* on Ri-TDNAtransformed roots of the carrot showed 80 % of the fungal infection units were produced during the period of root aging (Bécard an Fortin 1988; Diop et al. 1992). The use of Sunbags in in vitro system is another alternative to obtain large-scale AM fungal inoculum without contaminations. The axenic AM fungal propagules can be conserved at 4 °C in the dark for several months or used for fundamental or inoculation practices. The possibility of continuous culture and cryopreservation has resulted in an international collection of in vitro AMF (websites: http://www.mbla.uclac.be/ginco-beland; http://res2.agr.ca/ecorc/gino.can/).

5.6 Qualitative and Quantitative Production of Mycorrhizal Inocula

The industrial activity of inoculum producers has been developed using different AM fungi, which are quite often not well characterized in terms of ecological requirements and stability. Along with this, the lack of quality control for several marketed inocula is among the main reasons for the low acceptance of mycorrhizal technology in horticultural and agricultural practices (Gianinazzi and Vosátka 2004). This situation has led to the need for mycorrhizal inoculum production industry to develop, in its own interest, criteria that will satisfy minimum requirements of quality. Whatever the mode of inoculum production chosen and the formulation procedure adopted by the companies, the marketed product has to meet the expected requirements of end users. Although these objectives may vary according to the companies, they should all aim at the use of AM fungi as a natural plant health insurance (Gianinazzi and Gianinazzi-Pearson 1988). In this context, the following criteria should be fulfilled by the companies: (1) plants to be inoculated must be able to form mycorrhizas; (2) the AM fungal inoculum must be free of agents that could negatively affect normal plant growth and development; and (3) the shelf life of the inoculum should be sufficient to suit end-user markets. The introduction of such criteria by the inoculum producers could contribute to the definition of conditions for the registration of products at national or international levels (Von Alten et al. 2002). Furthermore, in the product description, inclusion of the following recommendations for quality standards may be considered: pH, nutrient carriers, and additives. Additives could be included if their primary aim is to support mycorrhizal development, but additives which are general fertilizers should not be included unless this is stated clearly.

For better quality and production, the relevant number of AM fungal propagules should be determined. Therefore, there is a need for an independent testing service that can be used by producers to check that batches of inocula meet baseline standards that have been established and agreed to by individual companies on the basis of a voluntary code of best practices (Gianinazzi and Vosátka 2004). The inoculum formulation procedure usually consists of placing fungal propagules (root fragments colonized with AM fungi, fragments of fungal mycelium, and spores) in a desired carrier (perlite, peat, inorganic clay, zeolite, vermiculite, sand, etc.) for a given application. Some companies producing AM fungal inocula have adopted the approach of one type of formulation (i.e., single fungal species) for all markets, while others produce a range of products for their target buyers.

The outcome of the AM symbiosis depends on environmental factors, AM fungal characteristics, and plant variables. However, present knowledge makes it difficult to predict the effectiveness of inocula. For example, the procedure called direct inoculum production process could help to improve predictability of AM fungal inoculum effectiveness (Feldmann and Grotkass 2002). Quality control of commercial inoculum must be dealt with, and a reference system for information concerning inoculum effectiveness based on the results of standard tests should be established for the buyers as well as a list of examples where the relevant inoculum had already been successfully used.

5.7 Ethical and Legal View of Using Inoculum

Suitable legislation based on quality control adapted to AM fungal inocula is essential for the development of mycorrhizal technology. At present, registration procedures for AM fungal inocula vary between countries, with some having very strict regulations (e.g., France and Canada) and others being less demanding or even without regulations. No regulation or lack of adherence to strict regulation will encourage the marketing of ineffective products. Overregulation could also destroy the market by preventing the development of small and medium enterprises and inoculum producers and distributors (Gianinazzi and Vosátka 2004). In France, beneficial microbes such as *Rhizobium* are considered as biofertilizers, and their registration requires a complex and expensive procedure that implies detailed description of the biological properties of the relevant microbes (identification, dissemination, toxicity, etc.). Furthermore, demonstration of the beneficial effects of the microbe via several controlled field trials (three to five per year) was performed, and finally, tests of the lack of toxicity or allergenicity of the formulated products on humans, animals, and plants were done. At the EU level, there is no registration for biofertilizers. However, the directive 91/414/EEC regulates the use of microbial products for plant protection. The data requirements for approval of plant protection products focus on possible unacceptable impacts on plants or the environment, harmful effects on human or animal health, and contamination of groundwater. The cost of such a process would handicap attempts to introduce mycorrhizal technology into plant production systems. Because of attempts to apply this directive to AM fungi, the European network on AM fungi, Cost Action 8.38 (2001), initiated discussions within the EU on the need for a registration procedure for AM fungal inocula. Because AM fungal inocula do not produce toxins, they should be regarded as a natural part of the plant, and the guidelines for approval of microbial plant protection products are not directly applicable to them, and the "risk assessment" criteria are also inappropriate (http://www.dijon.inra.fr/

cost838/index.html). However, unexpected consequences of transport of inocula should be considered (Schwartz et al. 2006).

5.8   Conclusion

Intensive agricultural practices are currently being reevaluated and are coming under increased scrutiny as the awareness of the consequences of excessive use of fertilizers and chemical pesticide usage improves. The concept of biofertilizers is to domesticate some of these microorganisms in agricultural production systems, so that additional natural reservoirs of nutrients in the atmosphere, hydrosphere, and pedosphere can be tapped to meet the requirements of sustainable agriculture. This approach also augments yield and monetary returns to the farmers, particularly to the small landholders, for which the incremental input cost is low.

The conventional difficulty of keeping low cost compatible with high quality of final products is always a point of concern of cultivating AM fungi using root-organ culture. The cheaper and lower technical methods of the abovementioned potted and hydroponic ones lead to difficulties in product control and product quality which discourages adoption. Soilless culture methods such as aeroponics have been demonstrated to be a promising way of inoculum production with potential for intensive production systems such as in horticulture. On-farm production of inocula combined with strategic management practices may be more practical for a large-scale agricultural production.

References

Barea JM (2000) Rhizosphere and mycorrhiza of field crops. In: Balazs E, Galante E, Lynch JM, Schepers JS, Toutan JP, Werner D, Werry PA (eds) Biological resource management: connecting science and policy (OECD). Springer, Berlin, pp 110–125

Bécard G, Fortin JA (1988) Early events of vesicular-arbuscular mycorrhiza formation on Ri T-DNA transformed roots. New Phytol 108:211–218

Bécard G, Piché Y (1992) Establishment of vesicular-arbuscular mycorrhiza in root organ culture: review and proposed methodology. In: Norris J, Read D, Varma A (eds) Techniques for the study of mycorrhiza. Academic, New York, pp 89–108

Brundrett MC, Abbott LK, Jasper DA (1999) Glomalean mycorrhizal fungi from tropical Australia. 1. Comparison of the effectiveness and specificity of different isolation procedures. Mycorrhiza 8:305–314

Butcher DN (1980) The culture of isolated roots. In: Ingram DS, Helgelson JP (eds) Tissue culture methods for plant pathologists. Blackwell, Oxford, pp 13–17

Butcher DN, Street HE (1964) Excised root culture. Bot Rev 30:513–586

Carruthers S (1992) Aeroponics system review. Practical Hydroponics, July/August issue, pp 18–21

Chabot S, Bécard G, Piché Y (1992) Life cycle of *Glomus intraradices* in root organ culture. Mycologia 84:315–321

Chellappan P, Anitha Christy SA, Mahadevan A (2002) Multiplication of arbuscular mycorrhizal fungi on roots. In: Mukerji KG, Manoharachary C, Chamola BP (eds) Techniques in mycorrhizal studies. Kluwer, Dordrecht, pp 285–297

Cost Action 8.38 (2001) Managing arbuscular mycorrhizal fungi for improving soil quality and plant health in agriculture. In: Gianinazzi S (ed) Report of 1999 Activity – EUR 19687. European Commission, Directorate-General for Research, Luxembourg

Cress WA, Throneberry GO, Lindsey DL (1979) Kinetics of phosphorus absorption by mycorrhizal and nonmycorrhizal tomato roots. Plant Physiol 64:484–487

Cress WA, Johnson GV, Barton LL (1986) The role of endomycorrhizal fungi in iron uptake by *Hilaria jamesii*. J Plant Nutr 9:547–556

Declerk S, Strullu DG, Plenchette C (1996) *In vitro* mass production of the arbuscular mycorrhizal fungus *Glomus versiforme* associated with Ri T-DNA transformed carrot roots. Mycol Res 100:1237–1242

Diop TA (1990) Méthodes axéniques de production d'inocula endomycorhiziens à vésicules et à arbuscules: étude avec le *Gigaspora margarita*. MSc Thesis, University of Laval, Quebec

Diop TA (1995) Ecologie des champignons mycorhiziens à vésicules et à arbuscules associés à *Acacia albida* (del) dans les zones sahéliennes et soudano-guinéennes du Sénégal. Thèse de docteur en Biologie et Physiologie végétales, Angers, France

Diop TA (2003) *In vitro* culture of arbuscular mycorrhizal fungi: advances and future prospects. Afr J Biotechnol 2(12):692–697

Diop TA, Becard G, Piché Y (1992) Long-term *in vitro* culture of an endomycorrhizal fungus, *Gigaspora margarita*, on Ri T-DNA transformed root of carrot. Symbiosis 12:249–259

Dodd JC, Arias I, Kooman I, Hayman DS (1990) The management of populations of vesicular-arbuscular mycorrhizal fungi in acid-infertile soils of a savanna ecosystem II. The effects of pre-crops on the spore populations of native and introduced VAM fungi. Plant Soil 122:241–248

Douds DD, Gadkar V, Adholeya A (2000) Mass production of VAM fungus biofertilizer. In: Mukerji KG, Chamola BP, Singh J (eds) Mycorrhizal biology. Kluwer, New York, pp 197–215

Douds DD, Nagahashi G Jr, Pfeffer PE, Kayser WM, Reider C (2005) On farm production and utilization of arbuscular mycorrhizal fungus inoculum. Can J Plant Sci 85:15–21

Dugassa DG, Grunewaldt-Stöcker G, Schönbeck F (1995) Growth of Glomus intraradices and its effect on linseed (Linum usitatissimum L.) in hydroponic culture. Mycorrhiza 5:279–282

Elmes RP, Mosse B (1984) vesicular arbuscular endomycorrhizal inoculum production. II. Experiments with maize (Zea mays) and other hosts in nutrient flow culture. Can J Bot 62:1531–1536

Elmes RP, Hepper CM, Hayman DS, O'Shea J (1984) the use of vesicular arbuscular mycorrhizal root by the nutrient film technique as inoculum for field sites. Ann Appl Biol 104:437–441

Elsen A, Declerck S, De Waele D (2001) Effects of *Glomus intraradices* on the reproduction of the burrowing nematode (*Radopholus similis*) in dixenic culture. Mycorrhiza 11:49–51

Feldmann F, Grotkass C (2002) Directed inoculum production – shall we be able to design populations of arbuscular mycorrhizal fungi to achieve predictable symbiotic effectiveness? In: Gianinazzi S, Schüepp H, Barea JM, Haselwandter K (eds) Mycorrhizal technology in agriculture: from genes to bioproducts. Birkhäuser, Basel, pp 223–233

Fortin JA, Bécard G, Declerck S, Dalpé Y, St-Arnaud M, Coughan AP, Piché Y (2002) Arbuscular mycorrhiza on root-organ cultures. Can J Bot 80:1–20

Ganesan V, Mahadevan A (1998) The role of mycorrhizae in the improvement of tuber crops in pot and field conditions. In: Prakash A (ed) Fungi in biotechnology. CBS, New Delhi, pp 51–58

Gaur A (1997) Inoculum production technology development of vesicular-arbuscular mycorrhizae. PhD thesis, University of Delhi, Delhi, India

Gianinazzi S, Gianinazzi-Pearson V (1988) Mycorrhizae: a plant's health insurance. Chim Oggi Oct:56–58

Gianinazzi S, Vosátka M (2004) Inoculum of arbuscular mycorrhizal fungi for production systems: science meets business. Can J Bot 82:1264–1271

Gryndler M (2000) Interactions of arbuscular mycorrhizal fungi with other soil organisms. In: Kapulnik Y, Douds DD (eds) Arbuscular mycorrhizas: physiology and function. Kluwer, Dordrecht, pp 239–262

Heinzemann J, Weritz J (1990) Rockwool: a new carrier system for mass multiplication of vesicular-arbuscular mycorrhizal fungi. Angew Bot 64:271–274

Howeler RH (1985) Mineral nutrition and fertilization of cassava (*Manihot esculenta Crantz*). CIAT, Cali, Colombia

Howeler RH, Asher CJ, Edwards DG (1982) Establishment of an effective endomycorrhizal association on cassava in flowing solution and its effects on phosphorus nutrition. New Phytol 90:229–238

Hung LL, Sylvia DM (1987) VAM inoculum production in aeroponic culture. In: Sylvia DM, Hung LL, Graham JH (eds) Mycorrhizae in the next decade practical applications and research priorities. Proceedings of the 7th NACOM, IFAS, University of Florida, Gainesville, pp 272–273

Hung LL, Sylvia DM (1988) Production of vesicular-arbuscular mycorrhizal fungus inoculum in aeroponic culture. Appl Environ Microbiol 54:353–357

Hung LL, O'Keefe DM, Sylvia DM (1991) Use of hydrogel as a sticking agent and carrier for vesicular–arbuscular mycorrhizal fungi. Mycol Res 95:427–429

Janos DP (1980) Mychorrhizae influence on tropical succession. Biotropica 12:56–64

Karunaratne S, Baker JH, Barker AV (1986) Phosphorus uptake by mycorrhizal and nonmycorrhizal roots of soybean. J Plant Nutr 9:1303–1313

Leyval C, Joner EJ, del Val C, Haselwandter K (2002) Potential of arbuscular mycorrhizal fungi for bioremediation. In: Gianinazzi S, Schüepp H, Barea JM, Haselwandter K (eds) Mycorrhizal technology in agriculture. Birkhäuser, Basel, pp 175–186

MacDonald RM (1981) Routine production of axenic vesicular arbuscular mycorrhizal. New Phytol 89:87–93

Mallesha BC, Bagyaraj DJ, Pai G (1992) Perlite-soilrite mix as a carrier for mycorrhizal and rhizobia to inoculate Leucaena leucocephala. Leucaena Res Rep 13:32–33

Martin-Laurent F, Lee SK, Tham FY, He J, Diem HG, Durand P (1997) A new approach to enhance growth and nodulation of *Acacia mangium* through aeroponic culture. Biol Fertil Soils 5:7–12

Mathew J, Johri BN (1988) Propagation of vesicular arbuscular mycorrhizal fungi in moong (*Vigna radiata* L.) through nutrient film technique. Curr Sci 57:156–158

Mosse B (1962) The establishment of vesicular arbuscular mycorrhizal under aseptic conditions. J Gen Microbiol 27:509–520

Mosse B, Thompson JP (1984) Vesicular-arbuscular endomycorrhizal inoculum production. I. Exploratory experiments with beans (*Phaseolus vulgaris*) in nutrient flow culture. Can J Bot 62:1523–1530

Nopamornbodi O, Thamsurakul S, Vasuvat Y (1987) Effect of VAM on growth, yield and phosphorus absorption of soybean and mungbean in Thailand. In: Sylvia DM, Hung LL, Graham JH (eds) Mycorrhizae in the next decade practical applications and research priorities. Proceedings of the 7th NACOM, IFAS, University of Florida, Gainesville, pp 52

Peuss H (1958) Untersuchungen zur Ökologie und Bedeutung der Tabakmycorrhiza. Arch Microbiol 29:112–142

Potty VP (1985) Cassava as alternate host for multiplication of VAM fungi. Plant Soil 88:135–137

Raja P, Mahadevan A (1991) Axenic cultivation of VAM fungi – a review. J Plant Res 7:1–6

Redecker D, Thierfelder H, Werner D (1995) A new cultivation system for arbuscular-mycorrhizal fungi on glass beads. Angewandte Botanic 69:189–191

Reider C, Herdman WR, Drinkwater LE, Janke R (2000) Yields and nutrient budgets under composts, raw dairy manure and mineral fertilizer. Compost Sci Util 8:328–339

Schwartz MW, Hoeksema JD, Gehring CA, Johnson NC, Klironomos JN, Abbott LK, Pringle A (2006) The promise and the potential consequences of the global transport of mycorrhizal fungal inoculum. Ecol Lett 9:501–515

Selvaraj T, Kim H (2004) Use of sucrose-agar globule with root-exudates for mass production of vesicular arbuscular mycorrhizal fungi. J Microbiol 42:60–63

Sharma AK, Singh C, Akhauri P (2000) Mass culture of Arbuscular mycorrhizal fungi and their role in biotechnology. Proc Natl Acad Sci India 4 & 5:223–238

Sieverding E (1987) On-farm production of VAM inoculum. In: Sylvia DM, Hung LL, Graham JH (eds) Proceedings of the 7th North American conference on mycorrhiza, Gainesville, FL, pp 284

Sieverding E (1991) Inoculum production. Vesicular–arbuscular mycorrhiza management in tropical agroecosystems. Deutsche Gesellschaft fur Technische Zusammenarbeit, Bremer, Germany, pp 223–246

Singh S, Srivastava K, Badola JC, Sharma AK (2012) Aeroponic production of AMF inoculum and its application for sustainable agriculture. Wudpecker J Agric Res 1(6):186–190

Srivastava K, Sharma AK (2011) Arbuscular mycorrhizal fungi in challenging environment – a prospective. In: Fulton SM (ed) Mycorrhizal fungi: soil, agriculture, and environmental implications. Nova, New York, pp 1–35

Sreenivasa MN (1992) Selection of an efficient vascular arbuscular mycorhizzal fungus for chilli. Sci Hortic 50:515–519

Strullu DG, Romand C (1987) Culture axénique de vésicules isolées à partir d'endomycorhizes et ré-association in vitro à des racines de tomate. C R Acad Sci Paris Sér III 305:15–19

Sylvia DM (1990) Inoculation of native woody plants with vesicular- arbuscular fungi for phosphate-mine land reclamation. Agric Ecosyst Environ 31:253–261

Sylvia DM, Hubbel DH (1986) Growth and sporulation of vesicular-arbuscular mycorrhizal fungi in aeroponic and membrane systems. Symbiosis 1:259–267

Sylvia DM, Jarstfer AG (1992a) Sheared-root inocula of vesicular-arbuscular mycorrhizal fungi. Appl Environ Microbiol 58:229–232

Sylvia DM, Jarstfer AG (1992b) Sheared roots as a VA-mycorrhizal inoculum and methods for enhancing plant growth. US Patent 5,096,481, Mar 17

Tarafdar JC (1995) Role of a VA mycorrhizal fungus on growth and water relation in wheat in presence of organic and inorganic phosphorus. J Ind Soc Soil Sci 43:200–204

Tommerup IC (1987) Physiology and ecology of VAM spore germination and dormancy in soil. In: Sylvia DM, Hung LL, Graham JH (eds) Mycorrhizae in the next decade practical applications and research priorities. Proceedings of the 7th NACOM, IFAS, University of Florida, Gainesville, pp 175–177

Turnau K, Haselwandter K (2002) Arbuscular mycorrhizal fungi, an essential component of soil microflora in ecosystem restoration. In: Gianinazzi S, Schüepp H, Barea JM, Haselwandter K (eds) Mycorrhizal technology in agriculture. Birkhäuser, Basel, pp 137–150

Von Alten H, Blal B, Dodd JC, Feldmann F, Vosátka M (2002) Quality control of arbuscular mycorrhizal fungi inoculum in Europe. In: Gianinazzi S, Schüepp H, Barea JM, Haselwandter K (eds) Mycorrhizal technology in agriculture. Birkhäuser, Basel, pp 223–234

White PR (1943) A handbook of plant tissue culture. J Cattel, Lancaster, PA

Wood T (1985) Commercial pot culture inoculum production: quality control and other headaches. In: Molina R (ed) Proceedings of the 6th North American conference on mycorrhizae, Bend, OR. Forest Research Laboratory, pp 84

Zobel RW, Del Tredici P, Torry JC (1976) Method for growing plants aeroponically. Plant Physiol 57:344–346

Chapter 6
Use of Arbuscular Mycorrhizal Fungal Inocula for Horticultural Crop Production

Keitaro Tawaraya

6.1 Introduction

Rock phosphate is the raw material of phosphate fertilizer and the global reserves are limited. The expected global peak of phosphate production has been predicted to occur around 2030 (Cordell et al. 2009) but this is a complex issue. The sudden emergence of the concept of peak phosphorus within the debate on global phosphorus scarcity in the international arena may have raised more questions than it has resolved (Cordell and White 2011). Excessive application of phosphate fertilizer is common practice on horticultural crops (Mishima et al. 2010; Reijneveld et al. 2010); this means that phosphorus use efficiency is usually low. If phosphorus is applied in excess of plant requirement, it can accelerate phosphorus enrichment of water leading to eutrophication of rivers, lakes, and marshes (Maguire et al. 2005). It is necessary to respond to the problem of depletion of the resource of rock phosphate by (1) reducing application of phosphate fertilizer to agricultural crops to a level related to that which is required for plant requirement within one cropping cycle, (2) selecting crop plants that are more efficient at acquiring and using soil phosphorus, and (3) recycling organic phosphorus for agricultural use. Inoculation with arbuscular mycorrhizal (AM) fungi is a promising technique in the horticultural industry, especially for plants exposed to diverse abiotic stresses. Zhang et al. (2014) suggested that inoculation with AM fungi increases the tolerance of loquat seedlings to drought stress, and that improved nutrient uptake by AM fungi greatly contribute to this tolerance. AM fungi can promote the growth of many plants by enhancing increasing the efficiency of use of phosphate and zinc fertilizers (Watts-Williams et al. 2014). However, growth responses of horticultural crops following inoculation with AM fungi have most commonly been investigated under pot culture conditions and because mycorrhizal dependency varies among

K. Tawaraya (✉)
Faculty of Agriculture, Yamagata University, Tsuruoka 997-8555, Japan
e-mail: tawaraya@tds1.tr.yamagata-u.ac.jp

© Springer-Verlag Berlin Heidelberg 2014   81
Z.M. Solaiman et al. (eds.), *Mycorrhizal Fungi: Use in Sustainable Agriculture and Land Restoration*, Soil Biology 41, DOI 10.1007/978-3-662-45370-4_6

plant species and cultivars, plants are not equally responsive. Ortas et al. (2013) observed in an inoculation experiment in tomato, with many AM fungi species, that plant response differed with AM fungi, and that there was no single inoculant species which showed superiority compared with the others examined. They also showed inoculation at seeding stage has higher mycorrhizal dependency than inoculation at seedling stage. But the effects of inoculation with AM fungi under field conditions are not widely demonstrated.

6.2   Application of Chemical Fertilizers to Horticultural Crops

Horticultural crops include vegetable crops, flower and ornamental plants, and fruit trees. Inoculant AM fungi have been mainly applied to vegetable crops and fruit trees. Vegetables crops are classified as leaf vegetables, fruit vegetables, and root vegetables. Leaf vegetables include onion (*Allium cepa*), garlic (*Allium sativum*), Welsh onion (*Allium fistulosum*), leek (*Allium porrum*), and lettuce (*Lactuca sativa*) all of which are mycorrhizal. Non-mycorrhizal leaf vegetables include cabbage (*Brassica oleracea*), rape (*Brassica campestris*), and spinach (*Spinacia oleracea*). Mycorrhizal fruit vegetables include tomato (*Lycopersicon esculentum*), eggplant (*Solanum melongena*), pepper (*Capsicum annuum*), pumpkin (*Cucurbita maxima*), and cucumber (*Cucumis sativus*) and mycorrhizal root vegetables include carrot (*Daucus carota*). Non-mycorrhizal root vegetables include radish (*Raphanus sativus*), turnip (*Brassica campestris*), and beet (*Beta vulgaris*). *Allium* plants including Welsh onion (*Allium fistulosum*), onion (*Allium cepa*), Chinese chive (*Allium tuberosum*), and garlic (*Allium sativum*) have coarse root systems and their capacity for phosphate uptake can be lower than for species with fibrous root systems (Greenwood et al. 1982). Therefore, mycorrhizas can be particularly beneficial in increasing the efficiency of phosphate uptake mostly of *Allium* species in phosphate deficient soil.

6.3   Inoculation of Horticultural Crops with Arbuscular Mycorrhizal Fungi

6.3.1   Mycorrhizal Dependency of Horticultural Crops

The degree of plant growth response associated with colonization of roots by AM fungi is expressed as mycorrhizal dependency and differs among plant species (Habte and Manjunath 1991; Howeler and Sieverding 1983; Planchette et al. 1983; Singh et al. 2012; Tawaraya 2003). Mycorrhizal dependency is generally high in the genus *Allium* because crop plants in Alliaceae such as Welsh onion,

Chinese chive, garlic, and leek have less well-developed root systems than do many other species (Greenwood et al. 1982). Mycorrhizal dependency can also differ among cultivars. For example, in a study of 16 Japanese cultivars of Welsh onion (*Allium fistulosum*) colonized with AM fungi *Glomus fasciculatum* (Tawaraya et al. 1999), 12 cultivars showed a positive response to AM colonization but one cultivar showed negative response. Furthermore, under glasshouse conditions, peanut grain production showed mycorrhizal dependency when inoculated with *Glomus rosea* in the presence of a low supply of P but for *G. clarum*, mycorrhizal dependency was only observed in the absence of applied P (Hippler and Moreira 2013).

It is necessary to check the mycorrhizal dependency of local cultivars and to select appropriate cultivars when AM fungi are inoculated in field conditions, or where the aim is to capitalize on contributions of indigenous AM fungi. The potential to increase the benefits from AM fungi will depend on the tradition of soil and agronomic management practices as well as plant cultivar used. For example, different degrees of dependency on the activity of AM fungi between native maize landraces and hybrids have been reported (Sangabriel-Conde et al. 2014), although the moderate level of fertilization used did not appear to have affected the mycorrhizal symbiosis, as all the maize types used this in experiment demonstrated mycorrhizal dependency. Cultivars of horticultural species bred in Japan are unlikely to be highly mycotrophic because of a history of heavy application of P fertilizer (Tawaraya et al. 2001). Breeding programs for plants used in intensive horticulture do not usually consider the AM fungi and their symbiotic associations with crop plants. Therefore, highly mycotrophic cultivars could be bred for use in horticulture in order to use phosphate resources more sustainably.

6.3.2 Inoculation Under Field Condition

Advanced scientific understanding of arbuscular mycorrhizal symbioses has recently demonstrated the potential for implementation of mycorrhizal biotechnology in horticultural plant production (Vosátka et al. 2012). Effects of AM fungal inoculation on nutrient uptake and growth of horticultural crops have been demonstrated in the field but most field experiments have been carried out at only one P level (Table 6.1). For example, inoculation with indigenous AM fungi increased the weight of marketable lettuce (*Lactuca sativa*) at one soil P level, 45 days after planting (Cimen et al. 2010b). In a different field experiment, inoculation with *Glomus intraradices* increased the bulb yield of onion (*Allium cepa*) at one soil P level 115 days after inoculation (Cimen et al. 2010a). Field inoculation with a selected inoculant of *Glomus intraradices* increased shoot dry weight of grapevine (*Vitis berlandieri* × *Vitis rupestris*) at 15 mg P kg^{-1} soil after 5 months (Camprubi et al. 2008).

Table 6.1 Growth response of horticultural crops with inoculation of AM fungi under field conditions

Crop species	AM fungi	Growth period	P levels	Growth responses	References
Allium fistulosum	*Glomus* R-10	109	4	+M > −M	Tawaraya et al. (2012)
Allium sativum	*Glomus fasciculatum*	145	4	+M > −M	Al-Karaki (2002)
Allium sativum	*Glomus mosseae*	–	4	No difference	Sari et al. (2002)
Colocasia esculenta	*Glomus, Gigaspora*	60	1	+M > −M	Li et al. (2005)
Lycopersicon esculentum	Indigenous	70	1	+M > −M	Cavagnaro et al. (2006)
Lycopersicon esculentum	*Glomus intraradices*	88	1	+M > −M, no difference	Subramanian et al. (2006)
Solanum tuberosum	*Glomus intraradices*	109	1	No difference	Douds et al. (2007)
Allium porrum	*Glomus 3 species*	36	1	+M.−M	Sorensen et al. (2008)

AM fungi can affect aspects of the quality of plant production as well as their yield. For example, inoculation with *Glomus mosseae* or *G. versiforme* increased survival rate and growth of tissue culture of taro plants (Li et al. 2005). In this study, the contents of nitrogen, phosphorus, potassium, calcium, copper, and zinc, the formation of corms, number of second and third branch corms and corm yield, and contents of protein, starch, and amino acids in the corms were also enhanced in response to mycorrhizas (Li et al. 2005). However, such responses are not always consisted. It was demonstrated that yields of potato tubers were increased by AM fungal inoculum (Douds et al. 2007) whereas inoculation with AM fungi did not affect shoot biomass of strawberry (*Fragaria × ananassa*) at 498 mg P kg^{-1} soil 14 weeks from transplanting (Stewart et al. 2005). In another example, the pre-inoculation of peach (*Prunus persica*) seedlings with AM fungi did not increase shoot growth at 85.3–95.8 mg P 100 g^{-1} soil after 2 years of transplanting (Rutto and Mizutani 2006).

Fruit yield of mycorrhizal tomato grown under drought stressed conditions was higher than that of non-mycorrhizal tomato at 0.10 g kg^{-1} soil 60 days after transplanting (Subramanian et al. 2006). On the other hand, when inoculation of garlic (*Allium sativum*) was investigated at four soil P levels, the AM fungal-inoculated garlic had higher mean bulb weight than did uninoculated plants at low P application levels (0 and 20 kg P ha^{-1}) under field conditions, 145 days after planting (Al-Karaki 2002). In another study, preinoculation with AM fungi increased shoot dry weight of leek (*Allium porrum*) 36 days at three soil P levels (20, 32, and 44 mg P kg^{-1}) after transplanting (Sorensen et al. 2008).

Growth improvement following inoculation of AM fungi has not been as clearly shown under field conditions as under pot culture conditions. For example, in spite of well-known mycorrhizal dependency of *Allium* spp., mycorrhizal inoculation did not increase yield and bulb weight of garlic (*Allium sativum* cv. Urfa local) at four P application levels (0, 40, 80, and 120 kg P_2O_5 ha^{-1}) under field conditions (Sari et al. 2002). Furthermore, shoot and fruit biomass of a wild-type tomato were not different from that of a mutant tomato although the wild type had higher AM colonization and higher concentrations of both P and Zn in shoot and fruits than did the mutant when grown in the field (Cavagnaro et al. 2006). Many factors such as temperature, moisture conditions, solar radiation, and pests can affect growth of plants under field conditions. It is difficult to claim that soil P is the most growth-limiting factor, especially in many horticultural field conditions; however, there are situations where P responses to mycorrhizal inoculation can be demonstrated even where P appears to be adequate for plant growth. Recently, we examined the effects of inoculating AM fungi on the growth, P uptake, and yield of Welsh onion (*Allium fistulosum* L.) under non-sterile field conditions (Tawaraya et al. 2012). Yield of inoculated plants grown in soil containing 300 mg P_2O_5 kg^{-1} soil was similar to that of non-inoculated plants grown in soil containing 1,000 mg P_2O_5 kg^{-1} soil. In this case, the cost of inoculation was US\$2,285 ha^{-1} which was lower than the cost of phosphate fertilizer (US\$5,659 ha^{-1}) added to soil containing 1,000 mg P_2O_5 kg^{-1} soil for non-inoculated plants. Thus, AM fungal inoculation can achieve marketable yield of *A. fistulosum* under field conditions with reduced application of P fertilizer.

6.4   Factors Affecting Contributions of AM Fungi to Horticultural Crops Under Field Conditions

There are many inconsistencies in studies of inoculation of horticulture plants with AM fungi in terms of the soil used, nutrient availability compared with plant requirement, plant species, and environmental conditions, so generalizations cannot be made about predicted responses in any one situation without local knowledge. Site-specific investigations are necessary to identify situations where there is potential for increasing benefits from AM fungi. These investigations should be incorporated into best management practice for individual horticulture crops at particular locations.

AM fungi should only be introduced if the population of indigenous AM fungi is low and/or the effectiveness of the indigenous AM fungi is low. If there is a high indigenous population of AM fungi with high ability to affect plant growth and nutrient uptake, heavy application of P fertilizer and fungicide should be avoided in order to maintain the population of indigenous AM fungi. However, it is difficult for farmers to determine the indigenous population of indigenous AM fungi and its capability in their fields. Therefore it is necessary to establish protocols and training workshops for farmers for evaluation of indigenous AM fungi.

Growth improvement following inoculation of AM fungi is not expected in soil with high concentrations of available P because plant root itself can take up sufficient amount of P from soil. Differences in plant growth between inoculated and non-inoculated plant are usually negligible in soil with high concentrations of available P. In some situations, AM fungi can be inoculated during the nursery stage for transplanting crops. Nursery soil may need to be sterilized either to remove indigenous AM fungi prior to inoculation with selected AM fungi, or to eliminate pathogens. Commercial soil media which contain high concentrations of available P need to be adjusted to reduce the P concentration if it is at a level that inhibits AM colonization during the nursery stage.

References

Al-Karaki GN (2002) Field response of garlic inoculated with arbuscular mycorrhizal fungi to phosphorus fertilization. J Plant Nutr 25:747–756

Camprubi A, Estaun V, Nogales A, Garcia-Figueres F, Pitet M, Calvet C (2008) Response of the grapevine rootstock Richter 110 to inoculation with native and selected arbuscular mycorrhizal fungi and growth performance in a replant vineyard. Mycorrhiza 18:211–216

Cavagnaro TR, Jackson LE, Six J, Ferris H, Goyal S, Asami D, Scow KM (2006) Arbuscular mycorrhizas, microbial communities, nutrient availability, and soil aggregates in organic tomato production. Plant Soil 282:209–225

Cimen I, Pirinc V, Sagir A (2010a) Determination of long-term effects of consecutive effective soil solarization with vesicular arbuscular mycorrhizal (VAM) on white rot disease (Sclerotium cepivorum Berk.) and yield of onion. Res Crops 11:109–117

Cimen I, Turgay B, Pirinc V (2010b) Effect of solarization and vesicular arbuscular mycorrhizal on weed density and yield of lettuce (Lactuca sativa L.) in autumn season. Afr J Biotechnol 9:3520–3526

Cordell D, White S (2011) Peak phosphorus: clarifying the key issues of a vigorous debate about long-term phosphorus security. Sustainability 3:2027–2049

Cordell D, Drangert J-O, White S (2009) The story of phosphorus: global food security and food for thought. Glob Environ Change 19:292–305

Douds DD, Nagahashi G, Reider C, Hepperly PR (2007) Inoculation with arbuscular mycorrhizal fungi increases the yield of potatoes in a high P soil. Biol Agric Hortic 25:67–78

Greenwood DJ, Gerwitz A, Stone DA, Barnes A (1982) Root development of vegetable crops. Plant Soil 68:75–96

Habte M, Manjunath A (1991) Categories of vesicular-arbuscular mycorrhizal dependency of host species. Mycorrhiza 1:3–12

Hippler FWR, Moreira M (2013) Mycorrhizal dependence of peanut plants on phosphorus levels. Bragantia 72:184–191

Howeler RH, Sieverding E (1983) Potentials and limitations of mycorrhizal inoculation illustrated by experiments with field grown cassava. Plant Soil 75:245–261

Li M, Liu RJ, Christie P, Li XL (2005) Influence of three arbuscular mycorrhizal fungi and phosphorus on growth and nutrient status of taro. Commun Soil Sci Plant Anal 36:2383–2396

Maguire RO, Chardon WI, Simard RR (2005) Assessing potential environmental impact of soil phosphorus by soil testing. In: Sims JT, Sharpley AN (eds) Phosphorus: agriculture and the environment. American Society of Agronomy, Madison, pp 145–180

Mishima S, Endo A, Kohyama K (2010) Nitrogen and phosphate balance on crop production in Japan on national and prefectural scales. Nutr Cycling Agroecosyst 87:159–173

Ortas I, Sari N, Akpinar C, Yetisir H (2013) Selection of arbuscular mycorrhizal fungi species for tomato seedling growth, mycorrhizal dependency and nutrient uptake. Eur J Hortic Sci 78:209–218

Planchette C, Fortin JA, Furlan V (1983) Growth responses of several plant species to mycorrhizae in a soil of moderate P fertility. I. Mycorrhizal dependency under field conditions. Plant Soil 70:199–209

Reijneveld JA, Ehlert PAI, Termorshuizen AJ, Oenema O (2010) Changes in the soil phosphorus status of agricultural land in the Netherlands during the 20th century. Soil Use Manage 26:399–411

Rutto KL, Mizutani F (2006) Peach seedling growth in replant and non-replant soils after inoculation with arbuscular mycorrhizal fungi. Soil Biol Biochem 38:2536–2542

Sangabriel-Conde W, Negrete-Yankelevich S, Maldonado-Mendoza IE, Trejo-Aguilar D (2014) Native maize landraces from Los Tuxtlas, Mexico show varying mycorrhizal dependency for P uptake. Biol Fertil Soils 50:405–414

Sari N, Ortas I, Yetisir H (2002) Effect of mycorrhizae inoculation on plant growth, yield, and phosphorus uptake in garlic under field conditions. Commun Soil Sci Plant Anal 33:2189–2201

Singh AK, Hamel C, DePauw RM, Knox RE (2012) Genetic variability in arbuscular mycorrhizal fungi compatibility supports the selection of durum wheat genotypes for enhancing soil ecological services and cropping systems in Canada. Can J Microbiol 58:293–302

Sorensen JN, Larsen J, Jakobsen I (2008) Pre-inoculation with arbuscular mycorrhizal fungi increases early nutrient concentration and growth of field-grown leeks under high productivity conditions. Plant Soil 307:135–147

Stewart LI, Hamel C, Hogue R, Moutoglis P (2005) Response of strawberry to inoculation with arbuscular mycorrhizal fungi under very high soil phosphorus conditions. Mycorrhiza 15:612–619

Subramanian KS, Santhanakrishnan P, Balasubramanian P (2006) Responses of field grown tomato plants to arbuscular mycorrhizal fungal colonization under varying intensities of drought stress. Sci Hortic 107:245–253

Tawaraya K (2003) Arbuscular mycorrhizal dependency of different plant species and cultivars. Soil Sci Plant Nutr 49:655–668

Tawaraya K, Imai T, Wagatsuma T (1999) Importance of root length in mycorrhizal colonisation of Welsh onion. J Plant Nutr 22:589–596

Tawaraya K, Tokairin K, Wagatsuma T (2001) Dependence of *Allium fistulosum* cultivars on the arbuscular mycorrhizal fungus, *Glomus fasciculatum*. Appl Soil Ecol 17:119–124

Tawaraya K, Hirose R, Wagatsuma T (2012) Inoculation of arbuscular mycorrhizal fungi can substantially reduce phosphate fertilizer application to *Allium fistulosum* L. and achieve marketable yield under field condition. Biol Fertil Soils 48:839–843

Vosátka M, Látr A, Gianinazzi S, Albrechtová J (2012) Development of arbuscular mycorrhizal biotechnology and industry: current achievements and bottlenecks. Symbiosis 58:29–37

Watts-Williams SJ, Turney TW, Patti AF, Cavagnaro TR (2014) Uptake of zinc and phosphorus by plants is affected by zinc fertiliser material and arbuscular mycorrhizas. Plant Soil 376:165–175

Zhang Y, Yao Q, Li J, Hu YL, Chen JZ (2014) Growth response and nutrient uptake of *Eriobotrya japonica* plants inoculated with three isolates of arbuscular mycorrhizal fungi under water stress condition. J Plant Nutr 37:690–703

Chapter 7
Management of the Arbuscular Mycorrhizal Symbiosis in Sustainable Crop Production

C. Yang, W. Ellouze, A. Navarro-Borrell, A. Esmaeili Taheri, R. Klabi, M. Dai, Z. Kabir, and C. Hamel

C. Yang • A. Navarro-Borrell • A.E. Taheri
Semiarid Prairie Agricultural Research Centre, 1 Airport Road, Box 1030, Swift Current, SK, Canada S9H 1X3

Department of Food and Bioproduct Sciences, University of Saskatchewan, 51 Campus Drive, Saskatoon, SK, Canada S9H 5A8

W. Ellouze
Institut de recherche en biologie végétale, 4101 Sherbrooke Est, Montréal, QC, Canada H1X 2B2

Faculté des Sciences de Tunis, Université Tunis El Manar, Campus Universitaire, Tunis 1060, Tunisia

R. Klabi
Semiarid Prairie Agricultural Research Centre, 1 Airport Road, Box 1030, Swift Current, SK, Canada S9H 1X3

Institut de recherche en biologie végétale, 4101 Sherbrooke Est, Montréal, QC, Canada H1X 2B2

M. Dai
Semiarid Prairie Agricultural Research Centre, 1 Airport Road, Box 1030, Swift Current, SK, Canada S9H 1X3

College of Horticultural and Landscape Architecture, Southwest University, 2# Tian Sheng Street, Bei Bei District, Chong Qing 400715, China

Z. Kabir
United States Department of Agriculture, 318 Cayuga Street, Salinas, CA 93901, USA

C. Hamel (✉)
Semiarid Prairie Agricultural Research Centre, 1 Airport Road, Box 1030, Swift Current, SK, Canada S9H 1X3

Department of Food and Bioproduct Sciences, University of Saskatchewan, 51 Campus Drive, Saskatoon, SK, Canada S9H 5A8

Institut de recherche en biologie végétale, 4101 Sherbrooke Est, Montréal, QC, Canada H1X 2B2
e-mail: hamelc@agr.gc.ca

© Springer-Verlag Berlin Heidelberg 2014
Z.M. Solaiman et al. (eds.), *Mycorrhizal Fungi: Use in Sustainable Agriculture and Land Restoration*, Soil Biology 41, DOI 10.1007/978-3-662-45370-4_7

7.1   Introduction

We live at a unique time in planet Earth's history where humans significantly impact the geochemical cycles that sustain life. Among the negative impacts of humans on the Earth's chemical equilibrium, loss of nitrogen (N) and phosphorus (P) and nitrous oxide (N_2O) emissions from cultivated fields affect air and water quality in addition to wasting non-renewable P and fossil fuel resources. This is not sustainable.

The challenges we face include the need to reduce the negative impacts of crop production on the environment, while further increasing land areas under cultivation to produce renewable fuels, food and industrial crops for a rapidly growing and developing world population. Clearly, sustainability will only be achieved by improving the efficiency of nutrient cycling in human-managed systems. Nutrients exported from production fields end up as organic wastes from urban and industrial sources. These nutrients should be recycled in crop production systems, and the contribution of biological N_2 fixation to crop production should be increased. Nutrient-efficient cropping systems should also be designed.

The improvement of nutrient-use efficiency in crops and concurrent reduction in the levels of labile N and P in cultivated soils are necessary to minimize undesirable losses of N and P to the environment. The natural processes of soil nutrient mobilization and cycling are largely driven by microorganisms about which very little is known. This makes managing soil bioresources a difficult task. Nevertheless, some 50 years of research has built an important body of knowledge on a group of microorganisms that are central to nutrient cycling in agroecosystems: the arbuscular mycorrhizal (AM) fungi. In the first part of this chapter, we show how the AM symbiosis improves the efficiency of nutrient use by crops. We review the agronomic practices influencing the AM symbiosis in the second part of the chapter and present ways to enhance the contribution of the symbiosis to cropping systems' efficiency. The agronomic practices discussed relate to soil tillage, fertilization, pesticide use, AM inoculation of crops and grazing management. We also discuss how bioactive molecules and crop genotypes creating effective AM symbioses may contribute to the sustainability of agroecosystems.

7.2   Arbuscular Mycorrhizas: A Component of Sustainable Crop Production

Nutrient-use efficiency in crop production is a function of two main factors, which are both influenced by the AM symbiosis. The first factor relates to the efficiency of the mechanisms involved in nutrient transformation in soils. The quality of crop plants as a sink for nutrients is another very important factor.

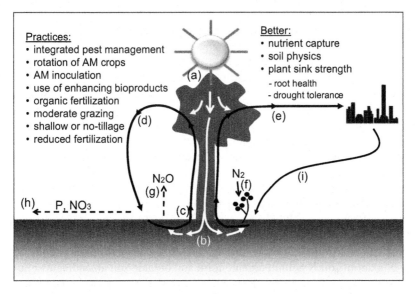

Fig. 7.1 Representation of the central role of the AM symbiosis in nutrient cycling in sustainable agroecosystem. Carbon and energy captured by the crop canopy (**a**) is transferred below ground and partly used to support the activity of AM fungi and associated microorganisms (**b**). Microbial activity mobilizes soil nutrients and the AM fungi facilitate their uptake by the crop (**c**). Part of the crop biomass is recycled in soil (**d**) whereas crop yield is exported to consumers' markets (**e**). The AM symbiosis favours the close cycling of N and P within the soil and plant components of the system, enhancing N_2 fixation in legumes (**f**) and reducing N and P losses through denitrification (**g**), leaching and run-off (**h**). The system is sustainable if the nutrients exported in yield are returned to the field (**i**). The agronomic practices listed in the *top left corner* leads to the AM-related benefits listed in the *top right corner* of the figure

7.2.1 Efficiency of Soil Nutrient Dynamics

Arbuscular mycorrhizas are the interface between the soil and those plants that can form this symbiosis. They connect roots to photosynthetic aerial plant parts and extensive networks of coenocytic hyphae that absorb water and nutrients, exude carbon (C)-rich compounds in the soil and interact with soil processes (Hamel and Strullu 2006). An AM symbiosis can be seen as a solar-powered soil management system where photosynthesis may increase in response to belowground demand (Fig. 7.1). The AM symbiosis, supported by plant photosynthesis, facilitates soil–plant processes that benefit crop production.

AM hyphae absorb and transport nutrients (Liu et al. 2007), which are translocated to plants through a symbiotic interface (Takeda et al. 2009). The symbiosis has been shown to improve plant nutrition by increasing plant uptake of nutrients present in growth-limiting amounts in soil, so the role of AM hyphae in plant P nutrition is particularly important to sustainability. The hyphae upload phosphate ions against a concentration gradient, concentrating phosphate in polyphosphate molecules (Takanishi et al. 2009; Tani et al. 2009) which are transported to the symbiotic interfaces located in root cortical cells, where P is downloaded into the plants tapping into the AM hyphal network (Karandashov and Bucher 2005; Bucher 2007).

The role of the AM symbiosis in efficient plant N nutrition may also be important. AM hyphae can translocate N to host plants (Govindarajulu et al. 2005; Rains and Bledsoe 2007; Cappellazzo et al. 2008; Whiteside et al. 2009), and the symbiosis can be important for N uptake from fertilizers (Azcón et al. 2008). However, the contribution of the this symbiosis to crop N nutrition is often thought to be small, as nitrate (NO_3^-), the main source of N for crops, is freely mobile in soil and, thus, transport of N in AM hyphae brings little advantages to crop N nutrition in conventional crop production systems. Nevertheless, in sustainable production, mineral N levels should be kept low as NO_3^- can be reduced to N_2O, a potent greenhouse gas, or lost from the soil–plant system through leaching (Herzog et al. 2008; van der Heijden 2010) and negatively impacting surface and ground water quality. Ammonium (NH_4^+), the other major soil mineral N form, can be lost from dry soils through volatilization or oxidized to NO_3^-. The large amount of soil N held in organic form and organic soil amendments must be mineralized by soil microorganisms before the N they contain can be used by plants. AM hyphal networks can stimulate the organisms involved in mineralization. The presence of AM fungi in the plant–soil system can enhance mineralization of N from organic residues in soil, and the N released can be better used by plants tapping AM networks located in the vicinity of mineralizing residues (Hodge et al. 2001). The amount of N from decomposing plant residues that is acquired by plants through AM networks may be important and was demonstrated to reach 25 % of residue N in 24 weeks and overcome Russian wild rye growth limitation (Atul-Nayyar et al. 2009). Furthermore, it accounted for about 20 % of *Plantago lanceolata* N content (Leigh et al. 2009). Plants and soil microorganisms compete for soil-available N, and large AM hyphal networks potentially offer the crop a pool of available N inaccessible to competing microorganisms (Whiteside et al. 2009).

The AM symbiosis is an asset particularly for mycorrhizal crops relying on organic N sources. An effective AM symbiosis should maximize the benefit of legume crops in rotations and the utilization of compost and other organic sources of nutrients, the use of which is necessary in a sustainable world. The gap in nutrient cycling created by the unidirectional movement of nutrients from production field to cities and industries must be removed and N fertilizer use must be reduced. Nitrous oxide is the most important greenhouse gas emitted from agriculture, with levels increasing with the abundance of plant-available soil N, and fertilizer manufacturing and transport is the other main source of greenhouse gas from agriculture (Snyder et al. 2009).

Phosphorus is required in large amount by plants, but phosphate deposits exploitable with current technologies are finite and rapidly disappearing (Gilbert 2009). The AM symbiosis allows efficient plant P uptake from soils with low levels of available P, reducing the risk of wasteful P loss, thus preserving the quality of water and aquatic ecosystems (van der Heijden 2010).

AM fungi can play an important role in forage systems, especially in mixtures of legume and grass species, as they improve biological N_2 fixation in legumes (Hayman 1986; Xavier and Germida 2002; Chalk et al. 2006). Their extraradical hyphal networks also facilitate the transfer of N from N_2-fixing plants to companion grasses (Sierra and Nygren 2006; Jalonen et al. 2009) and improve the efficiency of nutrient cycling, reducing nutrient losses from the soil–plant system (van der Heijden 2010). AM fungi can decrease the productivity of components of a plant mixture while promoting that of another (Gange et al. 1993; van der Heijden et al. 1998), giving a competitive advantage to AM-dependent species (Grime et al. 1987). AM fungi favoured the proliferation of perennial forbs (Gange et al. 1993) and of most legumes in plant mixtures, an effect attributed to the high P demand of these plants (Karanika et al. 2008). Increasing the proportion of legumes in pasture or mixed hay fields would improve stand productivity and reduce dependence on N fertilizers.

More efficient use of the AM symbiosis in crop production would improve the sustainability of crop production. The AM symbiosis can reduce nutrient loss from ecosystems in three main ways: (1) by improving crop nutrient extraction capacity (van der Heijden et al. 2008) allowing the production of good yield at lower levels of soil fertility; (2) by increasing soil aggregation via physical particle enmeshment and cementing with "sticky" exudates, which results in better soil nutrient storage and retention (Rillig and Mummey 2006) and in reduced erosion (Tisdall 1991; Wright and Upadhyaya 1996) and leaching losses (Querejeta et al. 2009); and (3) by promoting growth of host crops, thus increasing the size of this desirable nutrient sink (van der Heijden 2010).

7.2.2 Reduced Biotic and Abiotic Limitations Increase Nutrient Capture by AM Crops

Although the effect of AM fungi in field settings is complicated by plant-to-plant interactions (Koide and Dickie 2002), AM fungi generally have the potential to enhance nutrient supply and the productivity of plant stands (van der Heijden and Horton 2009). However, the effect of AM fungi on plant growth promotion goes beyond nutrition (Newsham et al. 1995; Finlay 2008). Plant disease and insufficient water are major plant growth-limiting factors which are also influenced by the AM symbiosis.

7.2.2.1 Arbuscular Mycorrhiza and Plant Bioprotection

The extraradical mycelium of AM fungi is a large structure often representing over 20 % of soil microbial biomass (Leake et al. 2004). This fungal structure is important in improving plant nutrition, but it also has a profound influence on the soil ecosystems. The extraradical AM mycelium derives its C and energy from plants and constitutes a sink similar in size to that of fine roots (Johnson 2010). Plants contribute considerable amounts of C and energy into soil systems, and this influences soil biodiversity and function (Finlay 2008). The AM symbiosis also influences soil microbial community structure through AM hyphal exudates, which have different effects on different organisms (Filion et al. 1999; Lioussanne et al. 2010), and through modification of root exudation (Sood 2003; Lioussanne et al. 2008). These processes influence the way that AM fungi interact with other soil microorganisms and can lead to protection against soilborne pathogens (St-Arnaud and Vujanovic 2007).

Several mechanisms have been proposed to explain how the AM symbiosis enhances plant tolerance to pests (see reviews by Azcón-Aguilar and Barea 1997, Harrier and Watson 2004, and St-Arnaud and Vujanovic 2007). They include the improvement of plant nutrition (especially P nutrition which modifies root exudation), increased cell wall lignification, competition with pathogens for C source and space in roots, antagonistic effects on pest especially in association with some bacteria (Bharadwaj et al. 2008; Siasou et al. 2009) and stimulation of the plant defence system (Pozo and Azcón-Aguilar 2007). The bioprotection conferred by the AM symbiosis to a plant probably results from the collective effects of more than one mechanism acting simultaneously (Pozo and Azcón-Aguilar 2007). Plant protection and other beneficial effects of AM fungi on host plants, such as tolerance to environmental stresses and improved soil physical properties, make AM fungi an important multifunctional component of sustainable agroecosystems.

7.2.2.2   Arbuscular Mycorrhiza and Plant Drought Stress

Limited water availability in agroecosystems is often a main cause of yield loss. The frequency of drought period is expected to increase in several regions of the world, with climate change. Severe drought may cause the xylem and rhizosphere to become air-filled and disrupt water flow, whereas milder water shortage leads to a state of C starvation exacerbated by photoinhibition (McDowell et al. 2008). The AM symbiosis can improve both plant and soil water relations in addition to increasing plant water use efficiency by mobilizing nutrients in dry soils (Augé 2004; Sheng et al. 2008). Mycorrhizal plants often contain more water and leaf chlorophyll than non-mycorrhizal plants (Colla et al. 2007; Al-Karaki and Clark 1999; Subramanian and Charest 1997, 1999; Srivastava et al. 2002) and have better gas exchange (Ruiz-Lozano and Azcón 1995; Aroca et al. 2009; Benabdellah et al. 2009) under drought.

The effect of the AM fungi on plant drought tolerance depends on the host–fungus combination (Davies et al. 2002; Pande and Tarafdar 2002). The better growth and water status of AM plant symbiosis is usually attributed to effective water extraction by an extraradical AM mycelium giving access to tightly held soil water and increasing soil–root hydraulic conductance (Gonzalez-Dugo 2010) and better osmotic adjustment and stomatal regulation (Augé 2001, 2004). Osmotic adjustment is an important drought-tolerance mechanism in plants (Martinez et al. 2004) that is influenced by the AM symbiosis (Wu et al. 2007). The symbiosis has also been shown to enhance antioxidant defence (Garg and Manchanda 2009; Hajiboland et al. 2010) protecting photosystem II against the reactive oxygen species (ROS) created by photoinhibition.

The AM symbiosis may also improve plant performance through drought periods by increasing the capacity of the soil to store water (Augé 2004). The amount of water a soil can hold depends on its structure, especially its porosity. The spaces between sand particles are large and water tends to drain through easily leaving a water film on sand grain surfaces and air-filled interstices. By contrast, gaps between silt or clay size particles are small, which restricts water flow (Brady and Weil 2001). Mycorrhizal roots promote the aggregation of soil particles improving soil porosity (Oades 1993; Rillig and Mummey 2006; Lambers et al. 2007), water infiltration (Kabir and Koide 2000, 2002) and storage (Augé 2004) in fine-textured soils. They stimulate the package of fine particles into microaggregates containing medium-sized pores with good water-holding capacity and macroaggregates of size suitable for the creation of interstitial macropores conducive to soil drainage and aeration.

The importance of the different mechanisms of protection operating in AM plants to provide protection against drought probably varies with the plant–fungus associations and conditions. However, Augé (2004) showed that soil hyphae abundance explained lethal leaf and soil water potential better than did root colonization, root density, soil aggregation and leaf phosphorus or osmotic potential (Augé 2004), indicating the importance of the extraradical AM mycelium in explaining the AM effect.

7.3   Managing the AM Symbiosis in Cropping Systems

The AM symbiosis can provide important benefits to the plant, the soil and the environment, and this makes the management of the symbiosis desirable in sustainable crop production. The management of the AM symbiosis is achievable through a variety of agronomic practices, in particular: (1) tillage, (2) crop nutrition, (3) grazing and (4) integrated pest management (IPM), as well as by (5) the selection of crop genotypes and crop rotation sequences, (6) the use of AM inoculants and (7) the use of biotechnologies that enhance the AM symbiosis of crop plants.

7.3.1 Understanding Tillage Effect to Optimize AM Symbiosis-Related Benefits

No-till or conservation tillage is practiced to reduce surface run-off and loss of sediments, nutrients and pesticides from topsoil to surface water. No-till (i.e., direct seeding into the standing stubbles of a previous crop) is an excellent way of conserving soil water resources in semiarid areas. Under moist climate, by contrast, soil inversion by mouldboard ploughing in fall and harrowing in spring is practiced to accelerate soil warming and remove excess moisture. Ploughing deeply disturbs soil and disrupts AM hyphal networks which are mostly located in the top 25 cm of the soil (Kabir et al. 1998). Tillage may also increase soil bulk density restricting root growth (Lampurlanés and Cantero-Martínez 2003; Yau et al. 2010) and AM colonization (Mulligan et al. 1985).

The terminology used to refer to different tillage managements is a source of confusion leading to the widespread belief that AM fungi contribute little to crop nutrition in developed countries. "Conservation tillage" refers to diverse practices including ridge tillage, reduced tillage, shallow tillage and strip tillage, which result in different patterns and extents of disturbance. "Conventional tillage" also refers to a range of different practices and has regionally defined meanings. Whereas inversion of the top 20–25 cm of the soil by fall ploughing and soil levelling by a few harrowing operations in spring is the conventional practice in temperate humid areas, it is conventional not to use any fall tillage and to harrow only once the top 7.0–7.5 cm on the soil before seeding, in dryer areas. Thus, conventional tillage under dryer climates corresponds to conservation tillage in wetter areas. It is important to consider the depth and intensity of soil disturbance resulting from a tillage system before concluding on its likely impact on AM fungi.

Tillage systems imposing disturbance only to the few top centimetres of the soil preserves AM hyphae networks and AM fungi functionality (Kabir 2005). However, the effects may depend on the farming system and/or on the inoculum level in the soil. Plants connected to fragmented AM extraradical hyphae networks can have reduced access to soil nutrients (O'Halloran et al. 1986) even if AM root colonization levels are not reduced (Evans and Miller 1990).

Low or no soil disturbance favours an abundance of active AM hyphae (Borie et al. 2006; Md González-Chávez et al. 2010; Roldán et al. 2007), and although the effect of tillage on AM fungi proliferation disappears with time, the restoration of the AM fungal biomass also takes time. A negative impact of severe soil disturbance on active AM fungal biomass may persist, even after 1 year (Wortmann et al. 2008), and can persist for 5 years after only one tillage event (Drijber 2002). Reduced AM fungal biomass in soil was associated with reduced soil mycorrhizal inoculum potential and reduced colonization of crop roots (Garcia et al. 2007; Lekberg et al. 2008; van Groenigen et al. 2010). Thus, low or no soil disturbance favours the abundance of AM fungi propagules (spores and hyphae) in soil (Evans and Miller 1990; Borie et al. 2006; Cornejo et al. 2009), which is particularly important for short-season annual crops.

Tillage not only affects AM fungal development and symbiosis functionality but can also change the structure of the AM fungal community. Richer AM fungal communities were found in no-till than in tilled soil and tillage was found to select for AM species of the genus *Glomus* (Alguacil et al. 2008; Boddington and Dodd 2000). The dominance of AM fungi species is also altered by tillage (Douds et al. 1995; Jansa et al. 2003), and the disappearance of some species, namely, *Glomus ambisporum* and *Glomus etunicatum*, following a disking-fallow treatment was reported (Rasmann et al. 2009). The selection of AM genotypes by tillage may be beneficial or detrimental to crops, depending on the species involved. Nevertheless, it is logical to expect that the loss of AM fungal diversity will at least result in reduced ability of the soil system to adapt to changes.

It may be advantageous to markedly disturb the soil despite the disruptive effect this has on extraradical AM hyphae. Crop residues remaining at the surface of soils under no-till or under reduced tillage systems are mixed in the few uppermost centimetres of the soil. The soil organic matter consequently accumulates at the soil surface resulting in nutrient stratification and enrichment of the top layer. Crop residues are relatively rich in nutrients, particularly in N and P. Concentrations of N and P in soil surface under conservation tillage may increase to inhibitory levels, as high fertility inhibits AM symbiotic development (Garcia et al. 2007; White and Weil 2010). Punctual ploughing operations may also reduce the risk of P loss from soils enriched by decades of P fertilization on which a layer of organic matter accumulates. Phosphate ions are very reactive in soil where they bind to surfaces. Organic molecules compete with phosphate ions for fixation sites in soil, and soluble P may not only be released through mineralization of crop residues, but P ions can also be displaced from fixation sites into the soil solution by organic molecules (Ouyang et al. 1999), raising available P to unhealthy levels. Mixing the organic layer accumulating on the surface of soils under conservation tillage with mineral soil increases the amount of binding sites available to fix both P ions and organic molecules, reducing the availability of these P ions (Guertal et al. 1991; Simard and Beauchenmin 2002). A host crop with abundant root production and low AM dependency such as wheat should follow a deep tillage operation that disrupts AM hyphae networks, in order to re-establish these networks without yield penalty.

7.3.2 Crop Nutrition Management in Sustainable Agroecosystems

Plants and AM fungi require a certain level of nutrients to growth and function, and the nutrients exported from the soil system through harvest and sale of products must be replaced somehow for the production system to be sustainable. In poor fertility soils, fertilization stimulates root colonization (Ramirez et al. 2009) and plant response to AM inoculation (Covacevich and Echeverría 2009). In contrast, cultivated soils are often nutrient rich as fertilization is a common component of crop management strategies. Although mineral N, P and K fertilizers and manures are sometimes abundantly used, they are not always necessary for higher crop production (McKenzie et al. 2003); fertilizers can be seen as cheap production insurance. A problem of using fertilization rate exceeding that exported in crop yield is P build-up in soil (Fixen 2006). The accumulation of residual P fertilizer in the soil creates unfavourable conditions for AM symbiotic development and can select for less mutualistic AM fungal strains (Johnson 1993).

Soil fertility management, particularly P fertility, is certainly a key element needing improvement in crop production. The soil P-testing methods usually used to determine fertilizer rates are quite coarse and inefficient tools. Plant-available P is estimated using different extraction protocols designed to dissolve easily soluble forms of P from soil calcium, aluminium and iron phosphates (Olsen and Sommers 1982). The different soil P-testing methods, such as Bray and Kurtz (1945) or (Olsen et al. 1954), are selected based on dominant regional soil type. They provide rudimentary indices of soil P fertility—"low", "medium" or "high"—that were calibrated to indicate the P requirement of each crop, often on varieties and in cropping systems that have long been replaced. Soil K testing is conducted in a similar way, and soil N may not be tested at all as current test results are largely unrelated to plant N nutrition (Ziadi et al. 1999).

Notwithstanding that nutrient calibration could be redone using new varieties and cropping practices, soil testing has important intrinsic limitations. It gives no indication on N and P availability from organic sources, yet these are major nutrient sources in soil. Most of the N and more than two-thirds of the P can occur in organic forms, and while soil microbial P is a small pool, it cycles rapidly and contributes importantly to plant nutrition (Stevenson and Cole 1999). Soil testing also gives no information on the contribution of AM fungi to crop nutrient uptake in the soils being tested.

Despite these important limitations, soil testing has been the tool available to manage crop nutrition since the mid-1900s, but as the environmental consequences of the poor nutrient-use efficiency of current agriculture are being unveiled, new tools are being proposed to better manage soil fertility. Models based on direct measurements of soil nutrient supply capacity and plant development models can be used to forecast the amount of nutrient needed to achieve a desirable crop yield (Greer et al. 2003). The precision of such models for P forecasting would be greatly improved by the inclusion of information on the AM hyphae network contributing to uptake. The information needed, i.e., the effective AM hyphae surface area for uptake, could be estimated using quantitative PCR (polymerase chain reaction) when reliable DNA markers are identified. Another approach to assessing the contribution of AM fungi to crop nutrition is to model the distribution of AM fungi in the landscape. Recent advances provide useful information towards the development of systems to monitor AM fungi and the function of the AM symbiosis in the field (Zimmerman et al. 2009; Johnson 2010; Tian et al. 2010).

Soil nutrient levels (Liu et al. 2000a), particularly N and P balance (Liu et al. 2000b), have a large influence on AM symbiosis development and function. Plants control AM symbiosis development and this influences their growth. As described by the functional equilibrium model developed by Johnson (2010), plants partition more C below ground under conditions of P limitation to develop capacity for soil P extraction. However, when the C cost to the plant exceeds the P uptake benefit derived from the AM symbiosis and becomes growth limiting, C export below ground is reduced. Plant growth may become limited by P availability when photosynthesis and N fertilizer rates are high. At this point, C supply to the AM symbiosis is increased to relieve P limitation, but high soil N levels increase the risk of harmful N losses to the environment (Snyder et al. 2009). Abundant P fertilization of AM crops reduces AM development, along with AM-related benefits to soils and crops, and increases the risk of nutrient loss from agroecosystems (van der Heijden 2010). Among the AM-derived benefits, the ability of AM fungi to link N mineralization from organic residues to plant demand is particularly relevant to nutrient cycling efficiency and sustainability of crop production. The presence of AM hyphae in patches of organic residues was shown to enhance mineralization (Hodge et al. 2001) with positive impact on host plant N nutrition (Atul-Nayyar et al. 2009; Leigh et al. 2009), as illustrated in Fig. 7.2.

The AM symbiosis enhances organic matter mineralization improving organic fertilizer availability to plants (Hodge et al. 2001; Amaya-Carpio et al. 2009; Atul-Nayyar et al. 2009). Organic fertilizers slowly release phosphorus; thus, they are less likely to raise soil P fertility to AM-inhibiting levels, as do readily available P sources. Organic fertilization can improve AM symbiosis development (Perner et al. 2006) even when mineral fertilization reduces it (Gryndler et al. 2006), but this depends on the amounts added. Compost may have a positive effect on the AM symbiosis (Valarini et al. 2009) and, used in combination with the AM symbiosis, can be as effective as mineral fertilizers (Perner et al. 2007; Salami 2007). The efficiency of sewage sludge can be increased by AM fungi (Arriagada et al. 2009), but these materials can be rich in P, explaining the negative impacts they have had on AM fungi (Tanu et al. 2004). Providing crops with a balanced nutrition will involve nutrient inputs from various sources, including N_2 fixation.

Rotation including N_2-fixing legumes reduces the dependence of cropping systems on N fertilizers and enhances AM symbiosis development and function (Bagayoko et al. 2000; Houngnandan et al. 2000; Alvey et al. 2001). Intercropping cereals and legumes may also increase mycorrhiza formation, with positive impacts on nodulation, N and P acquisition and use (Li et al. 2009), and is another agronomic practice enhancing efficient nutrient cycling in agroecosystems. The ability of AM fungi to improve plant N nutrition through better access to organic N source is advantageous in crops receiving N in organic form (He et al. 2003).

The AM symbiosis associated with different crops is influenced differently by soil fertility level (Johnson 2010). Similarly, some AM fungal strains stimulate plant growth best at relatively high soil fertility level, whereas others function best at lower level of fertility (Herrera-Peraza et al. 2011). The inclusion of detailed information in models will be needed to develop the most appropriate fertilization

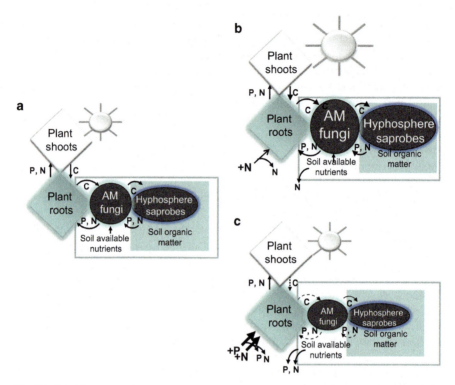

Fig. 7.2 Models representing dynamics of crop plant–AM fungi–saprobe relationships involving the distribution of C, P and N resources above and below ground, modified from Johnson (2010). (**a**) Mutualistic relationships in the distribution of fixed C and soil nutrients from inorganic and organic pools between crop plants, AM fungi and associated hyphosphere saprobes. (**b**) Increased photosynthesis or N fertilization stimulates plant growth, increasing plant P demand and AM symbiosis development and stimulating microbial mobilization of soil nutrients. Increased soil N availability increases the risk of N loss to the environment. (**c**) Fertilizer N and P applications beyond levels required to fulfil plant–AM fungi needs, by contrast, reduce C allocation to the AM symbiotic system. Fertilization increases the risk of N and P loss to the environment and may increase soil P saturation, with negative impact on the AM symbiosis in subsequent years

plan for each crop situation. Overall, slow release fertilizers generally favour AM symbiosis development and function when used in reasonable amounts, which varies with the requirements of different crops. The use of organic fertilizers tends to enhance the AM symbiosis, but care should be taken as their nutrient content may be unbalanced and it may be necessary to combine different fertilizing materials, organic or minerals. The use of crop rotations including N_2-fixing legumes is an excellent strategy both to enhance AM symbiosis-related benefits and reduce the footprint of crop production on the environment.

7.3.3 Pest Management: A Challenge and Opportunity in AM Resource Management

Pests and pathogens are responsible for about 45 % of all crop production losses worldwide (Agrios 2005). As the demand for food and bioproducts rises, pest management in agricultural systems is critical, and the use of agrochemicals, traditionally applied to control pests, is raising concerns. In addition to concern for human health (Lynch et al. 2009; Rull et al. 2009; Gilden et al. 2010), they can trigger the development of resistance in target organism populations (Sukhoruchenko and Dolzhenko 2008; Miyamoto et al. 2009; Hollomon and Brent 2009) and negatively impact beneficial organisms (Van Zwieten et al. 2004; Kinney et al. 2005). Although some cases of neutral and even positive effects of pesticides on the AM symbiosis have been reported (Reis et al. 2009), fungicides (Campagnac et al. 2008; Campagnac et al. 2009; Assaf et al. 2009; Zocco et al. 2008), herbicides (Vieira et al. 2007) and several insecticides (Veeraswamy et al. 1993) often have negative impacts on the performance and function of AM fungi. Pest control should be harmless to AM fungi because they are involved in many important ecological processes related to ecosystem function (Finlay 2008). Furthermore, AM fungi can enhance plant resistance to pests (Harrier and Watson 2004; Elsen et al. 2009), but pesticide application negatively impacting on AM fungi may increase crop vulnerability to pests.

Preserving the performance of the AM symbiosis with crops is aligned with the strategies and goals of IPM which is the road map of pest management in sustainable agroecosystems (Cuperus et al. 2004). IPM is defined as a comprehensive strategy for the control of pest populations in an environmental context to keep them below an economic threshold (FAO 1967). Sanitation, crop rotation, mechanical pest removal, use of chemical and biological control agents and host resistances are simultaneously used in IPM to control pests and minimize risks of environmental pollution and development of resistance in pests and to preserve the services provided by nontarget organisms such as AM fungi.

Improving AM fungi performance in agroecosystems could be a goal of IPM programmes, as AM fungi have important biocontrol activity in addition to improving plant nutrition and stress tolerance (St-Arnaud and Vujanovic 2007). The presence of AM hyphae in soil can have a sanitizing effect and reduce the abundance of pathogenic propagules (St-Arnaud et al. 1997; Fortin et al. 2002). The abundant extraradical AM hyphae leading to rapid and extensive AM root colonization could also reduce the incidence of diseases caused by parasitic nematodes (Pandey et al. 2009) which compete with AM fungi for occupation of plant roots and as food resource (Hol and Cook 2005). AM fungi are particularly important in the protection of plants against root diseases (Nogales et al. 2009; Azcón-Aguilar and Barea 1997), which may be symptomless and undetected in the field but decrease crop productivity and nutrient cycling efficiency.

Pesticides that are safe for the AM symbiosis (Wan and Rahe 1998; Reis et al. 2009) should be preferred (Wilson and Williamson 2008; Habte 1997;

Pimienta-Barrios et al. 2003), and alternative control measures should also be safe or beneficial to maximize the services of AM fungi in an IPM context. Well-planned crop rotations should always be practiced in order to break disease cycles. Interestingly, crop rotation can enhance AM symbiosis function, as mutualism in AM fungi appeared to be favoured by crop rotation (Johnson and Pfleger 1992). Host specificity level is low in the AM symbiosis (Smith and Read 1997), but often high in plant–pathogen interactions (van der Plank 1975) explaining how crop rotation with AM crop species can simultaneously reduce pathogenic infection and enhance AM symbiosis development and function.

The AM fungi are obligate biotrophs, making the presence of a host plant a mandatory condition for the symbiosis to exist. While weeds pose a real and serious risk of yield reduction, some weed species may benefit the agroecosystem by increasing biodiversity and abundance of beneficial AM fungi (Jordan et al. 2000). The removal of weed species hosting AM fungi from agroecosystems can alter the diversity, abundance and functioning of these important fungi, with an impact on crop growth (Feldmann and Boyle 1999; Kabir and Koide 2000). Weeds have been almost completely eliminated from cultivated fields by extremely effective technologies, in particular by glyphosate and glyphosate-resistant crops (Dewar 2009) although glyphosate-resistant weed genotypes are emerging (Duke and Powles 2009).

Sustainable crop production and IPM involve more planning and record keeping than conventional production. Easy-to-use agrochemicals simplify crop production but may reduce ecosystem functional efficiency with impact on the environment. Improving the sustainability of crop production is a realistic goal. The complexity of agroecosystems could be better monitored and managed through further application of knowledge in biology, meteorology, remote sensing and environmental modelling. A range of tools that support agronomic decisions could be further developed, but to reach a postmodern sustainable agriculture, silver bullet strategies must be put aside.

7.3.4   *Manipulating AM Fungi Through Plant Selection*

The host plant appears as a key element for the management of the AM symbiosis as it controls the development of the symbiosis in interaction with environmental conditions (An et al. 2010). Although most plants may host AM fungi, mycorrhizal development and plant response to AM colonization vary with the crop species (Saif 1987; Sudová 2009) and even with the genotypes of a same species (Krishna et al. 1985; Kaeppler et al. 2000; Eason et al. 2001; Karliński et al. 2010). Different plants are also associated with different levels of extraradical hyphae proliferation (Bingham and Biondini 2009) and different AM fungal community composition (Johnson et al. 1992; Mummey and Rillig 2006).

Variation in AM symbiosis formation has implications in cropping systems. Altering plant species in crop rotations can influence the AM potential of soils

(Johnson et al. 1991; Karasawa et al. 2002; Higo et al. 2009), an effect that is more marked in no-till systems (Gavito and Miller 1998). An AM population-mediated feedback effect of a previous crop on a following crop may occur as crop plants may modify AM fungal communities and as different host plants respond differently to different AM fungi (Saif 1987; Oliveira et al. 2006; Sudová 2009). A good crop rotation sequence should maximize the contribution of AM fungi to cropping systems in addition to minimizing pest population build-up, as mentioned in the previous section.

Plant response to AM fungal colonization varies interspecifically but also intraspecifically, revealing the possibility to breed plants which benefit from a more effective AM symbiosis either directly or indirectly. Kaeppler et al. (2000) identified one quantitative trait locus (QTL) for AM responsiveness in maize inbred lines, indicating the possibility to efficiently select maize varieties for AM symbiotic traits using genetic markers.

Different strategies could be used in the selection of plant genotypes for improved AM symbiosis. As plant response to AM colonization depends strongly on the plant–fungus combination involved (Facelli et al. 2009; Javaid et al. 2009), genotypes could be selected for effective association with highly effective AM (HE) strains. Such crop genotypes could be used with inocula containing these HE strains. Crop genotypes selected for their compatibility with HE strains would be an interesting tool for AM management in rotations involving non-host crop species or in biologically degraded soils. The advantage and disadvantage of this strategy would be the simplification of the systems, as it would make it easier to manage, but possibly less biodiverse. Whereas the negative impact of some desirable cropping practices (such as the production of desirable non-host crops) would be reduced by the possibility of rapidly restoring AM symbiosis-related beneficial effects on the agroecosystem, the performance of inoculated crops may vary geographically. Specific AM fungal strains appear to function well only within a range of environmental conditions beyond which their effect on plant productivity is reduced or even become negative (Herrera-Peraza et al. 2011). Thus, crop genotypes and HE strain-based inoculants should be developed specifically for different geographical locations. Alternatively, crop genotypes could be selected for their performance in association with AM fungi naturally occurring in cultivated soils. It would be important in this case to evaluate crop genotypes in different regions, as the dominant AM fungal genotypes naturally present in cultivated soils were shown to vary in a predictable manner with environmental conditions (Dai et al. 2012). This would be compatible with current practices in plant breeding. The evaluation of crop genotypes is normally done in multiple locations to assess yield stability (Tester and Langridge 2010). Thus, selecting plant genotypes for performance with local AM strains appears to be a feasible and ecologically sound option.

7.3.5 Inoculation with Highly Effective AM Strains

Tools for the management of the AM symbiosis in crop production are required (Facelli et al. 2009). Cultivars forming efficient AM symbioses are certainly important tools, but effective AM inoculants are also potentially useful where the ultimate goal of producing food and bioproducts sustainably sometimes conflicts with the maintenance of healthy AM fungal communities in cultivated soils. Despite their negative effect on AM fungi, disrupting tillage operations or the production of non-host crops may be desirable at times, and inoculation of AM-dependent crop species with selected AM fungal strains may restore ecosystem function where indigenous AM communities were negatively impacted.

Arbuscular mycorrhizal inoculants could also be used to stimulate plant growth beyond the capacity of an indigenous AM community. Inoculation with HE strains may have positive effects on cropping systems with little ecological consequence on indigenous AM communities, as shown by Antunes et al. (2009). A strain of *Glomus intraradices* inoculated onto maize positively interacted with the indigenous AM community in promoting plant growth, but did not influence the structure of the indigenous AM fungal community.

Host plant growth can be increased or decreased by AM inoculation (Johnson et al. 1997; Gosling et al. 2006). Different AM fungi have different abilities to build mutualistic relationships. They also have different abilities to compete for host plant roots, and competitive strains may be less effective mutualists (Bennett and Bever 2009). Highly effective strains are expected to enhance nutrient uptake and crop yield under field conditions. They only need to be competitive enough to produce a stable effect, and they should not be invasive. As the cost of production is an important consideration, HE strains with high propagule yield should be sought for the formulation of commercial inoculants. Species producing infective extraradical hyphae, such as *G. intraradices*, *G. etunicatum* and *Acaulospora spinosa*, will be preferred over species of *Gigaspora* and *Scutellospora* whose only propagules are spores (Klironomos and Hart 2002).

Inoculation of field crops has been shown to be effective (Sharma and Adholeya 2004; Stewart et al. 2005; Rivera et al. 2007; Higo et al. 2009; Mehraban et al. 2009). The effectiveness of an AM inoculation depends on the plant (Johnson et al. 1992; Gosling et al. 2006; An et al. 2010) but also on soil type (Herrera-Peraza et al. 2011). Different AM fungi are adapted to different soil types (Johnson et al. 1992; Herrera-Peraza et al. 2011; Dai et al. 2012). Highly effective strains often produce good effects on a range of crops, and in Cuba, crops are inoculated with HE strains selected based on the soil type where they will be introduced rather than on the crop species (Rivera et al. 2007).

7.3.6 Managing AM Fungi in Sustainable Crop Production Using Bioactive Molecules

Compatibility between crop plants and AM fungi can be enhanced through manipulation of signal molecules. The flavonoid "formononetin" was commercialized to enhance AM symbiosis development in plants (Koide et al. 1999) and to improve the yield of AM inoculum (de Novais and Siqueira 2009) by the biotech industry. Plant symbioses, including the AM symbiosis, are regulated by a crosstalk between plants and their microsymbionts and influence the formation and function of the AM symbiosis (Brachmann 2006).

Several plant compounds can influence the AM symbiosis (Brachmann 2006; Steinkellner et al. 2007). Strigolactones, a group of sesquiterpene lactones, can increase hyphae branching, mitochondrial density and respiration of AM fungi (Akiyama et al. 2005; Besserer et al. 2006a, b). Some plant flavonoids (Nair et al. 1991; Siqueira et al. 1991; Gianinazzi-Pearson et al. 1989; Tsai and Phillips 1991; Ishii et al. 1997; Fries et al. 1998; Aikawa et al. 2000; Cruz et al. 2004), ethylene (Ishii et al. 1996; Geil and Guinel 2002) and polyamines (El Ghachtouli et al. 1996) can stimulate AM fungi growth. Alginate oligosaccharide (Ishii et al. 2000) and nucleoside derivative (Kuwada et al. 2006) from a brown alga, *Laminaria japonica*, also stimulated AM hyphae growth, in vitro. The dipeptide Trp–Trp isolated from water-stressed bahia grass roots promoted growth and attracted the germ tube of germinating AM spores (Horii et al. 2009). The tryptophan dimer and Leu–Pro remarkably stimulated AM spore formation in *Glomus clarum*, *G. etunicatum* and *Gigaspora albida* grown in the absence of host roots or root exudates (Ishii and Horii 2009). Such signal molecules stimulating AM fungi growth and sporulation may facilitate the production of high-quality inoculants in bioreactors, with positive impact on the production costs of AM inoculants.

In natural ecosystems, plants manage their interactions with soil microorganisms using biochemical signals that selectively trigger responses in symbiotic partners (García-Garrido et al. 2009), but selectivity in signalling may not be perfect. Care should be taken as the application of signal molecules to crops or their inclusion in inoculant formulations may result in the attraction of undesirable organisms (Steinkellner et al. 2007) and reduce the ability of the crop to manage its rhizosphere. A good understanding of plant–microbe interaction is required to manipulate the rhizosphere exogenously. Crop genotypes with enhanced but regulated signalling systems selected in targeted plant-breeding programmes, appear as a safer avenue to AM symbiosis management in cropping systems. Rapid and early symbiosis development in crops would increase AM-derived benefits in short-season crops. Growing green manure plants with signalling properties highly promoting AM fungal development immediately following agronomic interventions unfavourable to the AM symbiosis, such as tillage or rotation with non-host crop species, would help counteract the negative impact of these cropping practices.

Plant signals may become useful bioproducts for sustainable production, but conversely, fungal signals may also lead to the development of useful biotechnologies. Plants recognize microbe-derived compounds and adjust their responses according to the type of microorganisms encountered (Mabood et al. 2006). AM fungal signals stimulating root receptivity to indigenous or introduced HE strains could be a useful symbiosis management tool. Such signal molecules could be formulated and applied to crops or introduced into AM inoculant formulations to improve symbiosis formation and function (Gianinazzi and Vosátkatka 2004).

7.3.7   AM Fungi in Pastures

Growing forage crops or pastures instead of annual crops is a sustainable alternative for the use of marginal lands. Perennial forage stands offer a permanent soil cover, are usually drought tolerant (Acuña et al. 2010), are a main source of feed for the livestock industry and maintain soil quality through their contribution to soil C (Shrestha and Lal 2010). Forage stands are typically low-disturbance and low-input systems, and thus they are good habitats for AM fungi.

Grazing management is the main way to manipulate and maintain pasture ecosystem health. Grazing increases the diversity of grassland plant species (Watkinson and Ormerod 2001) and that of AM fungi, although it decreases their proliferation (Murray et al. 2010). Defoliation reduces plant C export below ground (Ilmarinen et al. 2008), and reduced AM root colonization was attributed to grazing (Jirout et al. 2009). Grazing increases system heterogeneity, which is important to preserve biodiversity in grasslands (Klimek et al. 2008), but it may reduce AM function in intensive systems. Overgrazing, in turn, leads to reduced AM fungal diversity (Su and Guo 2007) and ecosystem degradation (Watkinson and Ormerod 2001).

The loss of AM diversity may negatively impact ecosystem function. Concurrent seasonal changes in AM fungi and plant community composition in native Canadian grasslands (Yang et al. 2010) suggested that seasonal shifts in above- and belowground diversity are necessary to maintain the function of ecosystems as the environmental conditions vary. The diversity of AM fungi creates a patchy soil environment (Mummey and Rillig 2008) conducive to the coexistence of different plant species. Different AM fungi influence different plants differently (Aldrich-Wolfe 2007), and variation in AM hyphae density and ribotype distribution in soil, with spatial structure at less than 30 cm (Mummey and Rillig 2008), indicates that seedling establishment occurs over a mosaic of AM fungal influence.

Marginal land conversion to native grasslands could be a sustainable alternative in dry areas at this time of climate change. Much of the brown soil zone of the Canadian Prairie, for example, is forecasted to become dryer and only marginally suitable for cereal production by mid-twenty-first century (Nyirfa and Harron 2002), a conclusion supported by trends in climate change (Cutforth 2000). In contrast, the productivity of native Canadian grasslands should be sustained under the conditions forecasted by climate models (Thorpe et al. 2008). The diversity and adaptation of native ecosystems gives them a resilience that tame forage species may not have. Ecovar seeds of several native plant species are commercially available and seeded native ecosystems could protect soils in marginal areas while supporting the cattle industry. It remains to be seen if the AM fungal diversity will also need to be restored in these soils.

7.4 Conclusion

Agrochemicals have increased the productivity of cultivated lands and simplified crop production. However, the application of agricultural technologies may reduce the efficiency of agroecosystems in the long term. Knowledge should be applied and technology developed that enhances ecosystem efficiency through the management of their complexity. In this context, it is particularly important to improve the precision and sustainability of crop nutrition management in a manner that protects and maximizes benefits from AM fungal resources. Technologies that mitigate negative effects of some necessary cropping practices on the functionality of the AM symbiosis in agriculture are also required.

References

Acuña H, Inostroza L, Sanchez MP, Tapia G (2010) Drought-tolerant naturalized populations of *Lotus tenuis* for constrained environments. Acta Agric Scand Sect B Soil Plant Sci 60:174–181

Agrios GN (2005) Plant pathology. Academic, California

Aikawa J, Ishii T, Kuramoto M, Kadoya K (2000) Growth stimulants for vesicular-arbuscular mycorrhizal fungi in *Satsuma Mandarin Pomace*. J Jpn Soc Hortic Sci 69:385–389

Akiyama K, Matsuzaki K, Hayashi H (2005) Plant sesquiterpenes induce hyphal branching in arbuscular mycorrhizal fungi. Nature 435:824–827

Aldrich-Wolfe L (2007) Distinct mycorrhizal communities on new and established hosts in a transitional tropical plant community. Ecology 88(3):559–566

Alguacil MM, Lumini E, Roldán A, Salinas-García JR, Bonfante P, Bianciotto V (2008) The impact of tillage practices on arbuscular mycorrhizal fungal diversity in subtropical crops. Ecol Appl 18:527–536

Al-Karaki GN, Clark RB (1999) Varied rates of mycorrhizal inoculum on growth and nutrient acquisition by barley grown with drought stress. J Plant Nutr 22:1775–1784

Alvey S, Bagayoko M, Neumann G, Buerkert A (2001) Cereal/legume rotations affect chemical properties and biological activities in two West African soils. Plant and Soil 231:45–54

Amaya-Carpio L, Davies FT, Fox T, He C (2009) Arbuscular mycorrhizal fungi and organic fertilizer influence photosynthesis, root phosphatase activity, nutrition, and growth of *Ipomoea carnea* ssp. *fistulosa*. Photosynthetica 47:1–10

An GH, Kobayashi S, Enoki H, Sonobe K, Muraki M, Karasawa T, Ezawa T (2010) How does arbuscular mycorrhizal colonization vary with host plant genotype? An example based on maize (*Zea mays*) germplasms. Plant and Soil 327:441–453

Antunes PM, Koch AM, Dunfield KE, Hart MM, Downing A, Rillig MC, Klironomos JN (2009) Influence of commercial inoculation with *Glomus intraradices* on the structure and functioning of an AM fungal community from an agricultural site. Plant and Soil 317:257–266

Aroca R, Bago A, Sutka M, Paz JA, Cano C, Amodeo G, Ruiz-Lozano JM (2009) Expression analysis of the first arbuscular mycorrhizal fungi aquaporin described reveals concerted gene expression between salt-stressed and nonstressed mycelium. Mol Plant Microbe Interact 9:1169–1178, doi:10.1094

Arriagada C, Pacheco P, Pereira G, Machuca A, Alvear M, Ocampo JA (2009) Effect of arbuscular mycorrhizal fungal inoculation on *Eucalyptus globulus* seedlings and some soil enzyme activities under application of sewage sludge amendment. R C Suelo Nutr Veg 9:89–101

Assaf TA, Turk MA, Hameed KM (2009) Impact of olive pomace wastes and fungicide treatment on indigenous arbuscular mycorrhizal fungi associated with chickpea (*Cicer arietinum* L.) under field conditions. Aust J Crop Sci 3:6–12

Atul-Nayyar A, Hamel C, Hanson K, Germida J (2009) The arbuscular mycorrhizal symbiosis links N mineralization to plant demand. Mycorrhiza 19:239–246

Augé RM (2001) Water relations, drought and vesicular-arbuscular mycorrhizal symbiosis. Mycorrhiza 11:3–42

Augé RM (2004) Arbuscular mycorrhizae and soil/plant water relations. Can J Soil Sci 84:373–381

Azcón R, Rodríguez R, Amora-Lazcano E, Ambrosano E (2008) Uptake and metabolism of nitrate in mycorrhizal plants as affected by water availability and N concentration in soil. Eur J Soil Sci 59:131–138

Azcón-Aguilar C, Barea JM (1997) Arbuscular mycorrhizas and biological control of soil-borne plant pathogens – an overview of the mechanisms involved. Mycorrhiza 6:457–464

Bagayoko M, Buerkert A, Lung G, Bationo A, Römheld V (2000) Cereal/legume rotation effects on cereal growth in Sudano-Sahelian West Africa: soil mineral nitrogen, mycorrhizae and nematodes. Plant and Soil 218:103–116

Benabdellah K, Ruiz-Lozano JM, Aroca R (2009) Hydrogen peroxide effects on root hydraulic properties and plasma membrane aquaporin regulation in *Phaseolus vulgaris*. Plant Mol Biol 70:647–661

Bennett AE, Bever JD (2009) Trade-offs between arbuscular mycorrhizal fungal competitive ability and host growth promotion in *Plantago lanceolata*. Oecologia 160:807–816

Besserer A, Puech-Pagès V, Kiefer P, Gomez-Roldan V, Jauneau A, Roy S, Portais JC, Roux C, Bécard G, Séjalon-Delmas N (2006) Strigolactones stimulate arbuscular mycorrhizal fungi by activating mitochondria. PLoS Biol 4:1239–1247

Bharadwaj DP, Lundquist PO, Alström S (2008) Arbuscular mycorrhizal fungal spore-associated bacteria affect mycorrhizal colonization, plant growth and potato pathogens. Soil Biol Biochem 40:2494–2501

Bingham MA, Biondini M (2009) Mycorrhizal hyphal length as a function of plant community richness and composition in restored northern tallgrass prairies (USA). Rangel Ecol Manag 62:60–67

Boddington CL, Dodd JC (2000) The effect of agricultural practices on the development of indigenous arbuscular mycorrhizal fungi. I. Field studies in an Indonesian Ultisol. Plant and Soil 218:137–144

Borie F, Rubio R, Rouanet JL, Morales A, Borie G, Rojas C (2006) Effects of tillage systems on soil characteristics, glomalin and mycorrhizal propagules in a Chilean Ultisol. Soil Tillage Res 88:253–261

Brachmann A (2006) Plant-fungal symbiosis en gros and en detail. New Phytol 171:242–246

Brady NC, Weil RR (2001) The nature and properties of soils, 13th edn. Prentice Hall, Upper Saddle River

Bray RH, Kurtz LT (1945) Determination of total, organic, and available forms of phosphorus in soils. Soil Sci 59:39–45

Bucher M (2007) Functional biology of plant phosphate uptake at root and mycorrhiza interfaces. New Phytol 173:11–26

Campagnac E, Fontaine J, Sahraoui ALH, Laruelle F, Durand R, Grandmougin-Ferjani A (2008) Differential effects of fenpropimorph and fenhexamid, two sterol biosynthesis inhibitor fungicides, on arbuscular mycorrhizal development and sterol metabolism in carrot roots. Phytochemistry 69:2912–2919

Campagnac E, Fontaine J, Sahraoui AL, Laruelle F, Durand R, Grandmougin-Ferjani A (2009) Fenpropimorph slows down the sterol pathway and the development of the arbuscular mycorrhizal fungus *Glomus intraradices*. Mycorrhiza 19:365–374

Cappellazzo G, Lanfranco L, Fitz M, Wipf D, Bonfante P (2008) Characterization of an amino acid permease from the endomycorrhizal fungus *Glomus mosseae*. Plant Physiol 147:429–437

Chalk PM, Souza RDF, Urquiaga S, Alves BJR, Boddey RM (2006) The role of arbuscular mycorrhiza in legume symbiotic performance. Soil Biol Biochem 38:2944–2951

Colla G, Rouphael Y, Cardarelli M, Tullio M, Rivera CM, Rea E (2007) Alleviation of salt stress by arbuscular mycorrhizal in zucchini plants grown at low and high phosphorus concentration. Biol Fertil Soils 44:501–509. doi:10.1007/s0037400702328

Cornejo P, Rubio R, Borie F (2009) Mycorrhizal propagule persistence in a succession of cereals in a disturbed and undisturbed andisol fertilized with two nitrogen sources. Chilean J Agric Res 69:426–434

Covacevich F, Echeverría HE (2009) Mycorrhizal occurrence and responsiveness of tall fescue and wheatgrass are affected by the source of phosphorus fertilizer and fungal inoculation. J Plant Interact 4:101–112

Cruz AF, Ishii T, Matsumoto I, Kadoya K (2004) Relationship between arbuscular mycorrhizal fungal development and eupalitin content in bahiagrass roots grown in a satsuma mandarin orchard. J Jpn Soc Hortic Sci 73:529–533

Cuperus GW, Berberet RC, Noyes RT (2004) The essential role of IPM in promoting sustainability of agricultural production systems for future generations. In: Koul O, Dhaliwal GS, Cuperus GW (eds) Integrated pest management: potential, constraints and challenges. CABI, Wallingford Oxfordshire

Cutforth HW (2000) Climate change in the semiarid prairie of southwestern Saskatchewan: temperature, precipitation, wind, and incoming solar energy. Can J Soil Sci 80:375–385

Dai M, Hamel C, St. Arnaud M, He Y, Grant CA, Lupwayi NZ, Janzen HH, Malhi SS, Yang X, Zhou Z (2012) Arbuscular mycorrhizal fungi assemblages in Chernozem great groups revealed by massively parallel pyrosequencing. Can J Microbiol 58:81–92

Davies WJ, Wilkinson S, Loveys B (2002) Stomatal control by chemical signaling and the exploitation of this mechanism to increase water-use efficiency in agriculture. New Phytol 153:449–460

de Novais CB, Siqueira JO (2009) Formononetin application on colonization and sporulation of arbuscular mycorrhizal fungi in Brachiaria. Pesq Agrop Brasileira 44:496–502

Dewar AM (2009) Weed control in glyphosate-tolerant maize in Europe. Pest Manag Sci 65:1047–1058

Douds DD, Galvez L, Janke RR, Wagoner P (1995) Effect of tillage and farming system upon populations and distribution of vesicular-arbuscular mycorrhizal fungi. Agr Ecosyst Environ 52:111–118

Drijber RA (2002) Microbial signatures for crop production systems. In: Lynch JM et al (ed) Proc OECD workshop on innovative soil-plant systems for sustainable agricultural practices, Izmir, Tübitak, Turkey, 3–7 June 2002, pp 132–146

Duke SO, Powles SB (2009) Glyphosate-resistant crops and weeds: now and in the future. AgBioForum 12:346–357

Eason WR, Webb KJ, Michaelson-Yeates TPT, Abberton MT, Griffith GW, Culshaw CM, Hooker JE, Dhanoa MS (2001) Effect of genotype of *Trifolium repens* on mycorrhizal symbiosis with *Glomus mosseae*. J Agric Sci 137:27–36

El Ghachtouli N, Paynot M, Martin-Tanguy J, Morandi D, Gianinazzi S (1996) Effect of polyamines and polyamine biosynthesis inhibitors on spore germination and hyphal growth of *Glomus mosseae*. Mycol Res 100:597–600. doi:10.1016/s0953-7562(96)80014-1

Elsen A, van der Veken L, De Waele D (2009) AMF-Induced bioprotection against migratory plant-parasitic nematodes in banana. Acta Horticult 828:91–100

Evans DG, Miller MH (1990) The role of the external mycelial network in the effect of soil disturbance upon vesicular-arbuscular mycorrhizal colonization of maize. New Phytol 114:65–71

Facelli E, Smith SE, Smith FA (2009) Mycorrhizal symbiosis overview and new insights into roles of arbuscular mycorrhizas in agro- and natural ecosystems. Australas Plant Pathol 38:338–344

FAO (1967) Report of the first session of the FAO panel of experts on integrated pest control. Food and Agriculture Organization of the United Nations, Rome

Feldmann F, Boyle C (1999) Weed-mediated stability of arbuscular mycorrhizal effectiveness in maize monocultures. J Appl Bot 73:1–5

Filion M, St-Arnaud M, Fortin JA (1999) Direct interaction between the arbuscular mycorrhizal fungus *Glomus intraradices* and different rhizosphere microorganisms. New Phytol 141:525–533. doi:10.1046/j.1469-8137.1999.00366.x

Finlay RD (2008) Ecological aspects of mycorrhizal symbiosis: with special emphasis on the functional diversity of interactions involving the extraradical mycelium. J Exp Bot 59:1115–1126

Fixen PE (2006) Soil test levels in North America. Better Crops 90:4–7

Fortin JA, Bécard G, Declerck S, Dalpé Y, St-Arnaud M, Coughlan AP, Piché Y (2002) Arbuscular mycorrhiza on root-organ cultures. Can J Bot 80:1–20

Fries LLM, Pacovsky RS, Safir GR (1998) Influence of phosphorus and formononetin on isozyme expression in the *Zea mays-Glomus intraradices* symbiosis. Physiol Plant 103:172–180

Gange AC, Brown VK, Sinclair GS (1993) Vesicular-arbuscular mycorrhizal fungi: a determinant of plant community structure in early succession. Funct Ecol 7:616–622

Garcia JP, Wortmann CS, Mamo M, Drijber R, Tarkalson D (2007) One-time tillage of no-till: effects on nutrients, mycorrhizae, and phosphorus uptake. Agron J 99:1093–1103

García-Garrido J, Lendzemo V, Castellanos-Morales V, Steinkellner S, Vierheilig H (2009) Strigolactones, signals for parasitic plants and arbuscular mycorrhizal fungi. Mycorrhiza 19:449–459

Garg N, Manchanda G (2009) Role of arbuscular mycorrhizae in the alleviation of ionic, osmotic and oxidative stresses induced by salinity in *Cajanus cajan* (L.) Millsp. (pigeonpea). J Agron Crop Sci 195:110–123

Gavito ME, Miller MH (1998) Early phosphorus nutrition, mycorrhizae development, dry matter partitioning and yield of maize. Plant and Soil 199:177–186

Geil RD, Guinel FC (2002) Effects of elevated substrate–ethylene on colonization of leek (*Allium porrum*) by the arbuscular mycorrhizal fungus *Glomus aggregatum*. Can J Bot 80:114–119

Gianinazzi S, Vosátkatka M (2004) Inoculum of arbuscular mycorrhizal fungi for production systems: science meets business. Can J Bot 82:1264–1271

Gianinazzi-Pearson V, Branzanti B, Gianinazzi S (1989) In vitro enhancement of spore germination and early hyphal growth of a vesicular-arbuscular mycorrhizal fungus by host root exudates and plant flavonoids. Symbiosis 7:243–255

Gilbert N (2009) The disappearing nutrient. Nature 461:716–718

Gilden RC, Huffling K, Sattler B (2010) Pesticides and health risks. J Obstet Gynecol Neonatal Nurs 39:103–110

Gonzalez-Dugo V (2010) The influence of arbuscular mycorrhizal colonization on soil-root hydraulic conductance in *Agrostis stolonifera* L. under two water regimes. Mycorrhiza 20 (6):365–373. doi:10.1007/s0057200902946

Gosling P, Hodge A, Goodlass G, Bending GD (2006) Arbuscular mycorrhizal fungi and organic farming. Agr Ecosyst Environ 113:17–35

Govindarajulu M, Pfeffer PE, Jin H, Abubaker J, Douds DD, Allen JW, Bucking H, Lammers PJ, Shacchar-Hill Y (2005) Nitrogen transfer in the arbuscular mycorrhizal symbiosis. Nature 435:819–823

Greer KJ, Sulewski C, Hangs R (2003) Applying PRS™ technology for nutrient management. In: Ellsworth JW (ed) Proceedings of the western nutrient management workshop, vol 5. Potash and Phosphate Institute, Brookings, pp 170–175

Grime JP, Mackey JM, Hillier SH, Read DJ (1987) Floristic diversity in a model system using experimental microcosms. Nature 328:420–422

Gryndler M, Larsen J, Hrselová H, Rezácová GH, Kubát J (2006) Organic and mineral fertilization, respectively, increase and decrease the development of external mycelium of arbuscular mycorrhizal fungi in a long-term field experiment. Mycorrhiza 16:159–166

Guertal EA, Eckert DJ, Traina SJ, Logan TJ (1991) Differential phosphorus retention in soil profiles under no-till crop production. Soil Sci Soc Am J 55:410–413

Habte M (1997) Use of benlate to obtain soil free of arbuscular mycorrhizal fungal activity for greenhouse investigations. Arid Soil Res Rehabil 11:151–161

Hajiboland R, Aliasgharzadeh N, Laiegh SF, Poschenrieder C (2010) Colonization with arbuscular mycorrhizal fungi improves salinity tolerance of tomato (*Solanum lycopersicum* L.) plants. Plant and Soil 331:313–327

Hamel C, Strullu DG (2006) Arbuscular mycorrhizal fungi in field crop production: potential and new direction. Can J Plant Sci 86:941–950

Harrier LA, Watson CA (2004) The potential role of arbuscular mycorrhizal (AM) fungi in the bioprotection of plants against soil-borne pathogens in organic and/or other sustainable farming systems. Pest Manag Sci 60:149–157

Hayman DS (1986) Mycorrhizae of nitrogen-fixing legumes. MIRCEN J Appl Microbiol Biotechnol 2:121–145

He X-H, Critchley C, Bledsoe C (2003) Nitrogen transfer within and between plants through common mycorrhizal networks (CMNs). Crit Rev Plant Sci 22:531–567

Herrera-Peraza RA, Hamel C, Fernández F, Ferrer RL, Furrazola E (2011) Soil–strain compatibility: the key to effective use of arbuscular mycorrhizal inoculants? Mycorrhiza 21(3):183–193

Herzog F, Prasuhn V, Spiess E, Richner W (2008) Environmental cross-compliance mitigates nitrogen and phosphorus pollution from Swiss agriculture. Environ Sci Pol 11:655–668

Higo M, Isobe K, Kang D-J, Ujiie K, Drijber RA, Ishii R (2009) Inoculation with arbuscular mycorrhizal fungi or crop rotation with mycorrhizal plants improves the growth of maize in limed acid sulfate soil. Plant Prod Sci 13:74–79

Hodge A, Campbell CD, Fitter AH (2001) An arbuscular mycorrhizal fungus accelerates decomposition and acquires nitrogen directly from organic material. Nature 413:297–299

Hol WHG, Cook R (2005) An overview of arbuscular mycorrhizal fungi-nematode interactions. Basic Appl Ecol 6:489–503

Hollomon DW, Brent KJ (2009) Combating plant diseases-the Darwin connection. Pest Manag Sci 65:1156–1163

Horii S, Matsumura A, Kuramoto M, Ishii T (2009) Tryptophan dimer produced by water-stressed bahia grass is an attractant for *Gigaspora margarita* and *Glomus caledonium*. World J Microbiol Biotechnol 25:1207–1215

Houngnandan P, Sanginga N, Woomer P, Vanlauwe B, Van Cleemput O (2000) Response of Mucuna pruriens to symbiotic nitrogen fixation by rhizobia following inoculation in farmers' fields in the derived savanna of Benin. Biol Fertil Soils 30:558–565

Ilmarinen K, Mikola J, Vestberg M (2008) Do interactions with soil organisms mediate grass responses to defoliation? Soil Biol Biochem 40:894–905

Ishii T, Horii S (2009) Importance of short molecular peptides to axenic culture of arbuscular mycorrhizal (AM) fungi, The 6th International conference on mycorrhiza, Belo Horizonte-Minas Gerais, Brazil, p 24

Ishii T, Shrestha YH, Matsumoto I, Kadoya K (1996) Effect of ethylene on the growth of vesicular-arbuscular mycorrhizal fungi and on the mycorrhizal formation of trifoliate orange roots. J Jpn Soc Hortic Sci 15:525–529

Ishii T, Narutaki A, Sawada K, Aikawa J, Matsumoto I, Kadoya K (1997) Growth stimulatory substances for vesicular-arbuscular mycorrhizal fungi in Bahia grass (*Paspalum notatum Flugge.*) roots. Plant and Soil 196:301–304

Ishii T, Kitabayashi H, Aikawa J, Matumoto I, Kadoya K, Kirino S (2000) Effects of alginate oligosaccharide and polyamines on hyphal growth of vesicular-arbuscular mycorrhizal fungi and infectivity on citrus roots. Proc Int Soc Citriculture 9:1030–1032

Jalonen R, Nygren P, Sierra J (2009) Transfer of nitrogen from a tropical legume tree to an associated fodder grass via root exudation and common mycelial networks. Plant Cell Environ 32:1366–1376

Jansa J, Mozafar A, Kuhn G, Anken T, Ruh R, Sanders IR, Frossard E (2003) Soil tillage affects the community structure of mycorrhizal fungi in maize roots. Ecol Appl 13:1164–1176

Javaid A, Ahmad S, Javaid A, Shad N, Jabeen K (2009) Screening of mungbean cultivars under rice allelopathic stress for best agronomic and symbiotic traits. Allelopath J 24:331–340

Jirout J, Triska J, Ruzickova K, Elhottova D (2009) Disturbing impact of outdoor cattle husbandry on community of arbuscular mycorrhizal fungi in upland pasture soil. Commun Soil Sci Plant Anal 40:736–745

Johnson NC (1993) Can fertilization of soil select less mutualistic mycorrhizae? Ecol Appl 3:749–757

Johnson NC (2010) Resource stoichiometry elucidates the structure and function of arbuscular mycorrhizas across scales. New Phytol 185:631–647

Johnson NC, Pfleger FL (1992) Vesicular-arbuscular mycorrhizae in sustainable agriculture. In: Bethlenfalvay GJ, Linderman RG (eds) Mycorrhizae in sustainable agriculture, vol 54, ASA special publication. ASA, CSSA, SSSA, Madison, pp 71–99

Johnson NC, Pfleger FL, Crookston RK, Simmons SR, Copeland PJ (1991) Vesicular-arbuscular mycorrhizas respond to corn and soybean cropping history. New Phytol 117:657–663

Johnson NC, Tilman D, Wedin D (1992) Plant and soil controls on mycorrhizal fungal communities. Ecology 73:2034–2042

Johnson NC, Graham JH, Smith FA (1997) Functioning of mycorrhizal associations along the mutualism-parasitism continuum. New Phytol 135:575–585

Jordan NR, Zhang J, Huerd S (2000) Arbuscular-mycorrhizal fungi: potential roles in weed management. Weed Res 40:397–410

Kabir Z (2005) Tillage or no-tillage: impact on mycorrhizae. Can J Plant Sci 85:23–29

Kabir Z, Koide RT (2000) The effect of dandelion or a cover crop on mycorrhiza inoculum potential, soil aggregation and yield of maize. Agr Ecosyst Environ 78:167–174

Kabir Z, Koide RT (2002) Effect of autumn and winter mycorrhizal cover crops on soil properties, nutrient uptake and yield of sweet corn in Pennsylvania, USA. Plant and Soil 238:205–215

Kabir Z, O'Halloran IP, Fyles JW, Hamel C (1998) Dynamics of the mycorrhizal symbiosis of corn (Zea mays L.): effects of host physiology, tillage practice and fertilization on spatial distribution of extra-radical mycorrhizal hyphae in the field. Agr Ecosyst Environ 68:151–163

Kaeppler SM, Parke JL, Mueller SM, Senior L, Stuber C, Tracy WF (2000) Variation among maize inbred lines and detection of quantitative trait loci for growth at low phosphorus and responsiveness to arbuscular mycorrhizal fungi. Crop Sci 40:358–364

Karandashov V, Bucher M (2005) Symbiotic phosphate transport in arbuscular mycorrhizas. Trends Plant Sci 10:22–29

Karanika ED, Mamolos AP, Alifragis DA, Kalburtji KL, Veresoglou DS (2008) Arbuscular mycorrhizas contribution to nutrition, productivity, structure and diversity of plant community in mountainous herbaceous grassland of northern Greece. Plant Ecol 199:225–234

Karasawa T, Kasahara Y, Takebe M (2002) Differences in growth responses of maize to preceding cropping caused by fluctuation in the population of indigenous arbuscular mycorrhizal fungi. Soil Biol Biochem 34:851–857

Karliński L, Rudawska M, Kieliszewska-Rokicka B, Leski T (2010) Relationship between genotype and soil environment during colonization of poplar roots by mycorrhizal and endophytic fungi. Mycorrhiza 20:315–324

Kinney CA, Mandernack KW, Mosier AR (2005) Laboratory investigations into the effects of the pesticides mancozeb, chlorothalonil, and prosulfuron on nitrous oxide and nitric oxide production in fertilized soil. Soil Biol Biochem 37:837–850

Klimek S, Marini L, Hofmann M, Isselstein J (2008) Additive partitioning of plant diversity with respect to grassland management regime, fertilisation and abiotic factors. Basic Appl Ecol 9:626–634

Klironomos JN, Hart MM (2002) Colonization of roots by arbuscular mycorrhizal fungi using different sources of inoculum. Mycorrhiza 12:181–184

Koide RT, Dickie IA (2002) Effects of mycorrhizal fungi on plant populations. Plant and Soil 244:307–317

Koide RT, Landherr LL, Besmer YL, Detweiler JM, Holcomb EJ (1999) Strategies for mycorrhizal inoculation of six annual bedding plant species. HortSci 34:1217–1220

Krishna KR, Shetty KG, Dart PJ, Andrews DJ (1985) Genotype dependent variation in mycorrhizal colonization and response to inoculation of pearl millet. Plant and Soil 86:113–125

Kuwada K, Kuramoto M, Utamura M, Matsushita I, Ishii T (2006) Isolation and structural elucidation of a growth stimulant for arbuscular mycorrhizal fungus from *Laminaria japonica* Areschoug. J Appl Phycol 18:795–800

Lambers H, Chapin FS III, Pons TL (2007) Plant physiological ecology, 2nd edn. Springer, Heidelberg

Lampurlanés J, Cantero-Martínez C (2003) Soil bulk density and penetration resistance under different tillage and crop management systems and their relationship with barley root growth. Agron J 95:526–536

Leake JR, Johnson D, Donnelly D, Muckle GE, Boddy L, Read DJ (2004) Networks of power and influence: the role of mycorrhizal mycelium in controlling plant communities and agroecosystem functioning. Can J Bot 82:1016–1045. doi:10.1139/B04-060

Leigh J, Hodge A, Fitter AH (2009) Arbuscular mycorrhizal fungi can transfer substantial amounts of nitrogen to their host plant from organic material. New Phytol 181:199–207

Lekberg Y, Koide RT, Twomlow SJ (2008) Effect of agricultural management practices on arbuscular mycorrhizal fungal abundance in low-input cropping systems of southern Africa: a case study from Zimbabwe. Biol Fertil Soils 44:917–923

Li Y, Ran W, Zhang R, Sun S, Xu G (2009) Facilitated legume nodulation, phosphate uptake and nitrogen transfer by arbuscular inoculation in an upland rice and mung bean intercropping system. Plant and Soil 315:285–296

Lioussanne L, Jolicoeur M, St-Arnaud M (2008) Mycorrhizal colonization with *Glomus intraradices* and development stage of transformed tomato roots significantly modify the chemotactic response of zoospores of the pathogen *Phytophthora nicotianae*. Soil Biol Biochem 40:2217–2224

Lioussanne L, Perreault F, Jolicoeur M, St-Arnaud M (2010) The bacterial community of tomato rhizosphere is modified by inoculation with arbuscular mycorrhizal fungi but unaffected by soil enrichment with mycorrhizal root exudates or inoculation with *Phytophthora nicotianae*. Soil Biol Biochem 42:473–483

Liu A, Hamel C, Hamilton RI, Ma BL, Smith DL (2000a) Acquisition of Cu, Zn, Mn and Fe by mycorrhizal maize (*Zea mays* L.) grown in soil at different P and micronutrient levels. Mycorrhiza 9:331–336

Liu A, Hamel C, Hamilton RI, Smith DL (2000b) Mycorrhizae formation and nutrient uptake of new corn (*Zea mays* L.) hybrids with extreme canopy and leaf architecture as influenced by soil N and P levels. Plant and Soil 221:157–166

Liu A, Plenchette C, Hamel C (2007) Soil nutrient and water providers: how arbuscular mycorrhizal mycelia support plant performance in a resource limited world. In: Hamel C, Plenchette C (eds) Mycorrhizae in crop production. Haworth, Binghampton, pp 37–66

Lynch SM, Mahajan R, Beane Freeman LE, Hoppin JA, Alavanja MCR (2009) Cancer incidence among pesticide applicators exposed to butylate in the agricultural health study (AHS). Environ Res 109:860–868

Mabood F, Gray EJ, Lee KD, Smith DL (2006) Exploiting inter-organismal chemical communication for improved inoculants. Can J Plant Sci 86:951–966

Martinez JP, Lutts S, Schanck A, Bajji M, Kinet JM (2004) Is osmotic adjustment required for drought stress resistance in the Mediterranean shrub *Atriplex halimus* L? J Plant Physiol 161:1041–1051

McDowell N, Pockman WT, Allen CD, Breshears DD, Cobb N, Kolb T, Plaut J, Sperry J, West A, Williams DG, Yepez EA (2008) Mechanisms of plant survival and mortality during drought: why do some plants survive while others succumb to drought? New Phytol 178:719–739

McKenzie RH, Bremer E, Kryzanowski L, Middleton AB, Solberg ED, Heaney D, Coy G, Harapiak J (2003) Yield benefit of phosphorus fertilizer for wheat, barley and canola in Alberta. Can J Soil Sci 83:431–441

Md González-Chávez CA, Aitkenhead-Peterson JA, Gentry TJ, Zuberer D, Hons F, Loeppert R (2010) Soil microbial community, C, N, and P responses to long-term tillage and crop rotation. Soil Tillage Res 106:285–293

Mehraban A, Vazan S, Naroui Rad MR, Ardakany AR (2009) Effect of vesicular-arbuscular mycorrhiza (VAM) on yield of sorghum cultivars. J Food Agric Environ 7:461–463

Miyamoto T, Ishii H, Seko T, Kobori S, Tomita Y (2009) Occurrence of *Corynespora cassiicola* isolates resistant to boscalid on cucumber in Ibaraki prefecture, Japan. Plant Pathol 58:1144–1151

Mulligan MF, Smucker AJM, Safir GF (1985) Tillage modifications of dry edible bean root colonization by VAM fungi. Agron J 77:140–144

Mummey DL, Rillig MC (2006) The invasive plant species *Centaurea maculosa* alters arbuscular mycorrhizal fungal communities in the field. Plant and Soil 288:81–90

Mummey DL, Rillig MC (2008) Spatial characterization of arbuscular mycorrhizal fungal molecular diversity at the submetre scale in a temperate grassland. FEMS Microbiol Ecol 64(2):260–270

Murray TR, Frank DA, Gehring CA (2010) Ungulate and topographic control of arbuscular mycorrhizal fungal spore community composition in a temperate grassland. Ecology 91:815–827

Nair MG, Safir GR, Siqueira JO (1991) Isolation and identification of vesicular-arbuscular mycorrhiza-stimulatory compounds from clover (*Trifolium repens*) roots. Appl Environ Microbiol 57:434–439

Newsham KK, Fitter AH, Watkinson AR (1995) Multi-functionality and biodiversity in arbuscular mycorrhizas. Trends Ecol Evol 10:407–411

Nogales A, Aguirreolea J, Santa María E, Camprubí A, Calvet C (2009) Response of mycorrhizal grapevine to *Armillaria mellea* inoculation: disease development and polyamines. Plant and Soil 317:177–187

Nyirfa W, Harron B (2002) Assessment of climate change on the agricultural resources of the Canadian Prairies. Agriculture and Agri-Food Canada, Prairie Farm Rehabilitation Administration, Regina, SK. Available via DIALOG. http://www.parc.ca/pdf/research_publications/agriculture4.pdf. Accessed 18 Aug 2010

O'Halloran IP, Miller MH, Arnold G, Sylvia DM (1986) Absorption of P by corn (*Zea mays* L.) as influenced by soil disturbance. Can J Soil Sci 66:287–302

Oades JM (1993) The role of biology in the formation, stabilization and degradation of soil structure. Geoderma 56:377–400

Oliveira RS, Castro PML, Dodd JC, Vosátka M (2006) Different native arbuscular mycorrhizal fungi influence the coexistence of two plant species in a highly alkaline anthropogenic sediment. Plant and Soil 287:209–221

Olsen SR, Sommers LE (1982) Phosphorus. In: Page AL, Miller RH, Keeney DR (eds) Methods of soil analysis. Part 2: chemical and microbiological properties, vol 9, 2nd edn, Agronomy. ASA-SSSA, Madison, pp 403–430

Olsen SR, Cole CV, Watanabe FS, Dean LA (1954) Estimation of available phosphorus in soils by extraction with sodium bicarbonate, vol 939. U.S. Department of Agriculture, Washington, DC

Ouyang DS, MacKenzie AF, Fan MX (1999) Availability of banded triple superphosphate with urea and phosphorus use efficiency by corn. Nutr Cycl Agroecosyst 53:237–247

Pande M, Tarafdar JC (2002) Effect of phosphorus, salinity and moisture on VAM fungal association in neem (*Azadirachta indica* L.). Symbiosis 32:195–209

Pandey R, Kalra A, Gupta ML (2009) Evaluation of bio-agents and pesticide on root-knot nematode development and oil yield of patchouli. Arch Phytopathol Plant Protect 42:419–423

Perner H, Schwarz D, George E (2006) Effect of mycorrhizal inoculation and compost supply on growth and nutrient uptake of young leek plants grown on peat-based substrates. HortSci 41:628–632

Perner H, Schwarz D, Bruns C, Mäder P, George E (2007) Effect of arbuscular mycorrhizal colonization and two levels of compost supply on nutrient uptake and flowering of pelargonium plants. Mycorrhiza 17:469–474

Pimienta-Barrios E, Gonzalez Del Castillo-Aranda ME, Muñoz-Urias A, Nobel PS (2003) Effects of benomyl and drought on the mycorrhizal development and daily net CO_2 uptake of a wild platyopuntia in a rocky semi-arid environment. Ann Bot 92:239–245

Pozo MJ, Azcón-Aguilar C (2007) Unraveling mycorrhiza-induced resistance. Curr Opin Plant Biol 10:393–398

Querejeta JI, Egerton-Warburton LM, Allen MF (2009) Topographic position modulates the mycorrhizal response of oak trees to interannual rainfall variability. Ecology 90:649–662

Rains KC, Bledsoe CS (2007) Rapid uptake of ^{15}N-ammonium and glycine-^{13}C, ^{15}N by arbuscular and ericoid mycorrhizal plants native to a Northern California coastal pygmy forest. Soil Biol Biochem 39:1078–1086

Ramirez R, Mendoza B, Lizaso JI (2009) Mycorrhiza effect on maize P uptake from phosphate rock and superphosphate. Commun Soil Sci Plant Anal 40:2058–2071

Rasmann C, Graham JH, Chellemi DO, Datnoff LE, Larsen J (2009) Resilient populations of root fungi occur within five tomato production systems in southeast Florida. Appl Soil Ecol 43:22–31

Reis MR, Tironi SP, Costa MD, Silva MCS, Ferreira EA, Belo AF, Barbosa MHP, Silva AA (2009) Mycorrhizal colonization and acid phosphatase activity in the rhizosphere of sugarcane cultivars after herbicide application. Planta Daninha 27:977–985

Rillig MC, Mummey DL (2006) Mycorrhizas and soil structure. New Phytol 171:41–53

Rivera R, Fernandez F, Fernandez K, Ruiz L, Sanchez C, Riera M (2007) Advances in the management of effective arbuscular mycorrhizal symbiosis in tropical ecosystems. In: Hamel C, Plenchette C (eds) Mycorrhizae in crop production. Haworth, Binghampton, pp 151–196

Roldán A, Salinas-García JR, Alguacil MM, Caravaca F (2007) Soil sustainability indicators following conservation tillage practices under subtropical maize and bean crops. Soil Tillage Res 93:273–282

Ruiz-Lozano JM, Azcón R (1995) Hyphal contribution to water uptake in mycorrhizal plants as affected by the fungal species and water status. Physiol Plant 95:472–478

Rull RP, Gunier R, Von Behren J, Hertz A, Crouse V, Buffler PA, Reynolds P (2009) Residential proximity to agricultural pesticide applications and childhood acute lymphoblastic leukemia. Environ Res 109:891–899

Saif SR (1987) Growth responses of tropical forage plant species to vesicular-arbuscular mycorrhizae—I growth, mineral uptake and mycorrhizal dependency. Plant and Soil 97:25–35

Salami AO (2007) Assessment of AM biotechnology in improving agricultural productivity of nutrient-deficient soil in the tropics. Arch Phytopathol Plant Protect 40:338–344

Sharma MP, Adholeya A (2004) Effect of arbuscular mycorrhizal fungi and phosphorus fertilization on the post vitro growth and yield of micropropagated strawberry grown in a sandy loam soil. Can J Bot 82:322–328

Sheng M, Tang M, Chen H, Yang B, Zhang F, Huang Y (2008) Influence of arbuscular mycorrhizae on photosynthesis and water status of maize plants under salt stress. Mycorrhiza 18:287–296

Shrestha RK, Lal R (2010) Carbon and nitrogen pools in reclaimed land under forest and pasture ecosystems in Ohio, USA. Geoderma 157:196–205

Siasou E, Standing D, Killham K, Johnson D (2009) Mycorrhizal fungi increase biocontrol potential of *Pseudomonas fluorescens*. Soil Biol Biochem 41:1341–1343

Sierra J, Nygren P (2006) Transfer of N fixed by a legume tree to the associated grass in a tropical silvopastoral system. Soil Biol Biochem 38:1893–1903

Simard RR, Beauchenmin S (2002) Relationship between soil phosphorus content and phosphorus concentration in drainage water in two agroecosystems. Rev Sci Eau 15:109–120

Siqueira JO, Safir GR, Nair MG (1991) Stimulation of vesicular-arbuscular mycorrhiza formation and growth of white clover by flavonoid compounds. New Phytol 118:87–93

Smith S, Read D (1997) Mycorrhizal symbiosis, 2nd edn. Academic, New York

Snyder CS, Bruulsema TW, Jensen TL, Fixen PE (2009) Review of greenhouse gas emissions from crop production systems and fertilizer management effects. Agr Ecosyst Environ 133:247–266

Sood SG (2003) Chemotactic response of plant-growth-promoting bacteria towards roots of vesicular-arbuscular mycorrhizal tomato plants. FEMS Microbiol Ecol 45:219–227

Srivastava AK, Singh S, Marathe RA (2002) Organic citrus, soil fertility and plant nutrition. J Sustain Agric 19:5–29

St-Arnaud M, Vujanovic V (2007) Effects of the arbuscular mycorrhizal symbiosis on plant diseases and pests. In: Hamel C, Plenchette C (eds) Mycorrhizae in crop production. Haworth, Bighamton, pp 68–122

St-Arnaud M, Hamel C, Vimard B, Caron M, Fortin JA (1997) Inhibition of Fusarium oxysporum f. sp. dianthi in the non-VAM species dianthus caryophyllus by co-culture with Tagetes patula companion plants colonized by Glomus intraradices. Can J Bot 75:998–1005

Steinkellner S, Lendzemo V, Langer I, Schweiger P, Khaosaad T, Toussaint JP, Vierheilig H (2007) Flavonoids and strigolactones in root exudates as signals in symbiotic and pathogenic plant-fungus interactions. Molecules 12:1290–1306

Stevenson JF, Cole MA (1999) The cycles of soils, 2nd edn. Wiley, New York

Stewart LI, Hamel C, Hogue R, Moutoglis P (2005) Response of strawberry to inoculation with arbuscular mycorrhizal fungi under very high soil phosphorus conditions. Mycorrhiza 15:612–619

Su Y-Y, Guo L-D (2007) Arbuscular mycorrhizal fungi in non-grazed, restored and over-grazed grassland in the inner Mongolia steppe. Mycorrhiza 17:689–693

Subramanian KS, Charest C (1997) Nutritional, growth, and reproductive responses of maize (*Zea mays* L.) to arbuscular mycorrhizal inoculation during and after drought stress at tasselling. Mycorrhiza 7:25–32

Subramanian KS, Charest C (1999) Acquisition of N by external hyphae of an arbuscular mycorrhizal fungus and its impact on physiological responses in maize under drought-stressed and well-watered conditions. Mycorrhiza 9:69–75

Sudová R (2009) Different growth response of five co-existing stoloniferous plant species to inoculation with native arbuscular mycorrhizal fungi. Plant Ecol 204:135–143

Sukhoruchenko GI, Dolzhenko VI (2008) Problems of resistance development in arthropod pests of agricultural crops in Russia. OEPP/EPPO Bull 38:119–126

Takanishi I, Ohtomo R, Hyatsu M, Saito M (2009) Short-chain polyphosphate in arbuscular mycorrhizal roots colonized by *Glomus* spp.: a possible phosphate pool for host plant. Soil Biol Biochem 41:1571–1573

Takeda N, Sato S, Asamizu E, Tabata S, Parniske M (2009) Apoplastic plant subtilases support arbuscular mycorrhiza development in *Lotus japonicus*. Plant J 58:766–777

Tani C, Ohtomo R, Osaki M, Kuga Y, Ezawa T (2009) ATP-dependent but proton gradient-independent polyphosphatase-synthesizing activity in extraradical hyphae of an arbuscular mycorrhizal fungus. Appl Environ Microbiol 75:7044–7050

Tanu A, Prakash A, Adholeya A (2004) Effect of different organic manures/composts on the herbage and essential oil yield of *Cymbopogon winterianus* and their influence on the native AM population in a marginal alfisol. Bioresour Technol 92(3):311–319

Tester M, Langridge P (2010) Breeding technologies to increase crop production in a changing world. Science 327:818–822

Thorpe J, Wolfe SA, Houston B (2008) Potential impacts of climate change on grazing capacity of native grasslands in the Canadian prairies. Can J Soil Sci 88:595–609

Tian C, Kasiborski B, Koul R, Lammers PJ, Bucking H, Shachar-Hill Y (2010) Regulation of the nitrogen transfer pathway in the arbuscular mycorrhizal symbiosis: gene characterization and the coordination of expression with nitrogen flux. Plant Physiol 153:1175–1187

Tisdall JM (1991) Fungal hyphae and structural stability of soil. Aust J Soil Res 29:729–743

Tsai SM, Phillips DA (1991) Flavonoids released naturally from alfalfa promote development of symbiotic *Glomus* spores in vitro. Appl Environ Microbiol 57:1485–1488

Valarini PJ, Curaqueo G, Seguel A, Manzano K, Rubio R, Cornejo P, Borie F (2009) Effect of compost application on some properties of a volcanic soil from central South Chile. Chilean J Agric Res 69:416–425

van der Heijden MGA (2010) Mycorrhizal fungi reduce nutrient loss from model grassland ecosystems. Ecology 91:1163–1171

van der Heijden MGA, Horton TR (2009) Socialisms in soil? The importance of mycorrhizal fungal networks for facilitation in natural ecosystems. J Ecol 97:1139–1150

van der Heijden MGA, Klironomos JN, Ursic M, Moutoglis P, Streitwolf-Engel R, Boller T, Wiemken A, Sanders IR (1998) Mycorrhizal fungal diversity determines plant biodiversity, ecosystem variability and productivity. Nature 396:69–72

van der Heijden MGA, Bardgett RD, Van Straalen NM (2008) The unseen majority: soil microbes as drivers of plant diversity and productivity in terrestrial ecosystems. Ecol Lett 11:296–310

van der Plank JE (1975) Principles of plant infection. Academic, New York

van Groenigen KJ, Bloem J, Bååth E, Boeckx P, Rousk J, Bodé S, Forristal D, Jones MB (2010) Abundance, production and stabilization of microbial biomass under conventional and reduced tillage. Soil Biol Biochem 42:48–55

Van Zwieten L, Rust J, Kingston T, Merrington G, Morris S (2004) Influence of copper fungicide residues on occurrence of earthworms in avocado orchard soils. Sci Total Environ 329:29–41

Veeraswamy J, Padmavathi T, Venkateswartu K (1993) Effect of selected insecticides on plant growth and mycorrhizal development in sorghum. Agr Ecosyst Environ 43:337–343

Vieira RF, Silva CMMS, Silveira APD (2007) Soil microbial biomass C and symbiotic processes associated with soybean after sulfentrazone herbicide application. Plant and Soil 300:95–103

Wan MT, Rahe JE (1998) Impact of azadirachtin on *Glomus intraradices* and vesicular-arbuscular mycorrhiza in root inducing transferred DNA transformed roots of *Daucus carota*. Environ Toxicol Chem 17:2041–2050

Watkinson AR, Ormerod SJ (2001) Grasslands, grazing and biodiversity: editors' introduction. J Appl Ecol 38(2):233–237

White CM, Weil RR (2010) Forage radish and cereal rye cover crop effects on mycorrhizal fungus colonization of maize roots. Plant and Soil 328:507–521

Whiteside MD, Treseder KK, Atsatt PR (2009) The brighter side of soils: quantum dots track organic nitrogen through fungi and plants. Ecology 90:100–108

Wilson GWT, Williamson MM (2008) Topsin-M: the new benomyl for mycorrhizal-suppression experiments. Mycologia 100:548–554

Wortmann CS, Quincke JA, Drijber RA, Mamo M, Franti T (2008) Soil microbial community change and recovery after one-time tillage of continuous no-till. Agron J 100:1681–1686

Wright SF, Upadhyaya A (1996) Extraction of an abundant and unusual protein from soil and comparison with hyphal protein of arbuscular mycorrhizal fungi. Soil Sci 161:575–586

Wu QS, Xia RX, Zou YN, Wang GY (2007) Osmotic solute responses of mycorrhizal citrus (*Poncirus trifoliata*) seedlings to drought stress. Acta Physiol Plant 29:543–549

Xavier LJC, Germida JJ (2002) Response of lentil under controlled conditions to co-inoculation with arbuscular mycorrhizal fungi and rhizobia varying in efficacy. Soil Biol Biochem 34:181–188

Yang C, Hamel C, Schellenberg MP, Perez JC, Berbara RL (2010) Diversity and functionality of arbuscular mycorrhizal fungi in three plant communities in semiarid Grasslands National Park, Canada. Microb Ecol 59:724–733

Yau SK, Sidahmed M, Haidar M (2010) Conservation versus conventional tillage on performance of three different crops. Agron J 102:269–276

Ziadi N, Simard RR, Allard G, Lafond J (1999) Field evaluation of anion exchange membranes as a N soil testing method for grasslands. Can J Soil Sci 79:281–294

Zimmerman E, St-Arnaud M, Hijri M (2009) Sustainable agriculture and the multigenomic model: how advances in the genetics of arbuscular mycorrhizal fungi will change soil management practices. In: Bouarab K, Brisson N, Daayf F (eds) Molecular plant-microbe interactions. CABI, Wallingford, pp 269–287

Zocco D, Fontaine J, Lozanova E, Renard L, Bivort C, Durand R, Grandmougin-Ferjani A, Declerck S (2008) Effects of two sterol biosynthesis inhibitor fungicides (fenpropimorph and fenhexamid) on the development of an arbuscular mycorrhizal fungus. Mycol Res 112:592–601

Chapter 8
Application of Arbuscular Mycorrhizal Fungi in Production of Annual Oilseed Crops

Mahaveer P. Sharma, Sushil K. Sharma, R.D. Prasad, Kamal K. Pal, and Rinku Dey

8.1 Introduction

Oilseed crops are the second most in importance after cereals and significantly contribute to the Indian economy. Oilseeds cover about 13 % of the total arable land and generate nearly 10 % of the total value of the agricultural products in India (Singh et al. 2006). The country grows nine dominant oilseed crops, with groundnut (*Arachis hypogaea* L.), soybean (*Glycine max* L. Merrill) and rapeseed-mustard (*Brassica juncea* L.) accounting for 87 % and 75 % of total oilseed production and acreage, respectively (Agricultural Statistics at a Glance 2004). In India, soybean is the premier oilseed crop and growing parallel with groundnut followed by rapeseed-mustard. When compared to other countries, the productivity of these oilseeds per unit area is very low in India and their productivity is declining due to the recurrence of drought, low nutrient use efficiency of crop, nutrient deficiency in soil and other biotic and abiotic stresses.

Microbial interactions with plant roots may involve either endophyte or free living microorganisms and can be symbiotic, associative or casual in nature.

M.P. Sharma (✉)
Microbiology Section, ICAR-Directorate of Soybean Research (ICAR-DSR), Khandwa Road, Indore 452001, India
e-mail: mahaveer620@gmail.com

S.K. Sharma
NAIMCC, ICAR-National Bureau of Agriculturally Important
Microorganisms (ICAR-NBAIM), Kusmaur, Mau Nath Bhanjan, UP 275101, India

R.D. Prasad
ICAR-Directorate of Oilseeds Research (ICAR-DOR), Rajendra Nagar, Hyderabad 500 030, India

K.K. Pal • R. Dey
ICAR-Directorate of Groundnut Research (ICAR-DGR), Ivnagar Road, Junagadh 362 001, India

© Springer-Verlag Berlin Heidelberg 2014   119
Z.M. Solaiman et al. (eds.), *Mycorrhizal Fungi: Use in Sustainable Agriculture and Land Restoration*, Soil Biology 41, DOI 10.1007/978-3-662-45370-4_8

Beneficial symbionts include N_2-fixing bacteria (e.g. rhizobia) in association with legumes and interaction of roots with AM fungi, with the latter being particularly important in relation to plant P uptake (Richardson et al. 2009). Legume crops are generally cultivated in nutrient poor environments in India and have a high P requirement for nodule formation, nitrogen fixation and optimum growth. The mycorrhizal condition in legume crops increases vegetative growth and seed yield in addition to improving nodulation (Mathur and Vyas 2000).

During the past 50 years, the widespread use of chemical fertilisers to supply N and P has had a substantial impact on food production and has become a major input in crop production around the world (Tilman et al. 2002). However, further increases in N and P application are unlikely to be as effective at increasing yields (Wang et al. 2011) as only 30–50 % of applied N fertiliser and 10–45 % of P fertiliser are taken up by crops (Adesemoye and Kloepper 2009; Garnett et al. 2009). In addition, the abundant use of chemical fertilisers in agriculture has had some deleterious environmental consequences and is a global concern (Bohlool et al. 1992; Tilman et al. 2002).

The scientific community must look for alternate technologies which can play a major role in sustaining and increasing the productivity of oilseed crops. One approach could be the use of combinations of plant growth-promoting microorganisms (PGPMs) that can fix atmospheric nitrogen and solubilise or mobilise phosphorus, zinc and other soil nutrients to stimulate plant growth and improve soil health (Babalola 2010; Sharma et al. 2010).

The rhizosphere is the dynamic environment where much interaction takes place and AM fungi are important biotrophic plant associates. These fungi colonise the root cortex and develop an extrametrical mycelium which is a bridge connecting the roots with the surrounding soil microhabitats (Barea et al. 2005). They are obligate symbionts and require a host plant to complete their life cycle (Wardle et al. 2004). AM fungi form a symbiotic association with most agricultural crops and are able to increase plant nutrition and plant health (Jansa et al. 2009). In addition, AM fungi establishment in the root causes changes in the microbial community of the rhizosphere (Meyer and Linderman 1986; Marschner et al. 2001) and increases plant tolerance to a wide range of biotic and abiotic stresses (Auge et al. 2004; Whipps 2004; Jansa et al. 2009). Many studies have demonstrated on field crops, including oilseeds, the benefits of AM inoculation on plant nutrition (Cardoso and Kuyper 2006; Hamel and Strullu 2006), nodulation (Meghvansia et al. 2008; Aryal et al. 2006), N-fixation (Peoples and Craswell 1992) and plant protection (Whipps 2004; Doley and Jite 2013a, b) under ideal conditions.

Certain cooperative microbial activities involving plant growth-promoting microorganisms can be exploited as a low-input biotechnology and form a basis for a strategy to help sustainable, environment-friendly practices fundamental to the stability and productivity of agricultural ecosystems (Kennedy and Smith 1995). The purpose of this review is to discuss (i) the current status of major oilseed crops in India; (ii) the application of AM fungi (single and dual inoculation) in the plant growth, nutrition and control of soil-borne diseases associated with major oil seeds; (iii) the strategies of manipulating soil and agricultural practices to manage indigenous AM fungi and quality performance and (iv) commercialisation possibilities of AM fungi.

8.2 Oilseed Crops of Global Importance

About one-third of the land area of the world comprises arid and semiarid climates. The increasing economic and agricultural utilisation of arid lands has emerged as a critical element in maintaining and improving the world's food supply (Zahran 1999). India plays a major role in global oilseeds and vegetable oil economy contributing about 15 % of the world's oil crops area of nine oilseeds (groundnut, soybean, rapeseed, mustard, sesame, sunflower, linseed, safflower and castor), 7 % of the world's oilseeds production and 6.7 % of vegetable oils production. However, the productivity in India is only 1,005 kg/ha as compared to the world average of 1,957 kg/ha (FAOSTAT). India has the largest area in groundnut, sesame, safflower and castor and ranks first in production of safflower, castor and sesame and ranked second in groundnut, third in rapeseed, fourth in linseed, fifth in soybean and tenth in sunflower (Table 8.1). In the domestic agricultural sector, oilseeds occupy a distinct position after cereals sharing 14 % of the country's gross cropped area and accounting for nearly 1.5 % of the gross national product and 7 % of the value of all agricultural products. India encompasses diverse agro-ecological conditions ideally suited for growing nine annual oilseed crops including groundnut, rapeseed-mustard, sunflower, sesame, soybean, safflower, castor, linseed and niger and two perennial oilseed crops (coconut and oil palm) and secondary oil crops such as maize and cotton. In addition to the above, more than 100 tree species of forest origin that have the potential to yield about one million tonnes of vegetable oil are grown in the country.

8.3 AM Fungi in the Production of Oilseed Crops

8.3.1 AM Fungi Inoculation Responses for Enhanced Growth and Nutrient Uptake

AM fungi are the most common type of association involved in agricultural systems. AM fungi are associated with improved growth of many plant species due to increased nutrient uptake, production of growth-promoting substances, induced tolerance to drought, salinity and transplant shock and synergistic interaction with other beneficial soil microorganisms such as N-fixers and P-solubilisers (Sreenivasa and Bagyaraj 1989). Symbiotic associations of plant roots with AM fungi can result in enhanced growth because of increased acquisition of P and nutrients with low mobility in soil. Effective nutrient acquisition by AM fungi is generally attributed to the extensive hyphal growth beyond the nutrient depletion

Table 8.1 Area, production and yield of oilseeds: global and Indian scenario (AICRPS 2009; Damodaram and Hegde 2010)

Oilseeds	Area ('000 ha)		Yield (kg/ha)		Production ('000 MT)		India's position in the world		Percent area to total oilseed area in India	Percent production to total oilseeds production in India
	World	India	World	India	World	India	Area	Production		
Groundnut	24,590	6,850	1,554	1,071	38,201	7,338	First	Second	22.37	25.86
Soybean	96,870	9,600	2,384	942	23,095	9,045	Fifth	Fifth	34.51	35.74
Safflower	691.44	350	890	643	615.21	225	First	First	1.07	0.68
Castor	1,524.7	880	1,037	1,276	1,580.6	1,123	First	First	3.14	4.22
Sunflower	25,024	2,050	1,424	542	35,643	1,112	Fourth	Tenth	6.58	4.18
Rapeseed and mustard	30,987	5,750	1,883	1,014	58,364	5,833	Third	Third	22.85	25.98
Linseed	2,437	550	903	296	2,200	163	Second	Fourth	1.48	0.61
Sesame	7,534	1,750	478	381	3,603	666	First	First	6.56	2.31

zone surrounding the root (Tisdale et al. 1995). In this way, AM fungi enable their host plants to gather mineral nutrients from a much larger volume of soil than the roots could reach on their own (Jansa et al. 2009).

8.3.1.1 AM Responses Under Glass House/Nursery and Field Conditions

AM fungi responses vary with AM fungal species used, soil pH, experimental conditions (Clark and Zeto 1996), root-geometry/architecture of the host plant which influences the nutrient uptake particularly soil supply of P and soil temperature (Raju et al. 1990). For example, in soybean, manganese (Mn) and iron (Fe), protection was more efficient when the plants were inoculated with *Glomus macrocarpum* than with *Glomus etunicatum*, whereas *Gigaspora margarita* was not effective with the inocula used (Cardoso and Kuyper 2006). Jalaluddin et al. (2008) found that the AM fungus *Scutellospora auriglobosa* increased the uptake of P in sunflower resulting in increased yield and reduced incidence of *Macrophomina phaseolina* which causes charcoal root rot in sunflower var. Helico-250 cultivated in Sindh region of Pakistan. Wang et al. (2011) while examining the tripartite symbiotic associations with rhizobia and AM fungi and correlating their relationships to root architecture as well as N and P availability of two soybean genotypes contrasting in root architecture grown in a field showed variable responses to AM fungi. The deep root soybean genotype had greater AM fungi colonisation at low P, but better nodulation with high P supply than the shallow root genotype. Co-inoculation with rhizobia and AM fungi significantly increased soybean growth under low P and/or low N conditions as indicated by increased shoot dry weight, along with plant N and P content. Moreover, the effects of co-inoculation were related to root architecture. The deep root genotype (HN112) benefited more from co-inoculation than the shallow root genotype (HN89).

AM fungal inoculation has been shown to reduce Mn and Fe toxicity in plants, and the concentration of Mn in shoots and roots of mycorrhizal plants can be lower than in non-mycorrhizal plants (Kothari et al. 1991; Nogueira et al. 2004). Mycorrhizal soybeans grew better and had lower shoot concentrations of Fe and Mn than did non-mycorrhizal soybeans under greenhouse conditions. In roots, the results were the same for Mn and the reverse for Fe. The decrease of Mn in shoots was attributed to reduced availability, while the decrease of Fe in the shoots was attributed to its retention in the roots. In excess, both Mn and Fe can be toxic to plants; thus, mycorrhizas may protect the plants from their toxicity (Nogueira et al. 2004).

Under field conditions, AM fungal inoculation enhanced biomass, nutrient uptake and yield of sesame applied with conventional P fertiliser (superphosphate) and slow release P source (rock phosphate) (Anil-Prakash and Tandon 2002). The influence of AM fungus on P and Fe uptake of mycorrhizal groundnut (*Arachis hypogaea* L.) and sorghum (*Sorghum bicolor* L.) plants was studied by Caris et al. (1998) using radiolabelled elements (^{32}P, ^{59}Fe). Plants possessing different strategies for the acquisition of Fe (Marschner 1995) were selected for this experiment. Groundnut is dicotyledonous and is a strategy I plant (Fe-deficiency response: enhanced net excretion of protons from the roots, increased Fe-reducing capacity), while sorghum is monocotyledonous (graminaceous) and is a strategy II plant (Fe-deficiency response: enhanced release of phytosiderophores from the roots). In both plant species, P uptake from the labelled soil increased more in shoots of mycorrhizal plants than in non-mycorrhizal plants. Mycorrhizal inoculation had no significant influence on the concentration of labelled Fe in shoots of peanut plants. In contrast, ^{59}Fe increased in shoots of mycorrhizal sorghum plants. The uptake of Fe from labelled soil by sorghum was particularly high under conditions producing a low Fe nutritional status of the plants providing evidence that hyphae of an AM fungus can mobilise and/or take up Fe from soil and translocate it to the plant.

Meghvansia et al. (2008) reported variations in efficacy of different treatments (involving AM fungal species and cultivar-specific bradyrhizobia) with different soybean cultivars indicating the specificity of the inoculation response. This provides a basis for selection of an appropriate combination of specific AM fungi and *Bradyrhizobium* which could further be utilised for verifying the symbiotic effectiveness and competitive ability of microsymbionts under particular agro-climatic conditions. Inoculation response of single or mixed species of AM fungi to soybean has shown enhanced growth, mineral nutrition and nutrient uptake (Sharma et al. 2012a, b; Ilbas and Sahin 2005; Meghvansia et al. 2008; Waceke 2003; Sanginga et al. 1999). The role of mycorrhiza-mediated *Rhizobium* symbiosis on soybean showed enhanced production of soybean under field conditions (Antunes et al. 2006). Synergistic effects of AM fungi and *B. japonicum* have a high potential to improve the nutrient supply of soybean including P and soil quality (Meghvansia et al. 2008). However, a much larger genetic variability of bradyrhizobia and AM fungi strains exist in different cultivar regions than was assumed previously (Taiwo and Adegbite 2001). Soybean can form tripartite symbiotic associations with nodule-inducing rhizobia and AM fungi, which may benefit both P and N efficiency (Lisette et al. 2003). Co-inoculation of soybean roots with *B. japonicum* 61-A-101 considerably enhanced colonisation by the AM fungus *Glomus mosseae* and increased N and P uptake (Xie et al. 1995). El-Azouni et al. (2008) studied the associative effect of AM fungi with *Bradyrhizobium* as biofertilisers on growth and nutrient uptake of *Arachis hypogaea*. The biomass and grain yield were significantly improved by using the dual bio-preparations of AM fungi and *Bradyrhizobium*. The bacterial mycorrhizal-legume symbiosis increased nodule number, nitrogenase activity, total pigments and carbohydrate, protein and lipid content. The N, P and K uptake was significantly increased due to the single or dual inoculation. Moreover, inoculation with AM fungi and *Rhizobium* enhanced nodulation and yield of groundnut applied with inorganic P fertiliser (Mandhare et al. 1995; Lekberg and Koide 2005) and organic amendments (Iyer et al. 2003).

Mostafavian et al. (2008) showed that besides *Rhizobium*, inoculation of AM fungi with *Thiobacillus* increased soybean yield. Jackson and Mason (1984) found positive relationships among P availability, mycorrhizal colonisation and pod yield in groundnut (*Arachis hypogaea* L.). Mirzakhani et al. (2009) indicated that seed yield and yield components of safflower were influenced by inoculation with *Azotobacter* and AM fungi. They showed that inoculation of seeds with

Table 8.2 Examples of AM fungi responses (applied singly or combined) to enhance growth and mineral nutrition of major oilseed crops

AM fungi species	Interaction/significant treatments	Crop	References
G. fasciculatum	Phosphorus levels	Soybean	Ilbas and Sahin (2005)
Indigenous *Glomus* sp.	Crop rotation and *Rhizobium*	Soybean	Sanginga et al. (1999)
Mixed AM fungi	Conventional, GM soybean and *Bradyrhizobium* sp.	Soybean	Powell et al. (2007)
G. mosseae	Root architecture and *Bradyrhizobium* sp.	Soybean	Wang et al. (2011)
Glomus intraradices, Acaulospora tuberculata *Gigaspora gigantea*	*Bradyrhizobium japonicum*	Soybean	Meghvansia et al. (2008)
Glomus fasciculatum	*Pseudomonas striata*, P sources	Soybean	Mahanta and Rai (2008)
Glomus etunicatum	Salt stress	Soybean	Sharifia et al. (2007)
Glomus intraradices	Glyphosate, *Bradyrhizobium japonicum*	Soybean	Powell et al. (2009)
G. fasciculatum	*Pseudomonas striata*, Rock phosphate	Soybean	Mahanta and Rai (2008)
Glomus mosseae, Glomus etunicatum, Gigaspora rosea	Phosphatic fertilisers	Soybean	Bethlenflavay et al. (1997)
G. mosseae	Heavy metals, phosphatic fertilisers	Soybean	Dev et al. (1997)
G. mosseae	*Bradyrhizobium japonicum*	Soybean	Shalaby and Hanna (2000)
G. intraradices	Phosphorus application	Groundnut	Lekberg and Koide (2005)
Glomus caledonium	Salt stress	Groundnut	Gupta and Krishnamurthy (1996)
G. fasciculatum	*Rhizobium* and phosphatic fertilisers	Groundnut	Mandhare et al. (1995)
G. fasciculatum	Phosphatic fertilisers	Groundnut	Singh and Chaudhari (1996)
Glomus sp.	*Bradyrhizobium*	Groundnut	Elsheikh and Mohamedzein (1998)
G. intraradices	*Azotobacter chroococcum*	Safflower	Mirzakhani et al. (2009)
G. intraradices	*Azotobacter chroococcum*	Sunflower	Mirzakhani et al. (2009)
Glomus mosseae, Glomus intraradices	Heavy metals	Sunflower	Adewole et al. (2010)

(continued)

Table 8.2 (continued)

AM fungi species	Interaction/significant treatments	Crop	References
G. fasciculatum	Phosphorus levels	Sunflower	Chandrashekara et al. (1995)
G. fasciculatum	–	Linseed, niger	Srinivasulu and Lakshman (2002)
AM fungi	Rock phosphate	Sesame	Anil-Prakash and Vandana (2002)
G. fasciculatum	–	Castor	Sulochana and Manoharachary (1989)
G. constrictum			
Gigaspora sp.			

Azotobacter and AM fungi (*G. intraradices*) at the time of planting increased the grain yield of safflower to about 38 % over control plants. Groundnut is an important food legume of Egypt, and to enhance the production of groundnut, new reclaimed soils were brought under cultivation. The lack of indigenous soil populations of AM fungi and rhizobia has restricted potential yields of groundnut cultivated in this area. A summary of AM fungi inoculation responses for enhanced growth and nutrient uptake is stated (Table 8.2).

8.3.2 AM Fungi Responses in the Stressed Environments (Drought, Heavy Metals and Salinity)

AM fungal responses have also been encouraging in stressed environments like acid/salt (Gupta and Krishnamurthy 1996; Sharifia et al. 2007), drought (Ruiz-Lozano 2003; Auge et al. 2004; Manoharan et al. 2010; Liu et al. 2007), heavy metals (Göhre and Paszkowski 2006; Nogueira et al. 2004) and modified micro-environmental conditions such as genetically modified soybean (Powell et al. 2009). AM fungi have also been observed to play a role in metal tolerance (Del Val et al. 1999) and accumulation (Zhu et al. 2001; Jamal et al. 2002). For example, groundnut is a major cash crop in the semiarid tropics where it is mainly grown under rainfed conditions. Poor soil fertility, drought and diseases are important factors causing low yields. Groundnut forms symbiotic associations with two types of soil microorganisms, one with *Bradyrhizobium* and another with AM fungi. The positive effect of AM fungi on plant growth and development make mycorrhiza a potentially very useful biological resource of assuring high plant productivity, with minimum application of chemical fertilisers or pesticides. Quilambo (2002) studied the effects of two AM inoculants on root colonisation, leaf growth and dry matter accumulation and distribution in two groundnut cultivars: Local and Falcon. The indigenous Soil Mozambique inoculants significantly increased root colonisation, leaf growth and dry matter in both cultivars under drought stress conditions. The commercial Hannover inoculant increased growth only under well-watered conditions. Drought stress effects could be alleviated by inoculation with Soil Mozambique inoculants. Therefore, peanut productivity, particularly under drought stress, may be improved by an adequate management of the AM symbiosis.

Most studies conducted on sunflower indicate that besides growth promotion, mycorrhizal colonisation of sunflower enhanced the ability to store more heavy metals in the roots. Adewole et al. (2010) found that AM inoculation to sunflower increased pollution tolerance to cadmium (Cd) and lead (Pb) and consequently increased the yield of sunflower. External mycelium of AM fungi provides a wider exploration zone (Khan et al. 2000; Malcova et al. 2003), thus providing access to greater volume of heavy metals present in the rhizosphere. However, the effectiveness of AM fungal isolates in improving plant growth also depends on the level of heavy metals in soil (Awotoye et al. 2009). Del Val et al. (1999) reported six AM fungal ecotypes showing consistent differences with regard to their tolerance to the presence of metals in soil. AM fungi may play a role in the protection of roots from heavy metal toxicity by mediating interactions between metals and plant roots (Leyval et al. 1997). Contaminated soils, which are often nutrient poor with low water-holding capacities, may provide an advantage to plants colonised by AM fungi by enabling them to act as pioneering species (Khan et al. 2000). Wu et al. (2004) used an intercropping system to examine the interactions of mycorrhizosphere and rhizosphere on metal uptake by growing mycorrhizal non-hyperaccumulator *Zea mays* and non-mycorrhizal hyperaccumulator *Brassica juncea* in a split-pot experiment. The intercropping system achieved higher phytoremediation efficiency in metal-contaminated soil, especially with dual inoculation of beneficial rhizobacteria and AM fungi. Similar studies were conducted by Zhang et al. (2004) who grew groundnut (leguminous crop) and maize (nonleguminous crop) and found that the iron-deficient maize released phytosiderophores which improved iron nutrition of groundnut through influencing its rhizosphere processes.

Among the biological approaches to enhance plant growth in saline conditions, the role of AM fungi is well established. Most native plants and crops of arid and semiarid areas are mycorrhizal, and it has been suggested that AM fungal colonisation might enhance salt tolerance of some plants (Tain et al. 2004). Under salt (base and acid) stress conditions, AM fungi response in terms of yield on groundnut was almost tripled in mycorrhizal plants compared with non-mycorrhizal control plants. Furthermore, they showed that AM inoculation promoted the establishment of groundnut plants under acid stress conditions (Gupta and Krishnamurthy 1996). Therefore, the additional beneficial effects of AM fungi in reducing salinity stress imposed on them (*Arachis hypogaea* var. *hypogaea* cv. Florunner) were studied by Al-Khaliel (2010) to understand the growth and physiological changes of groundnut plants under induced saline conditions. These investigations indicated that the AM fungi (*Glomus mosseae*) could improve growth of groundnut under salinity through enhanced nutrient absorption and photosynthesis. Chlorophyll content and leaf water content increased significantly under salinity stress by the inoculation with mycorrhizal fungi.

8.3.3 AM Fungi Inoculation Responses on the Control of Soil-Borne Diseases and Other Plant Pathogens

8.3.3.1 Influence of AM Fungi on Soil-Borne Diseases

The potential for AM fungi to suppress root diseases caused by soil-borne pathogens (Dehne 1982; Linderman 1994) has been intensively studied. *Sclerotium rolfsii* is an important soil-borne pathogen and causes disease in numerous crops including groundnut. The loss of yield caused by pathogen infection generally is 25 %, but it can be as high as 80–90 % (Grichar and Bosweel 1987). AM fungi have been shown to influence fungal diseases caused by root pathogens (Karagiannidis et al. 2002). Most studies concluded that disease severity could be reduced by root colonisation of AM fungi through several mechanisms including increasing the mineral absorption and plant growth (Smith and Read 1997), phenolic compounds (Devi and Reddy 2002) and pathogenesis-related proteins (Pozo et al. 1999). Ozgonen et al. (2010) studied the effects of AM fungi against stem rot caused by *Sclerotium rolfsii* Sacc. in groundnut. In field trials, the effect on disease locus of AM fungi ranged between 30 and 47 % with AM fungi differing in their benefit.

Disease and poor soils are considered to be the main causes of loss in the groundnut production. Rosette virus disease (RVD) and *Cercospora* leaf spots (CLS) are the major worldwide diseases that infect groundnut plants. In Cameroon, up to 53 % loss has been estimated (Fontem et al. 1996). CLS are caused by *Cercospora arachidicola* Hori (early leaf spot) and *Cercosporidium personatum* (Berk. and Curt.) Deighton (late leaf spot). Depending on the moment of contamination during the growing season, groundnut plants infected by RVD do not produce pods and, consequently, do not give any harvest (Savary 1991). Management against phytoviruses is very difficult because viral infection can be transmitted through seeds and also through some insect vectors (*Aphis* sp.). Strullu et al. (1991) showed that the symbiosis between mycorrhizas and roots of many crops has a positive influence on the plant's nutrition and in protection against some diseases. Zachee et al. (2008) determined the effect of mycorrhizal inoculation on the development of diseases (RVD and CLS) and on the physiology of groundnut plants (variety A-26) infected by RVD. A urea treatment and an absolute control were also used. It was observed that root colonisation rate was very low in control and urea plots compared to mycorrhiza-inoculated plots. Mycorrhizal applications reduced disease infection by almost 40 % and 54 %, respectively, for RVD and CLS. It was evident that mycorrhizal symbiosis with groundnut roots increased the resistance of plants to RVD and CLS and positively influenced the physiology of groundnut plants infected by RVD.

Fungal root pathogens can be reduced in crops by AM inoculation (Caron et al. 1986), including *Phytophthora* species (Davis and Menge 1980; Cordier et al. 1996), *Rhizoctonia solani* (Yao et al. 2002) and *Pythium ultimum* (Calvet et al. 1993). Bacterial diseases may also be reduced by mycorrhiza establishment on roots (Dehne 1982). Evidence of the suppression of nematode penetration and development following AM fungi inoculation has been reported by many workers (Elsen et al. 2001; Diedhiou et al. 2003). Harrier and Watson (2004) illustrated the role of AM fungi in organic and/or sustainable farming systems that rely on biological processes rather than agrochemicals to control plant diseases. However, the mechanisms by which AM fungi colonisation confer the protective effect are not well understood. Bio-protection within AM fungal-colonised plants is the outcome of complex interactions between plants, pathogens and AM fungi. These interactions have been shown to result in reductions in disease incidence (Matsubara et al. 2001), pathogen development (Cordier et al. 1996) and disease severity (Matsubara et al. 2001). The extent of AM fungi-induced protection of host plants against pathogens suppression ranges from complete protection (Torres-Barragan et al. 1996) to partial protection (Matsubara et al. 2001). The extent of partial protection is influenced by the AM fungal species and cultivar used (Yao et al. 2002). Information related to oilseed crops is summarised in Table 8.3. Effects may relate to direct interaction between mutualists and pathogens (Abdalla and Abdel-Fattah 2000), competition for infection sites (Abdel-Fattah and Shabanam 2002) and improved nutrition of AM fungi plants which offset the damage caused by the pathogen involved (masking effect). Inoculation with soil-based mixture of AM fungi (*Glomus fasciculatum*) decreased incidence of disease caused by *Macrophomina phaseolina* (Tassi) in groundnut and increased growth and

Table 8.3 Examples of AM fungi application providing protection to oilseed crops against soil-borne diseases and other plant pathogens

AM fungi	Pathogen	Plant	References
G. mosseae	*Rhizoctonia solani*	Groundnut	Abdalla and Abdel-Fattah (2000)
Glomus sp. *Gigaspora* sp.	Rosette virus disease (RVD), *Cercospora* leaf spot (CLS)	Groundnut	Zachee et al. (2008)
G. intraradices	*Fusarium oxysporum* f. sp. *lini*	Linseed	Dugassa et al. (1996)
G. mosseae	*Fusarium solani*	Groundnut	Abdalla and Abdel-Fattah (2000)
G. mosseae	*Macrophomina phaseolina, Rhizoctonia solani, Fusarium solani*	Soybean	Zambolim and Schenck (1983)
G. fasciculatum	*Sclerotium rolfsii*	Groundnut	Krishna and Bagyaraj (1982)
AM fungi	*Meloidogyne arenaria*		Carling et al. (1995)
AM fungi	*Meloidogyne incognita*	Soybean	Kellam and Schenck (1980)
Glomus sp., *Gigaspora* sp.	*Heterodera glycines*	Soybean	Tylka et al. (1991)
G. intraradices	*H. glycines*	Soybean	Price et al. (1995)
G. mosseae	*H. glycines*	Soybean	Todd et al. (2001)

production of defence-related enzymes (Doley and Jite 2013a, b). The various defence-related biochemical parameters such as protein, proline, total phenol, total chlorophyll content, acid and alkaline phosphatase activity, peroxidase and polyphenol activity showed marked increase in their content or activity in mycorrhizal healthy or diseased plants in comparison to non-mycorrhizal diseased or control ones (Doley and Jite 2013a, b). Zambolim and Schenck (1983) reported that *Glomus mosseae* reduced the influence of *Macrophomina phaseolina* (Tassi.), *Rhizoctonia solani* (Kuhn.) and *Fusarium solani* (Mart.) in soybean. The suppression of endoparasitic nematodes by AM fungi has been recently reported by many workers (Habte et al. 1999). Several mechanisms have been proposed to explain the nematode suppression by AM fungi (Pinochet et al. 1996). Carling et al. (1995) observed the individual and combined effects of two AM fungal species, *Meloidogyne arenaria* and P fertilisation on groundnut plant growth and pod yield. They found that the groundnut growth and yield were generally stimulated by AM fungi, which increased groundnut plant tolerance to the nematode and offset the growth reductions caused by *M. arenaria* at the two lower P levels. Price et al. (1995) investigated the effects of the AM fungi, *Glomus intraradices*, on the soybean cyst nematode (SCN), *Heterodera glycines*, on two soybean cultivars, cv. "Bragg" (nematode intolerant) and cv. "Wright" (moderately nematode tolerant) grown in the greenhouse in soils with low (35 µg/g) and high (70 µg/g) P. They found variable AM responses to cultivar. The cultivar "Wright" was more responsive than "Bragg" and exhibited greater nematode tolerance. Dugassa et al. (1996) demonstrated the effects of AM fungi on the health of *Linum usitatissimum* infected with wilt (*Fusarium oxysporum* f. sp. *lini*) and AM fungi showed increased resistance against the wilt pathogen; the level of these effects depended on the plant cultivars which all showed the same level of root colonisation by AM fungi.

8.3.3.2 Interaction Between AM Fungi and Other Plant Growth-Promoting Rhizobacteria (PGPR) Leading to Inhibition of Fungal Pathogens

Rhizosphere microorganisms can affect presymbiotic phases of mycorrhiza development (Barea et al. 1998). The bacteria have been found adhering to the AM fungi hyphae (Bianciotto et al. 1996) and as well as embedded within the spore walls (Walley and Germida 1996). Bacteria adhering to AM fungal mycelium may utilise hyphal exudates or use mycelium as vehicle for colonisation of rhizosphere (Bianciotto et al. 1996). Bacteria from genus *Paenibacillus*, which are antagonistic to a broad range of root pathogens and are able to stimulate mycorrhizal colonisation, were found frequently to be associated with *Glomus intraradices* mycelium (Mansfeld-Giese et al. 2002). Therefore, it should be mandatory to detect the cohesiveness of both AM fungi and PGPR participating in a particular rhizosphere while maintaining the healthy rhizosphere. The key step is to ascertain whether an antifungal biocontrol agent will negatively affect the AM fungi populations. Several studies have demonstrated that microbial antagonists of fungal pathogens, either fungi or PGPR, do not exert antimicrobial effect against AM fungi (Barea et al. 1998). There is a need to exploit the possibilities of dual (AM fungi and PGPR) inoculation to provide plant defence against root pathogens (Barea et al. 2005). Barea et al. (1998) conducted a series of experiments to evaluate the effect of *Pseudomonas* strains producing 2, 4-diacetylphloroglucinol (DAPG) on AM fungi formation and functioning. Three *Pseudomonas* strains producing DAPG were tested under in vitro and in situ for their effects on AM fungi; it was found that there was no negative impact on AM spore germination. Rather, there was stimulation of hyphal growth of *G. mosseae*. Under field conditions, none of the *Pseudomonas* strains affected the diversity of native AM fungi in the rhizosphere soil, root colonisation and AM functional symbiosis and rather improved plant growth and nutrient (N and P) acquisition by AM-mediated plants (Barea et al. 1998). Sanchez et al. (2004) showed that a fluorescent pseudomonad and *G. mosseae* had similar impacts on plant gene induction, supporting the hypothesis that some plant cell programmes may be shared during root colonisation by these beneficial microorganisms. Gram-positive and gamma-proteobacteria are more frequently associated with AM fungi than are gram-negative bacteria (Table 8.4), but their synergistic interaction is yet to be confirmed (Artursson et al. 2005).

Table 8.4 Examples of synergistic interactions between AM fungi and bacteria or PGPR leading to inhibition of fungal pathogens

Bacterial species	AM fungi species	Interaction effect	Inhibition of fungal pathogen	References
Bacillus pabuli	*Glomus clarum*	+	ND	Xavier and Germida (2003)
B. subtilis	*G. intraradices*	+	ND	Toro et al. (1997)
Paenibacillus validus	*G. intraradices*	+	ND	Hildebrandt et al. (2002)
Paenibacillus sp.	*G. mosseae*	+	+	Budi et al. (1999)
Paenibacillus sp.	*G. intraradices*	+	ND	Mansfeld-Giese et al. (2002)
Pseudomonas sp.	*G. versiformis*	+	ND	Mayo et al. (1986)
Pseudomonas sp.	*G. mosseae*	+	+	Barea et al. (1998)
Pseudomonas putida	Indigenous mixed AM fungi	+	ND	Meyer and Linderman (1986)
P. fluorescens	*G. mosseae*	+	+	Edwards et al. (1998)

Modified from Artursson et al. (2006)
+ positive, *ND* not determined

8.3.4   Soil and Agricultural Management Practices Influencing AM Fungi Response

To benefit from mycorrhizal associations (or more generally beneficial biological processes in the rhizosphere), emphasis has to be on agricultural practices that promote the occurrence and functioning of soil organisms, including AM fungi. The low host specificity of AM fungi may allow mycelial networks of a particular fungus in the soil to be connected directly to roots of plants of different species, forming hyphal links between their mycorrhizal roots. It has been shown that in fragile tropical agroecosystems, conventional agriculture, relying on tillage and external inputs (mineral fertilisers, biocides) for increase of productivity, may result in large ecological disturbances and may not be sustainable in the long term. Most of the cultivated plant species are able to form the mycorrhizas. However, the plant families *Brassicaceae* and *Chenopodiaceae* include species that do not usually form mycorrhizal symbiosis; among them, sugar beet and rape (Tester et al. 1987) are important. Growing these crops subsequently does not lead to multiplication of AM fungi, unless there are weeds that can act as hosts (Abbott and Robson 1991; Jansa et al. 2002).

8.3.4.1   Fertilisers, Manures, Fungicides and Tillage Practices

Application of farmyard manure can increase densities of AM fungal spores, although this depends on the soil types (Harinikumar and Bagyaraj 1989). Several studies indicated that cumulative P fertilisation decreases the spore density under Northern European field conditions (Martensson and Carlegren 1994; Kahiluoto et al. 2001). Another study showed that AM fungal colonisation was not affected by P addition when plants were deficient in N, but, when N was sufficient, P addition suppressed root colonisation (Sylvia and Neal 1990). Thus, there are agronomic soil management practices available for the farmer to regulate the AM fungi at the field site. An important measure, apart from the choice of cropping systems in conventional agriculture, is the use of fungicides particularly systemic fungicides applied in the field has shown to reduce the functioning of the AM fungi (Menge et al. 1978; Kling and Jakobsen 1997). AM fungi can be sensitive to certain but not all fungicides. Mancozeb, thiram and ziram are all dithiocarbamates and, as a group, appear to be deleterious to mycorrhizal fungi, at least when tested in groundnut (Sugavanam et al. 1994). Emisan (a mercuric treatment) and carbendazim (a benzimidazole) were both negative for AM fungi when tested in groundnut. Copper, however, appeared to provide a stimulus to mycorrhizae in groundnut. Application of fungicide to soil reduced sporulation and the root length colonised by AM fungi, although interaction of AM fungi and fungicide was observed to be highly variable depending on fungus-fungicide combination and on environmental conditions (Turk et al. 2006).

Fungicide seed treatments alter the microbial population dynamics in the rhizosphere by reducing root pathogen infection but may also affect nontarget organisms (Rodriguez-Kabana and Curl 1980; Trappe et al. 1984). Soil applications of metalaxyl have been reported to favour AM colonisation in corn and soybeans (Groth and Martinson 1983). Seed-applied captan had no effect on AM colonisation in studies conducted by Kucey and Bonetti (1988), and it reduced symptoms of *Fusarium solani* when applied along with AM inoculum in *Phaseolus vulgaris* plants (Gonçalves et al. 1991). Other fungicides such as benomyl, captan, pentachloronitrobenzene and emisan have been reported to also have negative effects on AM colonisation when applied as soil drenches (Kjoller and Rosendahl 2000; Schreiner and Bethlenfalvay 1997; Sugavanam et al. 1994). Murillo-Williams and Pedersen (2008) showed that under natural pathogen inoculum (non-fumigated soil), seed-applied fungicides with fludioxonil seemed to favour AM colonisation due to a reduced competition with aggressive pathogens like *Rhizoctonia* spp., an organism that is targeted by this fungicide.

Function of AM fungi and species composition may also be affected by farming systems. This is evidenced from a long-term field trial established in Switzerland designed to compare long-term effects of "conventional" vs. "organic" farming systems (Mäder et al. 2002). In this trial, about 40 % more roots were colonised by AM fungi in the organic systems than in the conventional system (Mäder et al. 2000). They suggested that AM fungal species differ in functional characteristics such as spore production and plant growth promotion (Van der Heijden et al. 1998). Moreover, less efficient AM fungal species might be selected by high-input farming (Scullion et al. 1998). Tillage affects the mycorrhizal hyphal network (Cardoso and Kuyper 2006). Mulligan et al. (1985) observed that excessive secondary tillage reduced AM colonisation of *Phaseolus vulgaris* L. Mycorrhizal root colonisation of corn growing in NT (no-tilled) and ridge till plots was greater than that in CT (conventional-tilled) plots (McGonigle and Miller 1993). AM hyphae and spores were more abundant in the top 0- to 15-cm layer of the soil profile and decreased dramatically below this depth (Kabir 2005). Similar results were reported for AM spores by An et al. (1990) in Kentucky, USA, under soybean. This suggests that tilling the soil to a depth of 15 cm would affect most of the AM fungi and that ploughing below this depth would dilute the AM propagules in the zone of seedling establishment (Kabir 2005). The role of glomalin in soil aggregation (Rillig 2004) was correlated with stabilisation of soil aggregates after a 3-year transition of a maize cropping system from conventional tillage to no tillage (Wright et al. 1999), and there are indications that some crop rotations favour glomalin production and aggregate stabilisation more than others (Wright and Anderson 2000). Thus, management of cropping systems to enhance soil stability and reduce erosion may benefit from consideration of the factors controlling production and maintenance of extraradical hyphae and glomalin (Cardoso and Kuyper 2006).

8.3.4.2 Crop Rotation and Sequences

AM fungi show only a limited degree of specificity; different plant species stimulate the amount and occurrence of different species of AM fungi; thus, through the management of plants, it is possible to modify mycorrhizal populations in the soil (Colozzi and Cardoso 2000; Hart et al. 2001). Mycorrhizal inoculum density declines when soils are kept fallow for extensive periods of time (Thompson 1987). The quantity of AM fungi in soils also differs between host species (Vivekanandan and Fixen 1991). Even the preceding crop in a crop rotation system affects the AM fungal spore densities in the field and thereby the yield of the following crop (Karasawa et al. 2001). Oehl et al. (2003) found that increased land use intensity was correlated with a decrease in AM fungal species richness and with a preferential selection of species that colonised roots slowly but formed spores rapidly. Soils used for agricultural production have a low diversity of AM fungi compared with natural ecosystems (Menendez et al. 2001) and are often dominated by *Glomus* species (Daniell et al. 2001; Oehl et al. 2003; Troeh and Loynachan 2003). One reason for this is the low diversity of hosts, which reaches an extreme in crop monoculture (Oehl et al. 2003). Monoculture may select for AM fungal species that provide limited benefits to the host plant. Johnson et al. (1992) found that maize yielded higher and had higher nutrient uptake on soils that had grown soybean continuously for the previous 5 years than on soil that had grown maize continuously for the previous 5 years. Conversely, soybean yielded higher and had higher nutrient uptake on soil which had grown 5 years of maize than 5 years of soybean. The most abundant AM fungal species in the continuous maize soil was negatively correlated with maize yield but positively correlated with soybean yield; there was a similar effect with soybean soil. They hypothesised that monocropping selects AM fungal species which grow and sporulate most rapidly and that these species will offer the least benefit to the plant because they divert more resources to their own growth and reproduction. The result can be reduced benefits of AM colonisation to the host plant while monocropping continues. Crop rotation effects on mycorrhizal functioning have repeatedly also been observed by other workers. Harinikumar and Bagyaraj (1988) observed a 13 % reduction in mycorrhizal colonisation after 1-year cropping with a non-mycorrhizal crop and a 40 % reduction after fallowing. Lack of inoculum or inoculum insufficiency after a long bare fallow (especially in climates with an extended, dry, vegetation less season) may result in low uptake of P and Zn and in plants with nutrient deficiency symptoms that have been described as long-fallow disorder. The use of mycorrhizal cover crops can overcome this disorder (Thompson 1996). Sanginga et al. (1999) found evidence for increased mycorrhizal colonisation of soybean if the preceding crop was maize and increased colonisation of maize if the preceding crop was *Bradyrhizobium*-inoculated soybean in the savanna of Nigeria. Similarly, Bagayoko et al. (2000) reported higher AM colonisation in cereals (sorghum, pearl millet when grown in rotation with legumes (cowpea, groundnut) than in continuous cropping. Osunde et al. (2003) reported that AM colonisation in maize benefited from previously grown soybean plants.

In a long-term experiment involving three tillage systems and four soybean-based crop rotations after six cropping seasons, rotation produced significantly higher grain yield and supported higher inoculum potential of AM fungi in the rhizosphere soil (Sharma et al. 2012a). On the other hand, irrespective of crop rotations, the tillage system did not all have the same effect. Moreover, the inoculum potential of resident AM fungi in soybean rotation involving maize in conservation tillage was highly correlated with grain yield of soybean implicating the resident AM fungi in enhancing the soybean yield.

8.3.5 Inoculation vs. Field Management of Indigenous AM Fungi

Selection of the appropriate AM fungi is among one of the critical issues for the application of AM technology in agriculture (Estaun et al. 2002). Ecologically sound selected strains of AM fungi inoculum are not presently available in large quantities at a low price. Alternatively, inoculum can be produced on site (on farm) under local agronomic conditions (Sieverding 1991). The successful introduction of a foreign microorganism into the soil depends on how well it adapts, develops and competes for nutrients. AM fungal consortia isolated from organic farms were more effective in plant growth promotion under conditions of low nutrient availability than were consortia from conventional farms (Scullion et al. 1998). Therefore, it is likely that on-farm selected strains (site specific) are better due to their adaptability to edaphic conditions than selected strains produced in vitro or in vivo under controlled conditions. Given limitations of bulk inocula requirements or instances where inoculation may not be feasible, the management of native and resident AM fungi through crop sequences and soil management practices (e.g. minimum tillage) could be a better option.

8.4 Production and Commercialisation of AM Fungi

8.4.1 Conventional Methods

The obligate biotrophic nature of AM fungi has complicated the development of cost-efficient large-scale production technologies to obtain high-quality AM fungal inoculum. This is one of the bottlenecks to commercial exploitation (IJdo et al. 2011). There are various techniques currently used to culture AM fungi on hosts such as on-farm production (Douds et al. 2005, 2006; Sharma and Sharma, 2006; Sharma and Sharma 2008; Sharma and Adholeya 2011), pot culture techniques using traps (Gaur and Adholeya 2000), nutrient film technique (Mosse and Thompson 1984) and aeroponics (Jarstfer and Sylvia 1995). The most frequently used technique for increasing propagule number has been the propagation of AM fungi on a suitable host in disinfested soil using pot cultures. Other factors for creating a favourable environment for culturing of AM fungi are a balance of light intensity, adequate moisture and moderate temperature without detrimental addition of fertilisers or pesticides (Jarstfer and Sylvia 1992; Al-Karaki et al. 1998). Cultures reaching high propagule density (e.g. 10 spores per gram) after a number of multiplication cycles can be stored using suitable methods after air-drying (Kuszala et al. 2001).

AM fungi have been cultured with plant hosts in different substrates such as sand, peat, expanded clay, perlite, vermiculite, soilrite (Mallesha et al. 1992), rockwool (Heinzemann and Weritz 1990) and glass beads (Redecker et al. 1995). They can also be produced aeroponically (Sylvia and Hubbell 1986). The aeroponic system was adopted for mycorrhiza production by the utilisation of seedlings with roots pre-colonised by an AM fungus and the use of modified Hoagland's nutrition with a very low P level (Hoagland and Arnon 1938). *Entrophospora kentinensis* was successfully propagated with bahia grass and sweet potato in an aeroponic system by Wu et al. (1995).

The nutrient film technique (NFT) was adapted for AM fungi inoculum production by Mosse and Thompson (1984). Further, Lee and George (2005) proposed a modified nutrient film technique for large-scale production of AM fungal biomass with the help of improved aeration by intermittent nutrient supply, optimum P supply and the use of glass beads as support materials.

8.4.2 In Vitro/Root Organ Culture (ROC) Method

In vitro culture of AM fungi was achieved for the first time in the early 1960s (Mosse 1962). Since then, various pioneering steps were aimed at axenic culturing of AM fungi. Continuous cultures of vigorous ROCs (Ri T-DNA-transformed) have been obtained through transformation of roots by the soil bacterium *A. rhizogenes* (Tepfer 1989) that provided the new way to obtain mass production of roots in a very short span of time. In most cases, purified and surface sterilised spores (Becard and Piche 1992) isolated from the field or from traps have been successful for establishing dual cultures under in vitro conditions. The root organ culture (ROC) is an attractive mass multiplication method for providing a pure, viable, rapid and contamination-free inoculum using less space and has an advantage over the pot culture multiplication/conventional system (Fortin et al. 2002; Cranenbrouck et al. 2005; Dalpe et al. 2005). Different production systems have been derived from the basic ROC in Petri plates. For example, root organs and AM fungi were cultured in small containers, by which large-scale production was obtained (Adholeya et al. 2005). Douds (2002) reported monoxenic culture of *G. intraradices* with Ri T-DNA transformed roots in two-compartment Petri dishes as a very useful technique for physiological studies and the production of clean fungal tissues. Various inocula based on inert or sterilised substrata, such as peat, expanded or calcined clays or lave, are used commercially and are less susceptible to contamination with pathogen (Whipps 2004). Various forms of AM fungi are commercially produced and available in various formulations for sale throughout the world.

8.4.3   On-Farm Production

As AM fungi are obligate symbionts, they require host plants to sporulate and colonise roots to complete their life cycles. Currently, AM fungi are multiplied in various ways like monoxenic/in vitro, pot culturing/greenhouse, aeroponic system and nutrient film technique (Fortin et al. 2002; Lee and George 2005). While inocula produced by these techniques are commercially available, the pot culture or conventional method is still widely used (Saito and Marumoto 2002). There are many steps including isolation of AM fungi, the use of substrate/potting mixture and subsequent maintenance and transportation which incur costs and limit commercialisation. On-farm multiplication of indigenous and resident AM fungi removes many steps, which reduce the cost and enhance the acceptability to the farmers (Douds et al. 2006). The on-farm technology is more appropriate since it uses the indigenous AM fungi already adapted to that site and environment. Apart from this, the technology can be used for producing introduced AM fungi (applied as starter culture in beds) using one or a succession of trap plants (Sieverding 1991). Under this method, the fungal inoculum is produced on raised/elevated beds in situ; in the farmer's own nursery or his kitchen garden, a space that he generally uses for growing seedlings for field transplantation (Sharma and Sharma 2008). The mycorrhizal roots can then be harvested and used in the field as inocula. The soil left in the nursery after removing the roots contains a many AM fungal propagules which will serve as the source of AM fungi for further multiplying the inocula in the subsequent cycles. This method can produce inoculum of the indigenous AM fungi already adapted to the site. This field-based method deals with preparing beds of sterilised (solarised by polythene) soils in which either the indigenous AM fungi community or introduced isolates are increased using one or a succession of trap plants (Sieverding 1991). An important consideration in producing AM fungi is the level of available phosphorus which is critical for inoculum production and needs to be analysed before multiplication. In general, under Indian conditions, the level of Olsen P (available P in tropical soils) is low (less than 10 ppm), but high available P level (beyond 20 ppm) could be detrimental to AM sporulation and hence should be determined prior to multiplication. A unique feature of such technique is that it will not only produce mycorrhizal spores, hyphae and highly colonised roots but at the same time beds can be used for preparing seedlings for field transplantation.

8.5   Need of Regulatory Mechanisms and Quality Assurance

Currently, large-scale production of AM fungi is not possible in the absence of a suitable host, and species cannot be identified in their active live stages (growing mycelium). As a consequence, quality control is often a problem, and tracing the organisms into the field to strictly relate positive effects to the inoculated AM fungus is nearly impossible (IJdo et al. 2011). Pringle et al. (2009) have also indicated the risks associated with the transport of AM fungi around the world and have detailed the problem that can arise with the introduction of exotic material. In India, registration of biofertiliser production units is compulsory and is being done by the Ministry of Agriculture and Cooperation through a nodal agency, National Centre of Organic Farming, Ghaziabad, India.

8.6   Conclusion

Oilseeds comprise both legumes and nonlegumes, and major oilseeds like groundnut, sesame and soybean are grown under rainfed conditions in the tropics and subtropics in the marginal lands with meagre amount of external application of fertilisers. Very often, the major oilseeds crop faces vagaries of weather conditions like erratic rainfall and mid- and end-of-season drought coupled with plethora of diseases and pests severely limiting the productivity. Thus, to enhance the productivity of the oilseed crops, management of nutrients is of utmost importance to enhance availability of nutrient in suboptimal conditions of cultivation. Therefore, there is great opportunity of application of microbes especially rhizobia, PGPMs and AM fungi alone or in combinations. Considering the plant genotype as a constant factor, microbial package should be developed based on climate, soil and microbe interactions. Furthermore, formulation of biofertiliser packages should be developed not only for enhancing nutrient availability and uptake but for managing soil-borne and foliar diseases, in addition to enhancing growth by production of plant growth regulators. Within the constraints of available resources, a large number of PGPMs and AM fungi have been identified with capability to enhance growth and yield of many oilseed crops, but effective strains tolerant to abiotic stresses are few. Therefore, ongoing effort is needed to identify efficient strains of PGPMs and AM fungi which can alleviate abiotic stresses and have potential biocontrol abilities, besides enhancing nutrient availability and uptake in suboptimal conditions of cultivation. Many studies have shown large amounts of hyphal biomass and higher indigenous AM fungi in crop rotations involving maize. The large-scale production of resident AM fungi is still in its infancy and the combined application of AM fungi and PGPMs are yet to be streamlined. Finally, potential commercial formulations need to be subjected to regulatory requirements and quality checks before they are eventually registered as a commercial formulation.

Acknowledgments The authors are thankful to the Director of ICAR-Directorate of Soybean Research, Indore, India, for his kind support during the compiling of this task. Authors are also thankful to Prof Hamel Chantal, Agriculture and Agri-Food Canada, for making critical comments and suggestions on this review chapter.

References

Abbott LK, Robson AD (1991) Factors influencing the occurrence of vesicular-arbuscular mycor-rhizas. Agric Ecosyst Environ 35:121–150

Abdalla ME, Abdel-Fattah GM (2000) Influence of the endomycorrhizal fungus *Glomus mosseae* on the development of peanut pod rot disease in Egypt. Mycorrhiza 10:29–35

Abdel-Fattah GM, Shabanam YM (2002) Efficacy of the arbuscular mycorrhizal fungus *Glomus clarum* in protection of cowpea plants against root rot pathogen *Rhizoctonia solani*. J Plant Dis Prot 109:207–215

Adesemoye AO, Kloepper JW (2009) Plant-microbes interactions in enhanced fertilizer-use efficiency. Appl Microbiol Biotechnol 85:1–12

Adewole MB, Awotoye OO, Ohiembor MO, Salami AO (2010) Influence of mycorrhizal fungi on phytoremediating potential and yield of sunflower in Cd and Pb polluted soils. J Agric Sci 55:17–28

Adholeya A, Tiwari P, Singh R (2005) Large-scale production of arbuscular mycorrhizal fungi on root organs and inoculation strategies. In: Declerck S, Strullu DG, Fortin JA (eds) In vitro culture of mycorrhizas. Springer, Heidelberg, pp 315–338

Agricultural Statistics at a Glance (2004) Agricultural Statistics Division, Directorate of Econom-ics & Statistics, Department of Agriculture & Cooperation Ministry of Agriculture, Govern-ment of India, India, p 221

AICRPS (2009) Annual progress report, All India coordinated research project on soybean (2009-10). Directorate of Soybean Research (ICAR), Indore, p 251

Al-Karaki GN, Al-Raddad A, Clark RB (1998) Water stress and mycorrhizal isolate effects on growth and nutrient acquisition of wheat. J Plant Nutr 21:891–902

Al-Khaliel AS (2010) Effect of salinity stress on mycorrhizal association and growth response of peanut infected by *Glomus mosseae*. Plant Soil Environ 56:318–324

An ZQ, Grove JH, Hendrix JW, Hershman DE, Henson GT (1990) Vertical distribution of endogonaceous mycorrhizal fungi associated with soybean, as affected by soil fumigation. Soil Biol Biochem 22:715–719

Anil-Prakash, Vandana T (2002) Exploiting mycorrhiza for oilseed crop production. In: Rajak RC (ed) Biotechnology of microbes and sustainable utilization pages. Scientific Publishers, India, p 370

Antunes PM, Rajcan I, Goss MJ (2006) Specific flavonoids as interconnecting signals in the tripartite symbiosis formed by arbuscular mycorrhizal fungi, *Bradyrhizobium japonicum* (Kirchner) Jordan and soybean (*Glycine max* L.) Merr. Soil Biol Biochem 38:533–543

Artursson V, Finlay RD, Jansson JK (2005) Combined bromodeoxyuridine immunocapture and terminal restriction fragment length polymorphism analysis highlights differences in the active soil bacterial metagenome due to *Glomus mosseae* inoculation or plant species. Environ Microbiol 7:1952–1966

Arturrson V, Finlay RD, Jansson JK (2006) Interactions between arbuscular mycorrhizal fungi and bacteria and their potential for stimulating plant growth. Environ Microbiol 8:1–10

Aryal UK, Shah SK, Xu HL, Fujita M (2006) Growth, nodulation and mycorrhizal colonization in Bean plants improved by rhizobial inoculation with organic and chemical fertilization. J Sustain Agric 29:71–83

Auge RM, Sylvia DM, Park S, Buttery BR, Saxton AM, Moore JL, Cho KH (2004) Partitioning mycorrhizal influence on water relations of *Phaseolus vulgaris* into soil and plant components. Can J Bot 82:503–514

Awotoye OO, Adewole MB, Salami AO, Ohiembor MO (2009) Arbuscular mycorrhiza contribution to the growth performance and heavy metal uptake of *Helianthus annuus* L. in pot culture. J Environ Sci Technol 3:157–163

Babalola OO (2010) Beneficial bacteria of agricultural importance. Biotechnol Lett 32:1559–1570

Bagayoko M, Buerkert A, Lung G, Bationo A, Romheld V (2000) Cereal/legume rotation effects on cereal growth in Sudano-Sahelian West Africa: soil mineral nitrogen, mycorrhizae and nematodes. Plant Soil 218:103–116

Barea JM, Andrade G, Bianciotto VV, Dowling D, Lohrke S, Bonfante P (1998) Impact on arbuscular mycorrhiza formation of *Pseudomonas* strains used as inoculants for biocontrol of soil-borne fungal plant pathogens. Appl Environ Microbiol 64:2304–2307

Barea JM, Pozo MJ, Azcon R, Azcon-Aguilar C (2005) Microbial co-operation in the rhizosphere. J Exp Bot 56:1761–1778

Becard G, Piche Y (1992) Establishment of AM in root organ cultures review and proposed methodology. In: Norris J, Read D, Verma A (eds) Techniques for the study of mycorrhiza. Academic, New York, pp 89–108

Bethlenflavay GJ, Schreiner RP, Mihara KL (1997) Mycorrhizal fungi effects on nutrient composition and yield of soybean seeds. J Plant Nutr 20:521–529

Bianciotto V, Bandi C, Minerdi D, Sironi M, Tichy HV, Bonfante P (1996) An obligately endosymbiotic mycorrhizal fungus itself harbors obligately intracellular bacteria. Appl Environ Microbiol 62:3005–3010

Bohlool BB, Ladha JK, Garrity DP, George T (1992) Biological N fixation for sustainable agriculture: a perspective. Plant Soil 141:1–11

Budi SW, Van Tuinen D, Martinotti G, Gianinazzi S (1999) Isolation from *Sorghum bicolor* mycorrhizosphere of a bacterium compatible with arbuscular mycorrhiza development and antagonistic towards soil-borne fungal pathogens. Appl Environ Microbiol 65:5148–5150

Calvet C, Pera J, Barea JM (1993) Growth response of marigold (*Tagetes erecta* L.) to inoculation with *Glomus mosseae*, *Trichoderma aureoviride* and *Pythium ultimum* in a peat-perlite mixture. Plant Soil 148:1–6

Cardoso IM, Kuyper TW (2006) Mycorrhizas and tropical soil fertility. Agric Ecosyst Environ 116:72–84

Caris C, Hördt W, Hawkins H, Römheld V, George E (1998) Studies of iron transport by arbuscular mycorrhizal hyphae from soil to peanut and sorghum plants. Mycorrhiza 8:35–39

Carling DE, Roncadori RW, Hussey RS (1995) Interactions of arbuscular mycorrhizae, *Meloidogyne arenaria* and phosphorus fertilization on peanut. Mycorrhiza 6:9–13

Caron M, Fortin JA, Richard C (1986) Effect of inoculation sequence on the interaction between Glomus intraradices and Fusarium oxysporum f. sp. radicis-lycopersici in tomatoes. Can J Plant Pathol 8:12–16

Chandrashekara CP, Patil VC, Sreenivasa MN (1995) VA-mycorrhiza mediated P effect on growth and yield of sunflower (*Helianthus annuus* L.) at different P levels. Plant Soil 176:325–328

Clark RB, Zeto SK (1996) Iron acquisition by mycorrhizal maize grown on alkaline soil. J Plant Nutr 19:247–264

Colozzi A, Cardoso EJBN (2000) Detecção de fungos micorrízicos arbusculares em raízes de cafeeiro e de crotalária cultivada na entrelinha. Pesqui Agropecu Bras 35:2033–2042

Cordier C, Gianinazzi S, Gianinazzi-Pearson V (1996) Colonization patterns of root tissues by *Phytophthora nicotianae* var. *parasitica* related to reduced diseases in mycorrhizal tomato. Plant Soil 185:223–232

Cranenbrouck S, Voets L, Bivort C (2005) Methodologies for *InVitro* cultivation of AM fungi with root organs. In: Declerck S, Strullu DG, Fortin A (eds) *In Vitro* culture of mycorrhizas. Springer, Berlin, pp 342–375

Dalpé Y, de Souza FA, Declerck S (2005) Life cycle of *Glomus* species in monoxenic culture. In: Declerck S, Fortin JA, Strullu DG (eds) Dans: biology of arbuscular mycorrhizal fungi under in vitro culture. Springer, Germany, pp 49–71

Damodaram T, Hegde DM (2010) Oilseeds situation: a statistical compendium 2010. Directorate of Oilseeds Research, Hyderabad, p 486

Daniell T, Husband R, Fitter AH, Young JPW (2001) Molecular diversity of arbuscular mycorrhizal fungi colonising arable crops. FEMS Microbiol Ecol 36:203–209

Davis RM, Menge JA (1980) Influence of *Glomus fasciculatus* and soil phosphorus on *Phytophthora* root rot of citrus. Phytolopathology 7:447–452

Dehne HW (1982) Interaction between vesicular-arbuscular mycorrhizal fungi and plant pathogens. Phytopathology 72:1114–1119

Del Val C, Barea JM, Azcon-Aguilar C (1999) Assessing the tolerance of heavy metals of arbuscular mycorrhizal fungi isolated from sewage-sludge contaminated soils. Appl Soil Ecol 11:261–269

Dev A, Gour RK, Jain RK, Bisen PS, Sengupta LK (1997) Effect of vesicular arbuscular mycorrhiza-Rhizobium inoculation interaction on heavy metal (Cu, Zn and Fe) uptake in soybean (*Glycine max*, var. JS-335) under variable P doses. Int J Tropic Agric 15:75–79

Devi MC, Reddy MN (2002) Phenolic acid metabolism of groundnut (*Arachis hypogaea* L.) plants inoculated with VAM fungus and Rhizobium. Plant Growth Regul 37:151–156

Diedhiou PM, Hallmann J, Oerke EC, Dehne HW (2003) Effects of arbuscular mycorrhizal fungi and a non-pathogenic *Fusarium oxysporum* on *Meloidogyne incognita* infestation of tomato. Mycorrhiza 13:199–204

Doley K, Jite PK (2013a) Effect of arbuscular mycorrhizal fungi on growth of groundnut and disease caused by *Macrophomina phaseolina*. J Exp Sci 4:11–15

Doley K, Jite PK (2013b) Disease management and biochemical changes in groundnut inoculated with *Glomus fasciculatum* and pathogenic *Macrophomina phaseolina* (Tassi) Goid. Plant Sci Feed 3:21–26

Douds DD Jr (2002) Increased spore production by Glomus intraradices in the split-plate monoxenic culture system by repeated harvest, gel replacement, and re-supply of glucose to the mycorrhiza. Mycorrhiza 12:163–167

Douds DD Jr, Nagahashi G, Pfeffer PE, Kayser WM, Reider C (2005) On-farm production and utilization of arbuscular mycorrhizal fungus inoculum. Can J Plant Sci 85:15–21

Douds DD Jr, Nagahashi G, Pfeffer PE, Reider C, Kayser WM (2006) On-farm production of AM fungus inoculum in mixtures of compost and vermiculite. Bioresour Technol 97:809–818

Dugassa GD, Vonalten H, Schnbeck F (1996) Effects of arbuscular mycorrhiza (AM) on health of *Linum usitatissimum* L infected by fungal pathogens. Plant Soil 185:173–182

Edwards SG, Young JPW, Fitter AH (1998) Interactions between *Pseudomonas fluorescens* biocontrol agents and *Glomus mosseae*, an arbuscular mycorrhizal fungus, within the rhizosphere. FEMS Microbiol Lett 116:297–303

El-Azouni IM, Hussein Y, Shaaban LD (2008) The associative effect of VA mycorrhizae with *Bradyrhizobium* as biofertilizers on growth and nutrient uptake of *Arachis hypogaea*. Res J Agric Biol Sci 4:187–197

Elsen A, Declerck S, De Waele D (2001) Effects of *Glomus intraradices* on the reproduction of the burrowing nematode *Radopholus similis* in dixenic culture. Mycorrhiza 11:49–51

Elsheikh EAE, Mohamedzein EMM (1998) Effect of Bradyrhizobium, VA mycorrhiza and fertilisers on seed composition of groundnut. Ann Appl Biol 132:325–330

Estaun V, Camprubi A, Joner EJ (2002) Selecting arbuscular mycorrhizal fungi for field application. In: Gianinazzi S, Schueep H, Barea JM, Haselwandter K (eds) Mycorrhiza technology in agriculture: from genes to bioproducts. Birkhauser, Basel, pp 249–259

Fontem DA, Iroume RN, Aloleko F (1996) Effet de la résistance variétale et des traitements fongicides sur la cercosporiose de l'arachide. Cahier Agric 5:33–38

Fortin JA, Becard G, Declerck S, Dalpe Y, St Arnaud M, Coughlan AP, Piche Y (2002) Arbuscular mycorrhiza on root-organ cultures. Can J Bot 80:1–20

Garnett T, Conn V, Kaiser B (2009) Root based approaches to improving nitrogen use efficiency in plants. Plant Cell Environ 32:1272–1283

Gaur A, Adholeya A (2000) Effects of the particle size of soil-less substrates upon AM fungus inoculum production. Mycorrhiza 10:43–48

Göhre V, Paszkowski U (2006) Contribution of the arbuscular mycorrhizal symbiosis to heavy metal phytoremediation. Planta 223:1115–1122

Gonçalves EJ, Muchojev JJ, Muchojev RMC (1991) Effect of kind and method of fungicidal seed treatment of bean seed on infections by VA-mycorrhizal fungus *Glomus macrocarpum* and by the pathogenic fungus *Fusarium solani*. I. Fungal and plant parameters. Plant Soil 132:41–46

Grichar VJ, Bosweel TE (1987) Comparison of lorsban and tilt with terrachlor for control of southern blight on peanut. The Texas Agriculture Experiment Station, PR-4534

Groth DE, Martinson CA (1983) Increased endomycorrhizal colonization of maize and soybeans after soil treatment with metalaxyl. Plant Dis 67:1377–1378

Gupta R, Krishnamurthy KV (1996) Response of mycorrhizal and nonmycorrhizal *Arachis hypogaea* to NaCl and acid stress. Mycorrhiza 6:145–149

Habte M, Zhang YC, Schmitt DP (1999) Effectiveness of *Glomus* species in protecting white clover against nematode damage. Can J Bot 77:135–139

Hamel C, Strullu DG (2006) Arbuscular mycorrhizal fungi in field crop production: potential and new direction. Can J Plant Sci 86:941–950

Harinikumar DM, Bagyaraj DJ (1988) Effect of crop rotation on native vesicular-arbuscular mycorrhizal propagules in soil. Plant Soil 110:77–80

Harinikumar KM, Bagyaraj D (1989) Effect of cropping sequence, fertilizers and farmyard manure on vesicular-arbuscular mycorrhizal fungi in different crops over three consecutive seasons. Biol Fertil Soils 7:173–175

Harrier LA, Watson CA (2004) The potential role of arbuscular mycorrhizal (AM) fungi in the bioprotection of plants against soil-borne pathogens in organic and/or other sustainable farming systems. Pest Manag Sci 60:149–157

Hart MM, Reader RJ, Klironomos JN (2001) Life strategies of arbuscular mycorrhizal fungi in relation to their successional dynamics. Mycologia 93:1186–1194

Heinzemann J, Weritz J (1990) Rockwoola new carrier system for mass multiplication of vesicular-arbuscular mycorrhizal fungi. Angew Bot 64:271–274

Hildebrandt U, Janetta K, Bothe H (2002) Towards growth of arbuscular mycorrhizal fungi independent of a plant host. Appl Environ Microbiol 68:1919–1924

Hoagland DR, Arnon DI (1938) The water-culture method for growing plants without soil. University of California, College of Agriculture, Agriculture Experiment Station Circular 347, Berkeley

Ijdo M, Cranenbrouck S, Declerck S (2011) Methods for large-scale production of AM fungi: past, present, and future. Mycorrhiza 21(1):1–16

Ilbas AI, Sahin S (2005) Glomus fasciculatum inoculation improves soybean production. Acta Agric Scand Sect B Soil Plant Sci 55:287–292

Iyer RM, Bhat N, Madhusudhanan K (2003) Effect of organic amendments on growth and colonization of *Glomus fasciculatum* native to oil palm. Int J Oil Palm Res 3(4):65–67

Jackson RM, Mason PA (1984) Mycorrhiza. Edward Arnold, London, p 60. ISBN 0-7131-2876-3

Jalaluddin M, Hamid M, Muhammad SE (2008) Selection and application of a AM-fungus for promoting growth and resistance to charcoal rot disease of sunflower var. Helico-250. Pak J Bot 40:1313–1318

Jamal A, Ayub N, Usman M, Khan AG (2002) Arbuscular mycorrhizal fungi enhance zinc and nickel uptake from contaminated soil by soya bean and lentil. Int J Phytoremed 4:205–221

Jansa J, Mozafar A, Anken T, Ruh R, Sanders IR, Frossard E (2002) Diversity and structure of AM fungi communities as affected by tillage in a temperate soil. Mycorrhiza 12(5):225–234

Jansa J, Hans-Rudolf O, Egli S (2009) Environmental determinants of the arbuscular mycorrhizal fungal infectivity of Swiss agricultural soils. Eur J Soil Biol 45:400–440

Jarstfer AG, Sylvia DM (1992) Inoculum production and inoculation strategies for vesicular-arbuscular mycorrhizal fungi. In: Meting B (ed) Soil microbial ecology; application in agriculture and environmental management. Marcel Dekker, New York, pp 349–377

Jarstfer AG, Sylvia DM (1995) Aeroponic culture of AM fungi. In: Varma A, Hock B (eds) Mycorrhiza structure, function, molecular biology and biotechnology. Springer, Heidelberg, pp 521–559

Johnson NC, Copeland PJ, Crookston RK, Pfleger FL (1992) Mycorrhizae: possible explanations for yield decline with continuous corn and soybean. Agron J 84:387–390

Kabir Z (2005) Tillage or no-tillage: impact on mycorrhizae. Can J Plant Sci 85:23–29

Kahiluoto H, Ketoja E, Vestberg M, Saarela I (2001) Promotion of AM utilization through reduced P fertilization II Field studies. Plant Soil 231:65–79

Karagiannidis N, Bletsos F, Stavropoulos N (2002) Effect of *Verticillium* wilt (*Verticillium dahlia* Kieb.) and mycorrhizae (*Glomus mosseae*) on root colonization, growth and nutrient in tomato and eggplant seedlings. Sci Hortic 94:145–156

Karasawa T, Kasahara Y, Takebe M (2001) Variable response of growth and arbuscular mycorrhizal colonization of maize plants to preceding crops in various types of soils. Biol Fertil Soils 33:286–293

Kellam MK, Schenck NC (1980) Interaction between a vesicular-arbuscular mycorrhizal fungus and root knot nematode on soybean. Phytopathology 72:293–296

Kennedy AC, Smith KL (1995) Soil microbial diversity and the sustainability of agricultural soils. Plant and Soil 170:75–86

Khan AG, Kuek C, Chaudhry TM, Khoo CS, Hayes WJ (2000) Role of plants, mycorrhizae and phytochelators in heavy metal contaminated land remediation. Chemosphere 41:197–207

Kjoller R, Rosendahl S (2000) Effects of fungicides in arbuscular mycorrhizal fungi: differential responses in alkaline phosphatases activity of external and internal hyphae. Biol Fertil Soils 31:361–365

Kling M, Jakobsen I (1997) Direct application of carbendazim and propiconazole at field rates to the external mycelium of three arbuscular mycorrhizal fungi species: effect on ^{32}P transport and succinate dehydrogenase activity. Mycorrhiza 7:33–37

Kothari SK, Marschner H, Romheld V (1991) Effect of vesicular arbuscular fungus and rhizosphere microorganisms on manganese reduction in the rhizosphere and manganese concentration in maize (*Zea mays* L.). New Phytol 117:649–655

Krishna KR, Bagyaraj DJ (1982) Influence of vesicular-arbuscular mycorrhiza on growth and nutrition of *Arachis hypogaea* L. Legume Res 5:18–22

Kucey RMN, Bonetti R (1988) Effect of vesicular-arbuscular mycorrhizal fungi and captan on growth and N$_2$-fixation by *Rhizobium*-inoculated field beans. Can J Soil Sci 68:143–149

Kuszala C, Gianinazzi S, Gianinazzi-Pearson V (2001) Storage conditions for the long-term survival of AM fungal propagules in wet sieved soil fractions. Symbiosis 30:287–299

Lee YJ, George E (2005) Development of a nutrient film technique culture system for arbuscular mycorrhizal plants. Hortic Sci 40:378–380

Lekberg Y, Koide RT (2005) Arbuscular mycorrhizal fungi, rhizobia, available soil P and nodulation of groundnut (*Arachis hypogaea*) in Zimbabwe. Agric Ecosyst Environ 110:143–148

Leyval C, Turnau K, Haselwandter K (1997) Effect of heavy metal pollution on mycorrhizal colonization and function: physiological, ecological and applied aspects. Mycorrhiza 7:139–153

Linderman RG (1994) Role of AM fungi in biocontrol. In: Pfleger FL, Linderman RG (eds) Mycorrhizae and plant health. APS Press, St. Paul, pp 1–26

Lisette J, Xavier C, Germida JJ (2003) Selective interactions between arbuscular mycorrhizal fungi and *Rhizobium leguminosarum* bv. *viceae* enhance pea yield and nutrition. Biol Fertil Soils 37:261–267

Liu A, Plenchette C, Hamel C (2007) Soil nutrient and water providers: how arbuscular mycorrhizal mycelia support plant performance in a resource limited world. In: Hamel C, Plenchette

C (eds) Mycorrhizae in crop production. Haworth Food & Agricultural Products Press, Binghamton, pp 37–66

Mäder P, Edenhofer S, Boller T, Wiemken A, Niggli U (2000) Arbuscular mycorrhizae in a long-term field trial comparing low-input (organic, biological) and high-input (conventional) farming systems in a crop rotation. Biol Fertil Soils 31:150–156

Mäder P, Fliessbach A, Dubois D, Gunst L, Fried P, Niggli U (2002) Soil fertility and biodiversity in organic farming. Science 296:1694–1697

Mahanta D, Rai RK (2008) Effects of sources of phosphorus and biofertilizers on productivity and profitability of soybean (*Glycine max*)–wheat (*Triticum aestivum*) system. Indian J Agron 53:279–284

Malcova R, Vosatka M, Gryndler M (2003) Effects of inoculation with *Glomus intraradices* on lead uptake by *Zea mays* L. and *Agrostis capillaries* L. Appl Soil Ecol 23:55–67

Mallesha BC, Bagyaraj DJ, Pai G (1992) Perlite-soilrite mix as a carrier for mycorrhiza and rhizobia to inoculate *Leucaena leucocephala*. Leucaena Res Rep 13:32–33

Mandhare VK, Kalbhor HB, Patil PL (1995) Effects of VA-Mycorrhiza, *Rhizobium* and phosphorus on summer groundnut. J Maharashtra Agric Univ 20:261–262

Manoharan PT, Shanmugaiah V, Balasubramanian N, Gomathinayagam S, Sharma MP, Muthuchelian K (2010) Influence of AM fungi on the growth and physiological status of E*rythrina variegata* Linn. grown under different water stress conditions. Eur J Soil Biol 46:151–156

Mansfeld-Giese K, Larsen J, Bodker L (2002) Bacterial populations associated with mycelium of the arbuscular mycorrhizal fungus *Glomus intraradices*. FEMS Microbiol Ecol 41:133–140

Marschner H (1995) Mineral nutrition of higher plants. Academic, London

Marschner P, Yang CH, Lieberei R, Crowley DE (2001) Soil and plant specific effects on bacterial community composition in the rhizosphere. Soil Biol Biochem 33:1437–1445

Martensson AM, Carlegren K (1994) Impact of phosphorus fertilization on VAM diaspores in two Swedish long-term field experiments. Agric Ecosyst Environ 47:327–334

Mathur N, Vyas A (2000) Influence of arbuscular mycorrhizae on biomass production, nutrient uptake and physiological changes in *Ziziphus mauritiana* Lam. under water stress. J Arid Environ 45:191–195

Matsubara Y, Ohba N, Fukui H (2001) Effect of arbuscular mycorrhizal fungus infection on the incidence of fusarium root rot in asparagus seedlings. J Jpn Soc Hortic Sci 70:202–206

Mayo K, Davis RE, Motta J (1986) Stimulation of germination of spores of *Glomus versiforme* by spore-associated bacteria. Mycologia 78:426–431

McGonigle TP, Miller MH (1993) Mycorrhizal development and phosphorus absorption in maize under conventional and reduced tillage. Soil Sci Soc Am J 57:1002–1006

Meghvansia MK, Prasad K, Harwani D, Mahna SK (2008) Response of soybean cultivars toward inoculation with three arbuscular mycorrhizal fungi and *Bradyrhizobium japonicum* in the alluvial soil. Eur J Soil Biol 44:316–323

Menendez AB, Scervino JM, Godeas AM (2001) Arbuscular mycorrhizal populations associated with natural and cultivated vegetation on a site of Buenos Aires province, Argentina. Biol Fertil Soils 33:373–381

Menge JA, Steirle D, Bagyaraj DJ, Johnson ELV, Leonard RT (1978) Phosphorus concentrations in plants responsible for inhibition of mycorrhizal infection. New Phytol 80:575–578

Meyer JR, Linderman RG (1986) Response of subterranean clover to dual inoculation with vesiculararbuscular fungi and a plant growth-promoting bacterium, *Pseudomonas putida*. Soil Biol Biochem 18:185–190

Mirzakhani M, Ardakani MR, Band AA, Rejali F, Shirani Rad AH (2009) Response of spring safflower to co-inoculation with *Azotobacter chroococum* and *Glomus intraradices* under different levels of nitrogen and phosphorus. Am J Agric Biol Sci 4:255–261

Mosse B (1962) The establishment of AM fungi under aseptic conditions. J Gen Microbiol 27:509–520

Mosse B, Thompson JP (1984) Vesicular-arbuscular endomycorrhizal inoculum production. I. Exploratory experiments with beans (*Phaseolus vulgaris*) in nutrient flow culture. Can J Bot 62:1523–1530

Mostafavian SR, Pirdashti H, Ramzanpour MR, Andarkhor AA, Shahsavari A (2008) Effects of mycorrhizae, *Thiobacillus* and sulphur nutrition on the chemical composition of soybean. Pak J Biol Sci 11:826–835

Mulligan MF, Smucker AJM, Safir GF (1985) Tillage modifications of dry edible bean root colonization by AM fungi. Agron J 77:140–144

Murillo-Williams A, Pedersen P (2008) Arbuscular mycorrhizal colonization response to three seed-applied fungicides. Agron J 100:795–800

Nogueira MA, Magelhaes GC, Cardoso EJBN (2004) Manganese toxicity in mycorrhizal and phosphorus-fertilized soybean plants. J Plant Nutr 27:141–156

Oehl F, Sieverding E, Ineichen K, Mader P, Boller T, Wiemken A (2003) Impact of land use intensity on the species diversity of arbuscular mycorrhizal fungi in agroecosystems of Central Europe. Appl Environ Microbiol 69:2816–2824

Osunde AO, Bala A, Gwam MS, Tsado PA, Sanginga N, Okugun JA (2003) Residual benefits of promiscuous soybean to maize (*Zea mays* L.) grown on farmers' fields around Minna in the southern Guinea savanna zone of Nigeria. Agric Ecosyst Environ 100:209–220

Ozgonen H, Akgul DS, Erkilic A (2010) The effects of arbuscular mycorrhizal fungi on yield and stem rot caused by *Sclerotium rolfsii* Sacc. in peanut. Afr J Agric Res 5:128–132

Peoples MB, Craswell ET (1992) Biological nitrogen fixation: investments, expectations and actual contributions to agriculture. Plant and Soil 141(13–39):1992

Pinochet J, Calvet C, Camprubí A, Fernandez C (1996) Interactions between migratory endoparasitic nematodes and arbuscular mycorrhizal fungi in perennial crops. Plant Soil 185:183–190

Powell JR, Gulden RH, Hart MM, Campbell RG, Levy-Booth DJ, Dunfield KE, Pauls KP, Swanton CJ, Trevors JT, Klironomos JN (2007) Mycorrhizal and rhizobial colonization of genetically modified and conventional soybeans. Appl Environ Microbiol 73:4365–4367

Powell JR, Campbell RG, Dunfield KE, Gulden RH, Hart MM, Levy-Booth DJ, Klironomos JN, Pauls KP, Swanton CJ, Trevors JT, Antunes PM (2009) Effect of glyphosate on the tripartite symbiosis formed by *Glomus intraradices*, *Bradyrhizobium japonicum*, and genetically modified soybean. Appl Soil Ecol 41:128–136

Pozo MJ, Azcon-Aguilar C, Dumas-Gaudot E, Barea JM (1999) β-1,3-Glucanase activities in tomato roots inoculated with arbuscular mycorrhizal fungi and/or *Phytophthora parasitica* and their possible involvement in bioprotection. Plant Sci 141:149–157

Price NS, Roncadori RW, Hussey RS (1995) The growth of nematode tolerant and intolerant soybeans as affected by phosphorus, *Glomus intraradices* and light. Plant Pathol 44:597–603

Pringle A, Bever JD, Gardes M, Parrent JL, Rillig MC, Klironomos JN (2009) Mycorrhizal symbioses and plant invasions. Annu Rev Ecol Evol Syst 40:699–715

Quilambo OA (2002) Minimising the effects of drought stress on growth of two peanut cultivars, using arbuscular mycorrhiza (AM) inoculants. J Trop Microbiol Biotechnol 1:22–28

Raju PS, Clark RB, Ellis JR, Maranville JW (1990) Effects of species of VA-mycorrhizal fungi on growth and mineral uptake of sorghum at different temperatures. Plant Soil 121:165–170

Redecker D, Thierfelder H, Werner D (1995) A new cultivation system for arbuscular-mycorrhizal fungi on glass beads. Angew Bot 69:189–191

Richardson AE, Barea JM, McNeill AM, Prigent-Combaret C (2009) Acquisition of phosphorus and nitrogen in the rhizosphere and plant growth promotion by microorganisms. Plant Soil 321:305–339

Rillig MC (2004) Arbuscular mycorrhizae, glomalin, and soil aggregation. Can J Soil Sci 84:355–363

Rodriguez-Kabana R, Curl EA (1980) Non-target effects of pesticides on soilborne pathogens and disease. Annu Rev Phytopathol 18:311–332

Ruiz-Lozano JM (2003) Arbuscular mycorrhizal symbiosis and alleviation of osmotic stress. New perspectives for molecular studies. Mycorrhiza 13:309–317

Saito M, Marumoto T (2002) Inoculation with arbuscular mycorrhizal fungi: the status quo in Japan and the future prospects. Plant Soil 244:273–279

Sanchez L, Weidmann S, Brechenmacher L, Batoux M, van Tuinen D, Lemanceau P (2004) Common gene expression in *Medicago truncatula* roots in response to *Pseudomonas fluorescens* colonization, mycorrhiza development and nodulation. New Phytol 161:855–863

Sanginga N, Carsky RJ, Dashiell K (1999) Arbuscular mycorrhizal fungi respond to rhizobial inoculation and cropping systems in farmers' fields in the Guinea savanna. Biol Fertil Soils 30:179–186

Savary S (1991) Approche de la pathologie des cultures tropicales: I exemple de I arachide en Afrique de I Ouest. Karthala-Orstom, Paris

Schreiner RP, Bethlenfalvay GJ (1997) Mycorrhizae, biocides, and biocontrol. 3. Effects of three different fungicides on developmental stages of three AM fungi. Biol Fertil Soils 24:18–26

Scullion J, Eason WR, Scott EP (1998) The effectivity of arbuscular mycorrhizal fungi from high input conventional and organic grassland and grass-arable rotations. Plant Soil 204:243–254

Shalaby AM, Hanna MM (2000) Interactions between VA mycorrhizal fungus *Glomus mosseae*, *Bradyrhizobium japonicum* and *Pseudomonas syringae* in soybean plants. Egypt J Microbiol 35:199–209

Sharifia M, Ghorbanli M, Ebrahimzadeh H (2007) Improved growth of salinity-stressed soybean after inoculation with salt pre-treated mycorrhizal fungi. J Plant Physiol 164:1144–1151

Sharma MP, Adholeya A (2011) Developing prediction equations and optimizing production of three AM fungal inocula under on-farm conditions. Exp Agric 47:529–537

Sharma MP, Sharma SK (2006) Arbuscular mycorrhizal fungi an emerging bio-inoculant for production of soybean. SOPA Digest III(IX):10–16

Sharma MP, Sharma SK (2008) On-farm production of arbuscular mycorrhizal fungi. Biofertilizer Newslett 16:3–7

Sharma MP, Sharma SK, Alok Dwivedi (2010) Liquid biofertilizer application in soybean and regulatory mechanisms. Agric Today April issue: 44–45

Sharma MP, Gupta S, Sharma SK, Vyas AK (2012a) Effect of tillage and crop sequences on arbuscular mycorrhizal symbiosis and soil enzyme activities in soybean (*Glycine max* L. Merril) rhizosphere. Indian J Agric Sci 82:25–30

Sharma MP, Jaisighani K, Sharma SK, Bhatia VS (2012b) Effect of native soybean rhizobia and AM fungi in the improvement of nodulation, growth, soil enzymes and physiological status of soybean under microcosm conditions. Agric Res 1:346–351

Sieverding E (1991) Vesicular–arbuscular mycorrhiza management in tropical agrosystems. Deutsche Gesellschaft für Technische Zusammansabeit (GT2) GmbH. Eschbon, Germany

Singh AL, Chaudhari V (1996) Use of zincated and boronated superphosphates and mycorrhizae in groundnut grown on a calcareous soil. J Oilseeds Res 13:61–65

Singh S, Basappa H, Singh SK (2006) Status and prospects of integrated pest management strategies in selected crops Oilseeds. In: Singh A, Sharma OP, Garg DK (eds) Integrated pest management principles and applications, vol 2. CBS Publishers and Distributors, New Delhi, p 656

Smith S, Read DJ (1997) Mycorrhizal symbiosis, 2nd edn. Academic, London

Sreenivasa MN, Bagyaraj DJ (1989) Use of pesticide for mass production of vesicular-arbuscular mycorrhizal inoculums. Plant Soil 119:127–132

Srinivasulu Y, Lakshman HC (2002) Response of *Guizotia abyssinica* Cass. (Niger) and *Linum usitattissimum* (Linseed) to VA-mycorrhizal inoculation in an unsterile soil. Karnataka J Agric Sci 15:405–407

Strullu DG, Romand C, Plenchette C (1991) Axenic culture and encapsulation of the intraradical forms of *Glomus* spp. World J Microbiol Biotechnol 7:292–297

Sugavanam V, Udaiyan K, Manian S (1994) Effect of fungicides on vesicular-arbuscular mycor-rhizal infection and nodulation in groundnut (*Arachis hypogeae* L.). Agric Ecosyst Environ 48:285–293

Sulochana T, Manoharachary C (1989) Vesicular-arbuscular mycorrhizal associations of castor and safflower. Curr Sci 58:459–461

Sylvia DM, Hubbell DH (1986) Growth and sporulation of vesicular-arbuscular mycorrhizal fungi in aeroponic and membrane systems. Symbiosis 1:259–267

Sylvia DM, Neal LH (1990) Nitrogen affects the phosphorus response of VA mycorrhiza. New Phytol 115:303–310

Tain CY, Feng G, Li XI, Zhang FS (2004) Different effects of arbuscular mycorrhizal fungal isolates from saline or non-saline soil on salinity tolerance of plants. Appl Soil Ecol 26:143–148

Taiwo LB, Adegbite AA (2001) Effect of arbuscular mycorrhiza and Bradyrhizobium inoculation on growth, N2 fixation and yield of promiscuously nodulating soybean (*Glycine max*). J Agric Res 2:110–118

Tepfer D (1989) Ri t-DNA from *Agrobacterium rhizogenes*. A source of genes having applications in rhizosphere biology and plant development, ecology and evolution. In: Kosuge T, Nester EW (eds) Plant-microbe interactions, vol 3. McGraw-Hill, New York, pp 294–342

Tester M, Smith SE, Smith FA (1987) The phenomenon of "nonmycorrhizal" plants. Can J Bot 65:419–431

Thompson JP (1987) Decline of vesicular-arbuscular mycorrhizae in long fallow disorder of field crops and its expression in phosphorus deficiency of sunflower. Aust J Agric Res 38:847–867

Thompson JP (1996) Correction of dual phosphorus and zinc deficiencies of linseed (*Linum usitatissimum* L.) with cultures of vesicular–arbuscular mycorrhizal fungi. Soil Biol Biochem 28:941–951

Tilman D, Cassman KG, Matson PA, Naylor R, Polasky S (2002) Agricultural sustainability and intensive production practices. Nature 418:671–677

Tisdale SI, Nelson WI, Baton JD (1995) Soil fertility and fertilizers. Macmillan, New York

Todd TC, Winkler HE, Wilson GWT (2001) Interaction of *Heterodera glycines* and *Glomus mosseae* on soybean. J Nematol 33:306–310

Toro M, Azcon R, Barea JM (1997) Improvement of arbuscular mycorrhizal development by inoculation with phosphate-solubilizing rhizobacteria to improve rock phosphate bioavailability (^{32}P) and nutrient cycling. Appl Environ Microbiol 63:4408–4412

Torres-Barragan A, Zavaleta-Mejia E, Gonzalez-Chavez C, Ferrera-Cerrato R (1996) The use of arbuscular mycorrhizae to control onion white rot (*Sclerotium cepivorum* Berk.) under field conditions. Mycorrhiza 6:253–258

Trappe JM, Molina R, Castellano M (1984) Reactions of mycorrhizal fungi and mycorrhiza formation to pesticides. Annu Rev Phytopathol 22:331–359

Troeh ZI, Loynachan TE (2003) Endomycorrhizal fungal survival in continuous corn, soybean, and fallow. Agron J 95:224–230

Turk MA, Assaf TA, Hameed KM, Al-Tawaha AM (2006) Significance of mycorrhizae. World J Agric Sci 2:16–20

Tylka GL, Hussey RS, Roncadori RW (1991) Interactions of vesicular-arbuscular mycorrhizal fungi, phosphorus, and *Heterodera glycines* on soybean. J Nematol 23:122–133

van der Heijden MGA, Klironomos JN, Ursic M, Moutoglis P, Streitwolf-Engel R, Boller T, Wiemken A, Sanders IR (1998) Mycorrhizal fungal diversity determines plant biodiversity, ecosystem variability and productivity. Nature 396:69–72

Vivekanandan M, Fixen PE (1991) Cropping system effects on mycorrhizal colonization, early growth and phosphorus uptake of corn. Soil Sci Soc Am J 55:136–140

Waceke JW (2003) A short communication: response of Soybean to concomitant inoculation with Arbuscular Mycorrhizal fungi and *Rhizobium*. J Tropic Microbiol 2:35–39

Walley FL, Germida JJ (1996) Failure to decontaminate *Glomus clarum* NT4 spores is due to spore wall-associated bacteria. Mycorrhiza 6:3–49

Wang X, Pan Q, Chen F, Yan X, Liao H (2011) Effects of co-inoculation with arbuscular mycorrhizal fungi and rhizobia on soybean growth as related to root architecture and availability of N and P. Mycorrhiza 21:173–181

Wardle DA, Bardgett RD, Klironomos JN, Setala H, van der Putten WH, Wall DH (2004) Ecological linkages between aboveground and belowground biota. Science 304:1629–1633

Whipps JM (2004) Prospects and limitations for mycorrhizas in biocontrol of root pathogens. Can J Bot 82:1198–1227

Wright SF, Anderson RL (2000) Aggregate stability and glomalin in alternative crop rotations for the central Great Plains. Biol Fertil Soils 31:249–253

Wright SF, Starr JL, Paltineanu IC (1999) Changes in aggregate stability and concentration of glomalin during tillage management transition. Soil Sci Soc Am J 63:1825–1829

Wu CG, Liu YS, Hung LL (1995) Spore development of *Entrophospora kentinensis* in an aeroponic system. Mycologia 87:582–587

Wu SC, Cheung KC, Luo YM, Wong MH (2004) Metal accumulation by *Brassica juncea* sharing rhizosphere with *Zea mays*: effect of mycorrhizal and beneficial bacterial inoculation. In: Proceedings of the fifth international conference on environmental geochemistry in the tropics, Haiko, Hainan, China, Nanjing, PR China: Institute Soil Science, Chinese Academy of Science, p 72

Xavier LJC, Germida JJ (2003) Bacteria associated with *Glomus clarum* spores influence mycorrhizal activity. Soil Biol Biochem 35:471–478

Xie Z, Staehelin C, Vierheili H, Wiemken A, Jabbouri S, Broughton WJ, Vogeli-Lange R, Boller T, Xie ZP (1995) Rhizobial nodulation factors stimulate mycorrhizal colonization of undulating and non-nodulating soybeans. Plant Physiol 108:1519–1525

Yao MK, Tweddell RJ, Desilets H (2002) Effect of two vesicular-arbuscular mycorrhizal fungi on the growth of micro-propagated potato plantlets and on the extent of disease caused by *Rhizoctonia solani*. Mycorrhiza 12:235–242

Zachee A, Bekolo N, Bime, Dooh N, Yalen M, Godswill N (2008) Effect of mycorrhizal inoculum and urea fertilizer on diseases development and yield of groundnut crops (*Arachis hypogaea* L.). Afr J Biotechnol 7:2823–2827

Zahran HH (1999) *Rhizobium*-legume symbiosis and nitrogen fixation under severe conditions and in an arid climate. Microbiol Mol Biol Rev 63:968–989

Zambolim L, Schenck NC (1983) Reduction of the effects of pathogenic, root-infecting fungi on soybean by the mycorrhizal fungus, *Glomus mosseae*. Phytopathology 73:1402–1405

Zhang F, Shen J, Li L, Liu X (2004) An overview of rhizosphere processes related with plant nutrition in major cropping systems in China. Plant Soil 260:89–99

Zhu YG, Smith SE, Barritt AR, Smith FA (2001) Phosphorus (P) efficiencies and mycorrhizal responsiveness of old and modern wheat cultivars. Plant Soil 237:249–255

Chapter 9
Arbuscular Mycorrhizal Diversity and Function in Grassland Ecosystems

Tomoko Kojima, Sasha Jenkins, Anjani Weerasekara, and Jing-Wei Fan

9.1 Introduction

Grasslands are widely distributed globally and occur on all continents except Antarctica. It is estimated that they make up one-fifth of the Earth's land surface (Parton et al. 1993). The majority of grasslands have resulted from anthropogenic activities where forests were cleared for domesticated animal grazing. Grasslands are important economically as they provide forage for livestock industry and a landscape for recreational and tourism activities. Grasslands are classified as natural, semi-natural or artificial. Natural and semi-natural grasslands are both grazed and unfertilised; natural systems do not receive any further agricultural improvements, whereas semi-natural grasslands are maintained by tillage, cutting, mowing or burning. Increased demand for grazing livestock has led to intensification and creation of artificial or improved grasslands that require regular reseeding and herbicide and fertiliser inputs. Some of the semi-natural grasslands are also used for hay and silage production. Artificial grasslands are generally more productive and profitable, whereas natural and semi-natural grasslands may have better soil quality and support a greater microbial biomass, species biodiversity and ecosystem function, including biogeochemical cycling, disease suppression and

T. Kojima (✉)
National Agriculture and Food Research Organization (NARO) Institute of Livestock and Grassland Science, 768 Senbonmatsu, Nasushiobara, Tochigi 329-2793, Japan
e-mail: kojima@affrc.go.jp

S. Jenkins • A. Weerasekara
Soil Biology and Molecular Ecology Group, School of Earth and Environment (M087) and UWA Institute of Agriculture, The University of Western Australia, Crawley, WA 6009, Australia

J.-W. Fan
State Key Laboratory of Grassland Agro-ecosystems, Institute of Arid Agroecology, School of Life Sciences, Lanzhou University, Lanzhou, Gansu Province 730000, China

© Springer-Verlag Berlin Heidelberg 2014
Z.M. Solaiman et al. (eds.), *Mycorrhizal Fungi: Use in Sustainable Agriculture and Land Restoration*, Soil Biology 41, DOI 10.1007/978-3-662-45370-4_9

carbon sequestration (Grayston et al. 2001; Oehl et al. 2003). Grasslands are under threat from the impacts of climate and land-use change (O'Donnell et al. 2007) due to an expanding human population with increased demand upon land resources for agricultural, residential, waste treatment, recreational and industrial development (Firbank 2005; Kan 2009). Meeting these demands requires a better understanding of soil functioning in grasslands and, ultimately, the ability to manipulate the diversity-function relationships and microbial interactions such as mycorrhizas (O'Donnell et al. 2007, 2001).

9.2   AM Fungal Ecology in Grasslands

Arbuscular mycorrhizal (AM) fungi are ubiquitous in grasslands (van der Heijden et al. 1998) where they are thought to play a major role in ecosystem functioning and services (Gianinazzi et al. 2010; Johnson et al. 2004) To date, AM fungi have been characterised in a wide range of grasslands across the world with varying soil and vegetation types including tropical grasslands (Zangaro et al. 2008), arid grasslands (Pezzani et al. 2006), boreal grasslands (Eriksen et al. 2002) and temperate grasslands (Barni and Siniscalco 2000). Succession in plant communities has been related to the changes in mycorrhizal type and their diversity (Allen 1996; Johnson et al. 1992). In particular, AM fungi play a vital role in facilitating plant nutrient uptake in nutrient-deficient soils of arid and semi-arid grasslands. In the case of succession in the Mexican Chihuahuan Desert, pioneer grasses were mycorrhiza-independent species, and late-successional grasses were more responsive to AM fungi and supported a higher spore density (Pezzani et al. 2006).

As a key link between above- and below-ground plant biomass, AM fungi play an important ecological role in shaping plant communities by influencing plant growth, plant diversity and competitive ability (Johnson et al. 2004; Klabi et al. 2014; McCain et al. 2011; van der Heijden et al. 1998). AM fungi are generalists and are able to colonise the majority of vascular plant species, including many important crop species such as maize, wheat, rice and potato (Roy-Bolduc and Hijri 2011). Thus, they do not display host specificity that is characteristic of other plant-microbe symbioses (Smith and Read 2008). As absolute symbionts, AM fungi provide a range of benefits (Roy-Bolduc and Hijri 2011) in return for plant-assimilated carbon from their hosts. They promote plant growth by mining soil pore spaces that are inaccessible to plant roots and by significantly increasing the total volume of soil explored for both nutrients and water (Garg and Chandel 2010; Kaya et al. 2003; Rillig et al. 2003). AM fungi also improve the efficiency of plant P uptake and a range of other nutrients, including organic N via various mechanisms (Al-Karaki and Clark 1998; Harrison et al. 2002; Subramanian and Charest 1999). AM fungi can contribute to improvement of soil structure through the formation of soil aggregates which in turn increases water-holding capacity (Andrade et al. 1998; Augé 2001; Augé et al. 2001) and tolerance of host plants to fungal pathogens, nematodes and water stress (Borowicz 2001; Forge et al. 2001; Newsham et al. 1995; Plenchette et al. 2005; Smith and Read 2008). They are also involved in alleviating effects of salinity on plant growth and ameliorating effects of heavy metal toxicity and pollution (Gianinazzi et al. 2010; Hildebrandt et al. 2001). These attributes enable AM fungi to have significant roles in maintaining sustainable levels of plant biomass in grasslands.

9.2.1 Host Plant Preferences

AM fungi can colonise a wide range of terrestrial plant species including grass species and do not display host specificity in the conventional sense of strict matching of host and symbiont (Smith and Read 2008). However, AM fungi can exhibit what has been termed 'ecological host specificity' (McGonigle and Fitter 1990) where roots of the same plant species may become preferentially colonised by some AM fungi in contrast to others when grown in the presence of the same community of AM fungi (Li et al. 2014; Scheublin et al. 2007; Vandenkoornhuyse et al. 2003). This may occur because of differences in root architecture or root proliferation. It may also arise due to variation in the infectivity of hyphae from spores of different AM fungi either arising from spores or from the common mycorrhizal network in soil (Fellbaum et al. 2014). Host preference for mycorrhiza formation can also be expressed in terms of the physiological effectiveness of the AM fungi (Klironomos 2003) that are present in the grassland community. Indeed, the indigenous assemblages of AM fungi associated with the native prairie grass *Andropogon gerardii* appeared to be functionally adapted to the local soil environment (Ji et al. 2013). However, species-specific interactions do not always occur (Santos et al. 2006), and this could be due to management and abiotic factors such as fertilisation and seasonality.

In experiments investigating interactions between AM fungi and plant species in grasslands compared with other agricultural systems, variation in the diversity and abundance of AM fungi associated with different plants has been demonstrated (Verbruggen et al. 2010). The same authors found that AM fungal species diversity was highest in grasslands compared to a cropping system. However, AM fungal communities of organically managed cropping systems were also more similar to those of grasslands compared to cropping systems where synthetic fertiliser had been applied (Verbruggen et al. 2010). For example, the spore number and diversity of AM fungi were both greater in soil collected from red clover or grassland than from under crops of barley or wheat (Menéndez et al. 2001). Hetrick and Bloom (1986) showed that there was higher spore production of AM fungi when sudangrass was grown as the host plant, compared with asparagus, tomato and red clover. Furthermore, in an experiment investigating the coexistence of two grass species, the distribution of phosphorus and nitrogen differed when plants were inoculated with different AM fungal species (van der Heijden 2003). Even when environmental conditions are similar, the AM fungal species composition or biomass can differ in rhizosphere soils of different plant species coexisting in the same soil. This has been demonstrated in semi-natural grasslands, where the AM fungal spore density in rhizosphere soil associated with *Miscanthus sinensis* was higher than in soil associated with *Pleioblastus chino* (Murakoshi et al. 1998).

Different cultivars of the same grass species can differ in response to colonisation by AM fungi, as was shown for growth of orchard grass with different combinations of AM fungal species (Tsuchida and Nonaka 2002, 2003). For the legume alfalfa, the effects of AM fungal species also varied with cultivar (Douds et al. 1998). Furthermore, AM fungi might alter plant species diversity by increasing competitive intraspecific suppression and decreasing interspecific suppression of small plants surrounded by larger neighbour plants (Moora and Zobel 1996). In another study, AM fungi were found to maintain grassland community stability by regulating plant diversity through increased interspecific competition between the grazing grass *Elymus nutans* and the poisonous plant, *Ligularia virgaurea* (Jin et al. 2011). Where there are dominant and subordinate plant species in a grassland community, the dominant species could be negatively affected depending on the combination of AM fungi present, especially if the growth of subordinates is enhanced by mycorrhizas to a greater extent than that of the dominant species (Klabi et al. 2014). Klabi et al. (2014) found that the competitiveness of the dominant cool-season grass *Elymus trachycaulus* ssp. *subsecundus* was reduced in the presence of *Glomus cubense*. These effects have been demonstrated in pot experiments where it has been suggested that less favourable AM fungi could reduce the dominance of some plant species in grassland (Mariotte et al. 2013). In addition, seasonal changes in the composition of AM fungi in roots can occur (Santos et al. 2006), and AM fungal diversity can be affected by plant species composition (Johnson et al. 2004). Furthermore, mycorrhizal dependency is different among plant species, and there are differences in mycorrhizal response between C4 and C3 grasses (Hetrick and Wilson 1991; Hetrick et al. 1990). Indeed, C4 grasses had higher mycorrhizal dependency than did C3 grasses (Lugo et al. 2003). Therefore, in grasslands, the composition of a plant community could affect the AM fungal community and *vice versa*. Consequently, factors that contribute to these processes are complex in space and time, and site-specific studies are needed to identify the mechanisms involved before confident generalisations can be made.

9.2.2 AM Fungal Diversity and Analytical Methods

AM fungal diversity in grasslands was initially characterised using bioassays and traditional taxonomic approaches that identified and quantified AM fungal spores extracted directly from the field or trap cultures (Leal et al. 2009; Oehl et al. 2003, 2009). However, it is often difficult to characterise spores below the family level using traditional methodology due to the lack of discriminating morphological characters (Redecker and Raab 2006) and because sporulation is species and environment dependent (Young 2012). As AM fungi are obligate symbionts (Franz Lang and Hijri 2009), selective isolation in artificial media remains a significant challenge. Consequently, the vast majority of the currently described taxa on the AM fungi database (www.amf-phylogeny.com) are uncultured morphospecies (Krüger et al. 2009, 2012). Since AM fungal communities influence plant competition, diversity and productivity, a greater taxonomic and phylogenetic resolution is needed for the understanding relationships between plant-fungi interaction and ecosystem functioning in grasslands (Klironomos 2003; McCain et al. 2011; Öpik et al. 2003; van der Heijden et al. 1998, 2003).

More recently, knowledge of AM fungi in grasslands has been based on the application of molecular methodologies with more rapid and precise identification of the fungi in both roots and soil (Montero Sommerfeld et al. 2013; Young 2012). PCR-based analysis of ribosomal RNA gene (rRNA) using primers that target the small subunit (SSU), large subunit (LSU) and the internal transcribed spacer (ITS) region has been extensively used as taxonomic biomarkers for characterising AM fungal communities in soil (Krüger et al. 2009, 2012; Lee et al. 2008). However, some primers (especially those targeting the SSU) have poor specificity and sensitivity for AM fungi making it difficult to distinguish between different species (Krüger et al. 2009; Redecker 2000; Schüßler et al. 2001). For better taxonomic resolution of AM fungi at the species level and a more unbiased determination community diversity, it is recommended that primers cover the ITS and LSU region (Stockinger et al. 2009). The ITS region is recognised as a general fungal bar-coding marker; however, it is difficult to differentiate between certain groups within *Glomeromycota* (Schocha et al. 2012). Recently, highly AM fungal-specific primers that amplify the SSU-ITS-LSU region (Krüger et al. 2009) have been described for better taxonomic resolution. Moreover, new phylogenetic reference data of AM fungi (Krüger et al. 2012) and databases (e.g. MaarjAM) (Öpik et al. 2010) containing representative sequences from published *Glomeromycota* sequence-based taxa and known morphospecies have been developed that improve DNA-based species characterisation of AM fungal communities in grasslands.

Initially, 'fingerprinting' technologies such as denaturing gradient gel electrophoresis (DGGE) and terminal restriction fragment length polymorphism (T-RFLP) were used to gain insights into how AM fungal community structure was shaped by environmental and management drivers (Lugo et al. 2003; Montero Sommerfeld et al. 2013; Yang et al. 2010). Microbial diversity was assessed by preparing clone libraries and analysing with Sanger sequencing methods (Chen et al. 2014; Liu et al. 2012; Santos-González et al. 2007; Yang et al. 2010). Data generated from these molecular techniques has revealed that the community structure and diversity of AM fungi in grassland are strongly influenced by seasonality, fertilisation and management practices (Chen et al. 2014; Montero Sommerfeld et al. 2013; Yang et al. 2010). With the advent of the next-generation sequencing (NGS), AM fungal diversity, function and distribution patterns are now being examined at levels that were unthinkable only a decade ago (Öpik et al. 2010; Schüßler and Walker 2010; Stürmer 2012) which has significantly improved our understanding of their role in grasslands and other environments (Dumbrell et al. 2011; Krüger et al. 2012; Redecker et al. 2013; Stockinger et al. 2010).

Glomus spp. are often dominant in grassland soil based on examination of spores. For example, sporocarpic *Glomus* spp. were dominant in a natural mountain grassland (Pampas) in Argentina (Lugo and Cabello 2002). *Glomus heterosporum* and *G. intraradices* were dominant in the tallgrass prairie of Kansas, USA (Eom et al. 2000), and unidentified *Glomus* spp. were dominant in grasslands of Central Europe (Oehl et al. 2003). *G. aggregatum* and *G. leptotichum* (*Ambicispora* sp.) were dominant in experimental plots in Minnesota, USA (Johnson et al. 1992), and *G. geosporum* was dominant in the Inner Mongolia steppe as determined using the method trap culture (Su and Guo 2007). In contrast, *Acaulospora colossica*, *Scutellospora calospora*, *Gigaspora gigantea*, *Archaeospora leptoticha* and several other species were common in grasslands of North Carolina, USA (Pringle and Bever 2002), and in several Japanese semi-natural grasslands, *Sclerocystis rubiformis* and an unidentified *Glomus* sp. (orange-dark red, about 100–150 μm) were the dominant species (Kojima et al. 2009; Fig. 9.1) along with Paris-type root colonisation (Fig. 9.2). Recently, DNA-based sequencing approaches have characterised AM fungal communities associated with the soil and roots of grassland plant species (Hiiesalu et al. 2014; Li et al. 2014). They revealed that grassland

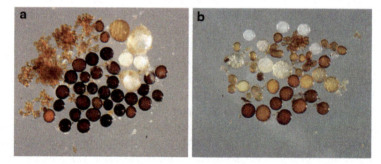

Fig. 9.1 AM fungal spores in the soil (10 g) collected from semi-natural: (**a**) *Zoysia japonica*-dominant or (**b**) *Miscanthus sinensis*-dominant grasslands

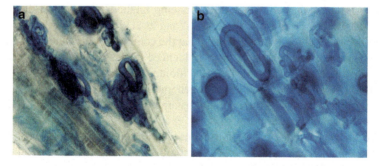

Fig. 9.2 AM fungal colonisation stained with Trypan blue in the roots of (**a**) *Zoysia japonica*, (**b**) *Pleioblastus chino* collected from semi-natural grasslands

communities are dominated by members of the families *Archaeosporaceae*, *Claroideoglomeraceae*, *Diversisporaceae*, *Glomeraceae*, *Gigasporaceae*, *Acaulosporaceae* and *Paraglomeraceae* (Hiiesalu et al. 2014; Li et al. 2014; Ohsowski et al. 2014). Members of the *Glomerales* (*Glomeraceae* and *Claroideoglomeraceae*) usually exhibit more r-selected traits, whereas members of the *Diversisporales* (*Gigasporaceae*, *Acaulosporaceae*, *Pacisporaceae* and *Diversisporaceae*) tend towards more K-selected traits such as large extraradical mycelium, spore size and density and increased hyphal diameter (Chagnon et al. 2013; Sýkorová et al. 2007). The most dominant AM fungal genus found in these grasslands was *Glomus*, followed by *Rhizophagus*, *Acaulospora*, *Scutellospora* and *Diversispora* which is in agreement with taxa recovered using traditional morphological methods (del Mar Alguacil et al. 2010; Hazard et al. 2014; Liu et al. 2012).

Conclusions about the composition of AM fungal species in grassland soils can differ depending on the methods used (Brundrett et al. 1999; Clapp et al. 2003). Sampling strategy, sampling intensity and the methodology used may further influence the detection and quantification of species diversity of AM fungi in roots or soils (Shi et al. 2012), including grasslands (Whitcomb and Stutz 2007). For example, when the AM fungal community colonising roots in a grassland was investigated using phylogenetic analysis, there was not a close correspondence with the morphotypes of AM fungi (Schnoor et al. 2011a). Furthermore, it has been shown that the treatment of root samples can influence conclusions related to AM fungal analysis (Renker et al. 2006). Molecular methods generally yield more information and have greater taxonomic resolution; however, assigning taxonomic affiliation is complicated by the fact that AM fungi reproduce asexually leading to higher diversity within species (Munkvold et al. 2004; Stockinger et al. 2009). This can be overcome by analysing molecular assays in combination with morphological-based methods that can identify different spore features (Wetzel et al. 2014). The integration of these methods will provide a greater insight into the AM fungal diversity and functional activities in grasslands.

9.3 The Dynamics of AM Fungi in Grasslands

In some grasslands, environmental factors such as altitude, nutrient availability, salinity, soil water content, temperature, pH as well as plant diversity were identified as key drivers of AM fungi diversity (An et al. 2008; Eason et al. 1999; Eom et al. 2001; Li et al. 2014). In other grasslands, light incidence, temperature and aluminium availability in the soil influenced AM fungal colonisation (Göransson et al. 2008; Zangaro et al. 2013) demonstrating the breadth of influence of local soil conditions. Intensification and changes in agricultural practice include fertilisation, manure application, pesticide usage, crop rotation, mowing and tillage (Barto et al. 2010; Binet et al. 2013; Birgander et al. 2014; Collins and Foster 2009; Kabir 2005; Mathimaran et al. 2007; Oehl et al. 2003), all of which could influence colonisation of roots by AM fungi. For example, long-term pesticide use resulted in an 80 % reduction in mycorrhizal root colonisation (Smith et al. 2000). Livestock grazing is an example of a major land-use practice which influences plant production, plant species composition and nutrient cycling in soil, all of which could alter root colonisation by AM fungi and their relative abundance.

Temporal changes and seasonal fluctuations in AM fungal colonisation have been observed (Bentivenga and Hetrick 1992; Birgander et al. 2014; Escudero and Mendoza 2005; Lingfei et al. 2005; Lugo et al. 2003). These changes are most likely to be a consequence of seasonal differences in rainfall and temperature that alter soil water content and metabolic activity, respectively (Lingfei et al. 2005; Lugo et al. 2003). In a number of studies, a distinct warm to cold seasonal shift in AM fungi community composition was observed, and root colonisation rates were greater during the growing season (Dumbrell et al. 2011; Hazard et al. 2014; Lugo et al. 2003; Montero Sommerfeld et al. 2013). In a seasonal investigation of the interaction between AM fungi and dark septate endophytic (DSE) fungi, AM fungi were most abundant during the peak growing season with dominant C4 plants, whilst DSE fungi were most abundant during the early part of the growing season (Mandyam and Jumpponen 2008). In contrast, higher AM fungi colonisation rates occurred during the growing season, but DSE colonisation was positively correlated with AM fungi colonisation suggesting they share an ecological niche in the grasslands of southwest China which were studied (Lingfei et al. 2005).

Interestingly, temporal variation in AM fungi colonisation or diversity is not always observed in grasslands. Saito et al. (2004) showed that the AM fungi in roots of *Zoysia japonica* remained relatively unchanged throughout a year. Other studies have shown that the composition and diversity of AM fungal communities do not change seasonally (Santos-González et al. 2007). Also, in grassland dominated by *Pleioblastus chino* and *Miscanthus sinensis*, the vegetation changed over a 4-year period, but the AM fungal community were relatively stable (Kojima et al. 2009). Sanders and Fitter (1992) showed colonisation in some plant species in a grassland in England changed both within and between years, but colonisation in other species was more constant. Thus, whilst the community structure of AM fungi in roots of some perennial plant species is relatively constant (McGonigle and Fitter 1990; Read et al. 1976), variability observed elsewhere may depend on location, soil characteristics or grassland composition.

9.3.1   Soil and Management Practices

In grasslands, colonisation of roots by AM fungi can occur through interception of hyphae from nearby roots, as well as from hyphae from germinated spores (Allen and Allen 1980). Thus, the network of AM fungi in grasslands is important not only for transfer of nutrients but also for colonisation of new roots (Wilson and Trinick 1983). Tilling or ploughing has been shown to reduce overall AM fungal abundance, spore density and species richness (Allison et al. 2005; Helgason et al. 2010;

Kabir 2005). Disking or tillage alters AM fungal dynamics in grasslands by disturbing the hyphal network and interacting with other soil organisms (Kabir et al. 1997; Murugan et al. 2013). In order to fully investigate these processes, many facets of AM fungi need to be considered. For example, AM colonisation was lower in a 3-year-old disked site compared to the natural prairie, but spore density was similar (Allen and Allen 1980). In another study, tillage increased the relative abundance of saprotrophic fungi at the expense of AMF and bacteria (Murugan et al. 2013). AM fungal biomass in roots and soil were both decreased by ploughing and rotavation in a sandy grassland (Schnoor et al. 2011a), and it was found that carbon allocation to the AM fungi decreased with soil disturbance. Furthermore, for a semi-natural grassland, soil disturbance with replicated ploughing reduced phylotype richness and changed the AM fungal community composition (Schnoor et al. 2011b). Disturbance of the soil structure is thought to make soil organic nitrogen accessible for mineralization which was otherwise protected from degradation (Kristensen et al. 2000). In another example, the reduction of woody species in a simulated grassland by burning decreased mycorrhizal colonisation (O'Dea 2007). Overall, management practices that disturb plant vegetation or AM fungal communities are likely to decrease AM fungal diversity in grasslands. Mechanical disturbances can also affect soil aggregation. Helgason et al. (2010) found that AM fungi were particularly affected by tillage disturbance with increases of 40–60 % among aggregate-size classes in non-tilled system compared to the conventional tilled system.

9.3.2 Soil Nutrients and Fertiliser Regime

Intensive management of grasslands is negatively correlated with AM fungal diversity, and this is thought to be due to management-induced changes in nutrient availability following the application of lime, synthetic fertiliser or manure (Jenkins et al. 2009). Moreover, since AM fungi play a role in mitigating nitrous oxide in soils, disruption of the mycorrhizal symbiosis through agricultural intensification may further contribute to increased N_2O emissions (Bender et al. 2014).

In general, increasing levels of fertilisation, whether they are synthetic or manures, is usually associated with reduced diversity of AM fungi compared to unfertilised grassland soils (Chen et al. 2014; Christie and Kilpatric 1992; Murugan et al. 2013; Wuen et al. 2002). However, the AM fungal community response to fertilisation is largely dependent upon the loading amount, fertiliser type and combination and the method of application (Chen et al. 2014; Lin et al. 2012). For example, AM fungal colonisation in two wild plant species was higher in the unfertilised compared to the grassland receiving phosphate fertilisation (Wuen et al. 2002). In another study, the genus *Acaulospora* dominated in the unfertilised treatment, whereas the genus *Glomus* prevailed in the fertilised soils, but this was largely dependent on the amount of P fertiliser applied (del Mar Alguacil et al. 2010). Overall, spore densities of AM fungi and mycorrhizal colonisation rates are higher in the unfertilised soil compared to fertilised soils. Nevertheless, fungal biomass and AM colonisation rates were higher in organically managed soils (receiving manure applications) compared to soils receiving synthetic fertiliser (Bittman et al. 2005; Eason et al. 1999; Kabir et al. 1997). Also, the age of the plant can influence the impact of fertilisation on AM fungal communities. For example, the formation of mycorrhizas in seedlings of *Plantago lanceolata* in managed grasslands were affected by fertilisation, but the adult plants were unaffected (Šmilauerová et al. 2012). In this case, fertilisation decreased total colonisation and relative arbuscular frequency in seedling roots. These differences might be dependent mainly on the phosphate and nitrate availability in the grassland.

The addition of nitrogen fertiliser has been reported to cause a reduction in species diversity of AM fungi (Chen et al. 2014; Egerton-Warburton and Allen 2000; Jumpponen et al. 2005), but AM fungi abundance is often unaffected (Chen et al. 2014). For example, Jumpponen et al. (2005) found that clades within *Glomus* spp. were associated with either the fertilised and unfertilised treatments in tallgrass prairie. The decreased species richness could reflect a reduction in translocation of C from the plant to the AM fungi under N-rich conditions and the roots becoming colonised by a few nitrophilic AM fungi taxa. The addition of phosphate fertiliser has been shown to reduce AM fungal abundance, in particular mycorrhizal colonisation rate, arbuscule formation and hyphal length density (Chen et al. 2014; Khaliq and Sanders 2000; Liu et al. 2012). Indeed, grasslands receiving high P fertiliser inputs can be less productive because under these non-limiting conditions, mycorrhizas can reduce plant growth when not contributing to the symbiosis (Kaeppler et al. 2000). Other examples of site-specific effects include studies where P fertilisation altered AM fungi community structure in some grasslands (del Mar Alguacil et al. 2010; Liu et al. 2012), but other grasslands were unaffected (Chen et al. 2014).

9.3.3 Grazing Pressure

Grazing pressure by domesticated animals in grasslands can alter the extent to which roots are colonised by AM fungi. In most studies, grazing or defoliation has been shown to decrease AM fungal colonisation in grasslands or in grassland species grown under controlled conditions in a glasshouse (Bethlenfalvay and Dakessian 1984; Eom et al. 2001; Grime et al. 1987; Saravesi et al. 2014). However, grazing has also been shown to increase AM fungal colonisation (Frank et al. 2003) or have no influence on colonisation of roots by AM fungi (Yang et al. 2013). A negative effect of defoliation on mycorrhizal colonisation has usually been ascribed to reduced photosynthetic capacity of plants which in turn, limits the carbon supply to roots and mycorrhizal fungi, particularly for heavy grazing conditions (Barto et al. 2010). However, different responses between AM fungi and their hosts during defoliation could also be associated with variation in plant genotype, soil and climatic factors, fertilisation and/or differences in relative growth rates of roots and hyphae of AM fungi inside roots and in the surrounding soil (Barto and Rillig 2010).

One possible reason for the inconsistent findings between studies investigating the effects of grazing on AM fungal community is due to methodological constraint. In a study of mountain grasslands in Argentina, grazing did not affect AM fungi when they were assessed as spores alone (Lugo et al. 2003). However, when the interaction between grazing management and AM fungi was investigated in a desert steppe in Inner Mongolia, China, using sporulation, colonisation and diversity measures, differences were found between control and grazed treatments (Bai et al. 2013). In another system, differences in plant species had more of an impact. Yamane et al. (1999) found that grazing pressure and the defoliation of shoots decreased AM fungi colonisation in roots of *Miscanthus sinensis* but not in roots of *Zoysia japonica*. Further, molecular analysis of AM fungi colonising roots of the same plant species showed that defoliation had a greater impact on the community composition of AM fungi in roots of *M. sinensis* than in roots of *Z. japonica* (Saito et al. 2004). Also, when the effect of overgrazing on AM fungal diversity was investigated in a grassland in Inner Mongolia by trap culture (Su and Guo 2007), both spore diversity and species richness of AM fungi were significantly decreased by long-term overgrazing. Grazing can alter root morphology, plant community composition and soil properties (Eom et al. 2001; Hiiesalu et al. 2014; Su and Guo 2007) which could all alter mycorrhiza formation to varying degrees. Hiiesalu et al. (2014) found that AM fungi species richness was associated with both above- and below ground plant species richness.

Increased grazing pressure can induce a plant succession; for example, in a Japanese semi-natural grassland, tallgrass species such as *Pleioblastus chino* and *Miscanthus sinensis* were replaced by turf grass species, *Zoysia japonica,* following long-term grazing (Numata 1961). In another study it was found the number of plant species and plant community cover in a grassland decreased with long-term overgrazing and affected AM mycorrhizal colonisation (Su and Guo 2007). Similarly, the changes in both AM fungal and plant community composition were more pronounced in a tallgrass prairie after agricultural disturbance than in a comparable natural system (Stover et al. 2012). Furthermore, AM fungal diversity was higher under light to moderate grazing pressure compared to heavy grazing, and these differences were attributed to changes in plant diversity, in particular the density of tillers of the dominant grass *Leymus chinensis* (Ba et al. 2012). Such a reduction in above-ground plant biomass would reduce the capacity of plants to supply carbon to roots. Reduced carbon supply to roots could fail to meet the demands of AM fungi, leading to reductions in mycorrhizal colonisation (Barto et al. 2010). Simultaneously, a change in root morphology induced by grazing (especially, hyphal density and area) could also influence root colonisation by AM fungi (Allsopp 1998; Lugo et al. 2003). Another factor that requires consideration is that species of AM fungi can differ in their tolerance of grazing pressure and in their competitive ability which is evident by the observation that non-*Glomus* species were particularly affected by overgrazing in a study in an Inner Mongolia steppe grassland (Su and Guo 2007).

Grazing has been shown to increase carbon storage in C4-dominated grasslands (Sanjari et al. 2008), whereas grazing decreased soil organic carbon in C3-dominated grasslands (Li et al. 2008). It has been suggested that changes in plant species composition as a result of grazing may be more important than direct or indirect effects of grazing on soil carbon (Yates et al. 2000). Indeed, sheep overgrazing for 17 years did not significantly alter soil organic carbon level and total soil N content (Raiesi and Asadi 2006) indicating that variation in decomposition rates of different plant species may be greater than direct impacts of grazing on carbon dynamics in the studied ecosystem. As AM fungi can make important contributions to soil aggregation, nutrient cycling and carbon sequestration (Rillig and Mummey 2006), grazing could influence a range of soil processes that vary in the extent to which they are mediated by mycorrhizas. Although the dynamics of soil carbon in grasslands is expected to influence AM fungi, this could be site specific and depend on differences in hyphal growth associated with soil organic matter, as well as on relationships between the various AM fungi present and the extent to which each one colonises roots of the plant species that are present (Allsopp 1998). In other words, the effects of grazing on AM fungi need to be interpreted within a very complex and dynamic soil-root environment.

9.3.4   Interactions with Other Soil Organisms

There are considerable interactions between AM fungi and other soil organisms in grasslands. For example, earthworms can be important agents in the distribution of mycorrhizal fungi and influence plant establishment in early succession (Gange 1993). Earthworm-AM fungal interactions affected plant diversity and productivity in a model grassland ecosystem, and different earthworm species were shown to have different influences (Zaller et al. 2011). Both earthworms and AM fungi play roles in the formation of soil aggregates (Rillig and Mummey 2006; Rillig et al. 2002; Spurgeon et al. 2013; Wright and Upadhyaya 1998) and that the amount of glomalin formed by AM fungi that is likely to contribute to soil aggregation changes with land use (Rillig et al. 2003). Although tillage can decrease the amount of glomalin in soil (Wright et al. 1999), it is an uncommon practice in most grasslands, so glomalin may have greater persistence in stable grassland ecosystems. Furthermore, Lutgen et al. (2003) showed that glomalin present in grassland soil changed seasonally with arbuscular colonisation in plant roots. There are multiple interactions between diverse communities of soil organisms and mycorrhizal fungi in grasslands, including those related to root-feeding nematodes. For example, it has been demonstrated that AM fungi can control nematodes feeding in the dune grass *Ammophila arenaria* (De La Peña et al. 2006) and protect plants from pathogens or parasites (Gworgwor and Weber 2003; Newsham et al. 1995).

9.4 Conclusions

Many studies on AM fungi in grassland ecosystems have been reported where species diversity, colonisation dynamics and biomass were investigated. Most studies have focused on the ecological aspects of AM fungi in grasslands, with less emphasis on their function. In addition to field observations, pot experiments have been used to simulate grassland environments to determine underlying processes affecting interactions between AM fungi and grassland plant species. Although it is difficult to investigate the function and effectiveness of unidentified AM fungal species, future studies need to establish new approaches for evaluating the function of unidentified and unculturable AM fungi which dominate grasslands.

AM fungal diversity in grasslands generally does not appear to change greatly within a few years after management practices are implemented. However, the full extent of how communities of AM fungi function in grasslands is not understood. In some situations, AM fungi in grasslands have been compared with those in crop fields, and usually AM fungal diversity was higher in the grassland than in the crop. Clarification of interactions between AM fungi and other soil organisms would improve understanding of how AM fungi contribute to grassland ecosystems. This information is important from both agricultural production and ecological perspectives. It is also important for protection of threatened plant species that occur in natural and semi-natural grasslands.

References

Al-Karaki G, Clark R (1998) Growth, mineral acquisition, and water use by mycorrhizal wheat grown under water stress. J Plant Nutr 21:263–276

Allen MF (1996) The ecology of arbuscular mycorrhizas: a look back into the 20th century and a peek into the 21st century. Mycol Res 100:769–782

Allen EB, Allen MF (1980) Natural re-establishment of vesicular-arbuscular mycorrhizae following stripmine reclamation in Wyoming. J Appl Ecol 17:139–147

Allison VJ, Miller RM, Jastrow JD, Matamala R, Zak DR (2005) Changes in soil microbial community structure in a tallgrass prairie chronosequence. Soil Sci Soc Am J 69:1412–1421

Allsopp N (1998) Effect of defoliation on the arbuscular mycorrhizas of three perennial pasture and rangeland grasses. Plant and Soil 202:117–124

An G-H, Miyakawa S, Kawahara A, Osaki M, Ezawa T (2008) Community structure of arbuscular mycorrhizal fungi associated with pioneer grass species Miscanthus sinensis in acid sulfate soils: habitat segregation along pH gradients. Soil Sci Plant Nutr 54:517–528

Andrade G, Mihara K, Linderman R, Bethlenfalvay G (1998) Soil aggregation status and rhizobacteria in the mycorrhizosphere. Plant and Soil 202:89–96

Augé RM (2001) Water relations, drought and vesicular-arbuscular mycorrhizal symbiosis. Mycorrhiza 11:3–42

Augé RM, Stodola AJ, Tims JE, Saxton AM (2001) Moisture retention properties of a mycorrhizal soil. Plant and Soil 230:87–97

Ba L, Ning JX, Wang DL, Facelli E, Facelli JM, Yang YN, Zhang LC (2012) The relationship between the diversity of arbuscular mycorrhizal fungi and grazing in a meadow steppe. Plant and Soil 352:143–156

Bai G, Bao Y, Du G, Qi Y (2013) Arbuscular mycorrhizal fungi associated with vegetation and soil parameters under rest grazing management in a desert steppe ecosystem. Mycorrhiza 23:289–301

Barni E, Siniscalco C (2000) Vegetation dynamics and arbuscular mycorrhiza in old-field successions of the western Italian Alps. Mycorrhiza 10:63–72

Barto EK, Rillig MC (2010) Does herbivory really suppress mycorrhiza? A meta-analysis. J Ecol 98:745–753

Barto EK, Alt F, Oelmann Y, Wilcke W, Rillig MC (2010) Contributions of biotic and abiotic factors to soil aggregation across a land use gradient. Soil Biol Biochem 42:2316–2324

Bender SF, Plantenga F, Neftel A, Jocher M, Oberholzer HR, Köhl L, Giles M, Daniell TJ, van der Heijden MG (2014) Symbiotic relationships between soil fungi and plants reduce N2O emissions from soil. ISME J 8(6):1336–1345

Bentivenga S, Hetrick B (1992) Seasonal and temperature effects on mycorrhizal activity and dependence of cool-and warm-season tallgrass prairie grasses. Can J Bot 70:1596–1602

Bethlenfalvay GJ, Dakessian S (1984) Grazing effects on mycorrhizal colonization and floristic composition of the vegetation on a semiarid range in northern Nevada. J Range Manage 37:312–316

Binet M et al (2013) Effects of mowing on fungal endophytes and arbuscular mycorrhizal fungi in subalpine grasslands. Fungal Ecol 6:248–255

Birgander J, Rousk J, Olsson PA (2014) Comparison of fertility and seasonal effects on grassland microbial communities. Soil Biol Biochem 76:80–89

Bittman S, Forge TA, Kowalenko CG (2005) Responses of the bacterial and fungal biomass in a grassland soil to multi-year applications of dairy manure slurry and fertilizer. Soil Biol Biochem 37:613–623

Borowicz VA (2001) Do arbuscular mycorrhizal fungi alter plant-pathogen relations? Ecology 82:3057–3068

Brundrett M, Abbott L, Jasper D (1999) Glomalean mycorrhizal fungi from tropical Australia. Mycorrhiza 8:305–314

Chagnon P-L, Bradley RL, Maherali H, Klironomos JN (2013) A trait-based framework to understand life history of mycorrhizal fungi. Trends Plant Sci 18:484–491

Chen YL, Zhang X, Ye JS, Han HY, Wan SQ, Chen BD (2014) Six-year fertilization modifies the biodiversity of arbuscular mycorrhizal fungi in a temperate steppe in inner Mongolia. Soil Biol Biochem 69:371–381

Christie P, Kilpatric D (1992) Vesicular-arbuscular mycorrhiza infection in cut grassland following long-term slurry application. Soil Biol Biochem 24:325–330

Clapp JP, Helgason T, Daniell TJ, Peter J, Young W (2003) Genetic studies of the structure and diversity of arbuscular mycorrhizal fungal communities. In: Sanders IR, Marcel GA, van der Heijden MGA (eds) Mycorrhizal ecology. Springer, Heidelberg, pp 201–224

Collins CD, Foster BL (2009) Community-level consequences of mycorrhizae depend on phosphorus availability. Ecology 90:2567–2576

De La Peña E, Echeverría SR, Van Der Putten WH, Freitas H, Moens M (2006) Mechanism of control of root-feeding nematodes by mycorrhizal fungi in the dune grass *Ammophila arenaria*. New Phytol 169:829–840

del Mar Alguacil M, Lozano Z, Campoy MJ, Roldan A (2010) Phosphorus fertilisation management modifies the biodiversity of AM fungi in a tropical savanna forage system. Soil Biol Biochem 42:1114–1122

Douds D Jr, Galvez L, Bécard G, Kapulnik Y (1998) Regulation of arbuscular mycorrhizal development by plant host and fungus species in alfalfa. New Phytol 138:27–35

Dumbrell AJ, Ashton PD, Aziz N, Feng G, Nelson M, Dytham C, Fitter AH, Helgason T (2011) Distinct seasonal assemblages of arbuscular mycorrhizal fungi revealed by massively parallel pyrosequencing. New Phytol 190(3):794–804

Eason W, Scullion J, Scott E (1999) Soil parameters and plant responses associated with arbuscular mycorrhizas from contrasting grassland management regimes. Agr Ecosyst Environ 73:245–255

Egerton-Warburton LM, Allen EB (2000) Shifts in arbuscular mycorrhizal communities along an anthropogenic nitrogen deposition gradient. Ecol Appl 10:484–496

Eom A-H, Hartnett DC, Wilson GW (2000) Host plant species effects on arbuscular mycorrhizal fungal communities in tallgrass prairie. Oecologia 122:435–444

Eom A-H, Wilson GW, Hartnett DC (2001) Effects of ungulate grazers on arbuscular mycorrhizal symbiosis and fungal community structure in tallgrass prairie. Mycologia 93:233–242

Eriksen M, Bjureke KE, Dhillion SS (2002) Mycorrhizal plants of traditionally managed boreal grasslands in Norway. Mycorrhiza 12:117–123

Escudero V, Mendoza R (2005) Seasonal variation of arbuscular mycorrhizal fungi in temperate grasslands along a wide hydrologic gradient. Mycorrhiza 15:291–299

Fellbaum CR, Mensah JA, Cloos AJ, Strahan GE, Pfeffer PE, Kiers ET, Bücking H (2014) Fungal nutrient allocation in common mycorrhizal networks is regulated by the carbon source strength of individual host plants. New Phytol 203:646–656

Firbank L (2005) Striking a new balance between agricultural production and biodiversity. Ann Appl Biol 146:163–175

Forge T, Muehlchen A, Hackenberg C, Neilsen G, Vrain T (2001) Effects of preplant inoculation of apple (*Malus domestica* Borkh.) with arbuscular mycorrhizal fungi on population growth of the root-lesion nematode, *Pratylenchus penetrans*. Plant and Soil 236:185–196

Frank DA, Gehring CA, Machut L, Phillips M (2003) Soil community composition and the regulation of grazed temperate grassland. Oecologia 137:603–609

Franz Lang B, Hijri M (2009) The complete Glomus intraradices mitochondrial genome sequence - a milestone in mycorrhizal research. New Phytol 183:3–6

Gange AC (1993) Translocation of mycorrhizal fungi by earthworms during early succession. Soil Biol Biochem 25:1021–1026

Garg N, Chandel S (2010) Arbuscular mycorrhizal networks: process and functions. A review. Agron Sustain Dev 30:581–599

Gianinazzi S, Gollotte A, Binet M-N, van Tuinen D, Redecker D, Wipf D (2010) Agroecology: the key role of arbuscular mycorrhizas in ecosystem services. Mycorrhiza 20:519–530

Göransson P, Olsson PA, Postma J, Falkengren-Grerup U (2008) Colonisation by arbuscular mycorrhizal and fine endophytic fungi in four woodland grasses - variation in relation to pH and aluminium. Soil Biol Biochem 40:2260–2265

Grayston S, Griffith G, Mawdsley J, Campbell C, Bardgett RD (2001) Accounting for variability in soil microbial communities of temperate upland grassland ecosystems. Soil Biol Biochem 33:533–551

Grime J, Mackey J, Hillier S, Read D (1987) Floristic diversity in a model system using experimental microcosms. Nature 328:420–422

Gworgwor NA, Weber HC (2003) Arbuscular mycorrhizal fungi-parasite-host interaction for the control of *Striga hermonthica* (Del.) Benth. in sorghum [*Sorghum bicolor* (L.) Moench]. Mycorrhiza 13:277–281

Harrison MJ, Dewbre GR, Liu J (2002) A phosphate transporter from *Medicago truncatula* involved in the acquisition of phosphate released by arbuscular mycorrhizal fungi. Plant Cell 14:2413–2429

Hazard C, Boots B, Keith AM, Mitchell DT, Schmidt O, Doohan FM, Bending GD (2014) Temporal variation outweighs effects of biosolids applications in shaping arbuscular mycorrhizal fungi communities on plants grown in pasture and arable soils. Appl Soil Ecol 82:52–60

Helgason BL, Walley FL, Germida JJ (2010) No-till soil management increases microbial biomass and alters community profiles in soil aggregates. Appl Soil Ecol 46:390–397

Hetrick BAD, Bloom J (1986) The influence of host plant on production and colonization ability of vesicular-arbuscular mycorrhizal spores. Mycologia 78(1):32–36

Hetrick B, Wilson G (1991) Effects of mycorrhizal fungus species and metalaxyl application on microbial suppression of mycorrhizal symbiosis. Mycologia 97–102

Hetrick B, Wilson G, Todd T (1990) Differential responses of C3 and C4 grasses to mycorrhizal symbiosis, phosphorus fertilization, and soil microorganisms. Can J Bot 68:461–467

Hiiesalu I, Pärtel M, Davison J, Gerhold P, Metsis M, Moora M, Öpik M, Vasar M, Martin Z, Wilson SD (2014) Species richness of arbuscular mycorrhizal fungi: associations with grassland plant richness and biomass. New Phytol 203:233–244

Hildebrandt U, Janetta K, Ouziad F, Renne B, Nawrath K, Bothe H (2001) Arbuscular mycorrhizal colonization of halophytes in central European salt marshes. Mycorrhiza 10:175–183

Jenkins SN, Waite IS, Blackburn A, Husband R, Rushton SP, Manning DC, O'Donnell AG (2009) Actinobacterial community dynamics in long term managed grasslands. Antonie Van Leeuwenhoek 95:319–334

Ji B, Gehring CA, Wilson GW, Miller R, Flores-Rentería L, Johnson NC (2013) Patterns of diversity and adaptation in Glomeromycota from three prairie grasslands. Mol Ecol 22:2573–2587

Jin L, Zhang GQ, Wang XJ, Dou CY, Chen M, Lin SS, Li YY (2011) Arbuscular mycorrhiza regulate inter-specific competition between a poisonous plant, *Ligularia virgaurea*, and a co-existing grazing grass, *Elymus nutans*, in Tibetan Plateau Alpine meadow ecosystem. Symbiosis 55:29–38

Johnson NC, Tilman D, Wedin D (1992) Plant and soil controls on mycorrhizal fungal communities. Ecology 73:2034–2042

Johnson D, Vandenkoornhuyse PJ, Leake JR, Gilbert L, Booth RE, Grime JP, Peter J, Young W, Read DJ (2004) Plant communities affect arbuscular mycorrhizal fungal diversity and community composition in grassland microcosms. New Phytol 161:503–515

Jumpponen A, Trowbridge J, Mandyam K, Johnson L (2005) Nitrogen enrichment causes minimal changes in arbuscular mycorrhizal colonization but shifts community composition - evidence from rDNA data. Biol Fertil Soils 41:217–224

Kabir Z (2005) Tillage or no-tillage: impact on mycorrhizae. Can J Plant Sci 85:23–29

Kabir Z, O'halloran I, Fyles J, Hamel C (1997) Seasonal changes of arbuscular mycorrhizal fungi as affected by tillage practices and fertilization: hyphal density and mycorrhizal root colonization. Plant and Soil 192:285–293

Kaeppler SM, Parke JL, Mueller SM, Senior L, Stuber C, Tracy WF (2000) Variation among maize inbred lines and detection of quantitative trait loci for growth at low phosphorus and responsiveness to arbuscular mycorrhizal fungi. Crop Sci 40:358–364

Kan A (2009) General characteristics of waste management: a review. Energy Educ Sci Technol Part A 23:55–69

Kaya C, Higgs D, Kirnak H, Tas I (2003) Mycorrhizal colonisation improves fruit yield and water use efficiency in watermelon (*Citrullus lanatus* Thunb.) grown under well-watered and water-stressed conditions. Plant Soil 253:287–292

Khaliq A, Sanders F (2000) Effects of vesicular–arbuscular mycorrhizal inoculation on the yield and phosphorus uptake of field-grown barley. Soil Biol Biochem 32:1691–1696

Klabi R, Hamel C, Schellenberg MP, Iwaasa A, Raies A, St-Arnaud M (2014) Interaction between legume and arbuscular mycorrhizal fungi identity alters the competitive ability of warm-season grass species in a grassland community. Soil Biol Biochem 70:176–182

Klironomos JN (2003) Variation in plant response to native and exotic arbuscular mycorrhizal fungi. Ecology 84:2292–2301

Kojima T, Saito M, Shoji A, Ando S, Sugawara K (2009) The diversity of arbuscular mycorrhizal fungi in Japanese grasslands. Jap J Grassland Sci 55:148–155

Kristensen H, McCarty G, Meisinger J (2000) Effects of soil structure disturbance on mineralization of organic soil nitrogen. Soil Sci Soc Am J 64:371–378

Krüger M, Stockinger H, Krüger C, Schüßler A (2009) DNA-based species level detection of Glomeromycota: one PCR primer set for all arbuscular mycorrhizal fungi. New Phytol 183:212–223

Krüger M, Krüger C, Walker C, Stockinger H, Schüßler A (2012) Phylogenetic reference data for systematics and phylotaxonomy of arbuscular mycorrhizal fungi from phylum to species level. New Phytol 193:970–984

Leal PL, Stürmer SL, Siqueira JO (2009) Occurrence and diversity of arbuscular mycorrhizal fungi in trap cultures from soils under different land use systems in the Amazon, Brazil. Braz J Microbiol 40(1):111–121

Lee J, Lee S, Young JPW (2008) Improved PCR primers for the detection and identification of arbuscular mycorrhizal fungi. FEMS Microbiol Ecol 65:339–349

Li C, Hao X, Zhao M, Han G, Willms WD (2008) Influence of historic sheep grazing on vegetation and soil properties of a Desert Steppe in Inner Mongolia. Agr Ecosyst Environ 128:109–116

Li XL, Gai JP, Cai XB, Li XL, Christie P, Zhang FS, Zhang JL (2014) Molecular diversity of arbuscular mycorrhizal fungi associated with two co-occurring perennial plant species on a Tibetan altitudinal gradient. Mycorrhiza 24:95–107

Lin X, Feng Y, Zhang H, Chen R, Wang J, Zhang J, Chu H (2012) Long-term balanced fertilization decreases arbuscular mycorrhizal fungal diversity in an arable soil in North China revealed by 454 pyrosequencing. Environ Sci Technol 46:5764–5771

Lingfei L, Anna Y, Zhiwei Z (2005) Seasonality of arbuscular mycorrhizal symbiosis and dark septate endophytes in a grassland site in southwest China. FEMS Microbiol Ecol 54:367–373

Liu Y, Shi G, Mao L, Cheng G, Jiang S, Ma X, An L, Du G, Johnson NC, Feng H (2012) Direct and indirect influences of 8 yr of nitrogen and phosphorus fertilization on Glomeromycota in an alpine meadow ecosystem. New Phytol 194:523–535

Lugo MA, Cabello MN (2002) Native arbuscular mycorrhizal fungi (AMF) from mountain grassland (Córdoba, Argentina) I. Seasonal variation of fungal spore diversity. Mycologia 94(4):579–586

Lugo MA, Maza MEG, Cabello MN (2003) Arbuscular mycorrhizal fungi in a mountain grassland II: seasonal variation of colonization studied, along with its relation to grazing and metabolic host type. Mycologia 95:407–415

Lutgen ER, Muir-Clairmont D, Graham J, Rillig MC (2003) Seasonality of arbuscular mycorrhizal hyphae and glomalin in a western Montana grassland. Plant and Soil 257:71–83

Mandyam K, Jumpponen A (2008) Seasonal and temporal dynamics of arbuscular mycorrhizal and dark septate endophytic fungi in a tallgrass prairie ecosystem are minimally affected by nitrogen enrichment. Mycorrhiza 18:145–155

Mariotte P, Meugnier C, Johnson D, Thébault A, Spiegelberger T, Buttler A (2013) Arbuscular mycorrhizal fungi reduce the differences in competitiveness between dominant and subordinate plant species. Mycorrhiza 23:267–277

Mathimaran N, Ruh R, Jama B, Verchot L, Frossard E, Jansa J (2007) Impact of agricultural management on arbuscular mycorrhizal fungal communities in Kenyan ferralsol. Agr Ecosyst Environ 119:22–32

McCain KNS, Wilson GWT, Blair JM (2011) Mycorrhizal suppression alters plant productivity and forb establishment in a grass-dominated prairie restoration. Plant Ecol 212:1675–1685

McGonigle T, Fitter A (1990) Ecological specificity of vesicular-arbuscular mycorrhizal associations. Mycol Res 94:120–122

Menéndez AB, Scervino JM, Godeas AM (2001) Arbuscular mycorrhizal populations associated with natural and cultivated vegetation on a site of Buenos Aires province, Argentina. Biol Fertil Soils 33:373–381

Montero Sommerfeld H, Díaz LM, Alvarez M, Añazco Villanueva C, Matus F, Boon N, Boeckx P, Huygens D (2013) High winter diversity of arbuscular mycorrhizal fungal communities in shallow and deep grassland soils. Soil Biol Biochem 65:236–244

Moora M, Zobel M (1996) Effect of arbuscular mycorrhiza on inter-and intraspecific competition of two grassland species. Oecologia 108:79–84

Munkvold L, Kjøller R, Vestberg M, Rosendahl S, Jakobsen I (2004) High functional diversity within species of arbuscular mycorrhizal fungi. New Phytol 64:357–364

Murakoshi T, Tojo M, Walker C, Saito M (1998) Arbuscular mycorrhizal fungi on adjacent semi-natural grasslands with different vegetation in Japan. Mycoscience 39:455–462

Murugan R, Loges R, Taube F, Joergensen RG (2013) Specific response of fungal and bacterial residues to one-season tillage and repeated slurry application in a permanent grassland soil. Appl Soil Ecol 72:31–40

Newsham K, Fitter A, Watkinson A (1995) Arbuscular mycorrhiza protect an annual grass from root pathogenic fungi in the field. J Ecol 83:991–1000

Numata M (1961) Ecology of grasslands in Japan. J Coll Arts Sci Chiba Univ 13:327–342

O'Dea M (2007) Influence of mycotrophy on native and introduced grass regeneration in a semiarid grassland following burning. Restor Ecol 15:149–155

O'Donnell AG, Seasman M, Macrae A, Waite I, Davies JT (2001) Plants and fertilisers as drivers of change in microbial community structure and function in soils. Plant and Soil 232:135–145

O'Donnell AG, Young IM, Rushton SP, Shirley MD, Crawford JW (2007) Visualization, modelling and prediction in soil microbiology. Nat Rev Microbiol 5:689–699

Oehl F, Sieverding E, Ineichen K, Mäder P, Boller T, Wiemken A (2003) Impact of land use intensity on the species diversity of arbuscular mycorrhizal fungi in agroecosystems of Central Europe. Appl Environ Microbiol 69:2816–2824

Oehl F, Sieverding E, Ineichen K, Mäder P, Wiemken A, Boller T (2009) Distinct sporulation dynamics of arbuscular mycorrhizal fungal communities from different agroecosystems in long-term microcosms. Agr Ecosyst Environ 134:257–268

Ohsowski BM, Zaitsoff PD, Öpik M, Hart MM (2014) Where the wild things are: looking for uncultured Glomeromycota. New Phytol 204:171–179

Öpik M, Moora M, Liira J, Kõljalg U, Zobel M, Sen R (2003) Divergent arbuscular mycorrhizal fungal communities colonize roots of *Pulsatilla* spp. in boreal Scots pine forest and grassland soils. New Phytol 160:581–593

Öpik M, Vanatoa A, Vanatoa E, Moora M, Davison J, Kalwij JM, Reier U, Zobel M (2010) The online database MaarjAM reveals global and ecosystemic distribution patterns in arbuscular mycorrhizal fungi (Glomeromycota). New Phytol 188:223–241

Parton WJ, Scurlock JMO, Ojima DS, Gilmanov TG, Scholes RJ, Schimel DS, Kirchner T, Menaut JC, Seastedt T, Moya EG, Kamnalrut A, Kinyamario JL (1993) Observations and modeling of biomass and soil organic matter dynamics for the grassland biome worldwide. Global Biogeochem Cycles 7:785–809

Pezzani F, Montaña C, Guevara R (2006) Associations between arbuscular mycorrhizal fungi and grasses in the successional context of a two-phase mosaic in the Chihuahuan Desert. Mycorrhiza 16:285–295

Plenchette C, Clermont-Dauphin C, Meynard J, Fortin J (2005) Managing arbuscular mycorrhizal fungi in cropping systems. Can J Plant Sci 85:31–40

Pringle A, Bever JD (2002) Divergent phenologies may facilitate the coexistence of arbuscular mycorrhizal fungi in a North Carolina grassland. Am J Bot 89:1439–1446

Raiesi F, Asadi E (2006) Soil microbial activity and litter turnover in native grazed and ungrazed rangelands in a semiarid ecosystem. Biol Fertil Soils 43:76–82

Read D, Koucheki H, Hodgson J (1976) Vesicular-arbuscular mycorrhiza in natural vegetation systems. New Phytol 77:641–653

Redecker D (2000) Specific PCR primers to identify arbuscular mycorrhizal fungi within colonized roots. Mycorrhiza 10:73–80

Redecker D, Raab P (2006) Phylogeny of the Glomeromycota (arbuscular mycorrhizal fungi): recent developments and new gene markers. Mycologia 98:885–895

Redecker D, Schüßler A, Stockinger H, Stürmer SL, Morton JB, Walker C (2013) An evidence-based consensus for the classification of arbuscular mycorrhizal fungi (Glomeromycota). Mycorrhiza 23:515–531

Renker C, Weißhuhn K, Kellner H, Buscot F (2006) Rationalizing molecular analysis of field-collected roots for assessing diversity of arbuscular mycorrhizal fungi: to pool, or not to pool, that is the question. Mycorrhiza 16:525–531

Rillig MC, Mummey DL (2006) Mycorrhizas and soil structure. New Phytol 171:41–53

Rillig MC, Wright SF, Eviner VT (2002) The role of arbuscular mycorrhizal fungi and glomalin in soil aggregation: comparing effects of five plant species. Plant and Soil 238:325–333

Rillig MC, Maestre FT, Lamit LJ (2003) Microsite differences in fungal hyphal length, glomalin, and soil aggregate stability in semiarid Mediterranean steppes. Soil Biol Biochem 35:1257–1260

Roy-Bolduc A, Hijri M (2011) The use of mycorrhizae to enhance phosphorus uptake: a way out of the phosphorus crisis. J Biofert Biopest 2:1

Saito K, Suyama Y, Sato S, Sugawara K (2004) Defoliation effects on the community structure of arbuscular mycorrhizal fungi based on 18S rDNA sequences. Mycorrhiza 14:363–373

Sanders I, Fitter A (1992) Evidence for differential responses between host-fungus combinations of vesicular-arbuscular mycorrhizas from a grassland. Mycol Res 96:415–419

Sanjari G, Ghadiri H, Ciesiolka CA, Yu B (2008) Comparing the effects of continuous and time-controlled grazing systems on soil characteristics in Southeast Queensland. Soil Res 46:348–358

Santos JC, Finlay RD, Tehler A (2006) Molecular analysis of arbuscular mycorrhizal fungi colonising a semi-natural grassland along a fertilisation gradient. New Phytol 172:159–168

Santos-González JC, Finlay RD, Tehler A (2007) Seasonal dynamics of arbuscular mycorrhizal fungal communities in roots in a seminatural grassland. Appl Environ Microbiol 73:5613–5623

Saravesi K, Ruotsalainen A, Cahill J (2014) Contrasting impacts of defoliation on root colonization by arbuscular mycorrhizal and dark septate endophytic fungi of *Medicago sativa*. Mycorrhiza 24:239–245

Scheublin TR, Van Logtestijn RSP, van der Heijden MGA (2007) Presence and identity of arbuscular mycorrhizal fungi influence competitive interactions between plant species. J Ecol 95:631–638

Schnoor TK, Lekberg Y, Rosendahl S, Olsson PA (2011a) Mechanical soil disturbance as a determinant of arbuscular mycorrhizal fungal communities in semi-natural grassland. Mycorrhiza 21:211–220

Schnoor TK, Mårtensson L-M, Olsson PA (2011b) Soil disturbance alters plant community composition and decreases mycorrhizal carbon allocation in a sandy grassland. Oecologia 167:809–819

Schocha CL, Seifertb KA, Huhndorf S, Robert V, Spouge JL, Levesque CA, Chen W (2012) Fungal barcoding consortium, nuclear ribosomal internal transcribed spacer (ITS) region as a universal DNA barcode marker for fungi. Proc Natl Acad Sci U S A 109:6241–6246

Schüßler A, Walker C (2010) The glomeromycota: a species list with new families and new genera 1–58 Libraries at The Royal Botanic Garden Edinburgh, The Royal Botanic Garden Kew, Botanische Staatssammlung Munich, and Oregon State University. www.amf-phylogeny.com

Schüßler A, Gehrig H, Schwarzott D, Walker C (2001) Analysis of partial Glomales SSU rRNA gene sequences: implications for primer design and phylogeny. Mycol Res 105:5–15

Shi P, Abbott L, Banning N, Zhao B (2012) Comparison of morphological and molecular genetic quantification of relative abundance of arbuscular mycorrhizal fungi within roots. Mycorrhiza 22:501–513

Šmilauerová M, Lokvencová M, Šmilauer P (2012) Fertilization and forb: graminoid ratio affect arbuscular mycorrhiza in seedlings but not adult plants of *Plantago lanceolata*. Plant and Soil 351:309–324

Smith SE, Read DJ (2008) Mycorrhizal symbiosis. Academic, London

Smith M, Hartnett D, Rice C (2000) Effects of long-term fungicide applications on microbial properties in tallgrass prairie soil. Soil Biol Biochem 32:935–946

Spurgeon DJ, Keith AM, Schmidt O, Lammertsma DR, Faber JH (2013) Land-use and land-management change: relationships with earthworm and fungi communities and soil structural properties. BMC Ecol 13:46. doi:10.1186/1472-6785-13-46

Stockinger H, Walker C, Schüßler A (2009) 'Glomus intraradices DAOM197198', a model fungus in arbuscular mycorrhiza research, is not *Glomus intraradices*. New Phytol 183:1176–1187

Stockinger H, Krüger M, Schüßler A (2010) DNA barcoding of arbuscular mycorrhizal fungi. New Phytol 187:461–474

Stover HJ, Thorn RG, Bowles JM, Bernards MA, Jacobs CR (2012) Arbuscular mycorrhizal fungi and vascular plant species abundance and community structure in tallgrass prairies with varying agricultural disturbance histories. Appl Soil Ecol 60:61–70

Stürmer SL (2012) A history of the taxonomy and systematics of arbuscular mycorrhizal fungi belonging to the phylum Glomeromycota. Mycorrhiza 22:247–258

Su Y-Y, Guo L-D (2007) Arbuscular mycorrhizal fungi in non-grazed, restored and over-grazed grassland in the Inner Mongolia steppe. Mycorrhiza 17:689–693

Subramanian KS, Charest C (1999) Acquisition of N by external hyphae of an arbuscular mycorrhizal fungus and its impact on physiological responses in maize under drought-stressed and well-watered conditions. Mycorrhiza 9:69–75

Sýkorová Z, Ineichen K, Wiemken A, Redecker D (2007) The cultivation bias: different communities of arbuscular mycorrhizal fungi detected in roots from the field, from bait plants transplanted to the field, and from a greenhouse trap experiment. Mycorrhiza 18:1–14

Tsuchida K, Nonaka M (2002) Effect of the indigenous arbuscular mycorrhizal fungi (AMF) on growth of the grass in a grassland. Jap J Soil Sci Plant Nutr 73:485–491

Tsuchida K, Nonaka M (2003) Effect of arbuscular mycorrhizal fungi (AMF) on growth of orchardgrass Japanese. Jap J Soil Sci Plant Nutr 74:23–29

van der Heijden MG (2003) Arbuscular mycorrhizal fungi as a determinant of plant diversity: in search of underlying mechanisms and general principles. In: Sanders IR, Marcel GA, van der Heijden MGA (eds) Mycorrhizal Ecology. Springer, Heidelberg, pp 243–265

van der Heijden MG, Klironomos JN, Ursic M, Moutoglis P, Engel RS, Boller T, Wiemken A, Sanders IR (1998) Mycorrhizal fungal diversity determines plant biodiversity, ecosystem variability and productivity. Nature 396:69–72

van der Heijden MG, Bardgett RD, Van Straalen NM (2008) The unseen majority: soil microbes as drivers of plant diversity and productivity in terrestrial ecosystems. Ecol Lett 11(3):296–310

Vandenkoornhuyse P, Ridgway KP, Watson IJ, Fitter AH, Young JPW (2003) Co-existing grass species have distinctive arbuscular mycorrhizal communities. Mol Ecol 12:3085–3095

Verbruggen E, Roling WFM, Gamper HA, Kowalchuk GA, Verhoef HA, van der Heijden MGA (2010) Positive effects of organic farming on below-ground mutualists: large-scale comparison of mycorrhizal fungal communities in agricultural soils. New Phytol 186:968–979

Wetzel K, Silva G, Matczinski U, Oehl F, Fester T (2014) Superior differentiation of arbuscular mycorrhizal fungal communities from till and no-till plots by morphological spore identification when compared to T-RFLP. Soil Biol Biochem 72:88–96

Whitcomb S, Stutz JC (2007) Assessing diversity of arbuscular mycorrhizal fungi in a local community: role of sampling effort and spatial heterogeneity. Mycorrhiza 17:429–437

Wilson J, Trinick M (1983) Infection development and interactions between vesicular-arbuscular mycorrhizal fungi. New Phytol 93:543–553

Wright S, Upadhyaya A (1998) A survey of soils for aggregate stability and glomalin, a glycoprotein produced by hyphae of arbuscular mycorrhizal fungi. Plant and Soil 198:97–107

Wright S, Starr J, Paltineanu I (1999) Changes in aggregate stability and concentration of glomalin during tillage management transition. Soil Sci Soc Am J 63:1825–1829

Wuen K, Saito S, Sato S, Sugawara K (2002) Arbuscular mycorrhizal colonization and sporulation in rhizosphere of common species on native and sown grasslands. Grassl Sci 48:248–253

Yamane Y, Nishiwaki A, Sugawara K, Saito M (1999) Arbuscular mycorrhizal colonization in the plant roots of *Miscanthus sinensis*, *Zoysia japonica*, and orchardgrass. Grassl Sci 45(appendix):196–197 (In Japanese)

Yang C, Hamel C, Schellenberg MP, Perez JC, Berbara RL (2010) Diversity and functionality of arbuscular mycorrhizal fungi in three plant communities in Semiarid Grasslands National Park. Microb Ecol 59:724–733

Yang W, Yong Z, Cheng G, Xinhua H, Qiong D, Yongchan K, Yichao R, Shiping W, Liang-Dong G (2013) The arbuscular mycorrhizal fungal community response to warming and grazing differs between soil and roots on the Qinghai-Tibetan Plateau. PLoS One 8:e76447. doi:10. 1371/journal.pone.0076447

Yates CJ, Norton DA, Hobbs RJ (2000) Grazing effects on plant cover, soil and microclimate in fragmented woodlands in south-western Australia: implications for restoration. Austral Ecol 25:36–47

Young JPW (2012) A molecular guide to the taxonomy of arbuscular mycorrhizal fungi. New Phytol 193:823–826

Zaller JG, Heigl F, Grabmaier A, Lichtenegger C, Piller K, Allabashi R, Frank T, Drapela T (2011) Earthworm-mycorrhiza interactions can affect the diversity, structure and functioning of establishing model grassland communities. PLoS One 6:e29293

Zangaro W, de Assis RL, Rostirola LV, de Souza PB, Gonçalves MC, Andrade G, Nogueira MA (2008) Changes in arbuscular mycorrhizal associations and fine root traits in sites under different plant successional phases in southern Brazil. Mycorrhiza 19:37–45

Zangaro W, Rostirola LV, de Souza PB, de Alves RA, Lescano LEAM, Rondina ABL, Nogueira MA, Carrenho R (2013) Root colonization and spore abundance of arbuscular mycorrhizal fungi in distinct successional stages from an Atlantic rainforest biome in southern Brazil. Mycorrhiza 23:221–233

Chapter 10
Application of AM Fungi to Improve the Value of Medicinal Plants

Ying Long Chen, Jun Xi Li, Lan Ping Guo, Xin Hua He, and Lu Qi Huang

10.1 Introduction

Medicinal herbs are known to be a source of phytochemical constituents or bioactive compounds (Toussaint et al. 2007). Unlike synthetic medicines, natural medicinal products are claimed to be safe to humans and the environment, and some can play a significant role in the treatment of cancer (Nema et al. 2013). The use of plants medicinally has a tradition in many cultures. In Europe, apothecaries stocked herbal ingredients for their medicines. Traditional Chinese medicine, Ayurvedic (Indian) medicine, and herbal medicine are examples of medical practices that incorporate the medical uses of plants. The well-known Chinese *Materia Medica* documented over 600 medicinal plants and recorded the first use of medicinal herbs in China as early as 1,100 BC (Cragg et al. 1997). It has been estimated that more than 30 % of known plant species have been or are being used medicinally in at least one medicinal tradition (Joy et al. 1998).

Y.L. Chen (✉)
State Key Laboratory of Soil Erosion and Dryland Farming on the Loess Plateau, Research Center of Soil and Water Conservation and Eco-environment, Chinese Academy of Sciences and Ministry of Education, Yangling 712100, China

School of Earth and Environment, The University of Western Australia, Crawley, WA 6009, Australia
e-mail: yinglongchen@hotmail.com

J.X. Li
Haidu College, Qingdao Agricultural University, Laiyang 265200, China

L.P. Guo • L.Q. Huang
Institute of Chinese Materia Medica, China Academy of Chinese Medical Sciences, Beijing 100700, China

X.H. He
School of Life Sciences, Yunnan Normal University, Kunming, Yunnan 650092, China

© Springer-Verlag Berlin Heidelberg 2014   171
Z.M. Solaiman et al. (eds.), *Mycorrhizal Fungi: Use in Sustainable Agriculture and Land Restoration*, Soil Biology 41, DOI 10.1007/978-3-662-45370-4_10

Like many other terrestrial plants, most medicinal plants are capable of forming mycorrhizal associations. Mycorrhizal symbioses play crucial roles in contributing to soil structure formation, plant nutrient acquisition, growth, productivity, and biodiversity in both agricultural and natural ecosystems (Smith and Read 2008). Among all kinds of mycorrhizas, arbuscular mycorrhizas are the most widely distributed association (Liu and Chen 2007; Smith and Read 2008). However, research on mycorrhizal relationships with medicinal plants began much later and varies among regions and countries. Colonization by AM fungi has been confirmed in many medicinal plants, and two major advances relate to (1) AM fungal community and diversity in the rhizosphere of medicinal plants (e.g., Kumar et al. 2010; Wubet et al. 2003; Zeng et al. 2013) and (2) improved medicinal values by AM fungal colonization (e.g., Copetta et al. 2006; Yuan et al. 2007; Morone-Fortunato and Avato 2008; Toussaint et al. 2008; Sasanelli et al. 2009).

Taber and Trappe (1982) pioneered and observed the presence of AM fungi in medicinal plants in Figi Islands and Hawaii, America. Waheed (1982) conducted field surveys on medicinal plants and their mycorrhizal status in Murree Hills and Kaghan Valley, Pakistan. Observations confirmed mycorrhizal associations were commonly present in medicinal plants in Pakistan (Gorsi 2002; Haq and Hussain 1995; Iqbal and Nasim 1986). The occurrence of AM associations in medicinal plants has also been reported from China (Wei and Wang 1989) and Japan (Udea et al. 1992). Many studies concerning AM fungi in medicinal plants emerged from the late 1980s.

With the development of Chinese, Indian, Arabian, and other traditional medicines, production systems have made extensive use of wild medicinal plants. At an earlier time, the traditional healthcare practice mainly depended on harvesting of wild medicinal plants. Increases in population, inadequate supply of drugs, side effects of several allopathic drugs, and development of resistance to the currently used drugs for infectious diseases have focused attention on the use of plant materials as a source of medicines for a wide variety of human ailments. Therefore, attempts to explore wild resources and to develop cultivation technologies for large-scale plantations have been made across Asian countries. In the traditional cultivation process, pests and diseases have direct impacts on the yield and quality of medicinal plants. Simultaneously, pesticide abuse involving harmful heavy metals affects the quality of plant-origin medicinal products and has led to environmental pollution. Development of innovative methods and technologies for cultivation and plantation of medicinal plants are needed.

Inoculation with AM fungi during an early stage of plant growth has become an alternative strategy for improved plant survival and growth (Kothamasi et al. 2001). During the establishment of AM symbioses, a range of chemical and biological reactions could occur in the rhizosphere and plant tissues, including producing plant secondary metabolites. AM associations have been reported to have functions in improving the growth of medicinal plants and the productivity of medicinal compounds (Chandra et al. 2010; Karthikeyan et al. 2009; Gupta and Janardhanan 1991).

10.2   Resource and Diversity of AM Fungi in Rhizosphere of Medicinal Plants

10.2.1   AM Fungal Species Diversity

Hundreds of plant species are reported for use as complementary medicines. Among these, a large proportion of medicinal plants have mycorrhizal associations. However, the diversity of AM fungal communities in rhizospheres of medicinal plants and the extent of colonization may vary depending on host plant species, sampling season, soil properties, and locality.

Gorsi (2002) reported that 76 medicinal plants were associated with AM fungi in the Azad Jammu and Kashmir areas in Pakistan. In Bangladesh, the AM fungal status of 40 medicinal plants was investigated on the Rajshahi University Campus (Zaman et al. 2008). Gautam and Roy (2009) investigated the seasonal variation of AM fungi associated with some medicinal plants in India, and Radhika and Rodrigues (2010) found that 30 out of 36 medicinal plant species were mycorrhizal in Goa region, India. Modern molecular approaches are now used to identify AM fungal diversity in rhizosphere soil and roots of medicinal plants. By analyzing phylogenetic data of 5.8S ribosomal DNA, Ethiopian researchers investigated the molecular diversity of AM fungi associated with *Prunus africana* (Wubet et al. 2003). They revealed 109 sequences belonging to members of the Glomeromycota. Subsequent 5.8S/ITS2 rDNA sequence analysis indicated high AM fungal diversity and dominance of *Glomus* type. Similarly, Appoloni et al. (2008) analyzed the community of AM fungi in roots of *Dichanthelium lanuginosum* and claimed that 16 rDNA phylotypes belonged to the genera *Archaeospora*, *Glomus*, *Paraglomus*, *Scutellospora*, and *Acaulospora*. The most diverse and abundant lineage was *Glomus* group A, with the most frequent phylotype corresponding to *Glomus intraradices*. Using nested PCR techniques, Cai et al. (2009) reported that the molecular diversity of the AM fungal community in the rhizosphere of *Phellodendron amurense* in Northeast China had three general groups in related to *Glomus*, *Scutellospora*, and *Hyponectria*. We list some most commonly used medicinal plant species capable of forming AM associations in Table 10.1.

10.2.2   AM Morphology

AM fungi in the rhizosphere of medicinal plants are abundant, but their colonization status is greatly influenced by plant species and environmental factors. Both biotic and abiotic factors affect the population of fungal species and distribution although abiotic/edaphic factors may be relatively more important than biotic factors for establishing and maintaining population pattern (Panwar and Tarafdar 2006). The vegetative stage of plants exhibits higher colonization rates compared to

Table 10.1 List of some commonly used medicinal plants and associated AM fungal species reported in the literature

Plant family	Plant species	AM fungal species	Source
Acanthaceae	*Andrographis paniculata*	*Acaulospora scrobiculata, Glomus aggregatum*	Radhika and Rodrigues (2010)
Amaranthaceae	*Amaranthus spinosus*	*A. denticulata, A. scrobiculata, A. tuberculata, G. claroideum, G. fecundisporum, G. monosporum*	Yang et al. (2002)
		Scutellospora pellucida, Sclerocystis clavispora	
Apocynaceae	*Hemidesmus indicus*	*Ambispora leptoticha, G. maculosum, G. geosporum, G. multicaule, G. fasciculatum*	Radhika and Rodrigues (2010)
Araliaceae	*Panax ginseng*	*A. cavernata, A. spinosa, G. fasciculatum, G. geosporum,*	Xing et al. (2000)
		G. macrocarpum, G. microaggregatum, G. mosseae	Cho et al. (2009)
Araliaceae	*Panax notoginseng*	*G. versiforme, G. monosporum, G. mosseae, G. constrictum,*	Ren et al. (2007)
		G. claroideum	Zhang et al. (2011)
Asteraceae	*Arnica montana*	*G. geosporum, G. constrictum, G. intraradices, G. mosseae,*	Jurkiewicz et al. (2010)
		G. macrocarpum, G. fasciculatum, G. versiforme	Chaudhary et al. (2008)
Asteraceae	*Echinacea purpurea*	*G. intraradices*	Araim et al. (2009)
Caprifoliaceae	*Lonicera japonica*	*G. constrictum, G. geosporum, G. mosseae, G. versiforme*	Gai et al. (2000)
Cercidiphyllaceae	*Cercidiphyllum japonicum*	*G. aggregatum, G. constrictum, G. dimorphicum, G. fasciculatum, G. flavisporum, G. intraradices, G. mosseae, S. aurigloba, Archaeospora leptoticha*	Wang et al. (2008)
Compositae	*Carthamus tinctorius*	*A. rehmii, G. claroideum*	Zhao (2006)
Compositae	*Xanthium sibirlcum*	*G. claroideum, G. mosseae*	Zhao (2006)
			Zhang and Tang (2006)
Elaeagnaceae	*Elaeagnus sarmentosa*	*G. etunicatum, G. mosseae, G. caledonium, G. constrictum*	Lin et al. (2003)
Elaeagnaceae	*Hippophae rhamnoides*	*G. albidum, G. claroideum, G. constrictum, G. coronatum, G. intraradices*	Tang et al. (2004)
Gentianaceae	*Gentiana scabra*	*G. Mosseae, G. geosporum*	Wang et al. (1998)

(continued)

Table 10.1 (continued)

Plant family	Plant species	AM fungal species	Source
Ginkgoaceae	*Ginkgo biloba*	*G. mosseae, G. aggregatum, G. geosporum, G. versiforme, G. caledonium, S. heterogama, Gigaspora gigantea, Gi. margarita*	Chen and Han (1999)
Labiatae	*Salvia miltiorrhiza*	*A. bireticulata, G. aggregatum, G. mosseae, G. clarum, G. reticulatum*	He et al. (2010)
Labiatae	*Scutellaria baicalensis*	*G. geosporum, G. versiforme*	Zhang and Tang (2006)
Leguminosae	*Robinia pseudoacacia*	*G. aggregatum, G. albidum, G. claroideum, G. constrictum, G. fasciculatum, G. mosseae, G. reticulatum*	Hu (2006)
Leguminosae	*Prosopis cineraria*	*G. fasciculatum, G. aggregatum, G. mosseae*	Verma et al. (2008)
Leguminosae	*Pueraria lobata*	*G. constrictum, G. geosporum, G. mosseae, G. reticulatum*	Gai et al. (2000)
Leguminosae	*Albizia julibrissin*	*G. mosseae, G. etunicatum*	Lin et al. (2003)
Liliaceae	*Allium macrostemon*	*G. caledonium*	Gai et al. (2000)
Liliaceae	*Aloe vera*	*G. maculosum, G. multicaule, G. geosporum*	Radhika and Rodrigues (2010)
Liliaceae	*Paris polyphylla*	*A. appendicula, A. brieticulata, A. excavate, G. albidum, G. ambisporum, G. luteum, Gi. albida, S. calospora, S. gilmorei*	Zhou et al. (2009)
Meliaceae	*Azadirachta indica*	*A. scrobiculata, G. fasciculatum, Gi. albida, S. calospora*	Radhika and Rodrigues (2010)
Meliaceae	*Naregamia alata*	*A. scrobiculata, Am. leptoticha, A. nicolsonii, G. rubiforme, G. maculosum, G. fasciculatum, S. verrucosa*	Radhika and Rodrigues (2010)
Plantaginaceae	*Plantago asiatica*	*G. intraradices*	Zhang and Tang (2006)
Rhamnaceae	*Ziziphus jujuba* Mill. var. *inermis*	*G. coronatum, G. intraradices, G. monosporum, G. reticulatum*	Tang et al. (2004)
Rosaceae	*Crataegus cuneata*	*G. constrictum, G. caledonium*	Di et al. (2006)
Rosaceae	*Prunus armeniaca*	*G. constrictum, G. geosporum*	Di et al. (2006)
Solanaceae	*Physalis minima*	*A. rehmi, G. fasciculatum, G. multicaule, G. maculosum, G. geosporum, G. rubiforme*	Radhika and Rodrigues (2010)

(continued)

Table 10.1 (continued)

Plant family	Plant species	AM fungal species	Source
Solanaceae	*Solanum nigrum*	*Gigaspora margarita, G. caledonium*	Gai et al. (2000)
			Zhang and Tang (2006)
Solanaceae	*Lycium barbarum*	*Gi. margarita, G. albidum*	Tang et al. (2004)
Taxaceae	*Taxus chinensis*	*G. aggregatum, G. ambisporum, G. clarum, G. constrictum, G. fasciculatum, G. geosporum, G. magnicaule, G. reticulatum, G. verruculosum, G. viscosum, A. denticulata*	Wang et al. (2008)
Trochodendraceae	*Euptelea pleiosperma*	*G. ambisporum, G. constrictum, G. fasciculatum, G. geosporum, G. hyderabadensis, G. intraradices, S. verrucosa*	Wang et al. (2008)
Umbelliferae	*Centella asiatica*	*G. multicaule, G. clarum, G. fasciculatum, A. delicate, S. scutata*	Radhika and Rodrigues (2010)
Umbelliferae	*Angelica dahurica*	*G. Mosseae, G. caledonium, G. constrictum, A. spinosa, S. calospara*	Cao et al. (2007)
Zingiberaceae	*Alpinia galanga*	*G. caledonium, G. mosseae, G. fasciculatum, G. geosporum, Am. leptoticha*	Radhika and Rodrigues (2010)

flowering and fruiting stages. In addition, herbaceous plants showed more entry points of hyphae into roots than did the shrubby and woody plants (Gorsi 2002). AM fungal spore density in soil around medicinal halophytes (*Suaeda fruticosa, Salsola baryosma, Haloxylon recurvum*) had a strong positive correlation with soil pH and organic carbon content but a negative correlation with soil phosphorus (Mathur et al. 2007). AM fungal spore density is often higher in association with wild medicinal plants compared with cultivated species, which may be due to the undisturbed nature of the ecosystem in wild habitats (Radhika and Rodrigues 2010). Experimental evidence of physical and functional selectivity in AM symbiosis has been demonstrated in field soils where diverse communities of AM fungi form associations with individual hosts (Helgason et al. 2002; Smith et al. 2000).

AM fungal morphology has been classified as either Arum type or the Paris type (Smith and Smith 1997). The physiological and functional disparity between Arum type and Paris type remains unclear. The development of Arum type is faster than that of Paris type (Cavagnaro et al. 2001). Zubeck and Blaszkowski (2009) reported that AM colonization rates in roots of *Mentha citrata, Origanum majorana, Salvia officinalis*, and *Thymus vulgaris* ranged from 67 to 100 % and were often Arum type. Both Arum and Paris types have been found in medicinal plants in Lamiaceae (Burni and Hussain 2011). Three wolfberry (*Lycium barbarum* L.) cultivars in arid

Table 10.2 AM morphologies and mycorrhizal status of medicinal plants

Plant family	Plant species	Morphology types	References
Acanthaceae	*Andrographis paniculata*	Arum	Muthukumar et al. (2006)
Amaranthaceae	*Amaranthus spinosus*	Arum	Muthukumar et al. (2006)
Araceae	*Pinellia ternata*	Intermediate	Cheng et al. (2009a)
Asteraceae	*Echinacea purpurea*	Arum	Zubek and Blaszkowski (2009)
Asteraceae	*Rhizoma Atractylodis Macrocephalae*	Paris	Cheng et al. (2009b)
Caesalpiniaceae	*Cassia siamea*	Arum	Muthukumar et al. (2006)
Campanulaceae	*Platycodon grandiflorus*	Arum	Cheng et al. (2009b)
Caprifoliaceae	*Lonicera Japonica*	Paris	Cheng et al. (2009b)
Compositae	*Vernonia cinerea*	Paris	Muthukumar et al. (2006)
Crassulaceae	*Sedum aizoon*	Arum	Cheng et al. (2009b)
Hypericaceae	*Hypericum perforatum*	Arum	Zubek and Blaszkowski (2009)
Labiatae	*Leucas aspera*	Arum	Muthukumar et al. (2006)
Lamiaceae	*Ocimum americanum*	Arum	Burni and Hussain (2011)
Lamiaceae	*Ocimum basilicum*	Paris	Burni and Hussain (2011)
Lamiaceae	*Rosmarinus officinalis*	Paris	Burni and Hussain (2011)
Lamiaceae	*Salvia lanata* Roxb	Intermediate	Burni and Hussain (2011)
Magnoliaceae	*Michelia champaca*	Paris	Panna and Highland (2010)
Solanaceae	*Lycium barbarum*	Paris	Zhang et al. (2010)
Umbelliferae	*Centella asiatica*	Intermediate	Muthukumar et al. (2006)
Umbelliferae	*Radix bupleuri*	Arum	Cheng et al. (2009b)

Northwestern China had the same Paris-type AM fungal associations, but colonization rate within a cultivar varied (Zhang et al. 2010). The Arum-type association was observed more often in medicinal plants than the Paris type (Table 10.2; Muthukumar et al. 2006).

10.3 Effect of AM Fungal Inoculation on Plant Growth

Several studies confirm that mycorrhizal medicinal plants generally have greater nutrient contents and grow better than non-mycorrhizal plants (e.g., Karagiannidisa et al. 2011; Nisha and Rajeshkumar 2010). For example, mycorrhizal inoculation increased dry matter accumulation in five medicinal plants (*Abelmoschus moschatus*, *Clitoria tematea*, *Plumbago zeylanica*, *Psoralea corylifolia*, and *Withania somnifera*) grown in five different soil types (Chandra et al. 2010). AM fungal associations improved the shoot height growth and root biomass of *Poncirus trifoliata*, *Piper longum*, *Salvia officinalis*, and *Plectranthus amboinicus* (Geneva et al. 2010; Gogoi and Singh 2011; Rajeshkumar et al. 2008; Wang et al. 2006).

10.3.1 Nutrient Uptake and Plant Biomass

The mycorrhizal hyphal network provides a larger absorptive surface than root hairs alone. The positive effects of AM fungal inoculation on plant growth are generally attributed to improved acquisition of nutrients of low mobility, especially phosphorus. It has been demonstrated that external hyphae of AM fungi are able to increase NH_4^+ and NO_3^- uptake and to assimilate these molecules into free amino acids (Johansen et al. 1996). Mycorrhizal symbioses stimulate plant uptake of nutrients such as P, Zn, Cu, Mn, and Fe in deficient soils (Chen and Zhao 2009), and mycorrhizal hyphae play an important role in nutrient uptake, convinced by labeling nutrient elements in controlled experiments (e.g., Hosamani et al. 2011). However, Zhao and Yan (2006) reported that leaf nitrogen contents were lower in mycorrhizal *Camptotheca acuminata* than in its non-mycorrhizal counterpart. Shoot dry weight was four times greater in *Withania somnifera* colonized by *Glomus fasciculatum* than in uninoculated plants (Hosamani et al. 2011). Glasshouse experiments showed that inoculation of palmarosa (*Cymbopogon martinii*) with *G. aggregatum* enhanced biomass production threefold compared to non-mycorrhizal plants (Gupta and Janardhanan 1991). Many studies of AM associations have shown that the effectiveness of AM fungi differs with plant species, soil nutrient level, and plant growth environments (e.g., Smith and Smith 2011).

10.3.2 Stress Tolerance

Compared to non-mycorrhizal plants, AM plants often show greater tolerance to several biotic and abiotic stresses, such as toxic metals, root pathogens, drought, high soil temperature, saline soils, adverse soil pH, and transplanting shock (Evelin et al. 2009; Lu et al. 2003; Tang et al. 1999; Turkmen et al. 2008). Where salt stress is a major threat to plant growth and productivity, AM fungi have been shown to promote plant growth and salinity tolerance. For example, AM fungi were able to colonize *Bacopa monnieri* roots effectively under high salinity levels (Khaliel et al. 2011). Inoculated plants significantly enhanced dry mass production, and this occurred to a greater extent when plants were grown at high salinity levels. AM fungi increased Na^+ and Cl^- uptake and reduced rhizosphere NaCl level. AM fungi can induce a buffering effect on the uptake of Na^+ when the content of Na^+ is within the permissible limit (e.g., Allen and Cunningham 1983), and mycorrhizal plants may employ mechanisms to promote plant tolerance to salinity. These may include enhanced nutrient acquisition, maintenance of the K^+/Na^+ ratio, biochemical changes (accumulation of proline, betaines, polyamines, carbohydrates, and anti-oxidants), physiological changes (photosynthetic efficiency, relative permeability, water status, abscisic acid accumulation, nodulation, and nitrogen fixation), molec-ular changes (the expression of genes: PIP, Na^+/H^+ antiporters, Lsnced, Lslea, and LsP5CS), and ultrastructural changes (Evelin et al. 2009).

High bicarbonate (HCO_3^-) content and associated high pH of irrigation water are detrimental to plant growth. Inoculation with AM fungi enhanced the tolerance of *Rosa multiflora* to HCO_3^- as indicated by greater nutrient uptake and leaf chlorophyll and lower root iron reductase activity and alkaline phosphatase activity (Cartmill 2004). When exposed to drought, AM plants exhibited a higher level of proline and activity of two antioxidant enzymes, superoxide dismutase and perox-idase. In addition, mRNA abundance of four genes involved in reactive oxygen species homeostasis and oxidative stress battling was higher in the mycorrhizal compared with non-mycorrhizal plants (Fan and Liu 2011). These findings illustrate the possible participation of drought-induced genes in the enhanced tolerance of AM plants to water deficit. It has been claimed that activation of physiological, biochemical, and molecular alterations may be involved in the improvement of the growth and drought tolerance of mycorrhizal *Poncirus trifoliata* seedlings (Fan and Liu 2011; Ruiz-Lozano et al. 2008).

Potential roles of AM associations in alleviating metal stress of plants have been demonstrated, but the mechanisms involved in the metal tolerance of AM fungi are still poorly understood (Hildebrandt et al. 2007). Heavy-metal-tolerant AM fungi isolated from polluted soils showed capability in binding heavy metals (Joner et al. 2000). *Datura metal* plants inoculated with AM fungi showed increased tolerance to heavy metals (Salvaraj and Kim 2004). A pot experiment with sweet basil (*Ocimum basilicum*) under heavy metal (Cr, Cd, Pb, and Ni) stress showed that the AM symbiosis could be used as a novel approach to enhance yield and maintain the quality of volatile oil under metal-contaminated soils (Prasad et al. 2011). Sweet basil has been traditionally used for the treatment of headaches, coughs, and diarrhea (Jayasinghe et al. 2003).

10.4 Effect of AM Fungi on Medicinal Composition

Secondary metabolites of medicinal plants are the critical resources in natural medicinal products used for pharmacological and therapeutical purposes. Terpenoids, phenolics, and alkaloids are the three major groups of secondary plant metabolites. There are a few attempts being made to study the relationship between the occurrence of AM fungi and improved secondary metabolite contents in medicinal plants. The improved phosphorus status or an altered hormonal balance of the plants may contribute to the AM effects on the secondary metabolites, but the reasons for such effects remain largely unknown (Toussaint 2007).

10.4.1 Terpenoids

The effects of AM fungi on terpenoid concentration in medicinal plants have received more attention in recent years. Essential oils are volatile, lipophilic mixtures of secondary plant compounds, mostly consisting of monoterpenes, sesquiterpenes, and phenylpropanoids, which are often used as flavors and fragrances, as antimicrobials and antioxidants, and as medicines (Deans and Waterman 1993). AM fungi increased the content of essential oil and alterations of its composition, such as in medicinal basil (*O. basilicum*) (Copetta et al. 2006).

Andrographis paniculata, commonly known as "king of bitters," has been used for centuries in Asia to treat gastrointestinal tract and upper respiratory infections, fever, herpes, sore throat, and other chronic and infectious diseases. The primary medicinal component of *A. paniculata* is andrographolide, a colorless diterpene lactone with bitter taste. The high concentration of andrographolide in the leaf extracts of *A. paniculata* inoculated with *Gi. albida* showed that the AM symbiosis can enhance the production of this secondary metabolite (Radhika and Rodrigues 2011). The concentration of andrographolide in mycorrhizal *A. paniculata* plants reached the highest level at the flowering growth stage.

10.4.2 Phenolic Compounds

Apart from terpenes and essential oil constituents, the relationship between mycorrhizal associations and other secondary plant metabolites such as phenolic compounds have been investigated. Phenolic compounds include phytoalexin, wallbound phenol, flavonoids, and isoflavonoids and their derivatives. A significant increase of total phenolic content was detected in leaves and flower heads of *Cynara cardunculus* inoculated with *Glomus intraradices*, either alone or in a mixture with *G. mosseae* under both greenhouse and field conditions (Ceccarelli et al. 2010). Isoflavone levels were altered in roots of legume plants locally and systemically when colonized by AM fungi (Catford et al. 2006). The shoot flavonoid levels in white clover (*Trifolium repens*) increased when roots were colonized by AM fungi (Ponce et al. 2004). An increased content of rosmarinic acid, a highly antioxidant phenolic compound, was detected in AM-colonized basil (Toussaint et al. 2008). The content of flavonoids in *Bupleurum chinense*, *Ginkgo biloba*, and *Astragalus membranaceus* (Meng and He 2011) and that of total coumarin and imperatorin in *Angelica dahurica* (Zhao and He 2011) were significantly higher in mycorrhizal compared with non-mycorrhizal plants of the same species. AM fungal colonization induced two different signaling pathways in the accumulation of phenylpropanoid metabolism: one through induction of phenylalanine ammonia-lyase and chalcone synthase and the other through suppression of isoflavone reductase (Zhao and Yan 2006).

There are controversial conclusions concerning mycorrhizal effects on phenolic contents in medicinal plants. Zeng et al. (2013) showed neutral effects of AM colonization on the composition of phenolic ingredients. The AM symbiosis did not alter the total concentrations of phenolic and rosmarinic acid in roots of *Salvia officinalis* (Nell et al. 2009) or the polyphenolic profile in leaves and stems of basil (Lee and Scagel 2009) after AM fungal inoculation.

10.4.3 Alkaloids

Vinca (*Catharanthus roseus*) is an important medicinal plant from which antineo-plastic alkaloids (e.g., vinblastine) are extracted. AM fungal inoculation significantly enhanced plant growth and the total content of vinblastine in Vinca leaves (Rosa-Mera et al. 2011). A positive correlation was found between mycorrhizal colonization and castanospermine content in field-grown *Castanospermum australe* seeds and in greenhouse-grown leaves in inoculated plants (Abu-Zeyad et al. 1999). This finding is interesting because castanospermine is effectively used in the treatments against AIDS and cancers (Spearman et al. 1991). Sweet basil has been traditionally used for the treatment of headaches, coughs, and diarrhea (Jayasinghe et al. 2003). AM fungal colonization increased the production of rosmarinic acid (antioxidant activity) in sweet basil shoots (Toussaint et al. 2007). Mycorrhizal *Ocimum basilicum* and *Coleus amboinicus* possessed higher amounts of the total phenols, ortho-dihydroxyphenols, flavonoids, alkaloids, and tannins in the root and leaf than non-mycorrhizal plants (Hemalatha 2002). The AM symbiosis also played a positive role in the accumulation of camptothecin in *Camptotheca acuminata* and vinca alkaloids in Vinca; both are important anticancer compounds (Rosa-Mera et al. 2011).

10.5   Conclusion

The quality of herbal materials in terms of active ingredients is largely influenced by abiotic and biotic factors, and AM fungal colonization can play an important role in improving the medicinal values of medicinal plants (Szakiel et al. 2011a, b). Their positive role in plant growth, disease resistance, and both yield and quality of medicinal materials make AM fungi potential alternatives to existing methods for promoting the growth of some important medicinal plants. With the increased demands in plant-oriented medicines, there is increasing scientific interest in the study of the interaction between medicinal plant production and their associated mycorrhizal symbioses. The advantages, prospects, and feasibility of introducing AM fungi into the process of cultivation of medicinal plants have been recognized (Xiao et al. 2011; Yang et al. 2008). AM fungal diversity and its significance in medicinal plant nutrient acquisition and secondary metabolite alteration have been investigated, and its application provides a sustainable method to enhance the agricultural and pharmaceutical outcomes of medicinal plants. AM fungi are not host specific, but their affinity to a particular host can be preferential (Rogers et al. 1994). Thus, selecting efficient AM fungi for a particular plant is essential for the cultivation of medicinal plants. Advanced mycorrhizal technologies in agriculture can be adopted to study medicinal plants and their production. To ensure a better understanding of the diversity and function of AM fungi and a wider application of AM fungi in the plantation of medicinal plants, the following research areas are recommended: (1) exploiting the diversity and distribution of AM fungi in important medicinal plants, (2) identifying the relationship between genetic structure and functional diversity of AM fungal species and mechanisms of signal perception and plant growth regulators in mycorrhizal establishment under diverse ecosystems, and (3) selecting and using efficient AM fungi for improved active secondary plant metabolites.

Acknowledgments This study was supported by National Natural Science Foundation of China (31471946) and Chinese Academy of Sciences "Hundred Talents Program".

References

Abu-Zeyad R, Khan AG, Khoo C (1999) Occurrence of arbuscular mycorrhiza in *Castanospermum australe* A. Cunn. & C. Fraser and effects on growth and production of castanospermine. Mycorrhiza 9:111–117
Allen EB, Cunningham GL (1983) Effects of vesicular-arbuscular mycorrhizae on *Distichlis spicata* under three salinity levels. New Phytol 93:227–236
Appoloni S, Lekberg Y, Tercek MT, Zabinski CA, Redecker D (2008) Molecular community analysis of arbuscular mycorrhizal fungi in roots of geothermal soils in Yellowstone National Park (USA). Microb Ecol 56:649–659

Araim G, Saleem A, Arnason JT, Charest AC (2009) Root colonization by an arbuscular mycorrhizal (AM) fungus increases growth and secondary metabolism of purple coneflower, *Echi-
nacea purpurea* L. Moench. J Agric Food Chem 57:2255–2258
Burni T, Hussain F (2011) Diversity in arbuscular mycorrhizal morphology in some medicinal
plants of family Lamiaceae. Pak J Bot 43(3):1789–1792
Cai BY, Ge QP, Jie WG, Yan XF (2009) The community composition of the arbuscular mycorrhizal fungi in the rhizosphere of *Phellodendron amurense*. Mycosystema 28(4):512–520
Cao DX, He XL, Zhao JL (2007) Ecological research of arbuscular mycorrhizal fungi in rhizosphere of *Angelica dahurica*. J Hebei Univ 27(5):525–529
Cartmill AD (2004) Arbuscular mycorrhizal fungi enhance tolerance to bicarbonate in *Rosa
multiflora* cv. Burr. Thesis for Master of Science. Texas A&M University
Catford JG, Staehelin C, Larose G, Piche Y, Vierheilig H (2006) Systemically suppressed
isoflavonoids and their stimulating effects on nodulation and mycorrhization in alfalfa split-
root systems. Plant and Soil 285:257–266
Cavagnaro TR, Smith FA, Lorimer MF, Haskard KA, Ayling FM, Smith SE (2001) Quantitative
development of Paris-type arbuscular mycorrhizas formed between *Asphodelus fistulosus* and
Glomus coronatum. New Phytol 149:105–113
Ceccarelli N, Curadi M, Martelloni L, Sbrana C, Picciarelli P, Giovannetti M (2010) Mycorrhizal
colonization impacts on phenolic content and antioxidant properties of artichoke leaves and
flower heads two years after field transplant. Plant and Soil 335:311–323
Chandra KK, Kumar N, Chand G (2010) Studies on mycorrhizal inoculation on dry matter yield
and root colonization of some medicinal plants grown in stress and forest soils. J Environ Biol
31(6):975–979
Chaudhary V, Kapoor R, Bhatnagar AK (2008) Effectiveness of two arbuscular mycorrhizal fungi
on concentrations of essential oil and artemisinin in three accessions of *Artemisia annua*
L. Appl Soil Ecol 40:174–181
Chen LQ, Han NL (1999) Identification of Ginkgo VA mycorrhizal fungi in Zhejiang Province.
For Res 12(6):581–584
Chen XH, Zhao B (2009) Arbuscular mycorrhizal fungi mediated uptake of nutrient elements by
Chinese milk vetch (*Astragalus sinicus* L.) grown in lanthanum spiked soil. Bio Fertil Soils
45:675–678
Cheng LT, Guo QS, Liu ZY (2009a) Infection pattern and dynamic change of arbuscular
mycorrhizal fungi in *Pinellia ternata*. Guizhou Agr Sci 37(2):37–39
Cheng LT, Liu ZY, Guo QS, Zhu GS (2009b) Advances in studies on arbuscular mycorrhizas in
medicinal plants. Chin Tradit Herbal Drugs 40(1):156–160
Cho EJ, Lee DJ, Wee CD, Kim HL, Cheong YH, Cho JS, Sohn BK (2009) Effects of AM fungi
inoculation on growth of *Panax ginseng* C.A. Meyer seedlings and on soil structures in
mycorrhizosphere. Sci Hortic 122(4):633–637
Copetta A, Lingua G, Berta G (2006) Effects of three AM fungi on growth, distribution of
glandular hairs, and essential oil production in *Ocimum basilicum* L. var. *Genovese*. Mycorrhiza 16:485–494
Cragg GM, Newman DJ, Sander KM (1997) Natural products in drug discovery and development.
J Nat Prod 60(1):52–60
Deans SG, Waterman PG (1993) Biological activity of volatile oils. In: Hay RKM, Waterman PG
(eds) Volatile oil crops. Longman Scientific and Technical, Harlow, UK, pp 97–109
Di LNE, Tang M, Wang YJ (2006) Arbuscular mycorrhizal fungi and their ecological distribution
of some kinds of wild plants in Yili Region of Xinjiang. J Northwest Sci-Tech Univ Agr
Forestry 34(6):96–100
Evelin H, Kapoor R, Giri B (2009) Arbuscular mycorrhizal fungi in alleviation of salt stress: a
review. Ann Bot 104(7):1263–1280
Fan QJ, Liu JH (2011) Colonization with arbuscular mycorrhizal fungus affects growth, drought
tolerance and expression of stress-responsive genes in *Poncirus trifoliata*. Acta Physiol Plant
33(4):1533–1542

Gai JP, Liu RJ, Meng XX (2000) Arbuscular mycorrhizal fungi on wild plants II. Mycosystema 19 (2):205–211

Gautam NK, Roy AK (2009) Seasonal diversity in arbuscular mycorrhizal fungi associated with some medicinal plants of Jharkhand. J Indian Bot Soc 88(1&2):56–59

Geneva MP, Stancheva IV, Boychinova MB, Mincheva NH, Yonova PA (2010) Effects of foliar fertilization and arbuscular mycorrhizal colonization on *Salvia officinalis* L. growth, antioxidant capacity, and essential oil composition. J Sci Food Agric 90:696–702

Gogoi P, Singh RK (2011) Differential effect of some arbuscular mycorrhizal fungi on growth of *Piper longum* L. (Piperaceae). Ind J Sci Technol 4(2):119–125

Gorsi MS (2002) Studies on mycorrhizal association in some medicinal plants of Azad Jammu and Kashmir. Asian J Plant Sci 1(4):383–387

Gupta ML, Janardhanan KK (1991) Mycorrhizal association of *Glomus aggregatum* with palmarosa enhances growth and biomass. Plant and Soil 131:261–263

Haq I, Hussain Z (1995) Medicinal plants of Palandri, District Poonch, Azad Jammu and Kashmir. Pak J Plant Sci 1:115–126

He XL, Wang LY, Ma J, Zhao LL (2010) AM fungal diversity in the rhizosphere of *Salvia miltiorrhiza* in Anguo City of Hebei Province. Biodivers Sci 18(2):175–181

Helgason T, Merryweather JW, Denison J, Wilson P, Young JPW, Fitter AH (2002) Selectivity and functional diversity in arbuscular mycorrhizas of co-occurring fungi and plants from a temperate deciduous woodland. J Ecol 90:371–384

Hemalatha M (2002) Synergistic effect of VA-mycorrhizae and azospirillum on the growth and productivity of some medicinal plants. Ph.D. Thesis, Bharathidasan University, Tamil Nadu, India, p. 108

Hildebrandt U, Regvar M, Bothe H (2007) Arbuscular mycorrhiza and heavy metal tolerance. Phytochemistry 68:139–146

Hosamani PA, Lakshman HC, Sandeepkumar K, Kadam MA, Kerur AS (2011) Role of arbuscular mycorrhizae in conservation of *Withania somnifera*. Biosci Discov 2(2):201–206

Hu B (2006) An investigation of arbuscular mycorrhizal fungi resources in Xianyang District of Shaan'xi. Shaanxi For Sci Tech 2:37–39

Iqbal SH, Nasim G (1986) Vesicular-arbuscular mycorrhiza in roots and other underground portions of *Curcuma longa*. Biologia 32:223–228

Jayasinghe C, Gotoh N, Aoki T, Wada S (2003) Phenolics composition and antioxidant activity of sweet basil (*Ocimum basilicum* L.). J Agric Food Chem 51:4442–4449

Johansen A, Finlay RD, Olsson PA (1996) Nitrogen metabolism of external hyphae of the arbuscular mycorrhizal fungus *Glomus intraradices*. New Phytol 133:705–712

Joner EJ, Briones R, Leyval C (2000) Metal-binding capacity of arbuscular mycorrhizal mycelium. Plant and Soil 226(2):227–234

Joy P, Thomos J, Mathew S, Skaria BP (1998) Medicinal plants. Kerala Agricultural University Press, Kerala

Jurkiewicz A, Ryszka P, Anielska T, Waligórski P, Białońska D, Góralska K, Michael MT, Turnau K (2010) Optimization of culture conditions of Arnica montana L.: effects of mycorrhizal fungi and competing plants. Mycorrhiza 20:293–306

Karagiannidisa N, Thomidisa T, Lazarib D, Panou-Filotheoua E, Karagiannidoua C (2011) Effect of three Greek arbuscular mycorrhizal fungi in improving the growth, nutrient concentration, and production of essential oils of oregano and mint plants. Sci Hortic 129(2):329–334

Karthikeyan B, Joe MM, Jaleel CA (2009) Response of some medicinal plants to vesicular arbuscular mycorrhizal inoculations. J Sci Res 1(1):381–386

Khaliel AS, Shine K, Vijayakumar K (2011) Salt tolerance and mycorrhization of *Bacopa monneiri* grown under sodium chloride saline conditions. Afr J Microbiol Res 5 (15):2034–2040

Kothamasi D, Kuhad RC, Babu CR (2001) Arbuscular mycorrhizae in plant survival strategies. Trop Ecol 42(1):1–13

Kumar A, Mangla C, Aggarwal A, Parkash V (2010) Arbuscular mycorrhizal fungal dynamics in the rhizospheric soil of five medicinal plants species. Middle-East J Sci Res 6(3):281–288

Lee J, Scagel CF (2009) Chicoric acid found in basil (*Ocimum basilicum* L.) leaves. Food Chem 115(2):650–656

Lin QH, Zeng XP, Zhang N, Huang ZH, Huang WN (2003) The isolation and identification of VA mycorrhizal fungi of nonleguminous tree of nitrogen fixation in Fujian. J Fujian College Forestry 23(3):270–273

Liu RJ, Chen YL (2007) Mycorrhizology. China Science, Beijing

Lu JY, Mao YM, Shen LY, Peng SQ, Li XL (2003) Effects of VA mycorrhizal fungi inoculated on drought tolerance of wild Jujube (*Zizyphus spinosus* Hu). Acta Hortic Sin 30(1):29–33

Mathur N, Singh J, Bohra S, Vyas A (2007) Arbuscular mycorrhizal status of medicinal halophytes in saline areas of Indian Thar Desert. Int J Soil Sci 2(2):119–127

Meng JJ, He XL (2011) Effects of AM fungi on growth and nutritional contents of *Salvia miltiorrhiza* Bge. under drought stress. J Agr Univ Hebei 34(1):51–61

Morone-Fortunato I, Avato P (2008) Plant development and synthesis of essential oils in micropropagated and mycorrhiza inoculated plants of *Origanum vulgare* L. ssp. *hirtum* (Link) Ietswaart. Plant Cell Tiss Org Cult 93:139–149

Muthukumar T, Senthikumar M, Rajangam M, Udaiyan K (2006) Arbuscular mycorrhizal morphology and dark septate fungal associations in medicinal and aromatic plants of western Ghats, Southern India. Mycorrhiza 17:11–24

Nell M, Vötsch M, Vierheilig H, Steinkellner S, Zitterl-Eglseer K, Franz C, Novak J (2009) Effect of phosphorus uptake on growth and secondary metabolites of garden (*Salvia officinalis* L.). J Sci Food Agric 89:1090–1096

Nema R, Khare S, Jain P, Pradhan A, Gupta A, Singh D (2013) Natural products potential and scope for modern cancer research. Am J Plant Sci 4:1270–1277

Nisha MC, Rajeshkumar S (2010) Effect of arbuscular mycorrhizal fungi on growth and nutrition of *Wedilia chinensis* (Osbeck) Merril. Ind J Sci Tech 3(6):676–678

Panna D, Highland K (2010) Mycorrhizal colonization and distribution of arbuscular mycorrhizal fungi associated with *Michelia champaca* L. under plantation system in northeast India. J For Res 21(2):137–142

Panwar J, Tarafdar JC (2006) Distribution of three endangered medicinal plant species and their colonization with arbuscular mycorrhizal fungi. J Arid Environ 65:337–350

Ponce MA, Scervino JM, Erra BR, Ocampo JA, Godeas A (2004) Flavonoids from shoots and roots of *Trifolium repens* (white clover) grown in presence or absence of the arbuscular mycorrhizal fungus *Glomus intraradices*. Phytochemistry 65:1925–1930

Prasad A, Kumar S, Khaliq A, Pandey A (2011) Heavy metals and arbuscular mycorrhizal (AM) fungi can alter the yield and chemical composition of volatile oil of sweet basil (*Ocimum basilicum* L.). Biol Fertil Soils 47(8):853–861

Radhika KP, Rodrigues BF (2010) Arbuscular mycorrhizal fungal diversity in some commonly occurring medicinal plants of Western Ghats, Goa region. J For Res 21(1):45–52

Radhika KP, Rodrigues BF (2011) Influence of arbuscular mycorrhizal fungi on andrographolide concentration in *Andrographis paniculata*. Aust J Med Herbalism 23(1):34–36

Rajeshkumar S, Nisha MC, Selvaraj T (2008) Variability in growth, nutrition and phytochemical constituents of *Plectranthus amboinicus* (Lour) Spreng. as influenced by indigenous arbuscular mycorrhizal fungi. Mj Int J Sci Tech 2(2):431–439

Ren JH, Liu RX, Li YL (2007) Study on arbuscular mycorrhizae of *Panax notoginseng*. Microbiology 34(2):224–227

Rogers HH, Runion GB, Krupa SV (1994) Plant responses to atmospheric CO_2 enrichment with emphasis on roots and the rhizosphere. Environ Pollut 83:155–189

Rosa-Mera CJDA, Ferrera-Cerrato R, Alarcón A, Sánchez-Colín MDJ, Muñoz-Muñiz OD (2011) Arbuscular mycorrhizal fungi and potassium bicarbonate enhance the foliar content of the vinblastine alkaloid in *Catharanthus roseus*. Plant and Soil 349:367–376

Ruiz-Lozano JM, Porcel R, Aroca R (2008) Evaluation of the possible participation of drought-induced genes in the enhanced tolerance of arbuscular mycorrhizal plants to water deficit. Mycorrhiza 18:185–205

Salvaraj T, Kim H (2004) Effect of vesicular-arbuscular mycorrhizal (VAM) fungi on tolerance of industrial effluent treatment in *Datura metal*. Agr Chem Biotechnol 47(2):106–109

Sasanelli N, Anton A, Takacs T, Addabbo TD, Biro I, Malov X (2009) Influence of arbuscular mycorrhizal fungi on the nematicidal properties of leaf extracts of *Thymus vulgaris* L. Helminthologia 46(4):230–240

Smith S, Read D (2008) Mycorrhizal symbiosis, 3rd edn. Academic, New York

Smith FA, Smith SE (1997) Structural diversity in VAM symbiosis. New Phytol 137:373–388

Smith FA, Smith SE (2011) What is the significance of the arbuscular mycorrhizal colonisation of many economically important crop plants? Plant and Soil 348:63–79

Smith FA, Jakobsen I, Smith SE (2000) Spatial differences in acquisition of soil phosphate between two arbuscular mycorrhizal fungi in symbiosis with *Medicago truncatula*. New Phytol 47:357–366

Spearman MA, Ballon BC, Gerrard JM, Greenberg AH, Wright JA (1991) The inhibition of platelet aggregation of metastatic H-ras-transformed 10T1/2 fibroblasts with castanospermine, an N-linked glycoprotein processing inhibitor. Cancer Lett 60:185–191

Szakiel A, Pączkowski C, Henry M (2011a) Influence of environmental abiotic factors on the content of saponins in plants. Phytochem Rev 10:471–491

Szakiel A, Pączkowski C, Henry M (2011b) Influence of environmental biotic factors on the content of saponins in plants. Phytochem Rev 10:493–502

Taber RA, Trappe JM (1982) Vesicular-arbuscular mycorrhiza in rhizomes, scale-like leaves, roots, and xylem of ginger. Mycologia 74(1):156–161

Tang M, Chen H, Shang HS (1999) Effects of arbuscular mycorrhizal fungi (AMF) on *Hippophae rhamnoides* drought-resistance. Sci Silvae Sin 35(3):48–52

Tang M, Xue S, Yang HP (2004) Vesicular arbuscular mycorrhizal (VAM) fungi of xerophyte in Gansu. J Yunnan Agri Univ 19(6):638–642

Toussaint JP (2007) Investigating physiological changes in the aerial parts of AM plants: what do we know and where should we be heading? Mycorrhiza 17:349–353

Toussaint JP, Smith FA, Smith SE (2007) Arbuscular mycorrhizal fungi can induce the production of phytochemicals in sweet basil irrespective of phosphorus nutrition. Mycorrhiza 17:291–297

Toussaint JP, Kraml M, Nell M, Smith SE, Smith FA, Steinkellner S, Schmiderer C, Vierheilig H, Novak J (2008) Effect of *Glomus mosseae* on concentrations of rosmarinic and caffeic acids and essential oil compounds in basil inoculated with *Fusarium oxysporum* f. sp. *basilici*. Plant Pathol 57:1109–1116

Turkmen O, Sensoy S, Demir S, Erdinc C (2008) Effects of two different AMF species on growth and nutrient content of pepper seedlings grown under moderate salt stress. Afr J Biotechnol 7:392–396

Udea T, Husoe T, Kubo S, Nakawashi I (1992) Vesicular arbuscular mycorrhizal fungi (Glomales) in Japan II: a field survey of vesicular arbuscular mycorrhizal association with medicinal plants in Japan. Trans Mycol Soc Jpn 33:77–86

Verma N, TarafdarCJ SKK, Panwa J (2008) Arbuscular Mycorrhizal (AM) diversity in *Prosopis cineraria* (L.) druce under arid agroecosystems. Agr Sci China 7(6):754–761

Waheed A (1982) Mycorrhizal and medicinal plants in Murree hills. M.Sc Thesis. The Punjab University, Lahore, Pakistan

Wang Q, Li HQ, Du YR, Lin Y, Li HW (1998) Isolation and identification of vesicular-mycorrhizal fungi in *Gentiana scabra*. Biotechnology 8(2):19–22

Wang CH, Yang XH, Li DY, Yu GB, Qin Q (2006) Effects of the different species of arbuscular mycorrhizal fungi on the vegetative growth and mineral contents in trifoliate orange seedlings. Chin Agr Sci Bull 22(12):199–203

Wang S, Tang M, Niu ZC, Zhang HQ (2008) Relationship between AM fungi resources of rare medicinal plants and soil factors in Lishan Mountain. Acta Bot Boreal-Occident Sin 28 (2):355–361

Wei GT, Wang HG (1989) Effects of VA mycorrhizal fungi on growth, nutrient uptake and effective compounds in Chinese medicinal herb *Datura stramonium* L. Sci Agric Sin 25 (5):56–61

Wubet T, Weib M, Kottke I, Teketay D, Oberwinkler F (2003) Molecular diversity of arbuscular mycorrhizal fungi in *Prunus africana*, an endangered medicinal tree species in dry Afromontane forests of Ethiopia. New Phytol 161:517–528

Xiao WJ, Yang G, Chen ML, Guo LP, Wang M (2011) AM and its application in plant disease prevention of Chinese medicinal herbs cultivation. Chin J Chin Med 36(3):252–257

Xing X, Li Y, Yolande D (2000) Ten species of VAM fungi in five ginseng fields of Jilin Province. J Jilin Agri Univ 22(2):41–46

Yang L, Wang GH, Ren LC, Zhao ZW (2002) Arbuscular mycorrhizae of the family *Amaranthaceae*. Acta Bot Yunnanica 24(1):37–40

Yang G, Guo LP, Huang LQ, Chen M (2008) Inoculation methods of AM fungi in medicinal plant. Resourc Sci 30(5):778–785

Yuan ZL, Dai CC, Chen LQ (2007) Regulation and accumulation of secondary metabolites in plant-fungus symbiotic system. Afr J Biotechnol 6(11):1266–1271

Zaman P, Roy AK, Khanum NS, Absar N, Yeasmin T (2008) Arbuscular mycorrhizal status of medicinal plants in Rajshahi University Campus. Mycosystema 27(4):543–553

Zeng Y, Guo LP, Chen BD, Hao ZP, Wang JY, Huang LQ, Yang G, Cui XM, Yang L, Wu ZX, Chen ML, Zhang Y (2013) Arbuscular mycorrhizal symbiosis and active ingredients of medicinal plants: current research status and prospectives. Mycorrhiza 23:253–265

Zhang WH, Tang M (2006) VA mycorrhiza fungi resources in the North of China. J Northwest For Univ 21(2):121–125

Zhang HH, Tang M, Chen H, Wang YJ, Ban YH (2010) Arbuscular mycorrhizas and dark septate endophytes colonization status in medicinal plant *Lycium barbarum* L. in arid Northwestern China. Afr J Microbiol Res 4(18):1914–1920

Zhang J, Liu DH, Guo LP, Jin H, Yang G, Zhou J (2011) Effects of arbuscular mycorrhizae fungi on biomass and essential oil in rhizome of *Atractylodes lancea* in different temperatures. Chin Tradi Her Drugs 42(2):372–375

Zhao SF (2006) The genera and species of AM fungi in Xinjiang and its application prospect. Xinjiang Agr Sci 43(S1):12–15

Zhao JL, He XL (2011) Effects of AM fungi on drought resistance and content of chemical components in *Angelica dahurica*. Acta Agr Bor Occi Sin 20(3):184–189

Zhao X, Yan XF (2006) Effects of arbuscular mycorrhizal fungi on the growth and absorption of nitrogen and phosphorus in *Camptotheca acuminata* seedlings. J Plant Ecol 30(6):947–953

Zhou N, Xia CL, Jiang B, Bai ZC, Liu GM, Ma XK (2009) Arbuscular mycorrhizae in *Paris polyphylla* var. *yunnanensis*. China J Chin Meter Med 34(14):1768–1772

Zubek S, Blaszkowski J (2009) Medicinal plants as hosts of arbuscular mycorrhizal fungi and dark septate endophytes. Phytochem Rev 8:571–580

Chapter 11
Arbuscular Mycorrhizas and Their Role in Plant Zinc Nutrition

Timothy R. Cavagnaro

11.1 Introduction

The majority of higher land plants form arbuscular mycorrhizas (AM). The formation of AM can result in improved plant nutrition, growth, disease resistance, and drought tolerance (Smith and Read 2008). It is for these reasons that AM are increasingly recognized as having an important role in many ecosystem processes and as an integral part of sustainable agroecosystems (Jackson et al. 2008). While it has been found that a large proportion of the P in plants can be delivered via the mycorrhizal pathway (Smith et al. 2004), AM also contribute significantly to plant acquisition of other nutrients including Zn, NH_4^+, NO_3^-, Cu, K, and others (Cavagnaro et al. 2006; Frey and Schuepp 1993; Johansen et al. 1993; Marschner and Dell 1994; Tanaka and Yano 2005). Despite the significance of AM in the uptake of nutrients other than P, our understanding of the underlying processes lags considerably behind that of P. Although significant advances have been made, this still remains a significant knowledge gap.

The importance of AM in plant nutrient acquisition, especially in low-fertility soils (Hetrick 1991; Menge 1983), is well recognized (Cardoso and Kuyper 2006; Jackson et al. 2008; Ryan and Angus 2003). This is especially true for nutrients with a low mobility in the soil (e.g., P, Zn, NH_4^+) (Tinker and Nye 2000). It has been estimated that up to 10 % of plant Zn is delivered by the extra-radical hyphae AM fungi (AMF) (Marschner and Dell 1994), although values vary considerably between studies (see Cavagnaro 2008 for review). Furthermore, the formation of AM by plants is particularly important where nutrients are distributed heterogeneously in the soil, due to the ability of arbuscular mycorrhizal (AM) fungi to forage for nutrients effectively (see Tibbett 2000 for review).

T.R. Cavagnaro (✉)

School of Agriculture, Food and Wine, University of Adelaide, Adelaide, SA 5005, Australia

e-mail: timothy.cavagnaro@adelaide.edu.au

© Springer-Verlag Berlin Heidelberg 2014   189

Z.M. Solaiman et al. (eds.), *Mycorrhizal Fungi: Use in Sustainable Agriculture and Land Restoration*, Soil Biology 41, DOI 10.1007/978-3-662-45370-4_11

AM have an important (and fundamentally interesting) dual role in plant Zn physiology. While the formation of AM can increase the capacity of plants to acquire Zn from soils with low Zn concentrations, they can also have a protective role against Zn accumulation to toxic levels in plant grown under high soil Zn conditions. Indeed it appears that there may be a critical soil Zn concentration below which Zn uptake is enhanced and above which it is reduced (Cavagnaro et al. 2010; Christie et al. 2004; Hildebrandt et al. 2007; Watts-Williams and Cavagnaro 2012). While of considerable interest, this latter protective role is not reviewed here (see Christie et al. 2004 for review).

Zinc is an essential element for plant growth. It is an important component of over 300 enzymes and plays important role in the catalytic activity and/or structure of many enzymes (Christie et al. 2004; Hacisalihoglu and Kochian 2003; Marschner 1995). While Zn is abundant in the Earth's crust, much of it occurs in forms unavailable to plants. Consequently, approximately 30 % of the world's soils are considered Zn deficient (Kochian 2000). Zinc deficiency in many important crops is, therefore, a common problem. This has flow on effects for consumers of those crops, with much of the world's human population not meeting its daily Zn requirements. This can have serious consequences for human health which can result in diminished human potential, felicity, and worker productivity (Brown and Wuehler 2000). Thus, understanding how plants acquire and utilize Zn is very important. While the importance of increasing the density of Zn in staple crops is widely recognized (Brown and Wuehler 2000; Burns et al. 2010), the role of AM in this has received little attention. If this is to change, we must develop a sound understanding how plants (and AM) acquire Zn. In addition to an understanding of the physiological and molecular aspects of plant uptake of Zn via the mycorrhizal pathway, an understanding of the effects of different land management practices on the formation and functioning of AM is also required.

This chapter will discuss the role of AM in plant Zn acquisition, with a strong emphasis on agroecosystems. To this end, the impacts of agricultural management on AM and their consequences to plant uptake of Zn via the mycorrhizal pathway will be explored. In order to do this, it is first necessary to revisit the physiological basis of Zn uptake, translocation and transfer to plants, by AMF; this is a subject that has been dealt with more depth in my earlier review (Cavagnaro 2008), and so only key details are represented here. Finally, emerging trends and future directions will be identified and discussed with a view to stimulating research on Zn-AM dynamics, especially where the aim is to capitalize on the benefits that AMF may afford crop plants, and those who depend upon them.

11.2   Zn Uptake by AM: Evidence from Isotope Studies

The importance of AMF in plant Zn nutrition has long been recognized (Cavagnaro 2008; Cavagnaro and Jackson 2007; Cavagnaro et al. 2006; Gao et al. 2007; Kothari et al. 1991; Marschner and Dell 1994; Ortas et al. 2002; Smith and Read 2008;

Watts-Williams and Cavagnaro 2012; Watts-Williams et al. 2013). Understanding how, and to what extent, plants acquire Zn via the mycorrhizal pathway will be important in studying AM/Zn interactions in agroecosystems. A wide range of experimental approaches has been used to demonstrate how AMF help plants acquire Zn, and indeed other nutrients. Studies that have used ^{65}Zn as a tracer have been particularly important. Unequivocal evidence for Zn uptake by AMF and its subsequent translocation and transfer to plants have come from studies where a ^{65}Zn tracer is supplied to the plant in compartments accessible to the AMF alone (Bürkert and Robson 1994; Cooper and Tinker 1978; Jansa et al. 2003). This approach makes it possible to partition root and AMF contributions to plant Zn acquisition. Using such an approach, it has been found that Zn supply to plants by AMF (via direct hyphal uptake, not effects of AMF on plant uptake) ranges widely among studies (two orders of magnitude). For example, Jansa et al. (2003) reported that almost 9 % of the Zn supplied to plants in a compartment accessible to hyphae only was transported to the plant over 25 days after Zn supply to the plant. While other studies have found smaller amounts of Zn supplied to the plant using a range of experimental systems (e.g., split-plate culture-based system Cooper and Tinker 1978), these studies demonstrate unequivocally the uptake, translocation, and transfer of Zn to plants by AMF.

11.3 Zn Uptake by AM: Field Studies

While the use of ^{65}Zn as a tracer has been highly valuable in demonstrating Zn uptake by AMF and transfer to plants, this approach has not, to my knowledge, been applied in a field setting, as with radioisotopes of P (e.g., Schweiger and Jakobsen 1999) and stable isotopes of N (Cavagnaro et al. 2012). Be that as it may, various other experimental methods have been used to demonstrate the role of AM in plant Zn nutrition, both in the laboratory and the field. These approaches include:

1. Crop rotations that reduce AMF inoculum potential (systems-based approach) and, hence, AMF colonization of target crops (Ryan and Angus 2003; Ryan and Ash 1999; Sorensen et al. 2005; Thompson 1987, 1996)
2. Mycorrhiza-defective mutants (genotypic approach) to establish AM controls (Cavagnaro et al. 2006; Ruzicka et al. 2010)
3. Fumigation of soils (fumigation-based approach) to eliminate AMF in situ

Emphasis here is placed on the first two of these approaches.

The inclusion of plants in crop rotations that do not form AM has been shown to be linked to a decrease in the inoculum potential of soils and, thence, levels of colonization in subsequent crops. This systems (or rotation)-based approach to reducing the inoculum potential of soils allows for identification of the role of AM in plant Zn nutrition under conditions that are commonly encountered in agricultural ecosystems. The growth and Zn and P acquisition of crops have been reported to be lower in crop rotations that include the non-mycorrhizal plant canola

(*Brassica napus* (L.)) or include long periods without plant cover (12–18 months fallow) (in southern Queensland, Australia) (Thompson 1987, 1996). In the case of the mycorrhizal crop linseed (*Linum usitatissimum* L.), inoculation of seedlings (in the glasshouse) with AMF ameliorated deficiencies in Zn and P when grown in soils that have been subjected to a long fallow period (and hence reduce AMF inoculum potential) (Thompson 1996). Also using a systems-based approach, increases in colonization of roots by AMF and Zn uptake in wheat (*Triticum aestivum* L.) (Ryan et al. 2002) and leek (*Allium porrum* L.) (Sorensen et al. 2005), following mycorrhizal cover crops, have been reported (see Cavagnaro 2008 for more detailed review). Together, these studies illustrate the importance of AMF in plant Zn acquisition in "real world" agricultural settings. Such an on-farm approach to research also permits study of the impacts of various agronomic practices (see below) on AM functioning and inoculum potential.

Irrespective of the ecosystem system, establishing non-mycorrhizal controls is one of the biggest challenges that we face in the study of AM. While a fumigation-based approach to establishing non-mycorrhizal controls in a field setting is effective at suppressing colonization of roots by AMF, and often the only practical approach, nontarget effects on other soil biota cannot be entirely discounted. In an attempt to overcome this issue, a genotypic approach to controlling for the formation of AM can be used. For example, we have compared the growth, nutrition, and soil ecology of a mycorrhiza-defective tomato (*Solanum lycopersicum* L.) mutant with reduced mycorrhizal colonization (Barker et al. 1998) (named *rmc*) with that of its mycorrhizal wild-type progenitor (cv. Peto 76R). This genotypic approach for controlling colonization of roots by AMF makes it possible to study the contribution of AM to plant Zn nutrition, with the wider soil biota "intact." For example, in tomato plants grown on an organically managed farm, we found that Zn contents in both the vegetative and edible aboveground biomass was up to 50 % higher in the mycorrhizal genotype (Cavagnaro et al. 2006). This highlights the importance of AM in plant Zn nutrition in an on-farm setting.

The establishment of non-mycorrhizal controls "lies at the heart of difficulties of experimenting with AMF at the ecosystem scale" (Rillig 2004). Thus, in this chapter I have provided selected examples that have tried to overcome some of these challenges. In particular, emphasis has been placed on approaches that allow studies to be undertaken with minimal deviations from normal farming practices or indirect effects of establishing non-mycorrhizal controls on other members of the soil biota. While further research is required, it is clear that AM have an important role to play in plant Zn nutrition in a range of agroecosystems.

11.4 Zn Uptake by AM: The Physiological Basis

Uptake of nutrients, including Zn, by AM involves three core processes:

1. Nutrient acquisition by the AMF

2. Nutrient translocation within the AMF to the intra-radical plant-fungal symbiotic interface
3. Nutrient transfer from the AMF into the interfacial apoplastic space from which it is taken up by the plant

Each of these processes, in the context of Zn acquisition, is now considered.

As is the case with other nutrients, AMF access nutrients not necessarily otherwise accessible to the roots by growing beyond the rhizosphere depletion zones that commonly from around roots (see Tinker and Nye 2000). Zinc uptake by AMF at distances of 40–50 mm from the root surface has been reported (Bürkert and Robson 1994; Jansa et al. 2003). While only a handful of studies have focused on the impacts of Zn addition on intra- and extra-radical growth of AMF, positive (Seres et al. 2006), negative (Liu et al. 2000), and neutral (Toler et al. 2005) responses have been reported (see Cavagnaro 2008 for detailed review). Each of these will likely impact the uptake of Zn by AMF in different ways. Be that as it may, as noted above, an increase in plant ^{65}Zn uptake has been related to the hyphal length density in the compartment containing the labeled Zn (Jansa et al. 2003). This suggests that AMF are able to "forage" for Zn, as is the case for other nutrients (Cavagnaro et al. 2005; Tibbett 2000). In addition to increasing hyphal length density in Zn "patches" in the soil, AMF may also increase Zn uptake via other mechanisms, such as enhanced levels of expression of genes implicated in Zn uptake (as with P Maldonado-Mendoza et al. 2001). This, however, is speculative and requires further consideration (see below).

The mechanisms underlying the long-distance translocation of Zn in the extra-radical hyphae of AMF in the soil, to the intracellular symbiotic interface within the root cortex, remain elusive (see below). Irrespective of how Zn is translocated within the extra-radical hyphae of AMF, any Zn delivered to the plant-fungal symbiotic interface needs to be unloaded into interfacial apoplastic space and thence taken up by the plant. To this end, some important insights have been gained. A cation diffusion facilitator (named *GintZnT1*) has been identified in the AMF *G. intraradices* (Gonzalez-Guerrero et al. 2005). *GintZnT1* has been suggested as having a role in Zn storage or efflux within hyphae (e.g., from an internal storage compartment involved in long-distance Zn translocation), or the efflux of Zn into to the plant/fungal interfacial apoplast (Gonzalez-Guerrero et al. 2005). Once inside the apoplast, the Zn then needs to be taken up across the plant plasma membrane. MtZIP2 is a plasma membrane-localized Zn transporter in *Medicago truncatula* (Gaertn); its expression is influenced by the effects of AM on the Zn status of the plant (Burleigh et al. 2003). Further studies of the molecular basis of Zn uptake via the mycorrhizal pathway will provide important insights into the functioning of AM; these factors are considered in more detail in an earlier review of this topic (Cavagnaro 2008). While such studies will be important, those that begin to integrate Zn uptake via the mycorrhizal pathway, with other aspects of plant and AM biology, are likely to be especially important. To this end, interactions between plant P and Zn nutrition and acquisition via the AM pathway seem to be an obvious, but little considered, starting point (Jansa et al. 2003; Marschner

1995; Watts-Williams and Cavagnaro 2012; Watts-Williams et al. 2014; Zhu et al. 2001a, b).

11.5   Zn Uptake by AM: Agroecosystems

It is clear that AM have a role to play in the Zn nutrition of plants. But do they have a place in the modern agricultural paradigm? There is no doubt that both the global population and pressures upon ecosystems are increasing. Indeed achieving global food security in a sustainable manner is one of the largest challenges society currently faces. To this end, we need to carefully consider the options available to us to increase both the yield and nutritive value of crops (Burns et al. 2010). AMF occur in the soils of most arable regions (see Read 1991; Treseder and Cross 2006) and, therefore, are essentially "freely available" to (most) agricultural producers. This, coupled with their role in improving plant growth, nutrition (including, but not limited to, Zn), plant disease resistance, and drought tolerance, suggests that AMF should be an important element of any such debate.

Agricultural yields have increased rapidly in recent decades due to a number of reasons, including the advent and widespread use of pesticides and fertilizers, the mechanization of agriculture, and the breeding of improved plant varieties. However, many of these factors can reduce the AMF inoculum potential of soils and/or the responsiveness of crops to AMF. For example, the application of fungicides can result in a decrease in AM colonization (see Cavagnaro and Martin 2011; Miller and Jackson 1998; Smith et al. 2000). Colonization of roots by AMF is often (but not always) reduced with high levels of fertilizer application. For example, Bolan et al. (1984) reported that at low soil P concentrations, colonization of roots by AMF was inhibited and that small additions of P to the soil increased colonization slightly. However, larger additions of P to the soil can result in a reduction in colonization. Importantly, the magnitude of the effect of soil P on colonization differs between plant and fungal combinations studied (Baon et al. 1992; Oliver et al. 1983). Similar reductions in colonization have been reported in response to N and Zn addition also (see Cavagnaro 2008). In addition to effects of agricultural practices on the formation of AM, plant breeding programs may have also inadvertently selected varieties with low mycorrhizal dependency and/or responsiveness by screening varieties in sterile growth media with high rates of nutrient addition. Be that as it may, these factors, alone and together, may have resulted in decreased mycorrhizal responsiveness and/or dependency upon AMF in many agricultural ecosystems.

Organically managed farms generally have higher AMF inoculum potential than conventionally managed farms (see Cavagnaro et al. 2006 and references therein). Furthermore, organic farming systems typically have higher soil organic matter, microbial biomass, and enhanced rates of N cycling and tend to support more diverse microbial communities (see Jackson et al. 2008 for recent review). These factors suggest that AMF may be more abundant and functionally important in these farming systems compared to conventionally managed farming systems. Many of the advances in agriculture mentioned above (pesticides and synthetic fertilizers) have not been readily available to farmers in the developing world (due largely to cost and access to distribution networks), where much of the future global increase in population is projected to occur. In other words, most farmers in the developing world can be considered to be using "organic" management practices (Cardoso and Kuyper 2006). Thus, AM are likely to be especially important in these agroecosystems (see Burns et al. 2012; Cardoso and Kuyper 2006). Disruption of hyphal networks due to soil cultivation, which is common place in most agroecosystems, may reduce AMF inoculum potential and, therefore, needs to be considered. These factors aside, there is a paucity of studies that have focused on Zn uptake by plants via the AM pathway in a field context. Given the benefits that plants can accrue due to forming AM, such studies should be of high priority, especially in the context of subsistence farming systems.

In response to widespread global inadequacies in dietary Zn intake (Brown and Wuehler 2000), much effort has focused on the development of Zn-efficient genotypes, that is, genotypes that can grow and yield well Zn-deficient soils. Although some studies have shown the Zn efficiency of some genotypes is the same irrespective of the presence of AMF, such responses are inconsistent across genotypes (see Hacisalihoglu and Kochian 2003 for review). Nevertheless, most crop species form AM, and AMF are found in the soils of most arable regions of the world (see work by Read 1991; Treseder and Cross 2006). This, coupled with the significant increases in plant Zn concentrations reported in many of the studies highlighted here, suggests that studies of Zn efficiency should consider (as has been the case in many, but not all, examples) the role of AM in enhancing the Zn efficiency of crops.

Much of our knowledge (with many important exceptions) on the physiology (including Zn nutrition) of AMF has come from studies using a reductionist approach (Johnson et al. 1997). For example, plants are often grown with single or limited numbers of isolates of AMF and (to a lesser extent) plant species. While important model systems, this must be balanced against the functional diversity that exists in AM (Cavagnaro et al. 2005; Smith et al. 2004), and the need to challenge plants with the AMF with which they naturally occur (Johnson et al. 2005). This point is especially relevant, but challenging, where new crop varieties are being screened on a large scale.

11.6   Zn Uptake by AM: Future Research

Our understanding of the uptake of Zn via the mycorrhizal pathway lags behind that of P. Given the widespread global deficiency of Zn in human diets, I argue here that there is an urgent need to better understand the role of AM in improving plant Zn nutrition. To this end, a number of future research opportunities are identified. These broadly fall under the themes of mechanisms, methods, and management.

This section is neither exhaustive nor complete, but it does, nevertheless, aim to stimulate further research in this area.

1. Molecular mechanisms. Identification of the molecular mechanisms by which plants take up Zn via the mycorrhizal pathway will be an important advance. Little is known about long-distance translocation of Zn in AM. For example, a motile vacuolar system as described for P (Uetake et al. 2002) may be important, especially if Zn can act as a counterion to polyP (Christie et al. 2004), but this is speculative (see Cavagnaro 2008 for more detailed discussion). Identification of additional, and further studies of already identified, genes involved in the Zn physiology of AM is also needed.
2. Functional diversity. Estimates of Zn uptake by AM (using ^{65}Zn) vary by approximately two orders of magnitude (see Jansa et al. 2003 and references therein). The underlying reasons for this variation remain unknown. Furthermore, studies that identify the proportion of plant Zn taken up via the mycorrhizal pathway, which in the case of P can be as high as 100 % with certain plant/fungal combinations (Smith et al. 2004), will be of particular value.
3. Isotopic studies. Studies that employ isotopic techniques in the field are likely to be important. Similar such studies have been important in helping to understand the role of AM in plant P nutrition in the field (e.g., Schweiger and Jakobsen 1999).
4. Field relevant studies. Since most people who experience Zn deficiency are in the developing world, their farming is essentially "organic," and the formation of AM can increase plant Zn nutrition, research on this topic should be of high priority. To this end, studies of agricultural management impacts on AM functioning (and other aspects of the biology/ecology of AM) are needed. The need for this is exemplified by the work on long fallow disorder in northern Australia (Thompson 1987, 1996). Genotypic and systems-based methods for the study of AM in the field may be of particular benefit. In studies using a genotypic-based approach, careful selection of mutant/wild-type genotypes should be undertaken so as to ensure that the pairs are as closely "matched," as is practicable. Where a systems-based approach is employed, appropriate crop rotations should be selected.
5. Pre-inoculation. The potential for pre-inoculation of crops with AMF inoculants also deserves further attention; high-value horticultural crops may benefit in particular. Although the importance of "matching" plants and AMF (Johnson et al. 2005), and edaphic and environmental conditions, need to be taken into account. Equally, the identification of land management practices that limit the formation and functioning of AM also need to be undertaken in parallel.

11.7 Conclusions

AM have an important role to play in plant Zn acquisition. Here emphasis has been placed on the role of AM under low soil Zn conditions. In this chapter I have sought to highlight some important advances that have been made and insights gained. If we are to capitalize on the benefits of AM in agroecosystems, we must have a strong knowledge base. Further, while there is a need for highly focused and detailed studies, there is also a need to integrate these findings in a wider context. This will necessarily involve studies undertaken under both laboratory and field conditions. While it is a significant challenge, it is sure to be rewarding.

Acknowledgments Unfortunately it is not possible to cite much of the excellent work on this topic within the confines of a single chapter. Thanks to Ms. Ansel Vies, Mr. Jack Stopher, and Dr. Vanessa Carne-Cavagnaro for their assistance in preparing this chapter. My research is also supported through the ARC Future Fellowship program (FT120100463).

References

Baon JB, Smith SE, Alston AM, Wheeler RD (1992) Phosphorus efficiency of three cereals as related to indigenous mycorrhizal infection. Aust J Agr Res 43:479–491

Barker SJ, Stummer B, Gao L, Dispain I, O'Connor PJ, Smith SE (1998) A mutant in *Lycopersicon esculentum* Mill. with highly reduced VA mycorrhizal colonization, isolation and preliminary characterisation. Plant J 15:791–797

Bolan NS, Robson AD, Barrow NJ (1984) Increasing phosphorus supply can increase the infection of plant roots by vesicular-arbuscular mycorrhizal fungi. Soil Biol Biochem 16:419–420

Brown KH, Wuehler SE (2000) Zinc and human health: results of recent trials and implications for program interventions and research. International Development Research Centre, Ottawa

Bürkert B, Robson A (1994) ^{65}Zn uptake in subterranean clover (*Trifolium subterraneum* L.) by 3 vesicular arbuscular mycorrhizal fungi in a root-free sandy soil. Soil Biol Biochem 26:1117–1124

Burleigh SH, Kristensen BK, Bechmann IE (2003) A plasma membrane zinc transporter from *Medicago truncatula* is up-regulated in roots by Zn fertilization, yet down-regulated by arbuscular mycorrhizal colonization. Plant Mol Biol 52:1077–1088

Burns AE, Gleadow RG, Cliff J, Zacarias AM, Cavagnaro TR (2010) Cassava: the drought, war and famine crop in a changing world. Sustainability 2:3572–3607

Burns AE, Gleadow RM, Zacarias AM, Cuambe E, Miller RE, Cavagnaro TR (2012) Variations in the chemical composition of cassava (*Manihot esculenta* Crantz) leaves and roots as affected by genotypic and environmental variation. J Agric Food Chem 60:4946–4956

Cardoso IM, Kuyper TW (2006) Mycorrhizas and tropical soil fertility. Agr Ecosyst Environ 116:72–84

Cavagnaro TR (2008) The role of arbuscular mycorrhizas in improving plant zinc nutrition under low soil zinc concentrations: a review. Plant and Soil 304:315–325

Cavagnaro TR, Jackson LE (2007) Isotopic fractionation of zinc in field grown tomato. Can J Bot 85:230–235

Cavagnaro TR, Martin AW (2011) Arbuscular mycorrhizas in southeastern Australian processing tomato farm soils. Plant and Soil 340:327–336

Cavagnaro TR, Smith FA, Smith SE, Jakobsen I (2005) Functional diversity in arbuscular mycorrhizas: exploitation of soil patches with different phosphate enrichment differs among fungal species. Plant Cell Environ 164:485–491

Cavagnaro TR, Jackson LE, Six J, Ferris H, Goyal S, Asami D, Scow KM (2006) Arbuscular mycorrhizas, microbial communities, nutrient availability, and soil aggregates in organic tomato production. Plant and Soil 282:209–225

Cavagnaro TR, Dickson S, Smith FA (2010) Arbuscular mycorrhizas modify plant responses to soil zinc addition. Plant and Soil 329:307–313

Cavagnaro TR, Barrios-Masias FH, Jackson LE (2012) Arbuscular mycorrhizas and their role in plant growth, nitrogen interception and soil gas efflux in an organic production system. Plant and Soil 353:181–194

Christie P, Li XL, Chen BD (2004) Arbuscular mycorrhiza can depress translocation of zinc to shoots of host plants in soils moderately polluted with zinc. Plant and Soil 261:209–217

Cooper KM, Tinker PB (1978) Translocation and transfer of nutrients in vesicular-arbuscular mycorrhizas. II Uptake and translocation of phosphorus, zinc and sulphur. New Phytol 81:43–52

Frey B, Schuepp H (1993) Acquisition of nitrogen by external hyphae of arbuscular mycorrhizal fungi associated with *Zea mays* L. New Phytol 124:221–230

Gao X, Kuyper TW, Zou C, Zhang F, Hoffland E (2007) Mycorrhizal responsiveness of aerobic rice genotypes is negatively correlated with their zinc uptake when nonmycorrhizal. Plant and Soil 290:283–291

Gonzalez-Guerrero M, Azcon-Aguilar C, Mooney M, Valderas A, MacDiarmid CW, Eide DJ, Ferrol N (2005) Characterization of a *Glomus intraradices* gene encoding a putative Zn transporter of the cation diffusion facilitator family. Fungal Genet Biol 42:130–140

Hacisalihoglu G, Kochian LV (2003) How do some plants tolerate low levels of soil zinc? Mechanisms of zinc efficiency in crop plants. New Phytol 159:341–350

Hetrick BAD (1991) Mycorrhizas and root architecture. Experientia 47:355–362

Hildebrandt U, Regvar M, Bothe H (2007) Arbuscular mycorrhiza and heavy metal tolerance. Phytochemistry 68:139–146

Jackson LE, Burger M, Cavagnaro TR (2008) Roots, nitrogen transformations, and ecosystem services. Annu Rev Plant Biol 59:341–363

Jansa J, Mozafar A, Frossard E (2003) Long-distance transport of P and Zn through the hyphae of an arbuscular mycorrhizal fungus in symbiosis with maize. Agronomie 23:481–488

Johansen A, Jakobsen I, Jensen ES (1993) External hyphae of vesicular-arbuscular mycorrhizal fungi associated with *Trifolium subterraneum*. 3. Hyphal transport of ^{32}P and ^{15}N. New Phytol 124:61–68

Johnson NC, Graham JH, Smith FA (1997) Functioning of mycorrhizal associations along the mutualism-parasitism continuum. New Phytol 135:575–586

Johnson NC, Wolf J, Reyes M, Panter A, Koch GW, Redman A (2005) Species of plants and associated arbuscular mycorrhizal fungi mediate mycorrhizal responses to CO_2 enrichment. Glob Chang Biol 11:1156–1166

Kochian LV (2000) Molecular physiology of mineral nutrients acquisition, transports, and utilization. In: Buchanan BB, Gruissem W, Jones RL (eds) Biochemistry and molecular biology of plants. American Society of Plant Biologists, Rockville

Kothari SK, Marschner H, Romheld V (1991) Contribution of the VA mycorrhizal hyphae in acquisition of phosphorus and zinc by maize grown in a calcareous soil. Plant and Soil 131:177–185

Liu A, Hamel C, Hamilton RI, Ma BL, Smith DL (2000) Acquisition of Cu, Zn, Mn and Fe by mycorrhizal maize (*Zea mays* L.) growth in soil at different P and micronutrient levels. Mycorrhiza 9:331–336

Maldonado-Mendoza IE, Dewbre GR, Harrison MJ (2001) A phosphate transporter gene from the extra-radical mycelium of an arbuscular mycorrhizal fungus *Glomus intraradices* is regulated in response to phosphate in the environment. Mol Plant Microbe Interact 14:1140–1148

Marschner H (1995) Mineral nutrition of higher plants. Academic, San Diego

Marschner H, Dell B (1994) Nutrient uptake in mycorrhizal symbiosis. Plant and Soil 159:89–102

Menge JA (1983) Utilization of vesicular-arbuscular mycorrhizal fungi in agriculture. Can J Bot 61:1015–1024

Miller RD, Jackson LE (1998) Survey of vesicular–arbuscular mycorrhizae in lettuce production in relation to management and soil factors. J Agric Sci 130:173–182

Oliver AJ, Smith SE, Nicholas DJD, Wallace W, Smith FA (1983) Activity of nitrate reductase in *Trifolium subterraneum*: effects of mycorrhizal infection and phosphate nutrition. New Phytol 94:63–79

Ortas I, Ortakei D, Kaya Z, Çinar A, Önelge N (2002) Mycorrhizal dependency of sour orange in relation to phosphorus and zinc nutrition. J Plant Nutr 26:1263–1279

Read DJ (1991) Mycorrhizas in ecosystems. Experientia 47:376–391

Rillig MC (2004) Arbuscular mycorrhizae and terrestrial ecosystem processes. Ecol Lett 7:740–754

Ruzicka DR, Barrios-Masias FH, Hausmann NT, Jackson LE, Schachtman DP (2010) Tomato root transcriptome response to a nitrogen-enriched soil patch. BMC Plant Biol 10:75–94

Ryan MH, Angus JF (2003) Arbuscular mycorrhizae in wheat and field pea crops on a low P soil: increased Zn-uptake but no increase in P-uptake or yield. Plant and Soil 250:225–239

Ryan MH, Ash A (1999) Effects of phosphorus and nitrogen on growth of pasture plants and VAM fungi in SE Australian soils with contrasting fertiliser histories (conventional and biodynamic). Agr Ecosyst Environ 73:51–62

Ryan MH, Norton RM, Kirkegaard JA, McCormick KM, Knights SE, Angus JF (2002) Increasing mycorrhizal colonisation does not improve growth and nutrition of wheat on Vertosols in south-eastern Australia. Aust J Agr Res 53:1173–1181

Schweiger PF, Jakobsen I (1999) Direct measurement of arbuscular mycorrhizal phosphorus uptake into field-grown winter wheat. Agron J 91:998–1002

Seres A, Bakonyi G, Posta K (2006) Zn uptake by maize under the influence of AM-fungi and Collembola *Folsomia candida*. Ecol Res 21:692–697

Smith SE, Read DJ (2008) Mycorrhizal symbiosis. Academic, Cambridge

Smith MD, Hartnett DC, Rice CW (2000) Effects of long-term fungicide applications on microbial properties in tallgrass prairie soil. Soil Biol Biochem 32:935–946

Smith SE, Smith FA, Jakobsen I (2004) Functional diversity in arbuscular mycorrhizal (AM) symbioses: the contribution of the mycorrhizal P uptake pathway is not correlation with mycorrhizal responses in growth or total P uptake. New Phytol 162:511–524

Sorensen JN, Larsen J, Jakobsen I (2005) Mycorrhiza formation and nutrient concentration in leeks (*Allium porrum*) in relation to previous crop and cover crop management on high P soils. Plant and Soil 273:101–114

Tanaka Y, Yano K (2005) Nitrogen delivery to maize via mycorrhizal hyphae depends on the form of N supplied. Plant Cell Environ 28:1247–1254

Thompson JP (1987) Decline of vesicular-arbuscular mycorrhizae in long fallow disorder of field crops and its expression in deficiency of sunflower. Aust J Agr Res 38:847–867

Thompson JP (1996) Correction of dual phosphorus and zinc deficiencies on Linseed (*Linum usitatissimum* L.) with cultures of vesicular-arbuscular mycorrhizal fungi. Soil Biol Biochem 28:941–951

Tibbett M (2000) Roots, foraging and the exploitation of soil nutrient patches: the role of mycorrhizal symbiosis. Funct Ecol 14:397–399

Tinker PB, Nye PH (2000) Solute movement in the rhizosphere. Oxford University Press, Oxford

Toler HD, Morton JB, Cumming JR (2005) Growth and metal accumulation of mycorrhizal sorghum exposed to elevated copper and zinc. Water Air Soil Pollut 164:155–172

Treseder KK, Cross A (2006) Global distributions of arbuscular mycorrhizal fungi. Ecosystems 9:305–316

Uetake Y, Kojima T, Ezawa T, Saito M (2002) Extensive tubular vacuole system in an arbuscular mycorrhizal fungus, *Gigaspora margarita*. New Phytol 154:761–768

Watts-Williams SJ, Cavagnaro TR (2012) Arbuscular mycorrhizas modify tomato responses to soil zinc and phosphorus addition. Biol Fertil Soils 48:285–294

Watts-Williams SJ, Patti AF, Cavagnaro TR (2013) Arbuscular mycorrhizas are beneficial under both deficient and toxic soil zinc conditions. Plant and Soil 371:299–312

Watts-Williams S, Turney TW, Patti AF, Cavagnaro TR (2014) Uptake of zinc and phosphorous is affected by zinc fertilizer material and arbuscular mycorrhizas. Plant and Soil 376:165–175

Zhu Y-G, Smith SE, Smith FA (2001a) Plant growth and cation composition of two cultivars of spring wheat (*Triticum aestivum* L.) differing in P uptake efficiency. J Exp Bot 52:1277–1282

Zhu Y-G, Smith SE, Smith FA (2001b) Zinc (Zn)-phosphorus (P) interactions in two cultivars of spring wheat (*Triticum aestivum* L.) differing in P uptake efficiency. Ann Bot 88:941–945

Chapter 12
Function of Mycorrhizae in Extreme Environments

Catherine A. Zabinski and Rebecca A. Bunn

12.1 Extreme Environments: What and Why

The most compelling story of mycorrhizae in extreme environments is that the symbiosis extends the ecological niche for host plants when environments are at their most limiting. And although research on symbioses in extreme environments is sparse, studies in marginal environments show that mycorrhizae can ameliorate harsh conditions for host plants (Bothe et al. 2010). However, mycorrhizal function is dependent on the identity of both symbionts, as well as environmental characteristics, and does not uniformly benefit host plants (Johnson et al. 1997). Further, when we consider the potential for a "file drawer" bias (Casada et al. 1996), that is, the publication of positive results (mycorrhizae benefiting host plants) while studies with negative results (mycorrhizae neutral or detrimental to host plants) get relegated to the file drawer, it makes the compelling story of mycorrhizal amelioration of extreme stresses tenuous or at least too simplistic. Here we clarify the current understanding of mycorrhizal effects on host plants in extreme environments and discuss why stress tolerance by mycorrhizal fungi is more likely due to acclimation rather than adaptation, but more research is required for conclusive proof. Finally, we suggest avenues for future research that could increase our understanding of the biology and ecology of this symbiosis.

Extreme environments are characterized by conditions that make survival difficult for most organisms (Rothschild and Mancinelli 2001). Research in extreme environments has increased dramatically as scientists study adaptations present in these environments (Gostinčar et al. 2010; Tiquia and Mormile 2010), search for

C.A. Zabinski (✉)
Land Resources and Environmental Sciences, Montana State University, Bozeman, MT, USA
e-mail: cathyz@montana.edu

R.A. Bunn
Department of Environmental Sciences, Western Washington University, Bellingham, WA, USA

© Springer-Verlag Berlin Heidelberg 2014   201
Z.M. Solaiman et al. (eds.), *Mycorrhizal Fungi: Use in Sustainable Agriculture and Land Restoration*, Soil Biology 41, DOI 10.1007/978-3-662-45370-4_12

organisms and molecules with potential biotechnology applications (Morozkina et al. 2010), and address the potential for life in other planetary environments (Canganella and Wiegel 2011; Harrison et al. 2013). While the harshest environments are primarily inhabited by prokaryotes, eukaryotes also colonize sites that are generally limiting for higher forms of life. Understanding mechanisms by which eukaryotes persist in extreme environments can inform fundamental science of stress physiology and evolutionary ecology, along with increasing the toolbox for managers restoring heavily disturbed sites (Smith et al. 2010). Furthermore, insight into plants' tolerance of heat, drought, and salinity is of critical import as agricultural food production strains to meet increasing demand under changing climate conditions.

Early terrestrial environments were extreme for the first land plants, which required a consistent water supply, a relatively narrow temperature range, and readily available nutrients. As plants transitioned into land-based habitats, they were exposed to damaging UV radiation, desiccating air, and high temperature fluctuations (Waters 2003). The mycorrhizal symbiosis evolved multiple times over the history of plant evolution, and evidence suggests that the ancestral form of this symbiosis, arbuscular mycorrhizae (AM), was present in the roots of the first plants colonizing terrestrial habitats (Wang and Qiu 2006). Given the increasing evidence that symbioses between plants and microorganisms, including mycorrhizae and fungal endophytes, can enhance survival and fitness in extreme environments (Rodriguez and Redman 2008; Chalk et al. 2010), it is likely that mycorrhizae have played an important role for plants' adaptations to extreme environments since the evolution of the symbiosis.

12.2 Symbiosis Function

Mycorrhizal function in non-extreme environments varies, but the most consistently observed phenomenon is enhanced nutrient uptake by the host plant in exchange for host carbon provided to the fungus. The mycorrhizal symbiosis has broad taxonomic and structural diversity, but in this chapter, we will focus on the two most commonly studied forms: ectomycorrhizae (EM), which form between fungi of the *Basidiomycota* and *Ascomycota* and woody plants that may be either angiosperms or gymnosperms, and arbuscular mycorrhizae (AM), between fungi from the *Glomeromycota* and primarily herbaceous plants (but also some woody gymnosperms and angiosperms). In addition to the taxonomic diversity, the structural differences between AM and EM mycorrhizae are most pronounced on the fungal fruiting bodies and at the fungal-root interface. EM fungi produce relatively large fruiting bodies, and EM hyphae form a fungal sheath surrounding the host plant roots, both of which require a large carbon investment from the host plant. The sheath provides a physical barrier between the root and the soil, thereby conferring some level of pathogen protection for the root. In contrast, AM fungi do not construct an external sheath, and the fruiting bodies are generally a small spore formed at the end of soil hyphae. AM fungal carbon requirements are more modest, and protection mechanisms occur biochemically, rather than physically.

The relationship between a host plant and its mycorrhizal fungi has been measured via a cost/benefit framework (Koide and Elliott 1989; Schwartz and Hoeksema 1998). Typically, the net benefit to the host plants is equated to biomass, which is used along with flowering, nutrient status, and seed production to estimate fitness of the host plants. With these measurements, the symbiosis varies from mutualistic to parasitic depending on the species and life stage of host plant, the fungal species, and the soil conditions (Johnson et al. 1997; Hoeksema et al. 2010). However, even without biomass differences, mycorrhizal fungal nutrient uptake pathways are utilized over those of the host plant (Smith et al. 2009), and there may be an advantage for the host plant not always evidenced by increased biomass (Smith et al. 2010) but with long-term implications for host plant fitness. From the perspective of the fungal partner, the symbiosis is clearly beneficial for AM fungi, as a result of their dependence on host plant carbon, and probably comparably so for EM fungi, although some species of EM fungi are able to obtain carbon from decomposing organic matter. An analysis of the costs and benefits of the symbiosis in extreme environments may be in simple terms of survival and the persistence of symbiont populations.

The relative costs of the mycorrhizal symbiosis can increase for the host plant in marginal or extreme environments if the plant has stress-induced constraints on photosynthesis. Alternatively, if a mycorrhizal symbiont supplies nutrients that are limiting in the extreme environment or reduces the host plant's contact with toxic compounds, the benefit of the symbiosis increases relative to benefits in non-extreme environments. Mechanisms of stress tolerance or avoidance conferred by the fungus include: (1) enhancing host plant access to resources, which indirectly affects the host plant's ability to tolerate a suboptimal environment, (2) protecting the host plant from the stress, and (3) altering the plant biochemistry apart from enhanced nutrient concentrations. Because enhancement of nutrient status has cascading effects for the host plant, the three mechanisms are not exclusive of one another.

12.2.1 Mycorrhizal Effects on Host Plant Access to Resources

Mycorrhizal fungal hyphae occur both inside roots and in the soil surrounding the roots. Fungal hyphae are more than two orders of magnitude smaller in diameter than are fine roots and thus able to penetrate soil pores that are inaccessible to roots (hyphal diameter is ~2–20 μm, while fine roots are ~2 mm diameter; Friese and Allen 1991). In this manner, mycorrhizal hyphae greatly increase the volume of soil explored and thereby increase the uptake of immobile nutrients, such as phosphorus. Uptake of other immobile soil ions, including zinc and copper, is also well documented in AM fungi (Smith and Read 2008). Some EM fungi are also capable of accessing nutrients in decomposing organic matter (Plassard and Dell 2010).

Besides enhancing nutrient uptake, mycorrhizal plants may increase host plant water uptake by maintaining a plant/soil continuum or accessing water otherwise unavailable to roots. The small diameter of fungal hyphae suggests that it is unlikely that mycorrhizae have a significant effect on large amounts of water movement, but it is possible that even a small enhancement of water uptake could be crucial to plants under drought conditions (Boyd et al. 1986; Stahl et al. 1998). Mycorrhizae can also improve water status directly via the increased surface area afforded by extraradical hyphae or through the regulation of aquaporins, the membrane proteins that control water intake (Porcel et al. 2006; Lehto and Zwiazek 2011). Indirectly, mycorrhizae can improve water access through their effects on soil structure and aggregate stability (Rillig and Mummey 2006).

In marginal and extreme environments, if mycorrhizal fungi can tolerate abiotic conditions that host plants cannot, the symbiosis can augment root function in those portions of the soil profile. For example, fungal hyphae appear to have a higher tolerance for high soil temperatures and are present in thermal soils where there are no roots, thus expanding the host plant's access to resources (Bunn et al. 2009). However, fungal tolerance for environmental extremes is not always higher than the host plant. For instance, AM hyphal growth is inhibited in saline soils with roots present, suggesting that the AM hyphae have lower tolerance to the saline solution than roots (McMillen et al. 1998).

12.2.2 Mycorrhizal Effects in Soils with Toxic Compounds

The mycorrhizal interface represents increased linkage between the root and the soil matrix. This increase allows for additional uptake of nutrients as discussed above but also could result in an increase in the uptake of toxic compounds. The role of mycorrhizae in toxic, and particularly metal-contaminated, soils has received intense interest with its immediate applicability to remediation and restoration projects. A significant body of literature reports mycorrhizal plants have greater metal tolerance than non-mycorrhizal plants, but these results are by no means universal (Smith and Read 2008), and rarely have studies separated the benefits of enhanced nutrient uptake from enhanced metal tolerance (Meharg and Cairney 1999; Meharg 2003). Additionally, the published literature includes reports of mycorrhizae both increasing and decreasing toxic element concentration in the host plant. From a remediation perspective, either result can be positive, depending on whether the management objective is to move the element of consideration out of the soil and into the biota or whether the objective is to stabilize sediments without introducing contaminants into the biota via trophic transfer.

EM plants (mainly conifers) are typically some of the first plant species to colonize mine spoils (Meharg and Cairney 1999). Physically, the fungal sheath encasing EM roots acts as a barrier between the roots and soil contaminants.

Because high metal concentration in the soil can damage root apical zones, this physical protection contributes to enhanced growth of mycorrhizal plants (Marschner 1995). Biochemically, EM fungi may protect the host plant from high concentrations of metals extracellularly, when metals are immobilized in exuded ligands or on fungal cell surfaces; intracellularly, when metals are immobilized in the cytosol also via ligands; or transcellularly, via increased efflux of metals from fungal cells or storage in vacuoles (Bellion et al. 2006).

AM hyphae can either increase or decrease the uptake of metals from the soil (Weissenhorn et al. 1995; Leyval et al. 1997; Neagoe et al. 2013) but generally reduce the transfer of metals to the shoots (Joner and Leyval 1997; Chen et al. 2005). Reduced element transfer to the plant in phytotoxic soils can be the result of enhanced binding, adsorption, or chelation of metals by mycorrhizal fungal tissues or fungal exudates (Evelin et al. 2009) and sequestration of metals into structures to minimize cellular damage (Ferrol et al. 2009) as well as increased efflux of metals from the cytoplasm (Colpaert et al. 2011). AM fungi can actually avoid metals by changing the direction of hyphal growth or growing through patches of higher metal concentration with reduced branching (Ferrol et al. 2009). The variable effects of the symbiosis on the host plant result from the net effect of enhanced element uptake in conjunction with binding, adsorption, chelation, sequestration, and efflux of excess elements (Audet and Charest 2009). A better understanding of the shifts in mycorrhizal function with changing element concentration would improve our ability to utilize these symbionts in remediation efforts (Hildebrandt et al. 2007).

Similarly, for organic contaminants, mycorrhizae can either increase or decrease host plant uptake (Gunderson et al. 2007). AM hyphae can take up polycyclic aromatic hydrocarbons (PAHs) in soils and transport them to the host plant (Gao et al. 2010), a positive result for those interested in phytoextraction, and can also increase the degradation of PAHs in the soil as a result of positive interactions with soil biota that occur in the rhizosphere and mycorrhizosphere (Yu et al. 2011).

12.2.3   Mycorrhizal Effects on Host Plant Biochemistry

The formation of the mycorrhizal symbiosis involves a complex array of communication between the host plant and the fungus (Maillet et al. 2011). Some of the biochemical responses between the plant and fungus may allow the host plant to better respond to stress. For example, plants produce reactive oxygen species as a signaling mechanism to regulate a number of cellular processes in response to stress (Nanda et al. 2010). These chemicals react with cell constituents in an irreversible way, especially when produced in high concentrations, and lead to aging and death of the cell. Reactive oxygen species are also produced in normal physiological processes, such as oxidative phosphorylation or photosynthesis. Cells regulate both low-level steady-state concentrations and high-level stress-response concentrations of reactive chemicals with antioxidants, a broad class of compounds that can neutralize chemically reactive molecules (Gill and Tuteja 2010).

Mycorrhizal enhancement of growth in environments with limiting conditions is correlated with higher levels of antioxidant enzymes in host plant tissues, suggesting improved oxidative stress regulation for plants in stressful conditions (Bressano et al. 2010). This phenomenon of elevated levels of antioxidants in mycorrhizal plants occurs following exposure to elevated metals (Schützendübel and Polle 2002; Azcón et al. 2010), temperature stress (Zhu et al. 2010), salinity (Garg and Manchanda 2008; Estrada et al. 2013), and drought (Alvarez et al. 2009). The benefit conferred by the mycorrhizal symbiosis may be the augmentation of the biochemical pathways to produce compounds that the plant uses in response to a variety of abiotic stress conditions.

12.3 Adaptation or Acclimation?

Discovering mycorrhizal fungi adapted to extreme environments could have wide ranging applications in agriculture and restoration. And, at first glance, extreme environments present seemingly ideal conditions for adaptation. Limiting environmental parameters promote strong evolutionary and ecological responses, which should enable organisms to adapt to marginal environments, expanding their ecological niche and geographic distribution (Kawecki 2008). But, while selection coefficients for adaptations that enhance survival and reproduction are high, small population size and low genetic variation limit evolutionary potential (Parsons 1991). Whether adaptation occurs depends on the balance between selection pressure and availability of genetic variation. In contrast, acclimation is the adjustment to a new environment that occurs without a change in the genetic profile of a population. It can be difficult to distinguish instances of acclimation from those of true adaptation. Adaptation requires that populations exhibit genetic variation that results in differential ability to occur in conditions particular to that site.

Species occurring in adverse environments may be found only in sites with those conditions, with a consequently narrow ecological distribution, or may occur across a wide range of site conditions with a resulting wider distribution. Mycorrhizal fungi appear to favor the second strategy, exhibiting a high degree of functional plasticity (Lekberg and Koide 2008) along with a high level of genetic variation (Ehinger et al. 2012). For AM fungi this variation is distributed within populations, as opposed to between populations (Koch et al. 2004; Rosendahl 2008). And, most of the variation within a population is represented within a single multinucleate spore (Pawlowska and Taylor 2004), which may give rise to genetically different variants (Ehinger et al. 2012). Similarly, EM fungi exhibit broad genetic variation within species nearly equal to that found between species (Smith and Read 2008).

Evidence for acclimation in AM fungi exists from studies of cold temperature tolerance (Addy et al. 1998) and Ni tolerance (Amir et al. 2008). Tolerance can be induced in a period of months (Amir et al. 2008) and can be lost in isolates that are not maintained under the same stress conditions (Sudova et al. 2007). For AM fungi, the combined evidence of the fungi's functional plasticity, genetic variation, and impermanent tolerance to stresses supports the hypothesis that mycorrhizae acclimate rather than adapt to extreme environments. There are several reports of ecotypic variation in EM fungi, relative to metal tolerance (Colpaert et al. 2011) and serpentine soils (Jourand et al. 2010). Further studies may reveal evidence that populations of EM fungi are adapted to site conditions and particularly to extreme sites.

Adaptation within a symbiosis is especially problematic to discern as the response of each symbiont is mediated by the presence of the other. To circumvent this issue, assessment of both host and fungal traits with combinations of potentially adapted and non-adapted symbionts is needed (Johnson et al. 2010). Patterns of response would serve as the basis for research on mechanisms resulting in the enhanced performance of either symbiont in harsh environments. Comparison across host plants is critical, since host plants adapted to marginal or extreme environments may not perceive that environment as extreme. A comparable study across a climatic gradient shows that host plants differentiated between AM fungi isolated from different climatic regimes only when growing under temperature regimes that they were not accustomed to (Antunes et al. 2011). While comparison across fungal isolates is also important, our understanding of fungal adaptations will be limited until we distinguish between induced responses versus evolutionary differentiation of races or ecotypes.

12.4 A Case Study in an Extreme Environment: Thermal Soils of Yellowstone National Park

Yellowstone National Park (YNP) is underlain by an area of high volcanic activity, where the heat from the subterranean magma chambers is transferred via the groundwater to the surface in the form of geysers, fumaroles, mudpots, hot springs, and thermal vents. The soils in the thermal areas are generally poorly developed, low in nutrients, and with temperatures that increase with depth. Plants growing in thermal areas are generally small in stature and shallow rooted to avoid high soil temperatures (Fig. 12.1), and AM are present in soils with temperatures in the rooting zone as high as 56 °C (Bunn and Zabinski 2003).

The thermotolerant grass, *Dichanthelium lanuginosum* (Elliott) Gould (hot springs panic grass), occurs in YNP thermal areas (Stout et al. 1997; Bunn and Zabinski 2003). We tested the effects of AM for this and two other host plants, *Agrostis scabra* Willd. (rough bent grass) and *Mimulus guttatus* DC. (yellow monkeyflower), both of which are widely distributed in the area, either on or off thermal soils. Plants were grown in the greenhouse with one of two sources of whole-soil AM inoculum, one from YNP thermal soils and the second from a nonthermal soil outside of YNP but in a similar climate regime. All combinations

Fig. 12.1 Thermal site in Yellowstone National Park, in the Midway Geyser Basin

of treatments were subjected to a soil temperature treatment, either ambient green-house temperature or with pots growing on a heat blanket, resulting in soil temper-atures near 50 °C at the base of the pot and 30 °C at the soil surface (Bunn et al. 2009).

Our first question was whether mycorrhizae affect host plant growth in high-temperature soils. The three host plants responded to the soil temperature treatment differently. *Dichanthelium lanuginosum*, which is only present on thermal sites, barely grew at ambient soil temperature and was far more robust at high soil temperature, whereas the biomass of the two species that occur on and off thermal sites was twice as high when growing on soils at ambient temperature as compared to high temperature (Fig. 12.2). Additionally, mycorrhizal effects for facultatively thermal plants were neutral in regard to biomass, at either ambient or high soil temperatures, while *D. lanuginosum* grew much better in the presence of mycor-rhizae. Therefore, mycorrhizal effects on host plant growth in high-temperature soils were dependent on the host species, and for *D. lanuginosum*, its growth on high-temperature soils was dependent on the symbiosis.

Our second question was whether symbiosis function in ambient versus high-temperature soils varies depending on the source of AM inoculum. We found no evidence for differences in mycorrhizal function with whole-soil inoculum from thermal versus conventional soils, when measuring both host plant response traits (biomass, flowering, and root characteristics) and fungal traits (internal coloniza-tion rates and extraradical hyphae). Those results are consistent with the general finding that mycorrhizal diversity is high and distributed within populations rather than between populations.

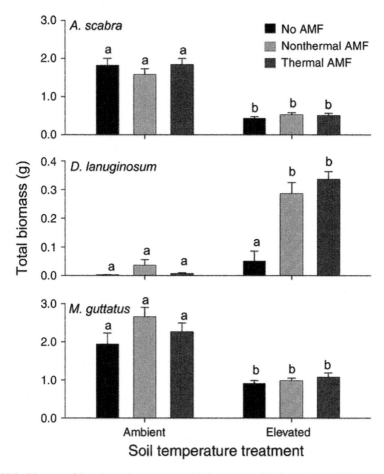

Fig. 12.2 Biomass of three host plants grown with three mycorrhizal treatments either at ambient soil temperature or elevated soil temperatures. *Lowercase letters* represent significant differences ($P < 0.05$) between least squares means of temperature treatments within each species. $n = 18$ for all factor combinations, except for *D. lanuginosum* in ambient temperatures, where $n = 10$, 10, and 8 for no, nonthermal, and thermal AMF, respectively

While plant species seem to be specifically adapted to high-temperature soils, AM fungi do not, which raises interesting questions in regard to the potential for adaptation of a symbiosis to extreme environmental gradients. The AM fungi present in thermal soils include both species that are widely distributed and species that are unique to thermal areas (Appoloni et al. 2008; Meadow and Zabinski 2012). While comparison of whole-soil inoculum from contrasting sites can tell us broadly about function of AM fungal communities, a comparison of isolates of the same species from thermal versus nonthermal sites would allow us to determine whether genetic differences exist between isolates. An additional limitation of this research is the difficulty of extrapolating greenhouse results to the field, due to the complexity of the natural environment relative to experimental conditions. We measured adaptation to a single environmental trait, soil temperature, when in fact organisms in the field are responding to a complex environment. In the greenhouse we used the same field soil and generated a temperature treatment with the use of heat blankets at the base of the pots. In the field, however, soil temperature differences are confounded with soil pH and accompanying chemical differences (Lekberg et al. 2011). Multifactor environmental gradients comparing isolate and host plant combinations become logistically difficult but may be necessary to measure symbiosis function and the potential for adaptation and acclimation of the symbionts to extreme environments.

12.5 Conclusions and Application to Management

The long history of the mycorrhizal symbiosis since plant colonization of terrestrial environments suggests that this symbiosis has functioned to reduce stress for the host plant, and as environmental conditions ameliorate, to increase host plant fitness via enhanced nutrient uptake. If mycorrhizae contribute to plant species' ability to tolerate fluctuating and extreme environmental conditions, as evidenced by enhanced antioxidant levels, contaminant sequestration, and enhanced nutrient status, then managing for changing environments should include recognition of the importance of soil biota, including mycorrhizal fungi, on plant growth.

From an application perspective, land managers would like to know how best to use mycorrhizae for revegetation of disturbed environments that have characteristics in common with extreme environments. Addition of inoculum to restoration sites or managed lands could potentially benefit both aboveground and belowground community developments if the site had depleted mycorrhizal fungal communities or if the inoculum was specifically adapted to site conditions. While there is some evidence for EM fungal ecotypes, there is a need for more studies across a wider taxonomic range of EM fungi to be able to estimate under what conditions site-adapted fungi should be used. For AM fungi, future research should address environment effects on gene expression and the potential for AM fungal species to acclimate to novel environmental conditions. Research that measures symbiont function across time and contrasting gradients, along with gene regulation relative to environmental conditions, will help to elucidate the potential for adaptation and acclimation of fungal symbionts to extreme environments.

References

Addy HD, Boswell EP, Koide RT (1998) Low temperature acclimation and freezing resistance of extraradical VA mycorrhizal hyphae. Mycol Res 102(5):582–586

Alvarez M, Huygens D, Fenandez C, Gacitúa Y, Olivares E, Saavedra I, Alberdi M, Valenzuela E (2009) Effects of ectomycorrhizal colonization and drought on reactive oxygen species metabolism of *Nothofagus dombeyi* roots. Tree Physiol 29:1047–1057

Amir H, Jasper DA, Abbott LK (2008) Tolerance and induction of tolerance to Ni of arbuscular mycorrhizal fungi from New Caledonian ultramafic soils. Mycorrhiza 19:1–6

Antunes PM, Koch AM, Morton JB, Rillig MC, Klironomos JN (2011) Evidence for functional divergence in arbuscular mycorrhizal fungi from contrasting climatic origins. New Phytol 189:507–514

Appoloni S, Lekberg Y, Tercek M, Zabinski C, Redecker D (2008) Molecular community analysis of arbuscular mycorrhizal fungi in roots of geothermal soils in Yellowstone National Park (USA). Microb Ecol 56(4):649–659

Audet P, Charest C (2009) Contribution of AM symbiosis to *in vitro* root metal uptake: from trace to toxic metal conditions. Botany 87:913–921

Azcón R, del Carmen PM, Roldán Barea JM (2010) Arbuscular mycorrhizal fungi, *Bacillus cereus*, and *Candida parapsilosis* from a multicontaminated soil alleviate metal toxicity in plants. Microb Ecol 59:668–677

Bellion M, Courbot M, Jacob C, Blaudez D, Chalot M (2006) Extracellular and cellular mechanisms sustaining metal tolerance in ectomycorrhizal fungi Issue. FEMS Microbiol Lett 254:173–181

Bothe H, Turnau K, Regvar M (2010) The potential role of arbuscular mycorrhizal fungi in protecting endangered plants and habitats. Mycorrhiza 20:445–457

Boyd R, Furbank RT, Read DJ (1986) Ectomycorrhiza and the water relations of trees. In: Gianinazzi-Pearson V, Gianinazzi S (eds) Physiological and genetic aspects of mycorrhizae. INRA, Paris, pp 689–694

Bressano M, Curetti M, Glachero L, Gil SV, Cabello M, DA March D, Luna CM (2010) Mycorrhizal fungi symbiosis as a strategy against oxidative stress in soybean plants. J Plant Physiol 167:1622–1626

Bunn RA, Zabinski CA (2003) Arbuscular mycorrhizae in thermal-influenced soils in Yellowstone National Park. West N Am Nat 63:406–415

Bunn R, Lekberg Y, Zabinski C (2009) Arbuscular mycorrhizal fungi ameliorate temperature stress in thermophilic plants. Ecology 90:1378–1388

Canganella F, Wiegel J (2011) Extremophiles: from abyssal to terrestrial ecosystems and possibly beyond. Naturwurwissenschaften 98:253–279

Casada RD, James PC, Espie RHM (1996) The "file drawer problem" of non-significant results: does it apply to biological research? Oikos 76:591–593

Chalk PM, Alves BJR, Boddey RM, Urquiaga S (2010) Integrated effects of abiotic stresses on inoculant performance, legume growth and symbiotic dependence estimated by N-15 dilution. Plant Soil 328:1–16

Chen B, Roos P, Borggaard OK, Zhu Y-G, Jakobsen I (2005) Mycorrhiza and root hairs in barley enhance acquisition of phosphorus and uranium from phosphate rock but mycorrhiza decreases root to shoot uranium transfer. New Phytol 165:591–598

Colpaert JV, Wevers JHL, Krznaric E, Adriaensen K (2011) How metal-tolerant ecotypes of ectomycorrhizal fungi protect plants from heavy metal pollution. Ann For Sci 68:17–24

Ehinger MO, Croll D, Koch AM, Sanders IR (2012) Significant genetic and phenotypic changes arising from clonal growth of a single spore of an arbuscular mycorrhizal fungus over multiple generations. New Phytol 196:853–861

Estrada B, Aroca R, Barea JM, Ruiz-Lozano JM (2013) Native arbuscular mycorrhizal fungi isolated from a saline habitat improved maize antioxidant systems and plant tolerance to salinity. Plant Sci 201–202:42–51

Evelin H, Kapoor R, Giri B (2009) Arbuscular mycorrhizal fungi in alleviation of salt stress: a review. Ann Bot 104:1263–1280

Ferrol N, González-Guerrero M, Valderas A, Benabdellah K, Azcón-Aguilar C (2009) Survival strategies of arbuscular mycorrhizal fungi in Cu-polluted environments. Phytochem Rev 8:551–559

Friese CF, Allen MF (1991) Spread of VA mycorrhizal fungal hyphae in the soil: inoculum types and external hyphal architecture. Mycologia 83:409–418

Gao Y, Chen Z, Ling W, Huang J (2010) Arbuscular mycorrhizal fungal hyphae contribute to the uptake of polycyclic aromatic hydrocarbons by plant roots. Bioresour Technol 101:6895–6901

Garg N, Manchanda G (2008) Effect of arbuscular mycorrhizal inoculation on salt-induced nodule senescence in *Cajanus cajan* (pigeonpea). J Plant Growth Regul 27:115–124

Gill S, Tuteja N (2010) Reactive oxygen species and antioxidant machinery in abiotic stress tolerance in crop plants. Plant Physiol Biochem 48:909–930

Gostinčar C, Grube M, de Hoog S, Zalar P, Gunde-Cimerman N (2010) Extremotolerance in fungi: evolution on the edge. FEMS Microbiol Ecol 71:2–11

Gunderson JJ, Knight JD, Van Rees KCJ (2007) Impact of ectomycorrhizal colonization of hybrid poplar on the remediation of diesel-contaminated soil. J Environ Qual 36:927–934

Harrison JP, Gheeraert N, Tsigelnitskiy D, Cockell CS (2013) The limits for life under multiple extremes. Trends Microbiol 21:204–212

Hildebrandt U, Regvar M, Bothe H (2007) Arbuscular mycorrhiza and heavy metal tolerance. Phytochemistry 68:139–146

Hoeksema JD, Chaudhary VB, Gehring CA, Johnson NC, Karst J, Koide RT, Pringle A, Zabinski C, Bever JD, Moore JC, Wilson GWT, Klironomos JN, Umbanhowar J (2010) A meta-analysis of context-dependency in plant response to inoculation with mycorrhizal fungi. Ecol Lett 13:394–407

Johnson NC, Graham JH, Smith FA (1997) Functioning of mycorrhizal associations along the mutualism–parasitism continuum. New Phytol 135:575–585

Johnson NC, Wilson GWT, Bowkers MA, Wilson JA, Miller RM (2010) Resource limitation is a driver of local adaptation in mycorrhizal symbioses. Proc Natl Acad Sci U S A 107:2093–2098

Joner EJ, Leyval C (1997) Uptake of ^{109}Cd by roots and hyphae of a *Glomus mosseae/Trifolium subterraneum* mycorrhiza from soil amended with high and low concentrations of cadmium. New Phytol 135:353–360

Jourand P, Ducousso M, Loulergue-Majorel C, Hannibal L, Santoni S, Prin Y, Lebrun M (2010) Ultramafic soils from New Caledonia structure *Pisolithus albus* in ecotype. FEMS Microbiol Ecol 72:238–249

Kawecki TJ (2008) Adaptation to marginal habitats. Annu Rev Ecol Evol Syst 39:321–342

Koch AM, Kuhn G, Fontanillas P, Fumagalli L, Goudet I, Sanders IR (2004) High genetic variability and low local diversity in a population of arbuscular mycorrhizal fungi. Proc Natl Acad Sci U S A 101:2369–2374

Koide R, Elliott G (1989) Cost, benefit, and efficiency of the vesicular-arbuscular mycorrhizal symbiosis. Funct Ecol 3:252–255

Lehto T, Zwiazek JJ (2011) Ectomycorrhizas and water relations of trees: a review. Mycorrhiza 21:71–90

Lekberg Y, Koide RT (2008) Effect of soil moisture and temperature during fallow on survival of contrasting isolates of arbuscular mycorrhizal fungi. Botany 86:1117–1124

Lekberg Y, Meadow J, Rohr JR, Redecker D, Zabinski CA (2011) Importance of dispersal and thermal environment for mycorrhizal communities: lessons from Yellowstone National Park. Ecology 92:1292–1302

Leyval C, Turnau K, Haselwandter K (1997) Effect of heavy metal pollution on mycorrhizal colonization and function: physiological, ecological and applied aspects. Mycorrhiza 7:139–153

Maillet F, Poinsot V, André O, Peuch-Pagès V, Haouv A, Guenier M, Cromer L, Giraudet D, Formey D, Niebel A, Andres Martinez E, Driguez H, Bécard G, Dénarié J (2011) Fungal lipochitooligosaccharide symbiotic signals in arbuscular mycorrhiza. Nature 469:58–63

Marschner H (1995) Mineral nutrition of higher plants, 2nd edn. Academic, San Diego, p 889

McMillen BG, Juniper S, Abbott LK (1998) Inhibition of hyphal growth of a vesicular–arbuscular mycorrhizal fungus in soil containing sodium chloride limits the spread of infection from spores. Soil Biol Biochem 30:1639–1646

Meadow JF, Zabinski CA (2012) Linking symbiont community structures in a model arbuscular mycorrhizal system. New Phytol 194:800–809

Meharg AA (2003) The mechanistic basis of interactions between mycorrhizal associations and toxic metal cations. Mycol Res 107:1253–1265

Meharg AA, Cairney JWG (1999) Co-evolution of mycorrhizal symbionts and their hosts to metal contaminated environments. Adv Ecol Res 30:70–112

Morozkina EV, Slutskaya ES, Fedorova TV, Tugay TI, Golubeva LI, Koroleva OV (2010) Extremophilic microorganisms: biochemical adaptation and biotechnological application (review). Appl Biochem 46:1–14

Nanda AK, Andrio E, Marino D, Pauly N, Dunand C (2010) Reactive oxygen species during plant-microorganism early interactions. J Integr Plant Biol 52:195–204

Neagoe A, Iordache V, Bergmann H, Kothe E (2013) Patterns of effects of arbuscular mycorrhizal fungi on plants grown in contaminated soil. J Plant Nutr Soil Sci 176:273–286

Parsons PA (1991) Evolutionary rates: stress and species boundaries. Annu Rev Ecol Syst 22:1–18

Pawlowska TE, Taylor JW (2004) Organization of genetic variation in individuals of arbuscular mycorrhizal fungi. Nature 427:733–737

Plassard C, Dell B (2010) Phosphorus nutrition of mycorrhizal trees. Tree Physiol 30:1129–1139

Porcel R, Aroca R, Azcón R, Ruiz-Lozano JM (2006) PIP aquaporin gene expression in arbuscular mycorrhizal Glycine max and Lactuca sativa plants in relation to drought stress tolerance. Plant Mol Biol 60:389–404

Rillig MC, Mummey DL (2006) Mycorrhizas and soil structure. New Phytol 171:41–53

Rodriguez R, Redman R (2008) More than 400 million years of evolution and some plants still can't make it on their own: plant stress tolerance via fungal symbiosis. J Exp Bot 59:1109–1114

Rosendahl S (2008) Communities, populations and individuals of arbuscular mycorrhizal fungi. New Phytol 178:253–266

Rothschild LJ, Mancinelli RL (2001) Life in extreme environments. Nature 409:1092–1101

Schützendübel A, Polle A (2002) Plant responses to abiotic stresses: heavy metal-induced oxidative stress and protection by mycorrhization. J Exp Bot 53:1351–1365

Schwartz MW, Hoeksema JD (1998) Specialization and resource trade: biological markets as a model of mutualisms. Ecology 79:1029–1038

Smith SE, Read DJ (2008) Mycorrhizal symbiosis, 3rd edn. Elsevier, Amsterdam

Smith FA, Grace EJ, Smith SE (2009) More than a carbon economy: nutrient trade and ecological sustainability in facultative arbuscular mycorrhizal symbioses. New Phytol 182:347–358

Smith SE, Christophersen HM, Pope S, Smith FA (2010) Arsenic uptake and toxicity in plants: integrating mycorrhizal influences. Plant Soil 327:1–21

Stahl PD, Schuman GE, Frost SM, Williams SE (1998) Arbuscular mycorrhizae and water stress tolerance of Wyoming big sagebrush seedlings. Soil Sci Soc Am J 62:1309–1313

Stout RG, Summers ML, Kerstetter T, McDermott TR (1997) Heat- and acid-tolerance of a grass commonly found in geothermal areas within Yellowstone National Park. Plant Sci 130:1–9

Sudova R, Jurkiewicz A, Turnau K, Vosátka M (2007) Persistence of heavy metal tolerance of the arbuscular mycorrhizal fungus Glomus intraradices under different cultivation regimes. Symbiosis 43:71–81

Tiquia SM, Mormile MR (2010) Extremophiles—a source of innovation for industrial and environmental applications. Environ Technol 31:823

Wang B, Qiu YL (2006) Phylogenetic distribution and evolution of mycorrhizas in land plants. Mycorrhiza 16:299–363

Waters ER (2003) Molecular adaptation and the origin of land plants. Mol Phylogenet Evol 29:456–463

Weissenhorn I, Leyval C, Belgy G, Berthelin J (1995) Arbuscular mycorrhizal contribution to heavy metal uptake by maize (*Zea mays* L.) in pot culture with contaminated soil. Mycorrhiza 5:245–251

Yu XZ, Wu SC, Wu FY, Wong MH (2011) Enhanced dissipation of PAHs from soil using mycorrhizal ryegrass and PAH-degrading bacteria. J Hazard Mater 28:1206–1217

Zhu X, Song F, Xu H (2010) Influence of arbuscular mycorrhiza on lipid peroxidation and antioxidant enzyme activity of maize plants under temperature stress. Mycorrhiza 20:325–332

Chapter 13
Alleviation of Soil Stresses by Arbuscular Mycorrhizal Fungi

Obed F. Madiba

13.1 Introduction

It is well known that arbuscular mycorrhizal (AM) fungi form symbiotic associations with a range of plants (Killham 1994; Marschner 1995; Smith and Read 2008). The extent of colonisation is not only controlled by P concentration in the soil solution but also by P content in the plant (Marschner 1995). A large amount of carbon from the host plant is needed by the fungi (Ryan et al. 2002). Mycorrhizal roots experience high respiration and thus have high carbon loss (Killham 1994; Marschner 1995; Calderon et al. 2012). Arbuscules (primarily for transfer of nutrients) and vesicles (primarily as storage organs) are crucial structures in the transfer of P and other nutrients, e.g. immobile nutrients (Zn, Cu), to the plant (Smith and Read 2008). The likely mechanism involves P (in the form of phosphate) uptake from the soil solution directly through the root epidermis and root hairs or via the AM fungi pathway (Marschner 1995; Solaiman and Saito 2001). Nutrients are then delivered to shoots and leaves by the transpiration stream.

Phosphorus (P) is generally a limiting nutrient in many agricultural systems especially in sandy soils (Evans et al. 2006). As a result, large application of P fertilisers has been administered to increase crop production, some of which results in P leaching (Fertiliser Working Group 2007) or fixation in soil in acid soils (Lujerdean et al. 2004) and alkaline soils (Aliasgharzad et al. 2010; Cardarelli et al. 2010). The consequences of fixation of P in soil result in the reduction of P availability to crops and consequently crop production will decrease (Lujerdean et al. 2004). Leaching of P from fertiliser does not only decrease nutrient availability, but can also cause algal blooms, thus polluting waterbodies (Lehmann et al. 2003). In addition, the raw material used to make P fertilisers (rock phosphate)

O.F. Madiba (✉)
School of Agricultural and Environmental Sciences, Faculty of Science and Agriculture, University of Limpopo, Sovenga 0727, South Africa
e-mail: freddy.madiba@ul.ac.za

© Springer-Verlag Berlin Heidelberg 2014   215
Z.M. Solaiman et al. (eds.), *Mycorrhizal Fungi: Use in Sustainable Agriculture and Land Restoration*, Soil Biology 41, DOI 10.1007/978-3-662-45370-4_13

Table 13.1 Examples of soil stresses effect on arbuscular mycorrhizal fungi

Soil stress	Mycorrhizal response	References
Soil disturbance	Disturbance affects AM fungal colonisation	Evans and Miller (1988), Miller (2000)
	Disturbance does not affect AM fungal colonisation	Duan et al. (2011)
Soil compaction	Increase in soil aggregation	Rillig et al. (2001), Schimel et al. (2007), Miransari et al. (2008)
Soil pH (alkalinity)	Reduce alkalinity effect	Smith and Read (2008), Cardarelli et al. (2010)
Soil pH (acidity)	Reduce acidity effect	Marschner (1995), Aliasgharzad et al. (2010)
Drought	Increase P uptake	Quilambo et al. (2005)
Salinity	Increase P uptake	Miransari and Smith (2007), Schimel et al. (2007), Daei et al. (2009)
Flooding	Low AM fungi colonisation	Solaiman and Hirata (1995), Sah et al. (2006)
	High AM fungi colonisation	Sivaprasad et al. (1990), Secilia and Bagyaraj (1992), Mendoza and Garcia (2005), Matsumara et al. (2008)

is a non-renewable resource and needs to be conserved. It has been estimated that mining of rock phosphate will reach 'peak P' around 2030 as rock phosphate reserves are predicted to be exhausted by 2060 (Cordell et al. 2009). Strategies are therefore required to increase crop production while at the same time protecting the environment.

Soil stresses, such as acidity and alkalinity, compaction, moisture, salinity, extreme soil temperatures and flooded soils, reduce plant growth and yield and hence reduce production (Miransari 2010). However, the production of rice increases with flooding (Sivaprasad et al. 1990; Secilia and Bagyaraj 1992) probably when colonisation by AM fungi increases. The presence of both commercial inocula of AM fungi and AM fungi indigenous to the agricultural system had potential to alleviate these stresses (Quilambo et al. 2005; Miransari 2010). However, examples of the impact of soil stresses on AM fungi show inconsistencies among studies (Table 13.1).

13.2 Soil Physical Constraints

No-tillage is defined as a cropping system in which there is minimal soil disturbance and can be practiced with diverse crop rotation to increase carbon sequestration for potential mitigation of climate change (Macvay et al. 2006). Additionally, no-tillage practices require that the soil must have a permanent plant cover for organic matter build-up (Diacona and Montemurro 2010). However, these requirements are not possible for many countries largely due to costs associated with the equipment needed, climate and other factors including government policies. No-tillage is an agricultural system which has been adopted in many countries, including Australia (D'Emden and Llewellyn 2006; Triplett and Dick 2008). For example, D'Emden and Llewellyn (2006) indicated that 86 % of farmers in Western Australia and 42 % in Southern Australia responded that they are using no-till in their cropping system. The main reason for this adoption is probably because no-tillage provides more benefits than negatives. The main benefit of no-tillage is that it reduces soil erosion either by water or wind erosion, thus increasing nutrients for crop growth (Diacona and Montemurro 2010). Furthermore, soil compaction, which prevents seedlings and water infiltration into the soil, is reduced by controlling traffic in the paddock. The retention of organic matter associated with no-tillage will enhance the ease of water movement and availability and the enhancement of soil biological activities (Diacona and Montemurro 2010). Where possible, weeds, pests and diseases can be controlled biologically (by using cover crops) and chemically using herbicides and by crop rotation (e.g. legumes), thus avoiding escalating costs of pesticides and nitrogen fertilisers. Again, by promoting diverse biological activities including those related to AM fungi, nutrient uptake, especially P, will be more efficient (Killham 1994; Marschner 1995; Smith and Read 2008; Verbruggen and Kiers 2010).

The role of AM fungi in nutrient uptake is influenced by crop rotation and agricultural practices, including tillage (Jasper et al. 1989; McGonigle and Miller 2000; Jansa et al. 2006). Evans and Miller (1988) reported that soil disturbance can cause destruction of the AM fungal hyphae network thus disrupting the flow of nutrients to roots (Miller 2000). However, in other circumstances, Duan et al. (2011) demonstrated that soil disturbance did not affect AM fungal colonisation as evidenced by high P uptake and increased growth but Jansa et al. (2003) reported that soil tillage affected the community structure of AM fungi in maize roots.

The timing of tillage can maintain the activity of AM fungi. For example, Maiti et al (2011) reported that although less soil disturbance results in high AM fungal colonisation of roots, the timing of tillage may play a role as well. Their findings indicate that no soil disturbance (no-tillage) in the off-season (fallow period) in rain-fed systems (in this case about 8 months after harvest) can have a positive effect on the indigenous AM fungal population. This suggests that the fallow period without tillage after harvest (in summer) and sowing another crop (in winter) is effective for AM fungal proliferation. Thus, although soil disturbance may disrupt AM fungal hyphae, the extent of destruction may depend on the soil type, the AM fungal species and the schedule of tillage. Agricultural practices such as cultivation exposes organic matter and aerates the soil which leads to oxidation of organic matter (Killham 1994). Subsequently, aggregate stability is reduced and soil bulk density increases, and this can lead to soil compaction (Miransari et al. 2008). Furthermore, the increased abundance of micro-aggregates can also contribute to soil compaction. If AM fungi are present in the soil in sufficient quantities, the hyphal network will contribute to binding micro-aggregates into macroaggregates (Miransari et al. 2008). Glomalin, a glycoprotein, as defined by Rillig et al. (2001)

is a 'glue-like' structure released from hyphae which protects hyphae of AM fungi from losing water due to their hydrophobicity. Due to the coating of AM fungal hyphae with glomalin, aggregates can be stabilised during wetting and drying cycles can occur whereas these cycles usually destroy aggregation (Schimel et al. 2007). Rillig et al. (2001) indicated that glomalin can exist in the soil from 6 months to 40 years, thus making it a crucial component in soil aggregation. Furthermore, AM fungi can reduce compaction impacts through their interactions with roots, providing strength in penetrating soil for exploration of nutrients and water (Miransari et al. 2008). AM fungal hyphae and glomalin contribute to soil organic carbon in grasslands, and due to their decomposition, more cementing agents, such as polysaccharides, are produced (Rillig et al. 2001). It is doubtful whether glomalin contributes significantly to aggregation in semiarid climates due to low production of biomass.

13.3 Soil Water Constraints

Stress caused by shortages of water can be challenging for microbial functioning and survival. In order to minimise water stress, microorganisms, including AM fungi, alter their physiology and cellular mechanisms (Schimel et al. 2007). Micro-organisms shift their resources from growth to survival mode as a strategy in water-limiting environments (Schimel et al. 2007). In addition, under dry conditions, water potential of AM fungi decreases and therefore accumulation of osmolytes is needed to prevent dehydration which usually comes at a cost to the environment (Auge 2001). However, upon re-wetting they have to dispose of the accumulated osmolytes and that usually occurs with respiration, releasing CO_2 from the ecosystem and thus contributing to large carbon losses. Furthermore, when the soil dries, substrate diffusion decreases due to discontinuation of pores, and this make substrates less available due to slow diffusion (Schimel et al. 2007).

Both indigenous and commercial AM fungi differ in their ability to increase crop growth under water stress conditions. Quilambo et al. (2005), working on low-P coastal soils of Mozambique, showed that indigenous AM fungi provided higher root colonisation compared to commercial AM fungi. Although AM fungi act as a pathway for increased uptake of nutrients and water to the plant, these are not the only requirements for the plant to grow.

By definition, waterlogged soils are those that most of the year or several months of the year are ponded with water (Marschner 1995). These soils can be submerged during and after heavy rain or excessive and frequent irrigation in poorly structured soils; paddy soils are good examples of such soils (Marschner 1995). In some cases, clearing forests for agriculture results in waterlogging (Cramer et al. 2004). The replacement of trees with shallow-rooted crops results in less interception of water by roots, and this increases ground water recharge, resulting in shallow water tables associated with waterlogging (Cramer et al. 2004).

AM fungi are thought to be aerobic, and any condition which results in O_2 deficiency (such as flooding) is detrimental to their development and survival (Atwell and Steer 1990). Less mycorrhizal colonisation occurs in reducing conditions where redox potential are lower compared to oxidising environments with high redox (Khan 1993). Lack of aeration in roots results in lower concentrations of N, K and P in shoots (Atwell and Steer 1990). Under waterlogged conditions, the reduction of iron Fe^{3+} to Fe^{2+} increases (Kirk et al. 1990), and in plant species such as rice, this reduction of Fe encourages the solubility and availability of phosphates to the rice plant (Kirk et al. 1990). Solaiman and Hirata (1996, 1997a) showed that rice grown in paddy soil had a higher plant dry matter compared to non-flooded soil. However, AM fungal colonisation in flooded soil was lower compared to non-flooded soil. Thus, under anaerobic conditions, AM fungal development is restricted due to O_2 unavailability (Atwell and Steer 1990).

Earlier research showed that AM fungi may be totally absent or temporarily absent in waterlogged conditions and become available in dry conditions (Solaiman and Hirata 1995, 1997b; Sah et al. 2006). In contrast, studies by Sivaprasad et al. (1990), Secilia and Bagyaraj (1992) and Matsumara et al. (2008) revealed that AM fungal colonisation and development was increased under flooded conditions. The findings of Matsumara et al. (2008) indicated that hyphal density and AM fungal colonisation of orange roots increased in a waterlogged soil where oranges were grown intercropped with bahia grass. This was attributed to the intercropping of oranges with bahia grass and inoculation with *G. margarita*. It was thought that the oranges were exposed to O_2 through the well-developed aerenchyma of bahia grass. Furthermore, Mendoza and Garcia (2005) indicated that the mycorrhizal fungal network increased in seedlings of a legume in a flooded soil.

13.4   Soil Chemical Constraints

Strongly saline soils (e.g. electrical conductivity (EC) >1.87 dS/m; McKenzie et al. 2004) can reduce crop production due to loss of water through osmosis (Daei et al. 2009). Under salty environments, water moves from plant cells to the soil solution and as a result, cells plasmolyse (shrink) and ultimately collapse and die (Brady and Weil 2008). Under saline conditions, P uptake is reduced and high ion toxicity from Na and Cl is experienced (Miransari and Smith 2007). Due to low uptake of these ions by AM fungi in salty conditions, their transportation to the plant will be restricted, thus alleviating salinity stress (Daei et al. 2009). Although AM fungi are capable of existing in saline soils, their development is inhibited in extreme saline conditions (Al-Karaki 2000). In addition, extreme salinity (EC of >1.87 dS/m) reduces crop growth by inhibiting spore germination and hyphae development for AM fungi (Juniper and Abbott 2006), thus reducing arbuscular development (Miransari 2010). These limitations will reduce the effectiveness of AM fungi in facilitating nutrient and water uptake by crops in saline conditions. However, under moderate salinity conditions (EC of 0.15–0.7 dS/m, McKenzie et al. 2004), AM fungal inoculation will alleviate saline stress (Miransari and Smith 2007). The mechanisms underlying salt-sensitive and salt-tolerant AM fungi are different depending on the species (Daei et al. 2009). Those species that are resistant enhance leaf respiration and transpiration, and as a result CO_2 and water exchange increases which will affect the water use efficiency (WUE) of the crop and consequently crop yield will increase (Daei et al. 2009; Miransari and Smith 2007). Unlike the case for water stress, AM fungi alleviate salinity stress by increasing osmolytes (carbohydrates) in host plants, and due to this mechanism, root and shoot growth can be reduced (Daei et al. 2009). Most soils in Mediterranean environments experience high evaporation during summer resulting in high salt accumulation on the surface of soils and reduced crop production. AM fungal inoculation could be beneficial in those situations, but not without other remediation practices.

Soil pH is an important soil characteristic, and AM fungi can differ substantially in their pH tolerance range. For example, Hayman and Mosse (1971) demonstrated enhanced colonisation and plant growth stimulation by AM fungi in soil of pH 5.6 and 7.0, but not in more acid soils of pH 3.3–4.4. Supporting this, soil pH was regarded as a constraint to the distribution of *A. laevis* and *Glomus* sp. (WUM 3) in southwestern Australia (Porter et al. 1987). In this environment, *Glomus* sp. (WUM 3) colonised roots well in the pH range of 5.3–7.5; *A. laevis* was tolerant of acidic soil up to pH 6.2, while *S. calospora* formed mycorrhizae at pH 5.3. Soil pH is one of the soil characteristics that is likely to determine the distribution and relative abundance of AM fungal species in soil (Robson and Abbott 1989).

Cardarelli et al. (2010) observed that inoculation of AM fungi in alkaline soil conditions increased P, K, Fe and Zn content in zucchini plants and decreased the toxic levels of Na with associated increased fruit yield and quality. Bacteria generally thrive well in soils with a high pH (Rousk et al. 2009) although Aliasgharzad et al. (2010) showed that AM fungi were more abundant in deeper soil layers (20–30 cm) with increasing pH.

Phosphorus is likely to be the soil nutrient that has the greatest influence on the extent of root colonisation by AM fungal species. Formation of hyphal entry points into the root, hyphal growth within the root, hyphal growth in soil and sporulation are highest at phosphorus levels suitable for optimal growth of particular hosts but are reduced at very low or very high phosphorus host levels (Thingstrup et al. 1998; Thomson et al. 1986). In high-P soil, the extent of colonisation varied substantially among plant families, genera and among closely related genotypes of the same species (Graham et al. 1991; Krishna et al. 1985; Manske 1989; Toth et al. 1990). Increasing nitrogen fertilisers of both ammonium and nitrate have also been reported to reduce root penetration and colonisation of subterranean clover by AM fungi (Chambers et al. 1980). The negative effect was more marked in the case of nitrogen fertilisers than that of P fertilisers (Jensen and Jakobsen 1980). However, these effects depend on the rate added and the available nutrient in the soil (Sharma and Adholeya 2000).

13.5 Conclusions

Phosphorus is a limiting nutrient in most agricultural systems especially in sandy soils. The practice of applying large amounts of P fertilisers to increase production has negative environmental consequences. Soil stresses, such as alkalinity and acidity, compaction, drought, salinity and extreme temperature and waterlogging, can reduce crop production. AM fungi are to some extent resistant to these stresses, but there may be negative influences on AM fungal development and survival.

References

Aliasgharzad N, Martensson LM, Olsson PA (2010) Acidification of a sandy grassland favours bacteria and disfavours fungal saprotrophs as estimated by fatty acid profiling. Soil Biol Biochem 42:1058–1064

Al-Karaki GN (2000) Growth of mycorrhizal tomato and mineral acquisition under salt stress. Mycorrhiza 10:51–54

Atwell BJ, Steer BT (1990) The effect of oxygen deficiency on uptake and distribution of nutrients in maize plants. Plant Soil 122:1–8

Auge RM (2001) Water relations, drought and vesicular-arbuscular mycorrhizal symbiosis. Mycorrhiza 11:3–42

Brady NC, Weil RR (2008) The nature and properties of soils. Pearson/Prentice Hall, Upper Saddle River

Calderon FJ, Schultz DJ, Paul EA (2012) Carbon allocation, belowground transfers, and lipid turnover in a plant-microbial association. Soil Sci Soc Am J 76:1614–1623

Cardarelli M, Rouphael Y, Rea E, Colla G (2010) Mitigation of alkaline stress by arbuscular mycorrhiza in zucchini plants grown under mineral and organic fertilization. J Plant Nutr Soil Sci 173:778–787

Chambers CA, Smith SE, Smith FA (1980) Effects of ammonium and nitrate ions on mycorrhizal infection, nodulation and growth of *Trifolium subterraneum*. New Phytol 85:47–62

Cordell D, Drangert J-O, White S (2009) The story of phosphorus: global food security and food for thought. Glob Environ Change 19:292–305

Cramer VA, Hobbs RJ, Atkins L, Hodgson G (2004) The influence of local elevation on soil properties and tree health in remnant eucalypt woodlands affected by secondary salinity. Plant Soil 265:175–188

D'Emden FH, Llewellyn RS (2006) No-tillage adoption decisions in southern Australian cropping and the role of weed management. Aust J Exp Agr 46:563–569

Daei G, Ardekani M, Rejali F, Teimuri S, Miransari M (2009) Alleviation of salinity stress on wheat yield, yield components, and nutrient uptake using arbuscular mycorrhizal fungi under field conditions. J Plant Physiol 166:217–225

Diacona M, Montemurro F (2010) Long-term effects on soil fertility. A review. Agron Sust Dev 30:411–422

Duan T, Facelli E, Smith SE, Smith FA, Nan Z (2011) Differential effects of soil disturbance and plant residue on function of arbuscular mycorrhizal (AM) symbiosis are not reflected in colonization of roots or hyphal development in soil. Soil Biol Biochem 43:571–578

Evans DG, Miller MH (1988) Vesicular-arbuscular mycorrhizas and the soil-disturbance-induced reduction of nutrient absorption in maize. I. Causal relations. New Phytol 110:67–74

Evans J, McDonald L, Price A (2006) Application of reactive phosphate rock and sulphur fertilisers to enhance the availability of soil phosphate in organic farming. Nutr Cycl Agroecosyst 75:233–246

Fertiliser Working Group (2007) Phasing-out the use of highly soluble phosphorus fertilisers in an environmentally sensitive areas of South west and Western Australia. Minister of the Environment, Western Australia, pp 5–6

Graham JH, Eissenstat DM, Drouillar DL (1991) On the relationship between a plant's mycorrhizal dependency and rate of vesicular-arbuscular mycorrhizal colonization. Funct Ecol 5:773–779

Hayman DS, Mosse B (1971) Plant growth responses to vesicular arbuscular mycorrhiza. I. Growth of endogone—inoculated plants in phosphate deficient soils. New Phytol 70:19–22

Jansa J, Mozafar A, Kuhn G, Anken T, Ruh R, Sanders IR, Frossard E (2003) Soil tillage affects the community structure of mycorrhizal fungi in maize roots. Ecol Appl 13:1164–1176

Jansa J, Weimken A, Frossard E (2006) The effects of agricultural practices on arbuscular mycorrhizal fungi. In: Frossard E, Blum W, Warkentin B (eds) Function of soils for human societies and the environment, vol 266, Special publication. Geological Society, London, pp 89–115

Jasper DA, Abbott LK, Robson AD (1989) Soil disturbance reduces the infectivity of external hyphae of vesicular-arbuscular mycorrhizal fungi. New Phytol 112:93–99

Jensen A, Jakobsen I (1980) The occurrence of vesicular-arbuscular mycorrhiza in barley and wheat grown in some Danish soils with different fertilizer treatments. Plant Soil 55:403–414

Juniper S, Abbott L (2006) Soil salinity delays germination and limits growth of hyphae from propagules of arbuscular mycorrhizal fungi. Mycorrhiza 16:371–379

Khan AG (1993) Occurrence and importance of mycorrhiza in aquatic trees of New South Wales, Australia. Mycorrhiza 3:31–38

Killham K (1994) Soil ecology. Cambridge University Press, Cambridge

Kirk GJD, Ahmad AR, Nye PH (1990) Coupled diffusion and oxidation of ferrous iron in soils. II. A model of diffusion and reaction of O_2, Fe^{2+}, H^+ and HCO_3^- in soils and a sensitivity analysis of the model. J Soil Sci Sci 41:411–431

Krishna KR, Shetty KG, Dart PJ, Andrews DJ (1985) Genotype dependent variation in mycorrhizal colonization and response to inoculation of pearl millet. Plant Soil 86:113–125

Lehmann J, da Silva JP, Steiner C, Nehls T, Zech W, Glaser B (2003) Nutrient availability and leaching in an archaeological Anthrosol and a Ferrasol of the Central Amazon basin: fertilizer, manure and charcoal amendments. Plant Soil 249:343–357

Lujerdean A, Rusu M, Marghitas M (2004) Effect of liming on soil phosphorus availability. In: Proceedings, Symposium on the prospects of the 3rd millennium agriculture, Romania, pp 39–41

Macvay KA, Budde JA, Fabrizzi K, Mikha MM, Rice CW, Schlegel AJ, Peterson DE, Sweeney DW, Thompson C (2006) Management effects on soil physical properties in a long-term tillage studies. Soil Sci Soc Am J 70:434–438

Maiti D, Variar M, Singh RK (2011) Optimizing tillage schedule for maintaining activity of the arbuscular mycorrhizal fungal population in a rainfed upland rice (*Oryza sativa* L.) agrosystem. Mycorrhiza 21:167–171

Manske GGB (1989) Genetical analysis of the efficiency of VA mycorrhiza with spring wheat. Agric Ecosyst Environ 29:273–280

Marschner H (1995) Mineral nutrition of higher plants, 2nd edn. Academic, London

Matsumara A, Horii S, Ishii T (2008) Observation of arbuscular mycorrhizal network system between trifoliate orange and some grasses under water-logged conditions. In: Proceedings, 27th International Conference, Seoul, South Korea, pp 69–75

McGonigle TP, Miller MH (2000) The inconsistent effect of soil disturbance on colonization of roots by arbuscular mycorrhizal fungi: a test of the inoculums density hypothesis. Appl Soil Ecol 14:147–155

McKenzie N, Jacquier D, Isbell R, Brown K (2004) Australian soils and landscapes: an illustrated compendium. CSIRO, Collingwood

Mendoza R, Escudero V, Garcia I (2005) Plant growth, nutrient acquisition and mycorrhizal symbioses of a water-logging tolerant legume (*Lotus glaber* Mill.) in saline-sodic soil. Plant Soil 275:305–315

Miller MH (2000) Arbuscular mycorrhizae and the phosphorus nutrition of maize: a review of Guelph studies. Can J Plant Sci 80:47–52

Miransari M (2010) Contribution of arbuscular mycorrhizal symbiosis to plant growth under different types of soil stress. Plant Biol 12:563–569

Miransari M, Smith DL (2007) Overcoming the stressful effects of salinity and acidity on soybean [*Glycine max* (L.) Merr.] nodulation and yields using signal molecule genistein under field conditions. J Plant Nutr 30:1967–1992

Miransari M, Bahrami HA, Rejali F, Malakouti MJ (2008) Using arbuscular mycorrhiza to reduce the stressful effects of soil compaction on wheat (Triticum aestivum L.) growth. Soil Biol Biochem 40:1197–1206

Porter WM, Robson AD, Abbott LK (1987) Factors controlling the distribution of vesicular-arbuscular mycorrhizal fungi in relation to soil pH. J Appl Ecol 24:663–672

Quilambo OA, Wiessenhorn I, Doddema H, Kuiper PJC, Stulen I (2005) Arbuscular mycorrhizal inoculation of peanut in low fertile tropical soil. II. Alleviation of drought stress. J Plant Nutr 28:1645–1662

Rillig MC, Wright SF, Nichols KA, Schmidt WF, Torn MS (2001) Large contribution of arbuscular mycorrhizal fungi to soil carbon pools in tropical forest soils. Plant Soil 233:167–177

Robson AD, Abbott LK (1989) The effect of soil acidity on microbial activity in soil. In: Robson AD (ed) Soil acidity and plant growth. Academic, Sydney, pp 139–165

Rousk J, Brooks PC, Baath E (2009) Contrasting soil pH effects on fungal and bacterial growth suggest functional redundancy in carbon mineralization. Appl Environ Microb 75:1589–1596

Ryan MH, Norton RM, Kirkegaard JA, McCormick KM, Knights SE, Angus JF (2002) Increasing mycorrhizal colonization does not improve growth and nutrition of wheat on Vertosols in south-eastern Australia. Aust J Agric Res 53:1173–1181

Sah S, Reed S, Jayachandran K, Dunn C, Fisher JB (2006) The effect of repeated short term flooding on mycorrhiza survival in Snap bean roots. HortScience 41:598–602

Schimel J, Balser TC, Wallenstein M (2007) Microbial stress-response physiology and its implications for ecosystem function. Ecology 88:1386–1394

Secilia J, Bagyaraj DJ (1992) Selection of efficient VA mycorrhizal fungi for wetland rice (Oryza sativa L.) plants. Biol Fertil Soils 13:108–111

Sharma MP, Adholeya A (2000) Response of Eucalyptus tereticornis to inoculation with indigenous AM fungi in a semiarid alfisol achieved with different concentrations of available soil P. Microbiol Res 154:349–354

Sivaprasad P, Sulochana KK, Salam MA (1990) Vesicular-arbuscular mycorrhiza (VAM) colonization in lowland rice roots and its effect on growth and yield. Int Rice Res Newslett 15:14–15

Smith SE, Read DJ (2008) Mycorrhizal symbiosis. Academic, London, 800pp

Solaiman MZ, Hirata H (1995) Effects of indigenous arbuscular mycorrhizal fungi in paddy fields on rice growth and N, P, K nutrition under different water regimes. Soil Sci Plant Nutr 41:505–514

Solaiman ZM, Hirata H (1996) Effectiveness of arbuscular mycorrhizal colonization at nursery-stage on growth and nutrition in wetland rice (*Oryza sativa* L.) after transplanting under different soil fertility and water regimes. Soil Sci Plant Nutr 42:561–571

Solaiman ZM, Hirata H (1997a) Effect of arbuscular mycorrhizal fungi inoculation of rice seedlings at the nursery stage upon performance in the paddy field and greenhouse. Plant and Soil 191:1–12

Solaiman ZM, Hirata H (1997b) Responses of directly seeded wetland rice to arbuscular mycor-
    rhizal fungi. J Plant Nutr 20:1479–1487
Solaiman ZM, Saito M (2001) Phosphate efflux from intraradical hyphae of *Gigaspora margarita*
    in vitro and its implication for phosphorus. New Phytol 151:525–533
Thingstrup I, Rubaek G, Sibbesen E, Jakobsen I (1998) Flax (*Linum usitatissimum* L.) depends on
    arbuscular mycorrhizal fungi for growth and P uptake at intermediate but not high soil P levels
    in the field. Plant Soil 203:37–46
Thomson BD, Robson AD, Abbott LK (1986) Effects of phosphorus on the formation of mycor-
    rhizas by *Gigaspora calospora* and *Glomus fasciculatum* in relation to root carbohydrates.
    New Phytol 103:751–765
Toth R, Toth D, Starke D, Smith DR (1990) Vesicular-arbuscular colonization in Zea mays
    affected by breeding for resistance to fungal pathogens. Can J Bot 68:1039–1044
Triplett GB Jr, Dick WA (2008) No-tillage crop production: a revolution in agriculture! Celebrate
    the centennial. A Suppl Agron J 100:S153–S165
Verbruggen E, Kiers ET (2010) Evolutionary ecology of mycorrhizal functional diversity in
    agricultural soils. Evol Appl 3:547–560

Chapter 14
Mechanisms for Alleviation of Plant Water Stress Involving Arbuscular Mycorrhizas

Bede Mickan

14.1 Introduction

Water is often a limiting factor in many dry land agricultural cropping systems, with many global climate change projections predicting an increase in drought frequency and duration. Furthermore, water deficits are now spreading to many agricultural regions where drought was uncommon in the past (Anwar et al. 2013). This poses a significant global challenge to maintain or increase food production from semiarid agricultural zones (Tscharntke et al. 2012). However, this reduction in rainfall has led to many innovative farm management practices such as no-tillage (Jemai et al. 2013), precision seeding, and reduced traffic (Boizard et al. 2012), which can increase soil water retention and water infiltration. In parallel with these developments, activities of arbuscular mycorrhizal (AM) fungi have been claimed to include mechanisms that increase both plant water relations directly (Manoharan et al. 2010) and indirectly increase the soil water retention and infiltration by the enhancement of soil aggregation (Rillig and Mummey 2006).

AM fungi are obligate symbionts. The fungi have an internal phase inhabiting roots and an external phase comprising the extra-radical hyphal network that forms a branching mass of hyphae with potential to acquire nutrients and water beyond the root depletion zone (Allen et al. 2003; Khalvati et al. 2010). The potential of AM fungi to enhance water relations of plants and soil is claimed to be related to a degree of alleviation of water stress, and the mechanisms involved have been widely investigated by a multidisciplinary group of specialists including ecologists, botanists, agricultural scientists, and more recently plant geneticists. Key reviews discussing the role of AM fungi in plant water relations include those of Augé

B. Mickan (✉)
Soil Biology and Molecular Ecology Group, School of Earth and Environment (M087), UWA
Institute of Agriculture, Faculty of Science, The University of Western Australia, Crawley,
WA, Australia 6009
e-mail: bedemickan@gmail.com

© Springer-Verlag Berlin Heidelberg 2014   225
Z.M. Solaiman et al. (eds.), *Mycorrhizal Fungi: Use in Sustainable Agriculture and
Land Restoration*, Soil Biology 41, DOI 10.1007/978-3-662-45370-4_14

(2001) and Ruiz-Lozano et al. (2012) and for AM fungal influences on soil structure a review by Rillig and Mummey (2006). Here I investigate how AM fungi can (1) directly alleviate water stress directly through nutrient and water acquisition and enhanced plant water physiology and (2) indirectly through enhancement of the structure of the soil/water interface.

14.2 Direct AM Fungi Benefits to Host Plants Under Water Stress

The ability of AM fungi to alleviate water stress directly in its most obvious mechanism is through an increase in the amount of nutrients and water acquired through the AM fungi hyphal network. However, there is also evidence to state that AM fungi are able to alter plant biochemical composition to increase antioxidant production and beneficial osmoregulator (proline) that directly alleviate a degree of water stress through drought-avoidance mechanisms (for key papers, see Table 14.1). Partitioning which mechanism is the largest contributor in alleviating water stress has not been determined, though it is most likely the combination of direct mechanisms in unity that are responsible for alleviating a degree of plant water stress.

14.2.1 Plant Nutritional Benefits from Mycorrhizal Symbiosis

Enhanced plant growth associated with the AM fungal symbiosis was first claimed to be through a direct increase in nutrient uptake, giving the host plant a greater tolerance to water stress in the early 1970s (Safir et al. 1971, 1972). Initially, these early authors claimed AM fungi were able to decrease water resistance from the root to leaves, and this was correlated with an increase in the growth of shoot mass (Safir et al. 1971). Later, they reported water resistance between AM-fungi- and non-AM-fungi-colonised plants was negligible once the addition of nutrients (Hoagland solution) was applied. Thus, non-AM fungi host plants are able to match water transpiration rate and shoot growth when there are sufficient nutrients for plant growth (Safir et al. 1972). Subsequently, it has been widely reported that colonisation of roots by AM fungi can increase the growth of the host plant through an enhanced acquisition of P and, more recently claimed, through the acquisition of N (Veresoglou et al. 2012). This increase can at least partly alleviate water stress of AM-fungi-colonised plants (Smith and Read 2008). Plant roots are able to acquire nutrients through a direct pathway, through the soil root interface, or through the extra-radical phase of AM fungi from the hyphal network path (Smith and Smith 2011). Thus, once the N and P concentrations in the soil have been exploited by a

Table 14.1 Key experimental papers on direct effects of AM fungi in plant water relations

Focal point	Experimental procedure	Experimental outcome	Reference
AM fungi effect on water resistance through soybean roots, stem, and leaves	Comparison of plant water resistance between non-AM-fungi- and AM-fungi-colonised soybeans	AM-fungi-colonised soybean reduced water resistance in roots compared to non-AM fungi soybean; this was related to an enhanced nutrient uptake proposed through the hyphal network	Safir et al. (1972)
AM fungi effect on leaf water potential, transpiration rates, and correlation with hyphal entry points into soybean	Allen MF calculated the plant water transpiration between AM-fungi- and non-AM-fungi-colonised soybeans	AM fungi soybean had 50 % lower leaf resistance with no change in leaf or root water potentials, thus AM fungi increased transpiration by 100 %. Allen correlated this increase in plant water to hyphal entry points to calculate direct water access through the hyphal network	Allen (1982)
Evaluated six AM fungi isolates' ability to alter rates of root water uptake under soil water-deficit conditions	Monitored soil-drying rates of non-AM fungi control plants of comparable size with nutritional status of AM-fungi-colonised lettuce	AM-fungi-colonised lettuce was able to deplete soil moisture significantly more than control non-AM-fungi-colonised lettuce. Furthermore some AM fungi species were able to deplete soil moisture than others, and this was directly correlated with hyphal mass production	Marulanda et al. (2003)
Quantification of plant water uptake through the hyphal network	Split chamber microcosm—using high-resolution on-line water content sensors	Plant water uptake through the hyphal network was estimated to be potentially up to 20 %. This value contradicts studies showing the contribution to be negligible	Ruth et al. (2011)
Antioxidant production and transpiration rates during drought stress on AM-fungi-/non-AM-fungi-colonised rice	Investigate drought tolerance mechanisms in rice induced through AM fungi colonisation	AM-fungi-colonised rice induced the accumulation of the antioxidant molecule glutathione whilst reducing oxidative damage to lipids. Combined with an AM fungi increase in leaf gas exchange aided to alleviate water stress above non-AM-fungi-colonised rice controls	Ruiz-Sánchez et al. (2010)

plant within the root depletion zone through the direct pathway, plants that have AM fungi associations are able to exploit nutrients through the hyphal network path because this pathway extends beyond the root depletion zone. Under these circumstances, when the exploitable limiting nutrient resource shifts from a direct pathway to a hyphal pathway, it is likely the AM fungi growth response will be positive (Allen et al. 2003). Under low soil-nutrient conditions, it is likely that the growth response associated with the AM fungi will be larger than that of a non-AM fungi counterpart, giving the AM-fungi-colonised plant the 'big plant, little plant advantage' arising from an increase in nutrient acquisition (Augé 2001).

AM fungal-associated increases in plant growth and phenology allow roots to develop deeper and more extensively, enabling greater access to exploitable water in soil. Furthermore, larger plants have an added advantage of a greater reserve of sugars to draw on under water-stressed conditions (Porcel and Ruiz-Lozano 2004). However, whilst there may be an increase in drought avoidance due to an increase in plant size attributed to AM fungi when the AM fungi growth response is positive, there is evidence that AM-fungi-colonised plants of similar size and nutrition status still show signs of enhanced plant tolerance to water stress (Augé 2001; Porcel and Ruiz-Lozano 2004).

14.2.2 Water Access Through the AM Fungi Hyphal Network

Pioneer studies by Allen (1982) claimed AM fungi hyphae were able to exploit water resources and transport water to the host plant. Allen reported mycorrhizal plants had 50 % lower leaf resistance with no change in leaf or root water potentials. More recently, the most widely claimed benefit of AM-fungi-colonised plants under water stress is through a superior water allocation mediated by the hyphal network of AM fungi, giving the plant access to water in a lower soil water potential as compared to comparative non-AM fungi plants (Marulanda et al. 2003; Ruiz-Lozano et al. 2012). It is currently widely accepted that AM fungi hyphae are able to exploit water and nutrients beyond the root depletion zone in both distance and space (Ruth et al. 2011; Ruiz-Lozano et al. 2012).

Investigations into the function and physiology of AM fungi hyphal network have incorporated split chamber microcosms, separated by fine nylon mesh (usually around 38 μm) allowing hyphae to pass whilst blocking roots (Al-Karaki et al. 2004). This allows empirical quantification of hyphal network services provided to the host plant, which is important, as AM fungi cannot survive without plant roots and they occupy the same volume of soil (Khalvati et al. 2005). Furthermore, natural abundance stable isotopes C^{13} and N^{15} along with deuterium have also been used to trace the movements of these nutrients from the soil to plant leaves, allowing definitive contributions of water, N, and P provided through the hyphal network (Egerton-Warburton et al. 2007).

The amount of AM fungi hyphae within pot experiments is variable with ranges from 1 to 40 m/g of soil, but this can be highly dependent on species identity (Smith and Smith 2011). In natural settings, the hyphal length can explore a much greater soil volume (e.g. 111 m/cm^3 of soil for one particular prairie community (Rillig et al. 2001)). The rate of hyphal spread through soil is also considerable. Jakobsen et al. (1992) in a glasshouse experiment showed *Acaulospora laevis* had extended through the soil about 80 mm by 28 days at a rate 3.0 mm/day from the host plant root. The quantity of hyphal network-derived water to the host plant values ranges from 0.1 μL/h (Allen 1982) for each hyphal penetration point to 0.37 μL/h (Faber et al. 1991); this represents in some experiments between 4 % (Khalvati et al. 2005) to 20 % of total water uptake through the hyphal network (Ruth et al. 2011). Marulanda et al. (2003) demonstrated that AM fungi *G. mosseae* when in association with *Lactuca sativa* under water-stressed conditions were able to deplete volumetric soil moisture by 0.95 %, equating to an additional 4.75 mL/plant/day compared to a comparative non-AM fungi control. Moreover, there was a direct correlation between the species (*G. intraradices*) of AM fungi that depleted the most volumetric soil moisture also produced the largest amount of hyphal network and frequency of root colonisation (Marulanda et al. 2003).

Alleviation of water stress has also been reported through mycorrhizal networks able to connect two or more plants through the hyphal network; this allows resource allocation by a source–sink relationship (Eason et al. 1991). Briefly, if plant A is deficient in water and plant B has sufficient water, then water is able to be transported from plant B via the hyphal network to plant A along the source–sink gradient (Simard et al. 2012). This has been shown to be especially pronounced in hydraulically lifted water from plants with deeper roots to plants with shallower roots. However, when Querejeta et al. (2012) quantified how much water can be transported from donor oak trees to receiver seedlings in shallower soil, they reported that a greater soil water potential gradient was more of a driving force in the vertical redistribution of water from deeper soil profiles to shallower soil profiles than the mycorrhizal network. The significance of mycorrhizal networks alleviating water stress from a host plant to another is still not fully ascertained and represents an exciting area of future research (Prieto et al. 2012).

Plant physiological evidence supporting AM fungi-associated alleviation of some degree of water stress in colonised plants has been shown by increases in (1) water uptake rate, (2) transpiration, and (3) stomatal conductance when compared to non-AM-fungi-colonised host plant controls (Marulanda et al. 2003; Khalvati et al. 2005). The widely claimed direct benefit to the host plant arises from AM fungi external hyphal network able to acquire both nutrients and water past the root depletion zone, and accessing microsites of water locked within the soil pores is probably the most substantive water-alleviating mechanism of AM fungi (Subramanian and Charest 1999; Khalvati et al. 2005; Ruth et al. 2011; Ruiz-Lozano et al. 2012).

14.2.3   AM Fungi Influence on Plant Biochemical Properties

There is evidence that AM fungi are able to alter biochemical properties of roots and shoots inside the host plant. AM fungi colonisation can increase in antioxidant and proline production, and this is claimed to be most beneficial to the host plant through a better osmotic adjustment potential and reduction in oxidative stress (Ruiz-Sánchez et al. 2010). Whether the claimed increase in beneficial biochemical compounds is directly related to the increased nutrient and water acquisition or if AM fungi are altering plant compounds independently of resource acquirement has not been determined.

14.2.4   AM Fungi Production of Antioxidant Compounds

Water-stressed plants display signs of oxidative damage as a result of increased production of degenerative free radical reactive oxygen species (ROS) (Ruiz-Lozano et al. 2012). These free radical ROS include superoxide (O_2^-), hydroxyl radicals, and others such as hydrogen peroxide (H_2O_2). The detrimental effects of O_2^- and H_2O_2 are in their ability to initiate reactions that cause the production of hydroxyl free radicals under water-stressed conditions (Porcel et al. 2003). Hydroxyl free radicals are among the most reactive species known to chemistry, indiscriminately causing oxidative damage to plant biomolecules and lipid membranes, denaturation of proteins, and mutation of DNA (Ruiz-Sánchez et al. 2010). It is claimed AM fungi are able to suppress oxidative damage by mediating antioxidant ROS scavenging enzymes into the host plant under water-stressed conditions (Porcel and Ruiz-Lozano 2004; Ruiz-Sánchez et al. 2010).

Experimental evidence in the antioxidant production in AM fungi plants has shown increases, decreases, and stabile concentrations in water-stressed plants. These variable results are highly dependent on plant, AM fungi species, and also the type of antioxidant (Kohler et al. 2008). Investigations by Porcel et al. (2003) showed that three out of four antioxidants within AM-fungi-colonised plant roots remained stable or slightly lower under water-stressed conditions between AM fungi and comparative non-AM fungi treatments. However, these authors reported antioxidant glutathione concentration increased 534 % in AM fungi plant roots under water stress compared to corresponding non-AM fungi treatment. Ruiz-Sánchez et al. (2010) reported similar results with the antioxidant glutathione reductase (reduced form). For water-stressed conditions, shoot concentration of antioxidant glutathione reductase in AM fungi plants increased by 436 % as compared to corresponding non-AM fungi plants (Ruiz-Sánchez et al. 2010). AM fungal stimulation of antioxidants in commercially grown lettuce (*Lactuca sativa* L.) is widely claimed to be beneficial for reduction of water stress to a certain degree via alleviation of oxidative stress (Porcel et al. 2003). Additionally, the increased nutritional status of the leaf has potential implications for increasing human health nutrient uptake (Baslam and Goicoechea 2012).

14.2.5 AM Fungi-Induced Accumulation of Proline

Water-stressed plants are also known to accumulate organic osmolytes such as proline (amino acid) and sugars that contribute to the host plant tolerance under water-stressed conditions through enhanced osmoregulation (Trotel-Aziz et al. 2000). Proline is a nonprotein amino acid that accumulates in plant tissues under water stress together with sugars, and after water stress recovery, it is readily metabolised (Singh et al. 2011). During water-deficit or saline conditions, plants accumulate proline to maintain osmotic balance under low water potentials (Ruiz-Lozano et al. 2012). During water stress, sugar and proline content in roots colonised by AM fungi has been shown to increase as compared to non-AM fungi roots, giving evidence that osmotic adjustment is occurring, enhancing the ability of the host plant to cope with water stress (Porcel et al. 2003; Kohler et al. 2008). Proline also acts as a reservoir of energy and N during water stress and has been found to increase when the plant is colonised by AM fungi. Whilst it is widely claimed proline accumulation in plant roots enhance plant tolerance to water stress through better osmoregulation at the root soil interface, there are also complex responses that show proline levels can be AM fungal species dependent (Ruiz-Lozano et al. 2012).

14.2.6 AM Fungi Influence on Plant Gas Exchange

There is growing evidence that shows AM-fungi-colonised plants maintain higher gas exchange rates to comparative non-AM fungi plants of similar size and plant nutrition status (see reviews by Augé 2001; Ruiz-Lozano et al. 2012). Colonisation of roots by AM fungi in response to water stress has been claimed through an observed increase in stomatal conductance (g_s), g_s maintenance (drought tolerance), and early g_s closure (drought avoidance) in soils with lower water potential (Khalvati et al. 2005). High plant g_s translates into higher rates of transpiration/ gas exchange which have often increased in AM fungi plants in well-watered, water-stressed, and also after exposure to salinity-affected water (Wu and Xia 2006; Sheng et al. 2008). Saline and water-deficit conditions have a common osmotic component as they reduce water uptake by roots causing dehydration of plant tissues, referred to as an osmotic stress. The observed higher g_s rates in AM-fungi-colonised plants have been in strong correlation with lower xylem-sap abscisic acid and lower abscisic acid fluxes to leaves in AM fungi host plants (Ruiz-Lozano and Aroca 2010). However, g_s in AM-fungi-colonised plant could be argued to be through an enhanced direct ability to deplete soil moisture through AM fungi hyphal network (Augé et al. 2008) and/or increased root branching caused by AM fungi, giving the plant greater access to water (Kothari et al. 1990). Interestingly, AM-fungi-colonised soils have the ability to increase the g_s of non-AM fungi plants under water-stressed conditions through the ability of the hyphal network to enhance soil moisture potential through increasing soil structure (Augé et al. 2007).

14.3 Indirect AM Fungi Benefits to Host Plants Through Enhanced Rhizosphere Processes

Whilst there is a direct role of AM fungi in alleviating water stress to the host plant, there is also an indirect benefit of AM fungi through buffering water stress within the rhizosphere (Audet 2012). There has been considerable interest in reports that AM fungi hyphal networks can enhance soil aggregation through direct physical effects or through AM fungi production of biochemicals (Martin et al. 2012). This indirect soil structure effect has potential to improve the water holding capacity of soil (Augé 2004). Soil aggregation is the arrangement or structure of soil particles held in a single mass or cluster, commonly defined using a hierarchal model. Soil organisms influence soil structure by physically binding soil particles together increasing the quantity and size of the aggregates improving the habitat for microfauna (Tisdall and Oades 1982). Whilst recognising the greatest influence on soil aggregation is probably through plant roots, AM fungi influence on soil aggregation can be seen at the plant community and plant hyphal levels and is influenced by AM fungi either directly through the hyphal network or indirectly through altering plant root physiology (Rillig and Mummey 2006). Moreover, non-AM fungi symbiont-forming plants have shown increased resistance to water stress in AM fungi-infected soil. This has been directly correlated with AM-fungi-colonised soil having a greater water retention capacity through enhanced soil aggregation as compared to non-AM-fungi-colonised soil (Augé et al. 2007).

14.3.1 AM Fungi Biochemical Influence on Soil Aggregation

Biochemical compounds including glomalin, mucilages, polysaccharides, and hydrophobins are exuded from hyphal tips and are also secreted on hyphal walls in the mycorrhizal rhizosphere (Singh et al. 2012). Glomalin is a stable glycoprotein that can be measured directly from the soil as a glomalin-related soil protein, which is also deposited on the hyphal walls of the hyphal network and on adjacent soil particles (Wilson et al. 2009). Glomalin-related soil protein acts to bind soil particles together, forming a stable soil aggregation that can enhance the C storage in soil. The benefit of high amounts of glomalin-related soil protein in soil is in the enhanced water retention capacity, as soil aggregation protects C-rich detritus from microbial decomposition (Rillig and Mummey 2006; Wilson et al. 2009; Verbruggen et al. 2012). The difficulty in quantifying AM fungi-produced glomalin-related soil protein is in the soil extraction process, as it is not clear whether glomalin-related soil protein is of AM fungi origin. In fact, there is evidence suggesting that glomalin-related soil protein also originates from other fungal species (Wilson et al. 2009). However, it is widely claimed AM fungi-produced glomalin, polysaccharides, and other related proteins exuded from the hyphal network act to bind soil particles of various sizes enhancing soil aggregation which correlates with increased water holding capacity of soil (Rillig and Mummey 2006; Wu et al. 2008; Hallett et al. 2009; Audet 2012).

14.3.2 AM Fungi Physical Influence on Soil Aggregation

Physical enmeshment and entanglement of AM fungi hyphae with soil particles to organic matter increase aggregate stability of soil (Daynes et al. 2013). Similar to plant roots, AM fungi hyphae form branching structures with glomalin acting to physically bind microaggregates with macroaggregates (Singh et al. 2012). Hyphal branching morphology is highly variable within and between AM fungi species, with dynamic hyphae lengths, chemistry, and thicknesses, which affects the enmeshment capability of hyphal network. Additionally, although the tensile strength of hyphae is unknown (owing to the narrowness of hypha), it may play an important role in stabilising soil aggregates under disturbance (Rillig and Mummey 2006). The hyphal network persistence in soil is also variable with turnover rates ranging from 5 to 6 days; hyphae runner have been recorded to last 32 days and also stabilise soil several months after plant death (Hallett et al. 2009). As for growing plant roots, AM fungi hyphae are also capable of aligning primary particles such as clay and organic matter together, exerting physical pressure on soil particles leading to a macroaggregate formation which has potential to increase the water holding capacity of soil (Singh et al. 2012). However, Daynes et al. 2013 reported the most pronounced influence on soil aggregation was the presence of plant roots, with AM fungi further stabilising soil structure. These authors reported the self-organising structure of soils to form aggregates in the absence of plant roots, AM fungi hyphae, and/or organic matter (Daynes et al. 2013). Thus, physical soil aggregate stability is dependent on soil characteristics and can also be influenced by soil fauna.

14.3.3 Biological Influence on Soil Aggregation

Biological alterations in soil by the AM fungi hyphal network have also been reported to alter soil aggregation through changes in the soil microbial food web. Changes to the prokaryotic communities induced by AM fungi alteration in soil

Table 14.2 Key papers on AM fungi-induced benefits to soil water through enhancing rhizosphere processes

Focal point	Procedure	Outcome	Reference
Developing a hierarchal model of soil structure, and its relationship with land management practices	Comprehensive review on soil management practices can alter soil aggregation	Soil management effect on soil structure through plant roots and AM fungi hyphae stabilise macro-aggregates, thus land management influences the growth of plant roots and the oxidation of organic carbon	Tisdall and Oades (1982)
		The water stability of microaggregates depends on the persistent organic binding agents and appears to be a characteristic of the soil, independent of management	
The role of AM fungi hyphae in soil's effect on non-AM fungi symbiosis plants	Experiments using non-AM forming plants in AM-fungi-colonised soil with a drought stress treatment	AM fungal colonisation of soil may play as important a role as colonisation of plant roots. AM fungi hyphae affect the water relations of host plant through enhancing soil water status or potential by increasing soil aggregation	Augé (2004)
AM fungi influence soil structure at the plant community, individual root, and the soil hyphal network	Comprehensive review on mechanisms of AM fungi influence on soil structure at various scales	AM fungi can influence soil aggregation at each of the plant community, individual root, and soil hyphal network levels. Through physical, chemical, and biological mechanisms to different degrees. Understanding these relationships will require analyses emphasising feedbacks between soil structure and AM fungi	Rillig and Mummey (2006)
The role of AM fungi in ecosystems using soil aggregate stability	Large-scale field manipulations, using fungicide application as non-AM fungi control	Field manipulations that increased AM fungi hyphae increased water stable aggregates and glomalin-related soil proteins. Fungicide	Wilson et al. (2009)

(continued)

Table 14.2 (continued)

Focal point	Procedure	Outcome	Reference
		application that decreased AM fungi, this correlated significantly with decreasing soil aggregate stability	
The role f AM fungi hyphae and Collembola in soil aggregation independent of plant roots	Split chamber microcosms, partitioning plant roots with hyphae and Collembola present	Collembola can enhance soil aggregation, which complement effects of AM fungi hyphae, and that these effects are independent of plant roots. Even though AM fungi hyphae food quality is regarded as lesser than saprobic fungi hyphae	Siddiky et al. (2012a)
Mechanisms that underpin the development and stabilisation of soil structure	Manipulative pot experiments of severely disturbed mine spoil soil, with additions of AM fungal isolates, organic matter, and plants	Organic matter, living plant roots, and AM fungi are required for stable soil structure in complex ways. The presence of adequate organic matter and plant roots as key contributors to the development of soil structure are further enhanced by AM fungi hyphae	Daynes et al. (2013)

may influence changes to soil aggregation at the microaggregate level. Exudates from the AM fungal hyphal network act as a substrate for bacterial growth, where bacteria such as *Paenibacillus* spp. have been reported to enhance microaggregate soil structure (Hildebrandt et al. 2002). Changes in rhizodeposits through AM fungi have also been shown to alter community composition of bacterial populations which have variable functional attributes (Toljander et al. 2007). AM fungi-induced alteration of soil structure leads to changes in available pore space in soil, which logically leads to changes in environments made habitable to soil organisms (Rillig 2004).

The AM fungi hyphal network also forms the basis of the soil food web by being a valuable food source for micro-arthropods even though this resource is a lesser quality than saprophytic fungi. Siddiky et al. (2012a) conducted an experiment investigating how collembola and AM fungi hyphae together increase water stable aggregation independent of plant roots. Interestingly AM fungi hyphae decreased 6 % in length from collembolan grazing, but the collembolan population increased by 20 %. It is claimed that collembola are able to increase soil aggregation through their faecal pellets. The understanding of mechanisms of how AM fungi hyphal network influences the soil food web and soil biological diversity and abundance is still in its infancy, but it presents an exciting area of future research (Siddiky et al. 2012b) (Table 14.2).

14.4 Conclusion

There is evidence that AM fungi colonisation in plants has the ability to directly alleviate water stress through physical, chemical, and biological drought-avoidance and tolerance mechanisms. Directly, AM fungi are able to exploit a greater amount of soil in both space due to the narrowness of hypha and distance through the hypha's ability to extend beyond the root depletion zone. Thus, separating traits of AM fungi that enable alleviation of water stress is difficult because variables of AM fungi are closely linked to one another. The conclusion most supported by evidence is that AM fungi are able to alleviate water stress to the host plant through multiple processes by enhancing (1) nutrient/water acquisition and antioxidant production, (2) proline accumulation, and (3) soil structure, thereby increasing soil water retention. The successful management of rainfed dry land agricultural production, whether it be cropping or grazing systems, will benefit from management practices that increase the diversity and abundance of AM fungi populations in water-limiting environments.

References

Al-Karaki G, McMichael B, Zak J (2004) Field response of wheat to arbuscular mycorrhizal fungi and drought stress. Mycorrhiza 14:263–269

Allen MF (1982) Influence of vesicular-arbuscular mycorrhizae on water movement through Bouteloua gracilis Lag ex Steud. New Phytol 91:191–196

Allen MF, Swenson W, Querejeta JI, Egerton-Warburton LM, Treseder KK (2003) Ecology of mycorrhizae: a conceptual framework for complex interactions among plants and fungi. Annu Rev Phytopathol 41:271–303

Anwar MR, Li Liu D, Macadam I, Kelly G (2013) Adapting agriculture to climate change: a review. Theor Appl Climatol 113(1-2):225–245

Audet P (2012) Arbuscular mycorrhizal symbiosis and other plant-soil interactions in relation to environmental stress. In: Ahmed P, Prasad MNV (eds) Environmental adaptations and stress tolerance of plants in the era of climate change. Springer, New York, pp 233–264

Augé RM (2001) Water relations, drought and vesicular-arbuscular mycorrhizal symbiosis. Mycorrhiza 11:3–42

Augé RM (2004) Arbuscular mycorrhizae and soil/plant water relations. Can J Soil Sci 84:373–381

Augé RM, Toler HD, Moore JL, Cho K, Saxton AM (2007) Comparing contributions of soil versus root colonization to variations in stomatal behavior and soil drying in mycorrhizal *Sorghum bicolor* and *Cucurbita pepo*. J Plant Physiol 164:1289–1299

Augé RM, Toler HD, Sams CE, Nasim G (2008) Hydraulic conductance and water potential gradients in squash leaves showing mycorrhiza-induced increases in stomatal conductance. Mycorrhiza 18:115–121

Baslam M, Goicoechea N (2012) Water deficit improved the capacity of arbuscular mycorrhizal fungi (AMF) for inducing the accumulation of antioxidant compounds in lettuce leaves. Mycorrhiza 22:347–359

Boizard H, Yoon SW, Leonard J, Lheureux S, Cousin I, Roger-Estrade J, Richard G (2012) Using a morphological approach to evaluate the effect of traffic and weather conditions on the structure of a loamy soil in reduced tillage. Soil Tillage Res 127:137–147

Daynes CN, Field DJ, Saleeba JA, Cole MA, McGee PA (2013) Development and stabilisation of soil structure via interactions between organic matter, arbuscular mycorrhizal fungi and plant roots. Soil Biol Biochem 57:683–694

Eason WR, Newman EI, Chuba PN (1991) Specificity of interplant cycling of phosphorus: the role of mycorrhizas. Plant Soil 137:267–274

Egerton-Warburton LM, Johnson NC, Allen EB (2007) Mycorrhizal community dynamics following nitrogen fertilization: a cross-site test in five grasslands. Ecol Monogr 77:527–544

Faber BA, Zasoski RJ, Munns DN, Shackel K (1991) A method for measuring hyphal nutrient and water uptake in mycorrhizal plants. Can J Bot 69:87–94

Hallett PD, Feeney DS, Bengough AG, Rillig MC, Scrimgeour CM, Young IM (2009) Disentangling the impact of AM fungi versus roots on soil structure and water transport. Plant Soil 314:183–196

Hildebrandt U, Janetta K, Bothe H (2002) Towards growth of arbuscular mycorrhizal fungi independent of a plant host. Appl Environ Microbiol 68:1919–1924

Jakobsen I, Abbott LK, Robson AD (1992) External hyphae of vesicular-arbuscular mycorrhizal fungi associated with *Trifolium subterraneum*. New Phytol 120:371–380

Jemai I, Ben Aissa N, Ben Guirat S, Ben-Hammouda M, Gallali T (2013) Impact of three and seven years of no-tillage on the soil water storage, in the plant root zone, under a dry subhumid Tunisian climate. Soil Tillage Res 126:26–33

Khalvati MA, Hu Y, Mozafar A, Schmidhalter U (2005) Quantification of water uptake by arbuscular mycorrhizal hyphae and its significance for leaf growth, water relations, and gas exchange of barley subjected to drought stress. Plant Biol 7:706–713

Khalvati M, Bartha B, Dupigny A, Schroder P (2010) Arbuscular mycorrhizal association is beneficial for growth and detoxification of xenobiotics of barley under drought stress. J Soils Sediments 10:54–64

Kohler J, Hernández JA, Caravaca F, Roldán A (2008) Plant-growth-promoting rhizobacteria and arbuscular mycorrhizal fungi modify alleviation biochemical mechanisms in water-stressed plants. Funct Plant Biol 35:141–151

Kothari SK, Marschner H, George E (1990) Effect of VA mycorrhizal fungi and rhizosphere microorganisms on root and shoot morphology, growth and water relations in maize. New Phytol 116:303–311

Manoharan PT, Shanmugaiah V, Balasubramanian N, Gomathinayagam S, Sharma MP, Muthuchelian K (2010) Influence of AM fungi on the growth and physiological status of *Erythrina variegata* Linn. grown under different water stress conditions. Eur J Soil Biol 46:151–156

Martin SL, Mooney SJ, Dickinson MJ, West HM (2012) The effects of simultaneous root colonisation by three *Glomus* species on soil pore characteristics. Soil Biol Biochem 49:167–173

Marulanda A, Azco´n R, Ruiz-Lozano JM (2003) Contribution of six arbuscular mycorrhizal fungal isolates to water uptake by Lactuca sativa plants under drought stress. Physiol Planta 119:526–533

Porcel R, Ruiz-Lozano JM (2004) Arbuscular mycorrhizal influence on leaf water potential, solute accumulation, and oxidative stress in soybean plants subjected to drought stress. J Exp Bot 55:1743–1750

Porcel R, Barea JM, Ruiz-Lozano JM (2003) Antioxidant activities in mycorrhizal soybean plants under drought stress and their possible relationship to the process of nodule senescence. New Phytol 157:135–143

Prieto I, Armas C, Pugnaire FI (2012) Water release through plant roots: new insights into its consequences at the plant and ecosystem level. New Phytologist 193:830–841

Querejeta JI, Egerton-Warburton LM, Prieto I, Vargas R, Allen MF (2012) Changes in soil hyphal abundance and viability can alter the patterns of hydraulic redistribution by plant roots. Plant Soil 355:63–73

Rillig MC, Wright SF, Nichols KA, Schmidt WF, Torn MS (2001) Large contribution of arbuscular mycorrhizal fungi to soil carbon pools in tropical forest soils. Plant Soil 233:167–177

Rillig MC (2004) Arbuscular mycorrhizae, glomalin, and soil aggregation. Can J Soil Sci 84:355–363

Rillig MC, Mummey DL (2006) Mycorrhizas and soil structure. New Phytol 171:41–53

Ruiz-Lozano JM, Aroca R (2010) Host response to osmotic stresses: stomatal behaviour and water use efficiency of arbuscular mycorrhizal plants. In: Koltai H, Kapulnik Y (eds) Arbuscular mycorrhizas: physiology and function. Springer, Dordrecht, pp 239–256

Ruiz-Lozano JM, Porcel R, Bárzana G, Azcón R, Aroca R (2012) Contribution of arbuscular mycorrhizal symbiosis to plant drought tolerance: state of the art. In: Aroca P (ed) Plant responses to drought stress. Springer, Berlin, pp 335–362

Ruiz-Sánchez M, Aroca R, Muñoz Y, Polón R, Ruiz-Lozano JM (2010) The arbuscular mycorrhizal symbiosis enhances the photosynthetic efficiency and the antioxidative response of rice plants subjected to drought stress. J Plant Physiol 167:862–869

Ruth B, Khalvati M, Schmidhalter U (2011) Quantification of mycorrhizal water uptake via high-resolution on-line water content sensors. Plant Soil 342:459–468

Safir GR, Boyer JS, Gerdemann JW (1971) Mycorrhizal enhancement of water transport in soybean. Science 172:581–583

Safir GR, Boyer JS, Gerdemann JW (1972) Nutrient status and mycorrhizal enhancement of water transport in soybean. Plant Physiol 49:700–703

Sheng M, Tang M, Chen H, Yang B, Zhang F, Huang Y (2008) Influence of arbuscular mycorrhizae on photosynthesis and water status of maize plants under salt stress. Mycorrhiza 18:287–296

Siddiky MRK, Kohler J, Cosme M, Rillig MC (2012a) Soil biota effects on soil structure: Interactions between arbuscular mycorrhizal fungal mycelium and collembola. Soil Biol Biochem 50:33–39

Siddiky MRK, Schaller J, Caruso T, Rillig MC (2012b) Arbuscular mycorrhizal fungi and collembola non-additively increase soil aggregation. Soil Biol Biochem 47:93–99

Simard SW, Beiler KJ, Bingham MA, Deslippe JR, Philip LJ, Teste F (2012) Mycorrhizal networks: mechanisms, ecology and modelling. Fungal Biol Rev 26:39–360

Singh LP, Gill SG, Tuteja N (2011) Unravelling the role of fungal symbionts in plant abiotic stress tolerance. Plant Signal Behav 6:175–191

Singh PK, Singh M, Tripathi BN (2012) Glomalin: an arbuscular mycorrhizal fungal soil protein. Protoplasma 250(3):663–669. doi:10.1007/s00709-012-0453-z

Smith SE, Read DJ (2008) Mycorrhizal symbiosis, 3rd edn. Academic, New York

Smith SE, Smith FA (2011) Roles of arbuscular mycorrhizas in plant nutrition and growth: new paradigms from cellular to ecosystem scales. Annu Rev Plant Biol 62:227–250

Subramanian KS, Charest C (1999) Acquisition of N by external hyphae of an arbuscular mycorrhizal fungus and its impact on physiological responses in maize under drought-stressed and well-watered conditions. Mycorrhiza 9:69–75

Tisdall JM, Oades JM (1982) Organic-matter and water-stable aggregates in soils. J Soil Sci 33:141–163

Toljander JF, Lindahl BD, Paul LR, Elfstrand M, Finlay RD (2007) Influence of arbuscular mycorrhizal mycelial exudates on soil bacterial growth and community structure. FEMS Microbiol Ecol 61:295–304

Trotel-Aziz P, Niogret M-F, Larher F (2000) Proline level is partly under the control of abscisic acid in canola leaf discs during recovery from hyper-osmotic stress. Physiol Planta 110:376–383

Tscharntke T, Clough Y, Wanger TC, Jackson L, Motzke I, Perfecto I, Whitbread A (2012) Global food security, biodiversity conservation and the future of agricultural intensification. Biol Conserv 151:53–59

Verbruggen E, Van Der Heijden MG, Weedon JT, Kowalchuk GA, Roeling WF (2012) Community assembly, species richness and nestedness of arbuscular mycorrhizal fungi in agricultural soils. Mol Ecol 21:2341–2353

Veresoglou SD, Chen B, Rillig MC (2012) Arbuscular mycorrhiza and soil nitrogen cycling. Soil Biol Biochem 46:53–62

Wilson GW, Rice CW, Rillig MC, Springer A, Hartnett DC (2009) Soil aggregation and carbon sequestration are tightly correlated with the abundance of arbuscular mycorrhizal fungi: results from long-term field experiments. Ecol Letters 12:452–461

Wu QS, Xia RX (2006) Arbuscular mycorrhizal fungi influence growth, osmotic adjustment and photosynthesis of citrus under well-watered and water stress conditions. J Plant Physiol 163:417–425

Wu QS, Xia RX, Zou YN (2008) Improved soil structure and citrus growth after inoculation with three arbuscular mycorrhizal fungi under drought stress. Eur J Soil Biol 44:122–128

Chapter 15
Role of Mycorrhizal Fungi in the Alleviation of Heavy Metal Toxicity in Plants

Hamid Amir, Philippe Jourand, Yvon Cavaloc, and Marc Ducousso

15.1 Introduction

Studies on the interactions between mycorrhiza and heavy metals are relatively recent. One of the first reports, by Gildon and Tinker (1981), highlighted a heavy metal tolerance of arbuscular mycorrhizal (AM) fungi in metal-polluted soil. These authors also suggested a possible role of AM fungi in increasing heavy metal uptake by plants. Since this report, about 500 articles have been published on this subject. Of these, about 150 articles deal with the influence of mycorrhizas on heavy metal absorption by plants and the alleviation of heavy metal toxicity; 78 % of these studies have been published during the last decade. Indeed, it is becoming more and more clear that mycorrhizal fungi can be used for the bioremediation of metal-polluted sites caused by industrial activities (Hildebrandt et al. 2007; Khade and Adholeya 2007; Orlowska et al. 2011a; Rajkumar et al. 2012). Several studies have suggested a role of these symbionts in the adaptation of plants to naturally metal-rich environments, i.e. mining areas and ultramafic (serpentine) soils (Ma et al. 2006; Leung et al. 2007; Jourand et al. 2010a, b; Lagrange et al. 2011; Amir et al. 2013).

Mycorrhizal symbioses occur in more than 80 % of the vascular plants (Brundrett 2009). Fungal symbionts constitute an important interface between the soil and the plant and induce physicochemical and biological changes in the rhizosphere (Hinsinger et al. 2009; Lambers et al. 2009; Smith and Smith 2010).

H. Amir (✉) • Y. Cavaloc
Laboratoire Insulaire du Vivant et de l'Environnement (LIVE), EA 4243, Université de la Nouvelle-Calédonie, BP R4, Nouméa Cedex 98851, New Caledonia
e-mail: hamid.amir@univ-nc.nc

P. Jourand • M. Ducousso
Laboratoire des Symbioses Tropicales et Méditerranéennes (LSTM), UMR 113 CIRAD/INRA/IRD/SupAgro/UMII, Centre IRD, BPA5, promenade Roger Laroque, Nouméa cedex 98848, New Caledonia

© Springer-Verlag Berlin Heidelberg 2014
Z.M. Solaiman et al. (eds.), *Mycorrhizal Fungi: Use in Sustainable Agriculture and Land Restoration*, Soil Biology 41, DOI 10.1007/978-3-662-45370-4_15

It is well known that they have a considerable influence on the mineral nutrition of plants. Generally, they improve the absorption of mineral elements in relation to the increase of the soil-plant interface which is due to the abundant mycelium colonizing a large volume of soil and to their weathering effects on minerals (Marschner and Dell 1994; Smith and Read 2008). However, the effects of mycorrhizal fungi on the absorption of mineral elements vary according to the type and concentration of the element, indicating that the mycorrhizal root is highly selective; this is especially the case for heavy metals which can be necessary at very low concentrations and toxic at higher levels (Leyval and Joner 2001).

Heavy metals such as Cd, Pb, Zn, Cu and Ni are naturally present in the soil solution as trace elements, but their concentrations can be considerably enhanced by industrial activities such as mining, automobiles, industrial wastes and pesticides (Joshi and Luthra 2000). There are also naturally metal-rich soils, such as ultramafic soils characterized by high contents of heavy metals, especially Ni, Co, Cr and Mn. The toxicity of these metals depends on their bioavailability which is influenced by physicochemical soil characteristics such as pH, clay and organic matter content and microbial activities, including that of mycorrhizal fungi (Berthelin et al. 1995; Leyval et al. 1995; Leyval and Joner 2001; Amir and Pineau 2003).

This review aims to synthesize research on the influence of arbuscular mycorrhizal (AM) and ectomycorrhizal (ECM) fungi on plant absorption of heavy metals and on the alleviation of their toxicity. Prior to discussing this central topic, a few points should be noted about heavy metal tolerance and adaptation of these symbionts to metal-rich soils.

15.2   Presence of Mycorrhizal Fungi in Metal-Rich Soils

Mycorrhizal fungi have been found in all heavy metal-polluted soils, even when these soils are highly contaminated (Vallino et al. 2006; Gamalero et al. 2009). They are also relatively abundant in naturally metalliferous soils, such as ultramafic soils, with high contents of Fe, Mn, Ni, Co and Cr (Amir et al. 1997; Turnau and Mesjasz-Przybylowicz 2003; Perrier et al. 2006; Gonçalves et al. 2007; Jourand et al. 2010b). In these environments, soils under nickel hyperaccumulating plants, with up to 1,500 μg g^{-1} of DTPA-extractable Ni, contained viable AM fungal spores; however, root colonization by AM fungi was partly, or (in rare cases) totally, inhibited (Amir et al. 2007).

The diversity of AM fungi in heavy metal-polluted soils is generally lower than in other soils (Pawlowska et al. 1996; Regvar et al. 2003; Hassan Sel et al. 2011). *Glomus* species are clearly the most abundant (Griffioen 1994; Khade and Adholeya 2007; Ortega-Larrocea et al. 2010). The most reported species of this genus in these soils are *G. fasciculatum*, *G. intraradices*, *G. etunicatum* and *G. mosseae*; some *Acaulospora* and *Gigaspora* species are also present at low frequencies (Khade and Adholeya 2007; Ortega-Larrocea et al. 2007, 2010; Wu et al. 2010). *Scutellospora* and *Sclerocystis* have been rarely reported (Khade and Adholeya 2007).

In ultramafic soils, *Glomus* species also seem to be highly dominant (Perrier et al. 2006; Gustafson and Casper 2006; Schechter and Bruns 2012), with *G. etunicatum* and *G. fasciculatum* being the most commonly observed (Gustafson and Casper 2006; Lagrange et al. 2011). Ji et al. (2012), using spore morphology, compared the AM fungal communities of two ultramafic and two non-ultramafic soils and reported differences in soil chemical characteristics, but without differences in AM fungal diversity. Perrier (2005) and Branco and Ree (2010) found diversity of ECM fungi was not limited in ultramafic soils.

Hrynkieiuicz et al. (2008) studied the structure of the ECM fungal community associated with *Salix caprea* in former silver-mining sites in Germany after 33 years of revegetation. Fungal diversity was represented by four families: Thelephoraceae, Cortinariaceae, Tricholomataceae and Tuberaceae, with Thelephoraceae the most frequent.

15.3 Heavy Metal Tolerance of Mycorrhizal Fungi

Several studies have indicated that isolates of AM fungi from heavy metal-polluted soils are more tolerant to metals than are those isolated from other soils (Leyval et al. 1995; Diaz et al. 1996; Gonzalez-Chavez et al. 2002; Tullio et al. 2003). The same conclusion has been reported for AM fungal isolates from ultramafic soils (Amir et al. 2008). This latter study showed that AM fungi isolated from roots of Ni-hyperaccumulating plants were more tolerant to Ni than were those isolated from ultramafic soils under non-hyperaccumulating plants, with these latter isolates being more tolerant than those from non-ultramafic soils. The same authors also found that AM fungal tolerance to Ni can be induced by the presence of high concentrations of this metal in the substrate where the symbiont has been grown.

The maximum metal concentrations tolerated by AM fungi vary greatly according to the type of metal, the type of substrate used and the type of propagule tested (Tullio et al. 2003; Khade and Adholeya 2007; Amir et al. 2008; Wu et al. 2009, 2010). For example, the spores of two *Glomus* spp. isolates were able to germinate in sand with up to 50 μg g^{-1} of Ni, whereas the less tolerant isolate did not germinate at 15 μg g^{-1} Ni (Amir et al. 2008). Wu et al. (2009, 2010) reported that spore germination of *G. mosseae* isolate from heavy metal-contaminated soils tolerated up to 5 μg g^{-1} Zn and 15.5 μg g^{-1} Pb. Considering the heterogeneity of heavy metal concentrations at the scale of soil aggregates and microsites in metal-rich soils, the tolerance to heavy metals by AM fungi is generally sufficient to colonize plant roots. However, colonization can be reduced by high concentrations of heavy metals in soil (Lingua et al. 2008; Amir et al. 2007; Gamalero et al. 2009).

The effects of heavy metals on ECM fungi have been widely reviewed (Leyval et al. 1997; Hartley et al. 1997; Jentschke and Goldbold 2000; Meharg 2003). Investigations at ECM fungal species and community levels have revealed wide inter- and intraspecific variation in sensitivity to metals (Hartley et al. 1997). Fungi belonging to genera *Amanita, Cenococum, Laccaria, Lactarius, Paxillus, Pisolithus, Scleroderma, Suillus* and *Thelephora* have been shown to tolerate metals such as Al, Cd, Cu, Ni, Pb and Zn. Ray et al. (2005) showed that fungi belonging to genus *Hysterangium* were able to tolerate Al, Cd, Cr and Ni. In vitro solid agar or liquid medium metal tolerance tests allowed determination of specific EC_{50} (effective concentration of metal which reduces growth by 50 %) or IC_{50} (concentration that inhibits rates of growth by 50 %) varying in a range from µM up to mM (Hartley et al. 1997; Blaudez et al. 2000; Ray et al. 2005).

Several studies on ultramafic soils have highlighted ECM fungal community tolerance to heavy metals. In such soils, ECM fungal communities presented a high biodiversity and an adaptive tolerance to heavy metals, especially Ni (Perrier et al. 2006; Gonçalves et al. 2007, 2009; Urban et al. 2008; Jourand et al. 2010a, b). Majorel et al. (2012) on *Pisolithus albus* found five genes acting as markers of Ni-tolerance.

15.4 Role of Mycorrhizal Fungi on the Alleviation of Heavy Metal Toxicity of Plants

The role of mycorrhizal fungi in heavy metal-rich soils was first suggested by Gildon and Tinker (1981) and has been investigated in the two last decades with two objectives and approaches. One group of studies focused on the phytoextraction of heavy metals and the phytoremediation of polluted soils. These studies included extraction of metals from soil by heavy metal-accumulating plants (Khade and Adholeya 2007; Hildebrandt et al. 2007; Gamalero et al. 2009). Other studies investigated the ecological restoration of degraded areas (Leung et al. 2006; Ma et al. 2006; Chen et al. 2005; Amir et al. 2013). The conclusions of these studies are complex and vary in relation to the type of approach, the experimental conditions, the group of the plants studied, the characteristics of the soils used and the type of metal concerned and its concentration (Weissenhorn et al. 1995; Hildebrandt et al. 2007; Gamalero et al. 2009).

It is now clear that AM and ECM generally induce adaptation of plants to high metal concentrations in soil, and this occurs despite the complexity of the results obtained and difficulties in comparing different studies. Indeed, to demonstrate that mycorrhizas alleviate metal toxicity of plants or improve plant tolerance, it must be shown that, in comparison with a control without heavy metals, the negative effects of heavy metals on plant growth and plant health are less important in the presence of mycorrhizal fungi than in the non-mycorrhizal treatment. However, some experiments did not use a control without heavy metals and only evaluate the effects of the symbionts on plant growth and nutrition in metal-rich soils.

Experiments cover a large number of plant species and families, different metals (Cd, Cu, Zn, Pb, Cr, Mn, Ni, Al and As) and different experimental conditions. As there were no clear conclusions specific to plant taxa, type of metal or particular conditions and considering the large number of publications, conclusions are synthesized here without detailed reference to these variables. Out of 44 publications, 43 % reported a better tolerance of the plant to heavy metals in the presence of AM fungi (Hildebrandt et al. 1999; Zhang et al. 2005; Lin et al. 2007; Lingua et al. 2008, Andrade et al. 2009; Cavagnaro et al. 2010; Orlowska et al. 2011b; Aloui et al. 2012; Amir et al. 2013, etc.), and 52 % showed an increase in growth and/or an improvement in mineral nutrition under the same conditions (Sadeque et al. 2006; Janouvska et al. 2007; Redon et al. 2008; Cabala et al. 2009; Dubkova et al. 2012; Neagoe et al. 2013, etc.). Overall, 86 % of studies showed a better adaptation of mycorrhizal plants to heavy metal-rich soils, and only 12 % did not report any positive effects of AM fungi on plant growth in the presence of heavy metals (Carvalho et al. 2006; Marques et al. 2006; Sudova et al. 2008, etc.). Boulet and Lambers (2005) showed that AM fungal inoculum from ultramafic soil did not improve the growth of *Hakea verrucosa*, but enhanced its mineral nutrition in an ultramafic soil.

The effects of ECM on heavy metal toxicity of plants are generally clear, although there are fewer publications than those concerning AM fungi. ECM protection against Cu, Cd, Zn and Ni toxicity of *Pinus sylvestris* has been demonstrated (Ahonen-Jonnarth and Finlay 2001; Adriaensen et al. 2006; Colpaert et al. 2011). *Eucalyptus globulus* plants inoculated with *P. albus* from ultramafic soils were clearly tolerant to Ni (Jourand et al. 2010a). Improvement of plant biomass and mineral nutrition by ECM in the presence of toxic heavy metal concentrations has been reported for Ni and Cr (Aggangan et al. 1998), Mn (Walker et al. 2004) and Zn (Adriaensen et al. 2006). Only one study (Dučić et al. 2008) did not show positive effects of ECM on plant growth when exposed to heavy metal stress (Mn), but the tolerance of the ECM fungi isolate to Mn was not tested.

To estimate the effects of ECM fungi on plant tolerance, Jentschke and Goldbold (2000) suggested considering the sensitivity of seedling growth and plant mineral nutrition, especially N and Ca/Mg uptake, which could be influenced by heavy metal toxicity. These effects depend on fungal species and, for the same species, on fungal isolates that can show different levels of tolerance to the metal. There is evidence that there is a relationship between fungal ecotype and amelioration of plant host heavy metal tolerance (Adriaensen et al. 2003, 2006; Jourand et al. 2010a).

15.4.1 Influence of Mycorrhizas on Heavy Metal Absorption by Plants

It is important to stress that the alleviation of plant heavy metal toxicity by AM fungi, or the improvement of plant growth in the presence of heavy metals by these symbionts, is not necessarily induced by a reduction of heavy metal absorption by the plant. About half the studies on this topic showed an increase in heavy metal absorption by mycorrhizal plants in comparison to non-mycorrhizal controls. Some studies reported an increase in heavy metal concentrations or heavy metal accumulation in roots and shoots (Marques et al. 2006; Deram et al. 2008; Tseng et al. 2009; Redon et al. 2009; Andrade et al. 2009). Other studies showed an increase in heavy metal concentrations or heavy metal accumulation only in the roots (Joner and Leyval 2001; Rufyikiri et al. 2004; Carvalho et al. 2006; Honglin et al. 2006; Redon et al. 2008; Li et al. 2009; Bissonnette et al. 2010, Orlowska et al. 2012). About a third of the studies reported a clear reduction of heavy metal concentrations in the whole plant (Vivas et al. 2005; Sadeque et al. 2006; Andrade et al. 2010; Amir et al. 2013), and about 20 % showed a variation (increase or reduction) in heavy metal concentrations in the plant, depending on heavy metal concentrations in soil (Diaz et al. 1996; Audet and Charest 2007; Janouskova et al. 2007; Wu et al. 2009, 2010), AM fungal species or isolates (Zhang et al. 2005; Janouskova et al. 2007; Redon et al. 2009) and the metal considered (Guo et al. 1996; Zhang et al. 2005; Dubkova et al. 2012). These complex relationships between the heavy metal accumulation in plant organs and the positive effects of mycorrhizas on plant tolerance to these metals can be easily understood when considering the diversity of physiological adaptations of plants to high heavy metal concentrations in soils. These include metal excluders, metal indicators and different types of metal accumulators (Whiting et al. 2004; Kazakou et al. 2008; Fernando et al. 2008), the diversity of heavy metal neutralization mechanisms (Khan et al. 2000; Meharg 2003; Hildebrandt et al. 2007;) and, more generally, the diversity of factors affecting these processes in the rhizosphere (Hinsinger et al. 2009; Lambers et al. 2009). Thus, high concentrations of heavy metals in roots and shoots do not indicate a high level of toxicity, as these metals are generally stored in inactive forms.

A few studies have dealt with the influence of ECM on heavy metal absorption by plants. Needles of *Picea abies* associated with *Laccaria laccata* showed a Cd content 2.5 times lower than for non-mycorrhizal plants (Galli et al. 1993). *Suillus bovinus* reduced Zn content in plant tissues of *P. sylvestris* (Adriaensen et al. 2006). Bojarczuk and Kieliszewska-Rokicka (2010) studied the effects of various ECM on Cu and Pb accumulation in leaves of *Betula pendula* grown in heavy metal-contaminated soil. Heavy metal concentrations in leaves varied inversely with the abundance of ECM fungi. Walker et al. (2004) reported lower concentrations of Mn in *Betula lenta* seedlings when inoculated with *Pisolithus tinctorius* on coal mine spoil and a Ni-tolerant isolate of *P. albus* from New Caledonian ultramafic soil significantly reduced Ni transfer into plant tissues of *E. globulus* (Jourand et al. 2010a). Baum et al. (2006) reported both a decrease and increase in different heavy metal contents in stems and roots of *Salix* plants, depending on heavy metal concentrations, type of heavy metal and fungal isolate.

15.4.2   Combined Effects of Mycorrhizal Fungi and Other Factors on the Alleviation of Heavy Metal Toxicity and Metal Accumulation in Plants

Most studies aimed at improving phytoextraction/phytoremediation have shown combined effects of AM fungi and other treatments. The combination of *G. mosseae* and a Plant Growth Promoting (PGP) bacterium (*Brevibacillus brevis*), both isolated from Cd-contaminated soil, increased AM colonization, plant growth and plant Cd tolerance (Vivas et al. 2005). These effects were related to an increase in P and K and a decrease in Cd, Cr, Mn, Cu, Mo, Fe and Ni in plant tissues. The co-inoculation of Eucalyptus plants with *Glomus deserticola* and *Trichoderma koningii* was more effective for Cd uptake and plant growth than was each treatment considered separately (Arriagada et al. 2007). Ma et al. (2006) tested the combined effects of AM fungi and earthworms on *Leucaena leucocephala* in topsoil amended mine tailings and showed additional positive effects on plant growth, plant nutrition and a reduction in Pb and Zn mobility. The combined effects of AM fungi and organic amendments were also tested. Inoculation of *Trifolium repens* with AM fungi in heavy metal-contaminated soil amended with *Aspergillus niger*-treated sugar beet stimulated bacterial diversity, plant growth and the phytoextraction process (Azcon et al. 2009). Medina et al. (2010) showed that the combination of AM fungi and *A. niger*-treated dry olive cake increased *T. repens* growth and its tolerance to Cd. The association of ECM fungi and bacteria can also improve the adaptation of pine to metal-polluted soils (Krupa and Kozdroj 2007), resulting in a higher accumulation of the metals, especially Zn, in the roots and a reduction of metal translocation to the shoots.

15.5   Mechanisms Involved in the Role of Mycorrhizas in Alleviation of Heavy Metal Toxicity of Plants

During the two last decades, a relatively large number of studies have focused on the mechanisms which can explain the influence of mycorrhizas on the alleviation of heavy metal toxicity of plants. Several processes have been highlighted. Direct mechanisms concern extracellular heavy metal inactivation, heavy metal binding in fungal wall, enhanced efflux of heavy metals through cellular membranes, intracellular inactivation and adaptive response to oxidative stress. Indirect mechanisms act through the improvement of mineral nutrition, which enhances the growth and influences plant tolerance to environmental stress. The plant and associated mycorrhizal fungi have different strategies to cope with heavy metal toxicity; some are common and act in together; others are different and operate independently (Meharg 2003). This text focuses mainly on fungal strategies.

15.5.1 Extracellular Heavy Metal Inactivation Mechanisms

Different mechanisms of heavy metal exclusion by mycorrhizal fungi have been suggested, among them extracellular chelation, cell wall binding and heavy metal accumulation in extraradical mycelium (Colpaert et al. 2011). Mycorrhizas can inactivate heavy metals through the exudation of complexing agents into the soil solution. According to Meharg (2003), organic acid exudation has a clear role in mycorrhizal adaptation to metal-contaminated sites. Citric, malic and oxalic acids are known to be produced by mycorrhizal fungi (Ahonen-Jonnarth et al. 2000; Meharg 2003); they can mobilize or immobilize metals by complexation, depending on various factors, especially rhizosphere pH (Gimmler et al. 2001; Hinsinger et al. 2009). Phenolic compounds produced by ECM are also involved in metal immobilization in soil (Schützendübel and Polle 2002). Machuka et al. (2007) highlighted different metal-chelating compounds in in vitro culture of ECM fungi collected from pine plantations (species of *Scleroderma*, *Suillus* and *Rhizopogon*). Oxalic, citric and succinic acids but also hydroxamate- and catecholate-type compounds were found in the liquid medium. Cabala et al. (2009) reported the presence in the rhizosphere of different AM and ECM mycorrhizal plants of metal-bearing aggregates formed during symbiotic action between mycorrhizas and bacteria. These structures enhanced the binding of Zn, Pb and Mn in the rhizosphere. More recently, different studies showed the role of glomalin, a very abundant AM fungal glycoprotein released into the soil where it participates in soil aggregation. Glomalin seems to be involved in heavy metal inactivation in soil (Ferrol et al. 2009; Gamalero et al. 2009). Glomalin extracted from polluted soil or from hyphae irreversibly sequesters metals such as Cu, Cd, Zn and As (Gonzalez-Chavez et al. 2002). Cornejo et al. (2008) showed that a glomalin-related soil protein was more abundant in polluted soils with high concentrations of Cu and Zn. Up to 27 % of the total Cu was bound by this protein, and in a highly polluted soil, with a low pH, up to 90 % of the soil organic carbon was represented by the glomalin-related protein. Similar results were obtained by Vodnik et al. (2008) for the sequestration of Pb and Zn.

15.5.2 Heavy Metal Binding in Fungal Wall

Some of the metals inactivated in mycorrhizal plants are retained by fungal walls. Joner et al. (2000) exposed extraradical mycelium of different *Glomus* spp. isolates to high concentrations of Cd and Zn and measured their capacities to bind these metals. The most tolerant isolate adsorbed more metals than the others. The fungal wall was responsive for 50 % of the metal retained. Orlowska et al. (2008), analysing the elemental distribution in mycorrhizal plants of the Ni-hyperaccumulator *Berkheya coddii*, also reported a high binding capacity of the extraradical mycelium for Zn, Cu and Ni. Using EDXS analyses, with monoxenic cultures of *G. intraradices*, Gonzalez-Guerrero et al. (2008) showed that Cu, Zn and Cd at toxic concentrations were partly localized in the fungal cell wall. Marques et al. (2007) and Zhang et al. (2009) reported that Zn and Cu were mainly deposited in the cell wall of the root cortex of the mycorrhizal plants, including the AM fungal wall. Several cell wall-binding molecules have been reported, such as glucan, chitin and galactosamine polymers, minor peptides and proteins, all presenting potential binding sites as free carboxyl, amino, hydroxyl, phosphate and mercapto groups (Bellion et al. 2006). Glomalin is also partly located at the AM fungal wall (Purin and Rillig 2008). In ECM, Cd and Zn are predominantly bound in cell wall of mantle hyphae, Hartig net hyphae and cortical cells (Meharg 2003).

15.5.3 Intracellular Heavy Metal Inactivation

After heavy metals have passed through the fungal wall, other avoidance mechanisms, which may be activated, include alteration of heavy metal influx transporter processes and an increase in heavy metal efflux through the cell membrane (Meharg 2003; Ouziad et al. 2005). Many metal protein transporters or metal permeases have been highlighted (Hildebrandt et al. 2007), but their role in cell detoxification is not well defined.

Intracellular compartmentalization strategies to inactivate the absorbed part of heavy metals are relatively well documented. The toxic elements are translocated into fungal vacuoles were they are stored away from the cytosol. According to Gonzalez-Guerrero et al. (2008), the highest metal content is localized in the spores. Ferrol et al. (2009) observed AM fungal mycelium developed in a Cu-enriched medium and showed that when spores appeared in clusters, only one or a few of them contained a high concentration of Cu, thus protecting the rest of the fungal colony. Vesicles of the intraradical mycelium may also serve for the storage of heavy metals (Orlowska et al. 2008).

What are the molecular mechanisms of this compartmentalization? The toxic elements must be bound to other molecules inside the cell to inactive them. Different metal chelators may be involved in this process: organic acids, amino acids, glutathione, phytochelatins (thiol-rich peptides) and metallothioneins (sulphur-rich proteins). Three glomeromycotan metallothioneins have been identified in *Gigaspora* and *Glomus* species (Stommel et al. 2001; Lanfranco et al. 2002; Gonzalez-Guerrero et al. 2007). Metallothioneins have also been found in ECM fungi (Courbot et al. 2004; Bellion et al. 2006, 2007). By contrast, to our knowledge, no specific metal-binding phytochelatin has been clearly identified in AM and ECM fungi. Hegedüs et al. (2007) highlighted the role of glutathione in heavy metal tolerance of ECM fungal isolates of *Paxillus involutus*, from HM-polluted soils. In addition, fungal colonization of roots may directly influence the expression of several plant genes coding for proteins involved in detoxification and plant tolerance to heavy metals, such as heavy metal transporter genes and plant metallothioneins (Repetto et al. 2003; Rivera-Becerril et al. 2005). More details on this topic are given by Hildebrandt et al. (2007).

Sequencing studies of transcript genomes of ECM fungi in symbiosis with their host plant have been reviewed by Arlt et al. (2009). Experiments have involved specific ECM/plant host models such as *Pisolithus* with *Eucalyptus* or *Quercus*, *Laccaria* with *Pinus*, *Tuber* with *Tilia* and *Paxillus* with *Betula*. However, the authors insist that the screening of patterns of RNA and consequently expressed sequence tags (EST) or identified regulated genes were all non-targeted and not based upon specific hypothesis. It now seems evident that such general transcriptomic approaches and their results need to be connected with the role of ECM fungi in improving plant host tolerance to heavy metals. This could allow the identification of fungal symbiotic genes involved in the metal tolerance mechanisms and expression level variations of these genes, in relation to the presence of the metal.

15.5.4 Response to Oxidative Stress

When toxic metal cations are not inactivated by the described mechanisms, they are generally redox active and can create oxidative stress. They induce high reactive radical hydroxyl and superoxide groups and then the alteration of cellular reactions. Adaptation mechanisms to this stress are not sufficiently understood in mycorrhizal plants, but according to Ferrol et al. (2009), they must include nonenzymatic antioxidant systems such as glutathione and vitamins C, E and B6 and enzymatic systems such as catalases, superoxide dismutases (SOD), thioredoxins and glutaredoxins. A reduction of SOD in plant shoots was found to be related to plant inoculation with AM fungi (Neagoe et al. 2013) and was explained as an effect of a lower oxidative stress in AM fungal inoculated plants. Jacob et al. (2001) highlighted the capacity of ECM fungi to synthesize SOD in reaction to cadmium toxicity. Vallino et al. (2009), studying the ericoid mycorrhizal fungus *Oidiodendron maius*, characterized a new SOD found both in the cell extract and in the growth medium of the fungal culture. They suggested that the presence of this enzyme in the extracellular environment may also protect the plant partner. A few genes involved in oxidative stress homeostasis have been identified in AM fungi: three SOD, ten genes encoding glutathione *S*-transferases, a glutaredoxin, a gene encoding a protein involved in vitamin B6 biosynthesis and a metallothionein (Ferrol et al. 2009). Aloui et al. (2012) quantified a group of isoflavonoids accumulated in the roots of *Medicago truncatula* in reaction to Cd toxicity and reported a strong reduction of three of these compounds in AM fungal inoculated plants, reinforcing the hypothesis that AM colonization buffered the effects of heavy metals in plant roots. In addition, a strong decrease of the transcripts of chalcone reductase, an enzyme involved in isoflavonoid production, was noticed.

15.6 Conclusions

Mycorrhizal fungi are present in all metal-rich soils, and their diversity allows them to adapt to the various concentrations of the different heavy metals present in these soils. Thus, it is now clear that mycorrhizas play an important role in plant tolerance to heavy metals, and this has been highlighted in heavy metal-polluted soils, in ultramafic soils and in mining degraded areas. However, the effects of these symbionts in the alleviation of heavy metal toxicity of plants and on heavy metal accumulation in plant organs are complex and vary with the diversity of physiological and molecular mechanisms involved in these processes and in relation to the diversity of the factors affecting the plant/fungi symbiosis.

This relatively new knowledge has important practical consequences, especially in the fields of phytoextraction, phytoremediation and ecological restoration of mining degraded areas (Khan et al. 2000; Khade and Adholeya 2007; Hildebrandt et al. 2007; Amir and Ducousso 2010). Different concepts and strategies have been proposed for these objectives (Meharg 2003; Audet and Charest 2007; Lebeau et al. 2008; Marques et al. 2008; Amir and Ducousso 2010). The complexity of the processes in plant/soil/microbial systems and the induced variations have to be taken into account. In particular the following factors have to be considered:

– Variations related to fungal isolates: Screening among a collection of fungi for their tolerance to the metals studied and their effects on plant is one of the conditions of the success (Meharg 2003).
– Variation related to plant factors: The plant species and even the plant clone (Sudova et al. 2008) must be screened depending on the objectives (phytoextraction, phytoremediation or ecological restoration). For phytoextraction, plants that accumulate metals are suitable, but generally produce a lower biomass, and a compromise has to be found between accumulation ability and growth rate (Audet and Charest 2007). For ecological restoration, maximal plant diversity is the best (L'Huillier et al. 2010), and mycorrhizal inoculum must be adapted to the plant species.
– Soil characteristics: Experiments with soil (or the substrate) to depollute or to revegetate are necessary, and the possibility of improving the efficiency of the method by amendments or other practices needs to be performed (Ma et al. 2006; Marques et al. 2008; Azcon et al. 2009).

Further studies should focus on the genetic determinism of mycorrhizal effects on plant tolerance to heavy metals and the control of the multivariate aspects of the metal/soil/plant/mycorrhizal fungus system interactions.

Acknowledgments We are grateful to Valerie Medevielle for her technical support.

References

Adriaensen K, van der Lelie D, Van Laere A, Vangronsveld J, Colpaert JV (2003) A zinc-adapted fungus protects pines from zinc stress. New Phytol 16:549–555

Adriaensen K, Vangronsveld J, Colpaert JV (2006) A zinc tolerant *Suillus bovinus* improves growth of Zn-exposed *Pinus sylvestris* seedlings. Mycorrhiza 16:553–558

Aggangan N, Dell B, Malajczuk N (1998) Effects of chromium and nickel on growth of the ectomycorrhizal fungus *Pisolithus* and formation of ectomycorrhizas on *Eucalyptus urophylla* S. T. Geoderma 84:15–27

Ahonen-Jonnarth U, Finlay RD (2001) Effects of elevated nickel and cadmium concentrations on growth and nutrient uptake of mycorrhizal and non-mycorrhizal *Pinus sylvestris* seedlings. Plant Soil 236:129–138

Ahonen-Jonnarth U, Finlay RD, Van Hees PAW, Lundstrom US (2000) Organic acids produced by mycorrhizal *Pinus sylvestris* exposed to elevated aluminium and heavy metal concentrations. New Phytol 146:557–567

Aloui A, Dumas-Gaudot E, Daher Z, van Tuinen D, Aschi-Smit S, Morandi D (2012) Influence of arbuscular mycorrhizal colonisation on cadmium induced *Medicago truncatula* root isoflavonoid accumulation. Plant Physiol Biochem 60:233–239

Amir H, Ducousso M (2010) Les bactéries et les champignons du sol sur roches ultramafiques. In: L'Huillier L, Jaffré T, Wulf A (eds) Mines et environnement en Nouvelle-Calédonie : les milieux sur substrats ultramafiques et leur restauration. IAC Ed, Noumea, pp 129–145

Amir H, Pineau R (2003) Release of Ni and Co by microbial activity in New Caledonian ultramafic soils. Can J Microbiol 49:288–293

Amir H, Pineau R, Violette Z (1997) Premiers résultats sur les endomycorhizes des plantes de maquis miniers de Nouvelle-Calédonie. In: Jaffre T, Reeves RD, Becquer T (eds) The ecology of ultramafic and metalliferous areas. ORSTOM Ed, Nouméa, pp 79–85

Amir H, Perrier N, Rigault F, Jaffré T (2007) Relationships between Ni-hyperaccumulation and mycorrhizal status of endemic plant species from New Caledonian ultramafic soils. Plant Soil 293:23–35

Amir H, Jasper DA, Abbott LK (2008) Tolerance and induction of tolerance to Ni of arbuscular mycorrhizal fungi from New Caledonian ultramafic soils. Mycorrhiza 19:1–6

Amir H, Lagrange A, Hassaïne N, Cavaloc Y (2013) Arbuscular mycorrhizal fungi from New Caledonian ultramafic soils improve tolerance to nickel of endemic plant species. Mycorrhiza 23:585–595

Andrade SAL, Gratao PL, Silveira APD, Schiavinato MA, Azevedo RA, Mazzafera P (2009) Zn uptake, physiological response and stress attenuation in mycorrhizal jack bean growing in soil with increasing Zn concentrations. Chemosphere 75:1363–1370

Andrade SAL, Gratao PL, Azevedo RA, Silvera APD, Schiavinato MA, Mazzafera P (2010) Biochemical and physiological changes in jack bean under mycorrhizal symbiosis growing in soil with increasing Cu concentrations. Environ Exp Bot 68:198–207

Arlt M, Schwarz D, Franken P (2009) Analysis of mycorrhizal functioning using transcriptomics. In: Azcon-Aguilar C, Barea JM, Gianinazzi S (eds) Mycorrhizas-functional processes and ecological impact. Springer, Berlin, pp 47–58

Arriagada CA, Herrera MA, Ocampo JA (2007) Beneficial effect of saprobe and arbuscular mycorrhizal fungi on growth of Eucalyptus globulus co-cultured with Glycine max in soil contaminated with heavy metals. J Environ Manage 84:93–99

Audet P, Charest C (2007) Dynamics of arbuscular mycorrhizal symbiosis in heavy metal phytoremediation: meta-analytical and conceptual perspectives. Environ Pollut 147:609–614

Azcon R, Peralvarez MDC, Biro B, Roldan A, Ruiz-Lozano JM (2009) Antioxidant activities and metal acquisition in mycorrhizal plants growing in a heavy-metal multicontaminated soil amended with treated lignocellulosic agrowaste. Appl Soil Ecol 41:168–177

Baum C, Hrynkiewicz K, Leinweber P, Meibner R (2006) Heavy-metal mobilization by mycorrhizal willows (*Salix dasyclados*). J Plant Nutr Soil Sci 169:516–522

Bellion M, Courbot M, Jacob C, Blaudez D, Chalot M (2006) Extracellular and cellular mechanisms sustaining metal tolerance in ectomycorrhizal fungi. FEMS Microbiol Lett 254:173–181

Bellion M, Courbot M, Jacob C, Blaudez D, Chalot M (2007) Metal induction of a *Paxillus involutus* metallothionein and its heterologous expression in *Hebeloma cylindrosporum*. New Phytol 174:151–158

Berthelin J, Munier-Lamy C, Leyval C (1995) Effect of microorganisms on mobility of heavy metals in soils. In: Huang PM, Berthelin J, Bollag JM, McGill WB, Page AL (eds) Environmental impact of soil component interactions. Metals, other inorganics and microbial activities. CRC Lewis, London, pp 3–17

Bissonnette L, St-Arnaud M, Labrecque M (2010) Phytoextraction of heavy metals by two Salicaceae clones in symbiosis with arbuscular mycorrhizal fungi during the second year of a field trial. Plant Soil 332:55–67

Blaudez D, Jacob C, Turnau K, Colpaert JV, Ahonen-Jonnarth U, Finlay R, Botton B, Chalot M (2000) Differential responses of ectomycorrhizal fungi to heavy metals *in vitro*. Mycol Res 104:1366–1371

Bojarczuk K, Kieliszewska-Rokicka B (2010) Effect of ectomycorrhiza on Cu and Pb accumulation in leaves and roots of silver birch (*Betula pendula* Roth) seedlings growth in metal contaminated soil. Water Air Soil Pollut 207:227–240

Boulet F, Lambers H (2005) Characterisation of arbuscular mycorrhizal fungi colonization in cluster roots of *Hakea verrucosa* F. Muell (Proteaceae), and its effect on growth and nutrient acquisition in ultramafic soil. Plant Soil 269:357–367

Branco S, Ree RH (2010) Serpentine soils do not limit mycorrhizal fungal diversity. PLoS One 5 (7):e11757. doi:10.1371/journal.pone.0011757

Brundrett MC (2009) Mycorrhizal associations and other means of nutrition of vascular plants: understanding the global diversity of host plants by resolving conflicting information and developing reliable means of diagnosis. Plant Soil 320:37–77

Cabala J, Krupa P, Misz-Kennan M (2009) Heavy metals in mycorrhizal rhizospheres contaminated by Zn-Pb mining and smelting around Olkusz in southern Poland. Water Air Soil Pollut 199:139–149

Carvalho LM, Cacador I, Martins-Loucao MA (2006) Arbuscular mycorrhizal fungi enhance root cadmium and copper accumulation in the roots of the salt marsh plant *Aster tripolium* L. Plant Soil 285:161–169

Cavagnaro TR, Dickson S, Smith FA (2010) Arbuscular mycorrhizas modify plant responses to soil zinc addition. Plant Soil 329:307–313

Chen B, Roos P, Borgaard OK, Zhu YG, Jakobsen I (2005) Mycorrhiza and root hairs in barley enhance acquisition of phosphorus and uranium from phosphate rock but mycorrhiza decreases root to shoot uranium transfer. New Phytol 165:591–598

Colpaert JV, Wevers JHL, Krznaric E, Adriaensen K (2011) How metal-tolerant ecotypes of ectomycorrhizal fungi protect plants from heavy metal pollution. Ann For Sci 68:17–24

Cornejo P, Meier S, Borie G, Rillig MC, Borie F (2008) Glomalin-related protein in a Mediterranean ecosystem affected by copper smelter and its contribution to Cu and Zn sequestration. Sci Total Environ 406:154–160

Courbot M, Diez L, Ruotolo R, Chalot M, Leroy P (2004) Cadmium-responsive thiols in the ectomycorrhizal fungus *Paxillus involutus*. Appl Environ Microbiol 70:7413–7417

Deram A, Languereau-Leman F, Howsam M, Petit D, Van Haluwyn C (2008) Seasonal patterns of cadmium accumulation in *Arrhenatherum elatius* (Poaceae): influence of mycorrhizal and endophytic fungal colonisation. Soil Biol Biochem 40:845–848

Diaz G, Azcon-Aguilar C, Honrubia M (1996) Influence of arbuscular mycorrhizae on heavy metal (Zn and Pb) uptake and growth of *Lygeum spartum* and *Anthyllis Cytisoides*. Plant Soil 180:241–249

Dubkova P, Suda J, Sudova R (2012) The symbiosis with arbuscular mycorrhizal fungi contributes to plant tolerance to serpentine edaphic stress. Soil Biol Biochem 44:56–64

Dučić T, Parladé J, Polle A (2008) The influence of the ectomycorrhizal fungus *Rhizopogon subareolatus* on growth and nutrient element localisation in two varieties of Douglas fir (*Pseudotsuga menziesii* var. *menziesii* and var. *glauca*) in response to manganese stress. Mycorrhiza 18:227–239

Fernando DR, Woodrow IE, Jaffré T, Dumontet V, Marshall AT, Baker AJM (2008) Foliar manganese accumulation by *Maytenus founieri* (Celastraceae) in its native New Caledonian habitats: populational variation and localization by X-ray microanalysis. New Phytol 177:178–185

Ferrol N, Gonzalez-Guerrero M, Valderas A, Benabdallah K, Azcon-Aguilar C (2009) Survival strategies of arbuscular mycorrhizal fungi in Cu-polluted environments. Phytochem Rev 8:551–559

Galli U, Meier M, Brunold C (1993) Effects of cadmium on nonmycorrhizal and mycorrhizal Norway Spruce seedlings *Picea abies* (L.) Karst and its ectomycorrhizal fungus *Laccaria laccata* (Scop ex Fr) Bk and Br, sulfate reduction, thiols and distribution of the heavy-metal. New Phytol 125:837–843

Gamalero E, Lingua G, Berta G, Glick B (2009) Beneficial role of plant growth promoting bacteria and arbuscular mycorrhizal fungi on plant responses to heavy metal stress. Can J Microbiol 55:501–514

Gildon A, Tinker PB (1981) A heavy metal-tolerant strain of a mycorrhizal fungus. Trans Br Mycol Soc 77:648–649

Gimmler H, de Jesus J, Greiser A (2001) Heavy metal resistance of the extreme acidotolerant filamentous fungus *Bispora* sp. Microb Ecol 42:87–98

Gonçalves SC, Portugal A, Goncalves MT, Vieira R, Martins-Loucao MA, Freitas H (2007) Genetic diversity and differential *in vitro* responses to Ni in *Cenococcum geophilum* isolates from serpentine soils in Portugal. Mycorrhiza 17:677–686

Gonçalves SC, Martins-Louçao MA, Freitas H (2009) Evidence of adaptive tolerance to nickel in isolates of *Cenococcum geophilum* from serpentine soils. Mycorrhiza 19:221–230

Gonzalez-Chavez C, Haen JD, Vangronsveld J, Dodd JC (2002) Copper sorption and accumulation by the extraradical mycelium of different *Glomus* spp. (arbuscular mycorrhizal fungi) isolated from the same polluted soil. Plant Soil 240:287–297

Gonzalez-Guerrero M, Cano C, Azcon-Aguilar C, Ferrol N (2007) GintMT1 encodes a functional metallothionein in *Glomus intraradices* that responds to oxidative stress. Mycorrhiza 17:327–335

Gonzalez-Guerrero M, Melville LH, Ferrol N, Lott JNA, Azcon-Aguilar C, Peterson RL (2008) Ultrastructural localization of heavy metals in the extraradical mycelium and spores of the arbuscular mycorrhizal fungus *Glomus intraradices*. Can J Microbiol 54:103–110

Griffioen WAJ (1994) Characterization of a heavy metal tolerant endomycorrhizal fungus from the surroundings of a zinc refinery. Mycorrhiza 4:197–200

Guo Y, George E, Marschner H (1996) Contribution of an arbuscular mycorrhizal fungus to the uptake of cadmium and nickel in bean and maize plants. Plant Soil 184:195–205

Gustafson DJ, Casper BB (2006) Differential host plant performance as a function of soil arbuscular mycorrhizal fungal communities: experimentally manipulating co-occurring *Glomus* species. Plant Ecol 186:257–263

Hartley J, Cairney JWG, Meharg AA (1997) Do ectomycorrhizal fungi exhibit adaptive tolerance to potentially toxic metals in the environment? Plant Soil 189:303–319

Hassan Sel D, Boon E, St-Arnaud M, Hijri M (2011) Molecular biodiversity of arbuscular mycorrhizal fungi in trace metal-polluted soils. Mol Ecol 20:3469–3483

Hegedüs N, Tamas E, Szilagyi J, Karanyi Z, Nagy I, Penninckx MJ, Pocsi I (2007) Effects of heavy metals on the glutathione status in different ectomycorrhizal *Paxillus involutus* strains. World J Microbiol Biotechol 23:1339–1343

Hildebrandt U, Kaldorf M, Bothe H (1999) The zinc violet and colonization by arbuscular mycorrhizal fungi. J Plant Physiol 154:709–717

Hildebrandt U, Regvar M, Bothe H (2007) Arbuscular mycorrhiza and heavy metal tolerance. Phytochemistry 68:139–146

Hinsinger P, Bengough AG, Vetterlein D, Young IM (2009) Rhizosphere: biophysics, biogeochemistry and ecological relevance. Plant Soil 321:117–152

Honglin H, Shuzhen Z, Chen BD, Wu N, Shan XQ, Christ P (2006) Uptake of atrazine and cadmium from soil by maize (*Zea mays* L.) in association with the arbuscular mycorrhizal fungus *Glomus etunicatum*. J Agric Food Chem 54:9377–9382

Hrynkieiuicz K, Haug I, Baum C (2008) Ectomycorrhizal community structure under willows at former ore mining sites. Eur J Soil Biol 44:37–44

Jacob C, Courbot M, Brun A, Steinman HM, Jaquot JP, Botton B, Chalot M (2001) Molecular cloning, characterizing and regulation by cadmium of a superoxide dismutase from the ectomycorrhizal fungus *Paxillus involutus*. Eur J Biochem 268:3223–3232

Janouskova M, Vosatka M, Rossi L, Lugon-Moulin N (2007) Effects of arbuscular mycorrhizal inoculation on cadmium accumulation by different tobacco (*Nicotiana tabacum* L.) types. Appl Soil Ecol 35:502–510

Jentschke G, Goldbold DL (2000) Metal toxicity and ectomycorrhizas. Physiol Plant 109:107–116

Ji B, Bentivenga SP, Casper BB (2012) Comparisons of AMF fungal spore communities with the same hosts but different soil chemistries over local geographic scales. Oecologia 168:187–197

Joner EJ, Leyval C (2001) Time-course of heavy metal uptake in maize and clover as affected by root density and different mycorrhizal inoculation regimes. Biol Fertil Soils 33:351–357

Joner EJ, Briones R, Leyval C (2000) Metal-binding capacity of arbuscular mycorrhizal mycelium. Plant Soil 226:227–234

Joshi UN, Luthra YP (2000) An overview of heavy metals: impact and remediation. Curr Sci 78:2–4

Jourand P, Ducousso M, Reid R, Majorel C, Richert C, Riss J, Lebrun M (2010a) Nickel-tolerant ectomycorrhizal *Pisolithus albus* ultramafic ecotype isolated from nickel mines in New Caledonia strongly enhance growth of a host plant at toxic nickel concentrations. Tree Physiol 30:1311–1319

Jourand P, Ducousso M, Loulergue-Majorel C, Hannibal L, Santoni S, Prin Y, Lebrun M (2010b) Ultramafic soils from New Caledonia structure *Pisolithus albus* in ecotype. FEMS Microbiol Ecol 72:238–249

Kazakou E, Dimitrakopoulos PG, Baker AJM, Reeves RD, Trumbis AY (2008) Hypotheses, mechanisms and trade-offs of tolerance and adaptation to serpentine soils: from species to ecosystem level. Biol Rev 83:495–508

Khade SW, Adholeya A (2007) Feasible bioremediation through arbuscular mycorrhizal fungi imparting heavy metal tolerance: a retrospective. Bioremediation J 11:33–43

Khan AG, Kuek C, Chauhry TM, Khoo CS, Hayes WJ (2000) Role of plants, mycorrhizae and phytochelators in heavy metal contaminated land remediation. Chemosphere 41:197–207

Krupa P, Kozdroj J (2007) Ectomycorrhizal fungi and associated bacteria provide protection against heavy metals in inoculated pine (*Pinus sylvestris* L.) seedlings. Water Air Soil Pollut 182:83–90

L'Huillier L, Wulf A, Gâteblé G, Fogliani B, Zongo C, Jaffré T (2010) La restauration des sites miniers. In: L'Huillier L, Jaffré T, Wulf A (eds) Mines et environnement en Nouvelle-Calédonie : les milieux sur substrats ultramafiques et leur restauration. IAC Ed, Noumea, pp 147–230

Lagrange A, Ducousso M, Jourand P, Majorel C, Amir H (2011) New insights into the mycorrhizal status of Cyperaceae from ultramafic soils in New Caledonia. Can J Microbiol 57:21–28

Lambers H, Mougel C, Jaillard B, Hinsinger P (2009) Plant-microbe-soil interactions in the rhizosphere: an evolutionary perspective. Plant Soil 321:83–115

Lanfranco L, Bolchi A, Ros S, Ottonello S, Bonfante P (2002) Differential expression of metallothionein gene during the presymbiotic versus the symbiotic phase of an arbuscular mycorrhizal fungus. Plant Physiol 130:58–67

Lebeau T, Braud A, Jezequel K (2008) Performance of bioaugmentation-assisted phytoextraction applied to metal contaminated soils: a review. Environ Pollut 153:497–522

Leung HM, Ye ZH, Wong MH (2006) Interactions of mycorrhizal fungi with *Pteris vittata* (As hyperaccumulator) in As-contaminated soils. Environ Pollut 139:1–8

Leung HM, Ye ZH, Wong MH (2007) Survival strategies of plants associated with arbuscular mycorrhizal fungi on toxic mine tailings. Chemosphere 66:905–915

Leyval C, Joner EJ (2001) Bioavailability of heavy metals in the mycorrhizosphere. In: Gobran GR, Wenzel WW, Lombi E (eds) Trace elements in the rhizosphere. CRC, New York, pp 165–185

Leyval C, Singh BR, Joner EJ (1995) Occurrence and infectivity of arbuscular mycorrhizal fungi in some Norwegian soils influenced by heavy metals and soils properties. Water Air Soil Pollut 84:203–216

Leyval C, Turnau K, Haselwandter K (1997) Effect of heavy metal pollution on mycorrhizal colonization and function: physiological, ecological and applied aspects. Mycorrhiza 7:139–153

Li Y, Peng J, Shi P, Zhao B (2009) The effect of Cd on mycorrhizal development and enzyme activity of *Glomus mosseae* and *Glomus intraradices* in *Astragalus sinicus* L. Chemosphere 75:894–899

Lin AJ, Zhang XH, Wong MH, Ye ZH, Lou LQ, Wang YS, Zhu YG (2007) Increase of multi-metal tolerance of three leguminous plants by arbuscular mycorrhizal fungi colonization. Environ Geochem Health 29:473–481

Lingua G, Franchin C, Todeschini V, Castiglione S, Biondi S, Burlando B, Parravicini V, Torrigiani P, Berta G (2008) Arbuscular mycorrhizal fungi differentially affect the response to high zinc concentrations of two registered poplar clones. Environ Pollut 153:137–147

Ma Y, Dickinson NM, Wong MH (2006) Beneficial effects of earthworms and arbuscular mycorrhizal fungi on establishment of leguminous trees on Pb/Zn mine tailings. Soil Biol Biochem 38:1403–1412

Machuka A, Pereira G, Aguiar A, Milagres AMF (2007) Metal-chelating compounds produced by ectomycorrhizal fungi collected from pine plantations. Lett Appl Microbiol 44:7–12

Majorel C, Hannibal L, Soupe M, Carriconde F, Ducousso M, Lebrun M, Jourand P (2012) Tracking nickel-adaptive biomarkers in *Pisolithus albus* from New Caledonia using a transcriptomic approach. Mol Ecol 21:2208–2223

Marques APGC, Oliveira RS, Rangel AOSS, Castro PML (2006) Zinc accumulation in *Solanum nigrum* is enhanced by different arbuscular mycorrhizal fungi. Chemosphere 65:1256–1263

Marques APGC, Oliveira RS, Samardjieva KA, Pissara J, Rangel AOSS, Castro PML (2007) *Solanum nigrum* grown in contaminated soil: effect of arbuscular mycorrhizal fungi on zinc accumulation and histolocalisation. Environ Pollut 145:691–699

Marques APGC, Oliveira RS, Rangel AOSS, Castro PML (2008) Application of manure and compost contaminated soils and its effect on zinc accumulation by *Solanum nigrum* inoculated with arbuscular mycorrhizal fungi. Environ Pollut 151:608–620

Marschner H, Dell B (1994) Nutrient uptake in mycorrhizal symbiosis. Plant Soil 159:89–102

Medina A, Vassilev N, Azcon R (2010) The interactive effect of an AM fungus and an organic amendment with regards to improving inoculum potential and the growth and nutrition of *Trifolium repens* in Cd-contaminated soils. Appl Soil Ecol 44:181–189

Meharg AA (2003) The mechanistic basis of interactions between mycorrhizal associations and toxic metal cations. Mycol Res 107:1253–1265

Neagoe A, Lordache V, Bergmann H, Kothe E (2013) Patterns of effects of arbuscular mycorrhizal fungi on plants grown in contaminated soil. J Plant Nutr Soil 176:273–286

Orlowska E, Mesjasz-Przybylowicz J, Przybylowicz W, Turnau K (2008) Nuclear macroprobe studies of elemental distribution in mycorrhizal and non-mycorrhizal roots of Ni-hyperaccumulator *Berkheya coddii*. X Ray Spectrom 37:129–132

Orlowska E, Orlowski D, Mesjasz-Przybylowicz J, Turnau K (2011a) Role of mycorrhizal colonization in plant establishment on an alkaline gold mine tailing. Int J Phytoremediation 13:185–205

Orlowska E, Przybylowicz W, Orlowski D, Turnau K, Mesjasz-Przybylowicz J (2011b) The effect of mycorrhiza on the growth and elemental composition of Ni-hyperaccumulating plant *Berkheya coddii* Roessler. Environ Pollut 159:3730–3738

Orlowska E, Godzik B, Turnau K (2012) Effect of different arbuscular mycorrhizal fungal isolates on growth and arsenic accumulation in Plantago lanceolata L. Environ Pollut 168:121–130

Ortega-Larrocea MP, Siebe C, Estrada A, Webster R (2007) Mycorrhizal inoculums potential of arbuscular mycorrhizal fungi in soils irrigated with wastewater for various lengths of time, as affected by heavy metals and available P. Appl Soil Ecol 37:129–138

Ortega-Larrocea MP, Xoconostle-Cazares B, Moldano-Mendoza IE, Carrillo-Gonzalez R, Hernández-Hernández J, Garduño MD, López-Meyer M, Gómez-Flores L, González-Chávez M (2010) Plant and fungal biodiversity from metal mine wastes under remediation at Zimapan, Hidalgo, Mexico. Environ Pollut 158:1922–1931

Ouziad F, Hildlebrandt U, Schmelzer E, Bothe H (2005) Differential gene expressions in arbuscular mycorrhizal-colonized tomato grown under heavy metal stress. J Plant Physiol 162:634–649

Pawlowska TE, Blaszkowski J, Ruhling A (1996) The mycorrhiza status of plants colonizing a calamine spoil mound in southern Poland. Mycorrhiza 6:499–505

Perrier N (2005) Bio-Géodiversité fonctionnelle des sols latéritiques miniers : application à la restauration écologique (massif du Koniambo, Nouvelle-Calédonie). PhD thesis, University of New Caledonia, Noumea

Perrier N, Amir H, Colin F (2006) Occurrence of mycorrhizal symbioses in the metal-rich lateritic soils of the Koniambo Massif, New Caledonia. Mycorrhiza 16:449–458

Purin S, Rillig MC (2008) Immuno-cytolocalization of glomalin in the mycelium of the arbuscular mycorrhizal fungus *Glomus intraradices*. Soil Biol Biochem 40:1000–1003

Rajkumar M, Sandhia S, Prasad MNV, Freitas H (2012) Perspectives of plant-associated microbes in heavy metal phytoremediation. Biotechnol Adv 30:1562–1574

Ray P, Tiwari R, Reddy GU, Adholeya A (2005) Detecting the heavy metal tolerance level in ectomycorrhizal fungi *in vitro*. World J Microbial Biotechnol 21:309–315

Redon PO, Béguiristain T, Leyval C (2008) Influence of *Glomus intraradices* on Cd partitioning in a pot experiment with *Medicago truncatula* in four contaminated soils. Soil Biol Biochem 40:2710–2712

Redon PO, Béguiristain T, Leyval C (2009) Differential effects of AM fungal isolates on *Medicago truncatula* growth and metal uptake in a multimetallic (Cd, Zn, Pb) contaminated agricultural soil. Mycorrhiza 19:187–195

Regvar M, Vogel K, Irgel N, Wraber T, Hildebrandt U, Wilde P, Bothe H (2003) Colonization of pennycresses (*Thlaspi* spp.) of the Brassicaceae by arbuscular mycorrhizal fungi. J Plant Physiol 160:615–626

Repetto O, Bestel-Corre G, Dumas-Gaudot E, Berta G, Gianinazzi-pearson V, Gianinazzi S (2003) Targeted proteomics to identify cadmium-induced protein modifications in *Glomus mosseae*-inoculated pea roots. New Phytol 157:555–567

Rivera-Becerril F, van Tuinen D, Martin-Laurent F, Metwally A, Dietz KJ, Gianinazzi S, Gianinazzi-pearson V (2005) Molecular changes in *Pisum sativum* L. roots during arbuscular mycorrhizal buffering of cadmium stress. Mycorrhiza 16:51–60

Rufyikiri G, Huysmans L, Wannijn J, Van Hees M, Leyval C, Jacobsen I (2004) Arbuscular mycorrhizal fungi can decrease the uptake of uranium by subterranean clover grown at high levels of uranium in soil. Environ Pollut 130:427–436

Sadeque HFR, Kilham K, Alexander I (2006) Influences of arbuscular fungus Glomus mosseae on growth and nutrition of lentil irrigated with arsenic contaminated water: rhizosphere: perspectives and challenges. Plant Soil 283:33–41

Schechter SP, Bruns TD (2012) Edaphic sorting drives arbuscular mycorrhizal fungal community assembly in a serpentine/non-serpentine mosaic landscape. Ecosphere 3(54):art 42

Schützendübel A, Polle A (2002) Plant responses to abiotic stresses: heavy metal-induced oxidative stress and protection by mycorrhization. J Exp Bot 53:1351–1365

Smith SE, Read DJ (2008) Mycorrhizal symbiosis, 3rd edn. Academic, San Diego

Smith SE, Smith FA (2010) Plant performance in stressful environments: interpreting new and established knowledge of the roles of arbuscular mycorrhizas. Plant Soil 326:3–20

Stommel M, Mann P, Franken P (2001) EST-library construction using spore RNA of the arbuscular mycorrhizal fungus *Gigaspora rosea*. Mycorrhiza 10:281–285

Sudova R, Doubkova P, Vosatka M (2008) Mycorrhizal association of *Agrostis capillaris* and *Glomus intraradices* under heavy metal stress: combination of plant clones and fungal isolates from contaminated substrates. Appl Soil Ecol 40:19–29

Tseng CC, Wang JY, Yang L (2009) Accumulation of copper, lead and zinc by *in situ* plants inoculated with AM fungi in multicontaminated soil. Commun Soil Sci Plant Anal 40:2122

Tullio M, Pierandrei F, Salerno A, Rea E (2003) Tolerance to cadmium of vesicular arbuscular mycorrhizae spores isolated from cadmium-polluted and unpolluted soil. Biol Fertil Soils 37:211–214

Turnau K, Mesjasz-Przybylowicz J (2003) Arbuscular mycorrhiza of *Berkheya coddii* and other Ni-hyperaccumulating members of Asteraceae from ultramafic soils in South Africa. Mycorrhiza 13:185–190

Urban A, Puschenreiter M, Strauss J, Gorfer M (2008) Diversity and structure of ectomycorrhizal and co-associated fungal communities in a serpentine soils. Mycorrhiza 18:339–354

Vallino M, Massa N, Lumini E, Bianciotto V, Berta G, Bonfante P (2006) Assessment of arbuscular mycorrhizal fungal diversity in roots of *Solidago gigantea* growing in a polluted soil in northern Italy. Environ Microbiol 8:971–983

Vallino M, Martino E, Boella F, Murat C, Chiapello M, Peretto S (2009) Cu, Zn superoxide dismutase and zinc stress in the metal-tolerant ericoid mycorrhizal fungus *Oidiodendron maius Zn*. FEMS Microbiol Lett 293:48–57

Vivas A, Barea JM, Azcon R (2005) Interactive effect of *Brevibacillus brevis* and *Glomus mosseae*, both isolated from Cd contaminated soil, on plant growth, physiological mycorrhizal fungal characteristics and soil enzymatic activities in Cd polluted soil. Environ Pollut 134:257–266

Vodnik D, Grcman H, Macek I, van Elteren JT, Kovačevič M (2008) The contribution of Glomalin-related protein to Pb and Zn sequestration in polluted soil. Sci Total Environ 392:130–136

Walker RF, McLaughlin SB, West DC (2004) Establishment of sweet birch on surface mine spoil as influenced by mycorrhizal inoculation and fertility. Restor Ecol 12:8–19

Weissenhorn I, Leyval C, Belgy G, Berthelin J (1995) Arbuscular mycorrhizal contribution to heavy metal uptake by maize (*Zea mays* L.) in pot culture with contaminated soil. Mycorrhiza 5:245–251

Whiting SN, Reeves RD, Richards D, Johnson MS, Cooke JA, Malaisse F, Paton A, Smith JAC, Angle JS, Ginocchio R, Jaffré T, Johns R, McIntyre T, Purvis OW, Salt DE, Schat H, Zhao FJ, Baker AJM (2004) Research priorities for conservation of metallophyte biodiversity and their potential for restoration and site remediation. Restor Ecol 12:106–116

Wu FY, Ye ZH, Wong MH (2009) Intraspecific differences of arbuscular mycorrhizal fungi in their impacts on arsenic accumulation by *Pteris vittata* L. Chemosphere 76:1258–1264

Wu FY, Be YL, Leung HM, Ye ZH, Lin XG, Wong MH (2010) Accumulation of As, Pb, Zn, Cd and Cu and arbuscular mycorrhizal status in populations of *Cynodon dactylon* grown on metal-contaminated soils. Appl Soil Ecol 44:213–218

Zhang XH, Zhu YG, Chen BD, Lin AJ, Smith SE, Smith FA (2005) Arbuscular mycorrhizal fungi contribute to resistance of upland rice to combined metal contamination of soil. J Plant Nutr 28:2065–2077

Zhang XH, Lin AJ, Gao YL, Reid RJ, Wong MH, Zhu YG (2009) Arbuscular mycorrhizal colonisation increases copper binding capacity of root cell walls of *Oryza sativa* L. and reduces copper uptake. Soil Biol Biochem 41:930–935

Chapter 16
Arsenic Uptake and Phytoremediation Potential by Arbuscular Mycorrhizal Fungi

Xinhua He and Erik Lilleskov

16.1 Introduction

Arsenic (As) contamination of soils and water is a global problem because of its impacts on ecosystems and human health. Various approaches have been attempted for As remediation, with limited success. Arbuscular mycorrhizal (AM) fungi play vital roles in the uptake of water and essential nutrients, especially phosphorus (P), and hence enhance plant performance and productivity (Smith and Read 2008). As uptake and tolerance to As toxicity in plants are also enhanced by AM fungi (Zhao et al. 2009; Smith et al. 2010; Gonzalez-Chavez et al. 2011). The use of AM fungi has thus been proposed as a potential contributor to enhance plant As uptake and accumulation and to develop plant-based As remediation. Here, we review the problem of As toxicity in terrestrial ecosystems and human health, examine the recent progress in understanding the roles of AM fungi in plant As tolerance and accumulation, and explore the promise and challenges of using AM fungi as phytoremediation approaches to tackle this environmental problem.

X. He (✉)
Elizabeth Macarthur Agricultural Institute, Department of Primary Industries, NSW Trade & Investment, Menangle, NSW 2568, Australia

School of Plant Biology, University of Western Australia, Crawley, WA 6009, Australia

Northern Research Station, USDA Forest Service, Houghton, MI 49931, USA

School of Forest Resources and Environmental Science, Michigan Technological University, Houghton, MI 49931, USA
e-mail: xinhua.he@dpi.nsw.gov.au

E. Lilleskov
Northern Research Station, USDA Forest Service, Houghton, MI 49931, USA

School of Forest Resources and Environmental Science, Michigan Technological University, Houghton, MI 49931, USA

© Springer-Verlag Berlin Heidelberg 2014
Z.M. Solaiman et al. (eds.), *Mycorrhizal Fungi: Use in Sustainable Agriculture and Land Restoration*, Soil Biology 41, DOI 10.1007/978-3-662-45370-4_16

16.2   Arsenic in the Environment and Its Toxicity

Arsenic is an odorless and tasteless semimetal element that occurs naturally in rocks ($\sim$3 mg kg^{-1} Earth crust) (Mandal and Suzuki 2002). It can be released into air, water, and soils through natural activities (volcanic action, rock and soil erosion) or agricultural and industrial practices (fertilizers, herbicides, pesticides, mining, semiconductors) (Mandal and Suzuki 2002; Adriano 2001). As accumulation, migration, and toxicity are related to its chemical speciation. Inorganic As species are the more reduced arsenite [H_3AsO_3, As(III)] and more oxidized arsenate [$HAsO_4^{2-}$, As(V)]. Generated from inorganic As via biomethylation, organic As species include mono- and di-methylarsenite [MMA(III) and DMA(III)] and mono- and di-methylarsenate [MMA(V) and DMA(V)] (Cullen and Reimer 1989). Arsenic compounds are the most notorious toxins in human history and linked to many forms of cancer, diarrhea, nausea, stomach pain, vomiting, numbness, partial paralysis, and blindness (Nriagu 2002). The toxicity order of As is as follows: MMA(III) > DMA(III) > As(III) > As(V) > MMA(V) > DMA(V) (Ali et al. 2009; Kim et al. 2009; Ralph 2008).

Arsenic exposure occurs primarily through drinking water and food. The standard for drinking water to prevent chronic effects is $\leq$0.01 mg As L^{-1} (0.01 ppm) (http://www.who.int/mediacentre/factsheets/fs210/en/). More than 150 million people worldwide get exposed to 0.01–0.05 mg As L^{-1} drinking water, including countries in Southeast Asia, North and South America, and Europe (Bhattacharjee 2007; Kim et al. 2009; Smith et al. 2000). At present, the World Health Organization and most countries have not established legal As limits in food, though the US FDA recommends a "tolerable daily intake" of 0.13 mg As in food (Stone 2008).

16.3   Arsenic Biogeochemistry in Soil and Its Uptake in Plants

Arsenic and P belong to the same V_A chemical group, and they thus display similar chemical properties and geochemical behaviors (Cullen and Reimer 1989; Adriano 2001). However, whereas P is an essential plant nutrient, As can be toxic to crops as well as to primary and secondary plant consumers (Stone 2008; Kim et al. 2009). Various forms of As exist in soils depending on pH and redox status. As(III) dominates in anaerobic substrates, while As(V) dominates in aerobic soils (see Wenzel et al. 2002; Raab et al. 2007; Williams et al. 2007). Typical concentrations of As(III) are 0.01–3.0 μM in contaminated soils, while As(V) concentrations are >2.3 μM in contaminated or <53 nM in uncontaminated soils (Wenzel et al. 2002). Plant roots primarily take up inorganic As(III) and As(V) and are also capable of taking up organic MMA(III) or DMA(III). The toxic limits to most plants are 5–20 mg As kg^{-1} soil (Mendez and Maier 2008), and the common symptoms of As toxicity include reduced root growth, leaf chlorosis, increased sterility, and yield reduction (Meharg and Hartley-Whitaker 2002; Raab et al. 2007; Smith et al. 2010 and references therein).

The known plant uptake pathways for reduced and oxidized inorganic As are via silicon (Si) and phosphate (PO_4^-) transporters (Pht), respectively. As(III) enters into rice (*Oryza sativa*) roots passively by sharing a Si transport pathway through nodulin 26-like intrinsic proteins (NIPs) (Maurel et al. 2008; Ali et al. 2009; Zhao et al. 2009 and references therein). As(III) uptake was inhibited by glycerol and antimonite (Sb), but not by P (Abedin et al. 2002a, b; Meharg and Jardine 2003). In contrast, As(V) is taken up actively through PO_4^- transporters (e.g., Pht1;1 and Pht1;4) (Shin et al. 2004), which have a lower affinity for As(V) than P (Meharg and Macnair 1990; Meharg and Hartley-Whitaker 2002). The rapid reduction of As (V) to As(III) was demonstrated in tomato (*Lycopersicon esculentum*) and rice roots (Xu et al. 2007).

The uptake competition between As(V) and P was exhibited by excised roots of barley (*Hordeum vulgare*), velvet grass (*Holcus lanatus*), or mouse-ear cress (*Arabidopsis thaliana*) in solution culture (Meharg and Macnair 1990; Meharg and Hartley-Whitaker 2002; Zhao et al. 2009), but not by medic (*Medicago truncatula*) or barley in soil/sand (2:8) media (Christophersen et al. 2009a). The influx of As(III) was generally comparable to that of As(V) under low (<50 μM, high-affinity transporter range) but considerably higher under high (>100 μM, low-affinity transporter range) concentrations (Meharg and Macnair 1990; Meharg and Jardine 2003).

16.4   Arsenic Transport and Hyperaccumulation in Plants: The Basis for Phytoremediation

Arsenic is primarily accumulated in roots of most plants because its low mobility restricts its root-to-shoot translocation, except in As hyperaccumulators (Raab et al. 2007). Brooks et al. (1977) defined "hyperaccumulators" as plants that could tolerate and accumulate >1 mg metal g^{-1} (0.1 %) dry mass. An As hyperaccumulator has greater antioxidant capacity and lower reactive oxygen concentration and thus greater As tolerance than a non-As hyperaccumulator (Srivastava et al. 2005; Singh et al. 2006). After uptake, As(V) is rapidly reduced by As(V) reductases in roots to As(III), which can then be detoxified by complexation with glutathione (GSH) or phytochelatins (PCs) (Raab et al. 2005; Zhao et al. 2009; Zhu and Rosen 2009). As(III) or the complexed As(III) is transported across tonoplasts and sequestered in vacuoles, loaded into xylem, and translocated to and accumulated in shoots (Xu et al. 2007; Su et al. 2008).

High As tolerance and accumulation capacity constitute the basis of exploring plant hyperaccumulators for As phytoremediation. Candidate plants for phytoremediation must tolerate and accumulate high levels of As in their tissues and possess high biomass production potential. At present, several fern species and

Table 16.1 Potential As hyperaccumulator plant species (grouped according to De Koe 1994; Bech et al. 1997; Tu et al. 2002; Baldwin and Butcher 2007; Tripathi et al. 2007; Zhao et al. 2009)

Plant	Species
Ferns	*Pityrogramma calomelanos* (L.) Link (silverback fern), *P. austroamericana* Domin (leatherleaf goldback fern), *Pteris aspericaulis* (tricolor fern), *P. biaurita* (thinleaf brake fern), *P. cretica* var. *albolineata* (table fern), *P. cretica* var. *nervosa* (Cretan brake fern), *P. cretica* cv Mayii (moonlight fern), *P. fauriei* (Faurie's brake fern), *P. longifolia* (longleaf brake fern), *P. multifida* Poir. and *P. multifida* f. *serrulata* (spider brake fern), *P. oshimensis* Hieron. (an Asian fern), *P. quadriaurita* (striped brake fern), *P. ryukyuensis* Tagawa. (an Asian fern), *P. umbrosa* (Australian jungle brake fern), *P. vittata* (Chinese brake or ladder fern)
Grasses and forbs	*Agrostis castellana* (bentgrass or dryland browntop), *A. delicatula* (bentgrass), *Bidens cynapiifolia* (West Indian beggarticks)

a number of grasses and forbs have been identified as As hyperaccumulators (De Koe 1994; Bech et al. 1997; Tu et al. 2002; Baldwin and Butcher 2007; Tripathi et al. 2007; Zhao et al. 2009; see Table 16.1). As(III) generally accounts for 60–90 % of the total As in the shoots of As hyperaccumulator *Pteris* species, and the ratio of shoot-to-root As accumulation (translocation factor (TF)) ranges between 5 and 25 in hyperaccumulators (Tu and Ma 2002; Tu et al. 2002; Zhao et al. 2009; Leung et al. 2010a, b, 2013). This high As accumulation in plants can lead to demonstrable reductions of soil As content via phytoremediation programs (Xie et al. 2009). For instance, *Pteris vittata* (Chinese brake fern) was capable of reducing As from 190 to 140 mg kg^{-1} soil after 2 years growing in an As-contaminated field (Kertulis-Tartar et al. 2006) or from 130 to 10 $\mu g\ L^{-1}$ after 4–6 weeks growing in an As-contaminated groundwater (Natarajan et al. 2008).

16.5 Mycorrhizal Symbiosis

The potential for mycorrhizal symbiosis to improve As tolerance and phytoremediation has been only partially explored. About 90 % of higher plants associate with mycorrhizal fungi (Wang and Qiu 2006; Smith and Read 2008; Brundrett 2009). There are about 200 AM fungal species, and all of them belong to the phylum Glomeromycota (Walker et al. 2007a, b; Palenzuela et al. 2008). AM fungi are asexual obligate symbionts, and most of them are widespread and not host specific. In AM associations, fungal hyphae penetrate inside the walls of root cortical cells to form either "little-tree-shaped" structures, called arbuscules, or hyphal coils, both of which serve as the main nutrient exchange sites between fungus and plant.

While aboveground plant structures are easily observed, mycorrhizal fungi and their activities are challenging to characterize. A single gram of soil may contain up to 50 m of AM hyphae, which can extend >9 cm beyond the roots and expand extensively throughout the soil matrix (Nasim 2005). The small 2–10 μm diameter of mycorrhizal fungal hyphae can efficiently explore soil volume and microsites inaccessible to plant roots. One important function of mycorrhizal fungi is to enhance host plant nutrient acquisition by increasing access to inorganic N and P by hyphae extending beyond depletion zones caused by direct uptake by roots and by access to organic N and P via their extracellular protease and phosphatase activity (Smith and Read 2008).

16.5.1   Roles of Mycorrhizal Fungi in Arsenic Tolerance

There are several hypothesized mechanisms by which mycorrhizal fungi could affect host plant As tolerance (Meharg and Hartley-Whitaker 2002; Zhao et al. 2009; Smith et al. 2010; Gonzalez-Chavez et al. 2011). First, it has been hypothesized that AM fungi increase plant P nutrition and growth and thus alleviate toxic effects of As on plants due to the dilution of As uptake because P shares chemical properties with As (Adriano 2001). Second, As-tolerant fungi could provide added functional benefits over non-tolerant fungi. Numerous AM studies have addressed these hypotheses, with most studies focused on P nutrition effects.

Hypothesis 1: P Nutrition Effects. Consistent with the hypothesis that mycorrhizally mediated improved P nutrition enhances As tolerance, plant growth and P nutrition were simultaneously improved under As stress conditions by AM in most studies. For instance, As uptake, As tolerance, and P nutrition in both shoots and roots of maize (*Zea mays*) (Xia et al. 2007; Bai et al. 2008; Wang et al. 2008; Yu et al. 2009, 2010), lettuce (*Lactuca sativa*; Cozzolino et al. 2010), and *Eucalyptus globulus* (Arriagada et al. 2009) were concurrently enhanced by AM fungi. In addition, the activity of peroxidase, superoxide dismutase, and As(V) reductase was suppressed by *Glomus mosseae* (now *Funneliformis mosseae*), indicating that AM colonization could inhibit the reduction of As(V) to As(III) and As toxicity to plants could hence be alleviated (Yu et al. 2009). By contrast, the phytotoxicity of arsenate (AsV, $Na_2HAsO_4 \cdot 7H_2O$) led to an increase in superoxide dismutase, catalase, and peroxidase activities in a 1-month-old pea (*Pisum sativum*) (Garg and Singla 2012). Similarly, P accumulation was significantly higher under all As levels of 10, 50, 100, and 200 mg kg^{-1} soil in a 2-month-old mycorrhizal medic inoculated with *G. mosseae* BEG167 (Xu et al. 2008). Both As and P uptake were higher in 3-month-old *G. mosseae* BEG167-inoculated tomatoes growing in 25, 50, and 75 As kg^{-1} spiked soil, but similar in 150 mg As kg^{-1} spiked soil (Liu et al. 2005a). A hydroponic study with a 1-month-old *Pennisetum clandestinum* Hochst (kikuyu grass) showed that As(V) uptake was competitively inhibited by P uptake because of a higher selectivity of membrane transporters with respect to P rather than As(V) (Panuccio et al. 2012). Smith et al. (2010) and Christophersen et al. (2012) recently detailed mechanisms of direct root and/or mycorrhizal Pi/As (V) uptake pathways, summarizing the physiological basis for the observed P-mediated effects on As accumulation of both AM-responsive and

AM-nonresponsive plants. Thus, there is relatively strong support for this hypothesis.

Hypothesis 2: *Fungal As Tolerance*. Some pure culture studies suggest that the extent of As(V) toxicity to mycorrhizal fungi could vary among fungal taxa, increasing the potential for the selection of appropriate fungi for remediation efforts. There is some evidence that AM fungal populations can develop tolerance to As and that this tolerance results in improved host performance. For instance, fungal isolates of *G. mosseae* and *G. caledonium* associated with velvet grass roots from the As-contaminated site were more As(V) tolerant than those from the non-As-contaminated site (Gonzalez-Chavez et al. 2002). Root high-affinity As(V)/PO_4^- transportation was suppressed by both tolerant and non-tolerant *G. mosseae* in both tolerant and non-tolerant velvet grasses. As(V) uptake in the tolerant velvet grass growing in the As-contaminated site was reduced by inoculating with the tolerant AM isolates. The authors concluded that AM fungi had evolved As (V) tolerance and conferred enhanced As tolerance on velvet grass (Gonzalez-Chavez et al. 2002).

In summary, mycorrhizal fungi have been consistently shown to confer As tolerance on their host plants. The possible mechanism of As tolerance in mycorrhizal plants might be one or a combination of the following. First, AM fungi enhance P nutrition and plant growth, resulting in a higher P/As ratio and a relative As dilution in tissues of mycorrhizal plants (Liu et al. 2005a, b; Ahmed et al. 2006; Chen et al. 2007; Ultra et al. 2007a, b). The corresponding reasons are the induction of HvPht1;8 (*H. vulgare* phosphate transporter) and downregulation of HvPht1;1 and HvPht1;2 (Christophersen et al. 2009b) and both the upregulated and downregulated expressions of up to 130 life proteins, particularly for some glycolytic enzymes including glyceraldehyde-3-phosphate dehydrogenase, phosphoglycerate kinase, and enolase (Bona et al. 2010, 2011). This could provide "protective effects" against As uptake or stress because P shares chemical properties with As. Second, As-tolerant mycorrhizal fungi enhance As(III) exudation to the external media and reduce As(V) uptake at the As-contaminated habitats and thus confer enhanced As tolerance on AM plants (Gonzalez-Chavez et al. 2002), since the induction of GiPT (*Glomus intraradices* high-affinity phosphate transporter) expression correlates with As(V) uptake in the extra-radical mycelium of *G. intraradices* (Gonzalez-Chavez et al. 2011). In addition, under 2 μM As(III), both the Lsi1 and Lsi2 As(III) transporters were significantly decreased by 0.7- and 0.5-fold in mycorrhizal than in non-mycorrhizal 2.5 month-old rice seedlings, leading to a decrease of As(III) uptake per unit of root dry mass (Chen et al. 2012).

16.5.2 Roles of Mycorrhizal Fungi in Arsenic Uptake

The chemical similarity of P and As, combined with the mycorrhizal role in P nutrition, provides the likelihood that mycorrhizal fungi may enhance As uptake. Given that mycorrhizas generally enhance P uptake, it is possible that mycorrhizal fungi will also increase the uptake of As in their plant hosts. However, depending on mycorrhizal specificity for P vs. As uptake (Li et al. 2011), a variety of outcomes are possible. For example, if mycorrhizal fungi have more specific P uptake mechanisms than their hosts, they may reduce the proportional uptake of As relative to P (Smith et al. 2010).

Arsenic Accumulations in Shoots of Herbaceous Plants. Consistent with an enhanced As uptake via mycorrhizal fungal symbiosis, AM fungi appear to increase As accumulation in their hosts. Compared to the non-AM seedlings, As accumulation (both concentration and content) was increased in shoots and roots of 2- or 3-month-old *G. mosseae*-inoculated maize seedlings growing in 75 and 150 mg As kg^{-1} soil/sand (3:1) media (Wang et al. 2008), in 100 mg As kg^{-1} soil (Yu et al. 2009), and even in 600 mg As kg^{-1} soil/sand (2:1) media (Xia et al. 2007). A mixed inoculum of indigenous AM isolates (*Glomus* spp. and *Acaulospora* spp.) from As-contaminated soils, not the nonindigenous *G. caledonium* 90036 from non-As-contaminated soils, increased As accumulation in shoots of a 2.5-month-old maize in 185 and 290 mg As kg^{-1} soil (Bai et al. 2008). Plant total As accumulations were significantly increased in a 3-month-old *G. mosseae*-inoculated white clover (*T. repens*) and ryegrass (*Lolium perenne*) growing in 600 mg As kg^{-1} soil/sand (1:1) media (Dong et al. 2008). Shoot As and toxicity symptoms were reduced in a 6-week-old *G. aggregatum*-inoculated sunflower (*Helianthus annuus*) growing in 620 mg As kg^{-1} contaminated soil (Ultra et al. 2007a, b). Arsenic accumulation was also significantly increased in a 2-month-old *G. mosseae* BEG167-inoculated medic growing in 200 mg As kg^{-1} soil but was similar between the non-mycorrhizal and mycorrhizal plants under 10, 50, or 100 mg As kg^{-1} soil (Xu et al. 2008). In addition, P accumulation and P/As ratio of both shoots and roots were always higher in all mycorrhizal plants than in their non-mycorrhizal counterparts in almost all these studies, suggesting that AM fungi may have more specific uptake of P relative to As when compared with non-mycorrhizal plants.

Higher As Accumulations in Roots than in Shoots of Herbaceous Plants. Although the increase in As accumulation in crop plants might be of concern from a food chain perspective, interestingly, As accumulation in roots, rather than in shoots, was much more enhanced by mycorrhizal fungi in most studies with herbaceous plants, as 80–90 % of As accumulated in roots of maize, ryegrass, and clover (Dong et al. 2008; Wang et al. 2008). Grown under a range between 100 and 600 mg As kg^{-1} soil/sand media, accumulations of As contents were 10–50 times higher in roots than in shoots in tomato (Liu et al. 2005a), sunflower (Ultra et al. 2007a, b), medic (Xu et al. 2008), white clover (*Trifolium repens*), ryegrass (*L. perenne*) (Dong et al. 2008), and maize (Xia et al. 2007; Wang et al. 2008; Yu et al. 2009). In contrast, P accumulations and P/As ratio were generally higher in shoots than in roots in all these studies. In addition to the food chain implications, the enhancement of As accumulation in roots has implications for mycorrhizal plant utility in bioremediation efforts, as we shall see below. Also of relevance to As accumulation in the food chain, As concentrations in pods were reduced, while P uptake was increased in a 9-week-old nodulated AM (*G. mosseae*) lentil (*Lens culinaris* cv. Titore) irrigated with 1, 2, 5, and 10 mg As(V) L^{-1} to the sand/terra (1:1) media (Ahmed et al. 2006). Lower As concentration in pods would most likely reduce As toxicity risk in the food chain. Further studies are required to understand if this is a general consequence of mycorrhizal colonization. In addition, the highest As accumulated in maize roots when inoculated with *Acaulospora* spp. or *Glomus* spp. and earthworm (*Eisenia foetida*) (Hua et al. 2009, 2010).

Arsenic Accumulation in Fronds of Ferns. The roles of AM fungi in As uptake and tolerance have also been investigated in the As hyperaccumulation ferns. Similar to studies summarized above, there was often an increase of As accumulation in mycorrhizal ferns (Liu et al. 2009), though there were intraspecific differences in AM fungi on As accumulation in *P. vittata* (Wu et al. 2009). In contrast, mycorrhization led to an increase in the relative proportion of As accumulated in fronds vs. roots. For example, compared to its non-mycorrhizal counterpart, the amounts of As accumulation were about five times higher in fronds, but similar in roots, in an 8-month-old mycorrhizal *P. vittata*, when grown in 100 mg As(V) with 25 or 50 mg P kg^{-1} soil and inoculated with an AM inoculum from an As-contaminated site (Al Agely et al. 2005), and in a 4-month-old *G. mosseae* BEG167-colonized *P. vittata* growing in 300 mg As kg^{-1} soil (Liu et al. 2005b). Arsenic accumulations in fronds and roots were 3.0–3.9 and 2.5–3.6 times higher, respectively, in a 2-month-old mycorrhizal (an indigenous soil inoculum) *P. vittata* than in non-inoculated plants growing in 50 or 100 mg As kg^{-1} soil (Leung et al. 2006). However, As accumulation in *P. vittata* was not affected by 2- or 3-month inoculation with either *G. mosseae*, *G. caledonium*, or *G. intraradices* growing in 106 mg As kg^{-1} soil (Chen et al. 2006). Compared to non-mycorrhizal plants, frond As accumulation was reduced, while similar in roots, in an 8-month-old AM *Pityrogramma calomelanos* (silverback fern) growing in 240 mg As kg^{-1} soil (Jankong and Visoottiviseth 2008). However, a commercial AM inoculum (a mixture of *G. mosseae*, *G. intraradices*, and *G. etunicatum*) was applied to this 8-month-old silverback fern for only 2 months, possibly reducing mycorrhizal effects on the outcome. Soil As concentration was reduced by 24 %, while tissue As accumulation was up to 0.2 % in *P. vittata* growing under a mixed inoculum [indigenous AM fungi (*G. intraradices*, *G. geosporum, and G. mosseae*) +nonindigenous *G. mosseae*] and the addition of phosphate rock (Leung et al. 2010a). The contrasting results may be derived from experimentation with different AM isolates, different host plants, or other experimental conditions. Further assessments of mycorrhizal effects on As accumulation are needed, particularly under field conditions. In general, most of these fern studies showed a higher ratio of frond/root As accumulation in the mycorrhizal ferns than in their non-mycorrhizal counterparts, suggesting that As translocation from root to shoot was enhanced by mycorrhizal fungi even in As hyperaccumulation ferns. The mycorrhizal-mediated enhancement of As tolerance and accumulation either in shoots of As hyperaccumulating ferns or in roots of herbaceous annuals and perennials offers potential for screening fungal species for As remediation purpose.

16.6   Potential of Mycorrhizal Fungi in Arsenic Phytoremediation

Phytoremediation is a promising alternative for As remediation from contaminated soils and water since the chemical and physical remediation technologies are quite expensive and limited to on-site applications (Mendez and Maier 2008; Mondal et al. 2006; Tripathi et al. 2007; Wenzel 2009; Garg and Singla 2011). Genetic manipulation of As hyperaccumulating traits could contribute our efforts to As phytoremediation (Zhu and Rosen 2009), though the traits and genes are largely unknown to date. Because aboveground plant parts are easier to harvest, most attention has been given to identify high shoot As accumulators for phytoextraction by aboveground harvesting, while less has been given to high root As accumulators by belowground harvesting. But all shoot As hyperaccumulation ferns require a tropical or subtropical climate and may not grow well in other habitats. As an alternative, if roots could be easily harvested, then root hyperaccumulators could be used for phytoremediation, especially in herbaceous plants with dense root systems in shallow soil profiles, though root removal technique is not available or currently impractical. Further testing of a broad suite of species is needed for screening both shoot and root hyperaccumulators, in addition to those listed in Table 16.1. The potential roles of AM fungi (Gaur and Adholeya 2004; Garg and Singla 2011) and plant-associated bacteria (Khan 2005; Weyens et al. 2009) in heavy metal phytoremediation have been respectively proposed. However, the potential for AM fungi to contribute to As (a semimetal element) tolerance and hyperaccumulation in their host plants is poorly explored, particularly under field conditions.

Can mycorrhizas potentially offer a more cost-effective, environmentally sound, and sustainable pathway to global As phytoremediation? As seen in the previous sections, mycorrhizal fungi can tolerate and perform well in high levels of As under laboratory conditions and contaminated field sites, and they also can facilitate As accumulation in host plant tissues or increase the transfer of As from roots to shoots by indigenous isolates in particular (Orlowska et al. 2012). This indicates that mycorrhizal fungi could confer both As tolerance and accumulation ability on their host plants. A range of 10 and 50 times higher As accumulations in roots than in shoots had been reported for some annuals or perennials, including lentil, maize, medic, ryegrass, sunflower, tomato, and white clover (Liu et al. 2005a, b; Ahmed et al. 2006; Xia et al. 2007; Bai et al. 2008; Dong et al. 2008; Wang et al. 2008; Xu et al. 2008; Yu et al. 2009; Ultra et al. 2007a, b; Garg and Singla 2012), or in shoots than in roots for a dozen ferns (Al Agely et al. 2005; Leung et al. 2006; Chen et al. 2006; Jankong and Visoottiviseth 2008; Zhao et al. 2009). If these phenomena are generally true, the selection of combinations of plant and fungal species with high As tolerance and accumulation ability would tap their potential for As phytoremediation, particularly for both phytoextraction and phytostabilization (Mendez and Maier 2008). At present, no one has identified either a woody As phytoremediation plant or a candidate with both high shoot and high root As accumulation capacity. Thus, the current phytoremediation strategies are focused on herbaceous shoot hyperaccumulators, and phytostabilization is focused on herbaceous root hyperaccumulators. Furthermore, almost all current As phytoremediation practices are limited to laboratory experiments and a few very small field trials, where plants are introduced into the soil without established mycorrhizal symbioses.

Mycorrhizal diversity is high and mycorrhizal symbiosis develops well with the shoot As hyperaccumulation ferns even on As-contaminated field sites. A field investigation on both As-contaminated and As-uncontaminated fields in Central, Southern, and Southeastern China showed that the As hyperaccumulator *P. vittata* was associated with the fungal genera *Acaulospora*, *Diversispora*, *Glomus*, *Paraglomus*, and *Scutellospora*, with the common species *Glomus brohultii*, *G. geosporum*, *G. microaggregatum*, and *G. mosseae* (Wu et al. 2007). This high mycorrhizal fungal diversity may have significant ecological and physiological contributions to their host plants in contaminated sites. The known root As hyperaccumulation annuals and perennials mentioned above are mycorrhizal (Brundrett 2009; Wang and Qiu 2006). Given that indigenous AM fungi from contaminated soils performed better in both accumulation of As and plant growth (see the above section), these adapted indigenous fungi are a promising tool for As phytoremediation from the contaminated soil, particularly when large-scale on-farm production of mycorrhizal inocula becomes available (Douds et al. 2005; Ijdo et al. 2011). The introduction of As-tolerant mycorrhizal fungi to sites with no, limited, or unadapted mycorrhizal fungi could speed up not only As remediation with the establishment of mycorrhizal symbiosis between plants and fungi but also soil reclamation and vegetation restoration. Therefore, there is great potential to screen and then to integrate fungal isolates that enhance both As tolerance and hyperaccumulation with a shoot or root hyperaccumulation plant. In addition, the combination of mycorrhizal fungi with N_2-fixing microorganisms (*Rhizobia* or *Frankia*), As(V)-reducing bacteria (*Comamonas* sp., *Delftia* sp., *Rhodococcus* sp., and *Streptomyces* sp.), and dual AM and ectomycorrhizal (EM) or the tripartite AM, EM, and N_2-fixing plant (He et al. 2005, 2009; Roy et al. 2007; Yang et al. 2012) would further extend our efforts to identify plants with high As tolerance and accumulation capacity capable of functioning under nutrient-poor conditions.

Hyphae of a single fungal individual can potentially interconnect many plants of the same or different species, and a single plant can form mycorrhizas with many fungi as well. As a consequence, a common mycorrhizal network (CMN) forms within and between plant roots to link plants together (Newman 1988; He and Nara 2007; He et al. 2009). CMNs provide pathways to shuttle nutrients, such as C, N, P, and water, from one plant to another between the same and different plant species (Newman 1988; He and Nara 2007; He et al. 2009). These extensive mycorrhizal mycelia and networks could enhance As uptake and accumulation in shoots and/or roots. The transfer of As from a plant to another via a CMN has evidenced this potential. Plants were grown in two separate chambers separated by 25 μm steel mesh with a 1.0 cm air gap between chambers to restrict root growth but allow hyphal linkages. After 1 week of 0.1 % Na_2HAsO_4 application to leaves of a 50-day-old donor (either a grass of *Bromus hordeaceus*, *B. madritensis*, *Nassella pulchra* or a forb of *Madia gracilis, Sanicula bipinnata, Trifolium microcephalum*), AM-mediated transfer of As occurred between grass donors and forb receivers, but not the other direction (Meding and Zasoski 2008). By growing plants with high biomass production but low As uptake capacity together with those having low biomass production but high As uptake capacity, As transfer between mycorrhizal plants via CMN may provide another plant-based phytoremediation strategy.

The current barriers to the adoption of mycorrhizal inoculation reside at several levels. First, there is the need to identify the best candidate fungi for both phytoremediation and phytostabilization. Inoculum sources for mycorrhizal fungi used in phytoremediation should be derived from a similar soil, climate, and geographic region as the phytoremediation site as possible. This will both increase the chances of success and minimize the likelihood of the transfer of unwanted invasive soil organisms with the fungal inoculum (Schwartz et al. 2006). In addition, screening sites with naturally high As or long-term As contamination will provide the highest likelihood for encountering As-tolerant mycorrhizal fungal populations. Given that it is likely that the best strains will be isolated from sites that have naturally high As, in these locations, mycorrhizal fungi native to the site may be sufficient as an inoculum source, greatly simplifying the process of inoculation for phytoremediation. Second, for cases where inoculation is necessary, there are existing biotechnological approaches to producing large quantities of fungal inoculum (Douds et al. 2005; Ijdo et al. 2011), but such approaches are limited at present to very few fungal strains. The magnitude of this limitation will depend on the tractability of otherwise suitable mycorrhizal fungal inocula. It may be that native soil inoculum from sites discussed above could be used when otherwise appropriate (e.g., when conforming to regulations regarding soil transportation). Third, in temperate regions, the barrier to mycorrhizal fungal use for phytoextraction is the lack of appropriate host plants, because most mycorrhizally enhanced As accumulation outside tropical ferns occurs in host roots, which are more challenging to harvest. Up to 1,400 or 1,600 mg As DW kg^{-1} was accumulated in mycorrhizal roots of tomato (Liu et al. 2005a), ryegrass, and clover (Dong et al. 2008) compared to 70 or 80 mg As kg^{-1} DW accumulated in shoots, when growing under 150 or 600 mg As kg^{-1} soil-like media. In addition, annual or perennial bentgrass (*Agrostis castellana* and *A. delicatula*) and West Indian beggarticks (*Bidens cynapiifolia*) could accumulate 1,000–1,800 mg As DW kg^{-1} in roots at As-contaminated mine sites (De Koe 1994; Bech et al. 1997), though their mycorrhizal status had not been reported. Considering that the *Agrostis* and *Bidens* genera have more than 100 or 200 species and almost all tested species are mycorrhizal (Wang and Qiu 2006), it is likely that these species are mycorrhizal. The potential range of plants for phytoremediation could thus greatly be expanded if root-harvesting technologies that are economically and environmentally appropriate could be explored in the near future. Pilot studies are urgently needed to determine whether root As accumulation is common in a magnitude sufficient to make root As harvesting feasible for those herbaceous plants with dense, sufficiently accessible root systems. Identification of such plants and appropriate root-harvesting technologies, such as those widely used for root or tuberous crops, would greatly expand the potential range of plants for extraction of As-enriched root systems. Furthermore, if these plants can be identified, then the incorporation of mycorrhizal inoculation with appropriate strains should greatly enhance phytoremediation efforts for a broad range of host plants.

16.7 Conclusion

Chronic As exposure through drinking water or food consumption has become a major global environmental problem. Cost-effectively and environmentally sound plant-based As remediation technologies are urgently required. A number of As hyperaccumulation plants have been identified. Mycorrhizal plants display much greater tolerance to As toxicity under high As levels and exhibit enhanced As accumulation even in high As soils. These results demonstrate that mycorrhizas may offer global potential in As phytoremediation. With an appropriate combination of fungal and plant species, mycorrhizal plants with strong As tolerance and As hyperaccumulation capacity could thus be screened, particularly from naturally As-enriched sites, for As phytoextraction or phytostabilization. Biotechnological developments in the important ecological and physiological functions of mycorrhizas relevant to phytotolerance and phytoremediation will enhance the potential of mycorrhizal fungi to contribute to our efforts to curb global As contamination in a more environmentally sound, effective, practical, and sustainable manner, particularly by large-scale application of mycorrhizal inocula through on-farm production (Douds et al. 2005; Ijdo et al. 2011).

Acknowledgments Many interesting papers could not be cited due to space limitations, for which we apologize.

References

Abedin MJ, Feldmann J, Meharg AA (2002a) Uptake kinetics of arsenic species in rice plants. Plant Physiol 12:1120–1128

Abedin MJ, Cresser M, Meharg AA, Feldmann J, Cotter-Howells J (2002b) Arsenic accumulation and metabolism in rice (*Oryza sativa* L.). Environ Sci Technol 3:962–968

Adriano DC (2001) Trace elements in the terrestrial environment: biogeochemistry, bioavailability, and risks of metals, 2nd edn. Springer, New York

Ahmed FRS, Killham K, Alexander I (2006) Influences of arbuscular mycorrhizal fungus *Glomus mosseae* on growth and nutrition of lentil irrigated with arsenic contaminated water. Plant Soil 28:33–41

Al Agely A, Sylvia DM, Ma LQ (2005) Mycorrhizae increase arsenic uptake by the hyperaccumulator Chinese brake fern (*Pteris vittata* L.). J Environ Qual 3:2181–2186

Ali W, Isayenkov SV, Zhao FJ, Maathuis JM (2009) Arsenite transport in plants. Cell Mol Life Sci 6:2329–2339

Arriagada C, Aranda E, Sampedro I, Garcia-Romera I, Ocampo JA (2009) Contribution of the saprobic fungi *Trametes versicolor* and *Trichoderma harzianum* and the arbuscular mycorrhizal fungi *Glomus deserticola* and *G. claroideum* to arsenic tolerance of *Eucalyptus globulus*. Bioresour Technol 24:6250–6257

Bai JF, Lin XG, Yin R, Zhang HY, Wang JH, Chen XM, Luo YM (2008) The influence of arbuscular mycorrhizal fungi on As and P uptake by maize (*Zea mays* L.) from As-contaminated soils. Appl Soil Ecol 3:137–145

Baldwin PR, Butcher DJ (2007) Phytoremediation of arsenic by two hyperaccumulators in a hydroponic environment. Microchem J 8:297–300

Bech J, Poschenrieder C, Llugany M, Barcelo J, Tume P, Toloias FJ (1997) As and heavy metal contamination of soil and vegetation around a copper mine in Northern Peru. Sci Total Environ 20:83–91

Bhattacharjee Y (2007) A sluggish response to humanity's biggest mass poisoning. Science 31:1659–1661

Bona E, Cattaneo C, Cesaro P, Marsano F, Lingua G, Cavaletto M, Berta G (2010) Proteomic analysis of Pteris vittata fronds: two arbuscular mycorrhizal fungi differentially modulate protein expression under arsenic contamination. Proteomics 10:3811–3834

Bona E, Marsano F, Massa N, Cattaneo C, Cesaro P, Argese E, di Toppi L, Cavaletto M, Berta G (2011) Proteomic analysis as a tool for investigating arsenic stress in Pteris vittata roots colonized or not by arbuscular mycorrhizal symbiosis. J Proteomics 74:1338–1350

Brooks RR, Lee J, Reeves RD, Jaffre T (1977) Detection of nickeliferous rocks by analysis of herbarium specimens of indicator plants. J Geochem Explor 7:49–57

Brundrett MC (2009) Mycorrhizal associations and other means of nutrition of vascular plants, understanding the global diversity of host plants by resolving conflicting information and developing reliable means of diagnosis. Plant Soil 32:37–77

Chen BD, Zhu YG, Smith FA (2006) Effects of arbuscular mycorrhizal inoculation on uranium and arsenic accumulation by Chinese brake fern (*Pteris vittata* L.) from a uranium mining–impacted soil. Chemosphere 6:1464–1473

Chen BD, Xiao XY, Zhu YG, Smith FA, Xie ZM, Smith SE (2007) The arbuscular mycorrhizal fungus *Glomus mosseae* gives contradictory effects on phosphorus and arsenic acquisition by *Medicago sativa* Linn. Sci Total Environ 37:226–234

Chen XW, Li H, Chan WF, Wu C, Wu FY, Wu SC, Wong MH (2012) Arsenite transporters expression in rice (Oryza sativa L.) associated with arbuscular mycorrhizal fungi (AMF) colonization under different levels of arsenite stress. Chemosphere 89:1248–1254

Christophersen HM, Smith FA, Smith SE (2009a) No evidence for competition between arsenate and phosphate for uptake from soil by medic or barley. Environ Int 3:485–490

Christophersen HM, Smith FA, Smith SE (2009b) Arbuscular mycorrhizal colonization reduces arsenate uptake in barley via downregulation of transporters in the direct epidermal phosphate uptake pathway. New Phytol 184:962–974

Christophersen HM, Smith FA, Smith SE (2012) Unraveling the influence of arbuscular mycorrhizal colonization on arsenic tolerance in medicago: *Glomus mosseae* is more effective than *G. intraradices*, associated with lower expression of root epidermal Pi transporter genes. Front Physiol 3:91

Cozzolino V, Pigna M, Di Meo V, Caporale AG, Violante A (2010) Effects of arbuscular mycorrhizal inoculation and phosphorus supply on the growth of *Lactuca sativa* L. and arsenic and phosphorus availability in an arsenic polluted soil under non-sterile conditions. Appl Soil Ecol 45:262–268

Cullen WR, Reimer KJ (1989) Arsenic speciation in the environment. Chem Rev 8:713–764

De Koe T (1994) *Agrostis castellana* and *Agrostis delicatula* on heavy metal and arsenic enriched sites in NE Portugal. Sci Total Environ 14:103–109

Dong Y, Zhu YG, Smith FA, Wang YS, Chen BD (2008) Arbuscular mycorrhiza enhanced arsenic resistance of both white clover (*Trifolium repens* Linn.) and ryegrass (*Lolium perenne* L.) plants in an arsenic-contaminated soil. Environ Pollut 15:174–181

Douds DD, Nagahashi G, Pfeffer PE, Kayser WM, Reider C (2005) On-farm production and utilization of arbuscular mycorrhizal fungus inoculum. Can J Plant Sci 8:15–21

Garg N, Singla P (2011) Arsenic toxicity in crop plants: physiological effects and tolerance mechanisms. Environ Chem Lett 9:303–321

Garg N, Singla P (2012) The role of *Glomus mosseae* on key physiological and biochemical parameters of pea plants grown in arsenic contaminated soil. Sci Hortic 143:92–101

Gaur A, Adholeya A (2004) Prospects of arbuscular mycorrhizal fungi in phytoremediation of heavy metal contaminated soils. Curr Sci 8:528–534

Gonzalez-Chavez C, Harris PJ, Dodd J, Meharg AA (2002) Arbuscular mycorrhizal fungi confer enhanced arsenate resistance on *Holcus lanatus*. New Phytol 15:163–171

Gonzalez-Chavez MDA, Ortega-Larrocea MD, Carrillo-Gonzalez R, Lopez-Meyer M, Xoconostle-Cazares B, Gomez SK, Harrison Maria J, Figueroa-Lopez AM, Maldonado-Mendoza IE (2011) Arsenate induces the expression of fungal genes involved in As transport in arbuscular mycorrhiza. Fungal Biol 115:1197–1209

He XH, Nara K (2007) Element biofortification, can mycorrhizas potentially offer a more effective and sustainable way to curb human malnutrition? Trends Plant Sci 1:331–333

He XH, Critchley C, Ng H, Bledsoe CS (2005) Nodulated N_2-fixing *Casuarina cunninghamiana* is the sink for net N transfer from non-N_2-fixing *Eucalyptus maculata* via an ectomycorrhizal fungus *Pisolithus* sp. supplied as ammonium nitrate. New Phytol 16:897–912

He XH, Xu MG, Qiu GY, Zhou JB (2009) Use of ^{15}N stable isotope to quantify nitrogen transfer between mycorrhizal plants. J Plant Ecol 2:107–118

Hua JF, Lin XG, Yin R, Jiang Q, Shao YF (2009) Effects of arbuscular mycorrhizal fungi inoculation on arsenic accumulation by tobacco (*Nicotiana tabacum* L.). J Environ Sci (China) 21:1214–1220

Hua JF, Lin XG, Bai JF, Shao YF, Yin R, Jiang Q (2010) Effects of arbuscular mycorrhizal fungi and earthworm on nematode communities and arsenic uptake by maize in arsenic-contaminated soils. Pedosphere 20:163–173

Ijdo M, Cranenbrouck S, Declerck S (2011) Methods for large-scale production of AM fungi: past, present, and future. Mycorrhiza 21:1–16

Jankong P, Visoottiviseth P (2008) Effects of arbuscular mycorrhizal inoculation on plants growing on arsenic contaminated soil. Chemosphere 7:1092–1097

Kertulis-Tartar GM, Ma LQ, Tu C, Chirenje T (2006) Phytoremediation of an arsenic–contaminated site using *Pteris vittata* L.: a two–year study. Int J Phytoremediation 8:311–322

Khan AG (2005) Role of soil microbes in the rhizospheres of plants growing on trace metal contaminated soils in phytoremediation. J Trace Elem Med Biol 1:355–364

Kim KW, Bang S, Zhu YG, Meharg AA, Bhattacharya P (2009) Arsenic geochemistry, transport mechanism in the soil–plant system, human and animal health issues. Environ Int 3:453–454

Leung HM, Ye ZH, Wong MH (2006) Interactions of mycorrhizal fungi with *Pteris vittata* (As hyperaccumulator) in As-contaminated soils. Environ Pollut 13:1–8

Leung HM, Wu FY, Cheung KC, Ye ZH, Wong MH (2010a) The effect of arbuscular mycorrhizal fungi and phosphate amendment on arsenic uptake, accumulation and growth of Pteris vittata in As-contaminated soils. Int J Phytoremediation 12:384–403

Leung HM, Wu FY, Cheung KC, Ye ZH, Wong MH (2010b) Synergistic effects of arbuscular mycorrhizal fungi and phosphate rock on heavy metal uptake and accumulation by an arsenic hyperaccumulator. J Hazard Mater 181:497–507

Leung HM, Leung AOW, Ye ZH, Cheung KC, Yung KKL (2013) Mixed arbuscular mycorrhizal (AM) fungal application to improve growth and arsenic accumulation of *Pteris vittata* (As hyperaccumulator) grown in As-contaminated soil. Chemosphere 92:1367–1374

Li H, Ye ZH, Chan WF, Chen XW, Wu FY, Wu SC, Wong MH (2011) Can arbuscular mycorrhizal fungi improve grain yield, As uptake and tolerance of rice grown under aerobic conditions? Environ Pollut 159:2537–2545

Liu Y, Zhu YG, Chen BD, Christie P, Li XL (2005a) Yield and arsenate uptake of arbuscular mycorrhizal tomato colonized by *Glomus mosseae* BEG167 in As spiked soil under glasshouse conditions. Environ Int 3:867–873

Liu Y, Zhu YG, Chen BD, Christie P, Li XL (2005b) Influence of the arbuscular mycorrhizal fungus *Glomus mosseae* on uptake of arsenate by the As hyperaccumulator fern *Pteris vittata* L. Mycorrhiza 15:187–192

Liu Y, Christie P, Zhang JL, Li XL (2009) Growth and arsenic uptake by Chinese brake fern inoculated with an arbuscular mycorrhizal fungus. Environ Exp Bot 66:435–441

Mandal BK, Suzuki KT (2002) Arsenic round the world: a review. Talanta 5:201–235

Maurel C, Verdoucq L, Luu DT, Santoni V (2008) Plant aquaporins: membrane channels with multiple integrated functions. Annu Rev Plant Biol 5:595–624

Meding SM, Zasoski RJ (2008) Hyphal-mediated transfer of nitrate, arsenic, cesium, rubidium, and strontium between arbuscular mycorrhizal forbs and grasses from a California oak woodland. Soil Biol Biochem 4:126–134

Meharg AA, Hartley-Whitaker J (2002) Arsenic uptake and metabolism in arsenic resistant and nonresistant plant species. New Phytol 15:29–43

Meharg AA, Jardine L (2003) Arsenite transport into paddy rice (*Oryza sativa*) roots. New Phytol 15:39–44

Meharg AA, Macnair MR (1990) An altered phosphate uptake system in arsenate–tolerant *Holcus lanatus*. New Phytol 11:29–35

Mendez OM, Maier RM (2008) Phytostabilization of mine tailings in arid and semiarid environments—an emerging remediation technology. Environ Health Perspect 11:278–283

Mondal P, Majumder CB, Mohanty B (2006) Laboratory-based approaches for arsenic remediation from contaminated water: recent developments. J Hazard Mater 13:464–479

Nasim G (2005) The role of symbiotic soil fungi in controlling roadside erosion and the establishment of plant communities. Caderno de Pesquisa serie Biologia 17:119–136

Natarajan S, Stamps RH, Saha UK, Ma LQ (2008) Phytofiltration of arsenic–contaminated groundwater using *Pteris vittata* L.: effect of plant density and nitrogen and phosphorus levels. Int J Phytoremediation 10:220–235

Newman EI (1988) Mycorrhizal links between plants: their functioning and ecological significance. Adv Ecol Res 18:243–271

Nriagu JO (2002) Arsenic poisoning through the ages. In: Frankenberger JWT (ed) Environmental chemistry of arsenic. Marcel Dekker, New York, pp 1–2

Orlowska E, Godzik B, Turnau K (2012) Effect of different arbuscular mycorrhizal fungal isolates on growth and arsenic accumulation in *Plantago lanceolata* L. Environ Pollut 168:121–130

Palenzuela J, Ferrol N, Boller T, Azcón-Aguilar C, Oehl F (2008) *Otospora bareai*, a new fungal species in the Glomeromycetes from a dolomitic shrub land in Sierra de Baza National Park (Granada, Spain). Mycologia 10:296–305

Panuccio MR, Logoteta B, Beone GM, Cagnin M, Cacco G (2012) Arsenic uptake and speciation and the effects of phosphate nutrition in hydroponically grown kikuyu grass (*Pennisetum clandestinum* Hochst). Environ Sci Pollut Res 19:3046–3053

Raab A, Schat H, Meharg AA, Feldmann J (2005) Uptake, translocation and transformation of arsenate and arsenite in sunflower (*Helianthus annuus*): formation of arsenic–phytochelatin complexes during exposure to high arsenic concentrations. New Phytol 16:551–558

Raab A, Williams PN, Meharg A, Feldmann J (2007) Uptake and translocation of inorganic and methylated arsenic species by plants. Environ Chem 4:197–203

Ralph SJ (2008) Arsenic-based antineoplastic drugs and their mechanisms of action. Met Based Drugs 200:Article ID 260146

Roy S, Khasa DP, Greer CW (2007) Combining alders, frankiae, and mycorrhizae for the revegetation and remediation of contaminated ecosystems. Can J Bot 85:237–251

Schwartz MW, Schwartz MW, Hoeksema JD, Gehring CA, Johnson NC, Klironomos JN, Abbott LK, Pringle A (2006) The promise and the potential consequences of the global transport of mycorrhizal fungal inoculums. Ecol Lett 9:501–515

Shin H, Shin HS, Dewbre GR, Harrison MJ (2004) Phosphate transport in *Arabidopsis*, Pht1;1 and Pht1;4 play a major role in phosphate acquisition from both low- and high-phosphate environments. Plant J 3:629–642

Singh N, Ma LQ, Srivastava M, Rathinasabapathi B (2006) Metabolic adaptations to arsenic-induced oxidative stress in *Pteris vittata* L and *Pteris ensiformis* L. Plant Sci 17:274–282

Smith SE, Read DJ (2008) Mycorrhizal symbiosis, 3rd edn. Academic, San Diego

Smith AH, Lingas EO, Rahman M (2000) Contamination of drinking-water by arsenic in Bangladesh: a public health emergency. Bull World Health Organ 7:1093–1103

Smith SE, Christophersen HM, Pope S, Smith FA (2010) Arsenic uptake and toxicity in plants: integrating mycorrhizal influences. Plant Soil 327:1–21

Srivastava M, Ma LQ, Singh N, Singh S (2005) Antioxidant responses of hyper-accumulator and sensitive fern species to arsenic. J Exp Bot 5:1335–1342

Stone R (2008) Food safety: arsenic and paddy rice: a neglected cancer risk? Science 32:184–185

Su YH, McGrath SP, Zhu YG, Zhao FJ (2008) Highly efficient xylem transport of arsenite in the arsenic hyperaccumulator *Pteris vittata*. New Phytol 18:434–441

Tripathi RD, Srivastava S, Seema M, Singh N, Tuli R, Gupta DK, Maathui FJM (2007) Arsenic hazards: strategies for tolerance and remediation by plants. Trends Biotechnol 2:158–165

Tu C, Ma LQ (2002) Effects of arsenic concentrations and forms on arsenic uptake by the hyperaccumulator ladder brake. J Environ Qual 3:641–647

Tu C, Ma LQ, Bondada B (2002) Arsenic accumulation in the hyperaccumulator Chinese brake and its utilization potential for phytoremediation. J Environ Qual 3:1671–1675

Ultra VU, Tanaka S, Sakurai K, Iwasaki K (2007a) Effects of arbuscular mycorrhiza and phosphorus application on arsenic toxicity in sunflower (*Helianthus annuus* L.) and on the transformation of arsenic in the rhizosphere. Plant Soil 29:29–41

Ultra VUY, Tanaka S, Sakurai K, Iwasaki K (2007b) Arbuscular mycorrhizal fungus (*Glomus aggregatum*) influences biotransformation of arsenic in the rhizosphere of sunflower (*Helianthus annuus* L.). Soil Sci Plant Nutr 5:499–508

Walker C, Vestberg M, Demircik F, Stockinger H, Saito M, Sawaki H, Nishmura I, Schüßler A (2007a) Molecular phylogeny and new taxa in the *Archaeosporales* (*Glomeromycota*): *Ambispora fennica* gen. sp. nov., *Ambisporaceae* fam. nov., and emendation of *Archaeospora* and *Archaeosporaceae*. Mycol Res 11:137–153

Walker C, Vestberg M, Schüßler A (2007b) Nomenclatural clarifications in *Glomeromycota*. Mycol Res 11:253–255

Wang B, Qiu YL (2006) Phylogenetic distribution and evolution of mycorrhizas in land plants. Mycorrhiza 16:299–363

Wang ZH, Zhang JL, Christie P, Li XL (2008) Influence of inoculation with *Glomus mosseae* or *Acaulospora morrowiae* on arsenic uptake and translocation by maize. Plant Soil 31:235–244

Wenzel WW (2009) Rhizosphere processes and management in plant-assisted bioremediation (phytoremediation) of soils. Plant Soil 32:385–408

Wenzel WW, Brandstetter A, Wutte H, Lombi E, Prohaska T, Stingeder G, Adriano DC (2002) Arsenic in field–collected soil solutions and extracts of contaminated soils and its implication to soil standards. J Plant Nutr Soil Sci 16:221–228

Weyens N, van der Lelie D, Taghavi S, Newman L, Vangronsveld J (2009) Exploiting plant-microbe partnerships to improve biomass production and remediation. Trends Biotechnol 27:591–598

Williams PN, Villada A, Deacon C, Raab A, Figuerola J, Green AJ, Feldmann J, Meharg AA (2007) Greatly enhanced arsenic shoot assimilation in rice leads to elevated grain levels compared to wheat and barley. Environ Sci Technol 41:6854–6859

Wu FY, Ye ZH, Wu SC, Wong MH (2007) Metal accumulation and arbuscular mycorrhizal status in metallicolous and nonmetallicolous populations of *Pteris vittata* L. and *Sedum alfredii* Hance. Planta 22:1363–1378

Wu FY, Ye ZH, Wong MH (2009) Intraspecific differences of arbuscular mycorrhizal fungi in their impacts on arsenic accumulation by *Pteris vittata* L. Chemosphere 76:1258–1264

Xia YS, Chen BD, Christie P, Smith FA, Wang YS, Li XL (2007) Arsenic uptake by arbuscular mycorrhizal maize (*Zea mays* L.) grown in an arsenic-contaminated soil with added phosphorus. J Environ Sci (China) 1:1245–1251

Xie QE, Yan XL, Liao XY, Li X (2009) The arsenic hyperaccumulator fern *Pteris vittata* L. Environ Sci Technol 43:8488–8495

Xu XY, McGrath SP, Zhao FJ (2007) Rapid reduction of arsenate in the medium mediated by plant roots. New Phytol 17:590–599

Xu PL, Christie P, Liu Y, Zhang JL, Li XL (2008) The arbuscular mycorrhizal fungus *Glomus mosseae* can enhance arsenic tolerance in *Medicago truncatula* by increasing plant phosphorus status and restricting arsenate uptake. Environ Pollut 15:215–220

Yang Q, Wang G, Tu S, Liao X, Yan X (2012) Effectiveness of applying arsenate reducing bacteria to enhance arsenic removal from polluted soils by *Pteris vittata* L. Int J Phytoremediation 14:89–99

Yu Y, Zhang S, Huang H, Luo L, Wen B (2009) Arsenic accumulation and speciation in maize as affected by inoculation with arbuscular mycorrhizal fungus *Glomus mosseae*. J Agric Food Chem 5:3695–3701

Yu Y, Zhang SZ, Hunag HL, Wu NY (2010) Uptake of arsenic by maize inoculated with three different arbuscular mycorrhizal fungi. Commun Soil Sci Plant Anal 41:735–743

Zhao FJ, Ma JF, Meharg AA, McGrath SP (2009) Arsenic uptake and metabolism in plants. New Phytol 18:777–794

Zhu YG, Rosen BP (2009) Perspectives for genetic engineering for the phytoremediation of arsenic–contaminated environments: from imagination to reality? Curr Opin Biotechnol 2:220–224

Chapter 17
Arbuscular Mycorrhizal Colonization and Agricultural Land Use History

Irnanda A.F. Djuuna

17.1 Introduction

Most plant species form symbiotic associations with mycorrhizal fungi (Newman and Reddell 1987) and arbuscular mycorrhizal (AM) fungi are widespread in natural and agricultural ecosystems (Brundrett 1991). AM fungi can contribute to plant growth by enhancing water and nutrient uptake, especially phosphorus (P) (Ortas 1996; Jacobson 1997; Watts-Williams et al. 2014). Although AM fungi colonize roots of most plant species (Harley and Harley 1987; Smith and Read 2008), plants differ in their growth response to mycorrhizal colonization. Furthermore, plant species can influence the population of AM fungi (Crush 1978; Hiiesalu et al. 2014). AM fungi may also contribute to soil fertility by enhancing soil structure and protecting crops from root pathogens (Douds and Johnson 2003; Sharma et al. 2013). The soil environment, particularly those factors that control mineral fertility, strongly influences mycorrhizal function (Abbott and Robson 1982; Sikes et al. 2014).

In agricultural fields, the status of AM fungi is influenced by soil conditions and management practices (Jansa et al. 2014). The diversity of AM fungi species can be lower in agricultural systems than in nearby natural fields (Helgason et al. 1998; Sieverding 1991) or forested areas (Boerner et al. 1996). However, factors such as crop and rotation history can also influence the abundance of AM fungi in agricultural soil (Douds and Johnson 2003; Helgason et al. 2014).

I.A.F. Djuuna (✉)
Faculty of Agriculture and Agriculture Technology, University of Papua, Gunung Salju
St. Kampus UNIPA Amban, Manokwari, Papua Barat 98314, Indonesia
e-mail: irnanda_afd@yahoo.com

© Springer-Verlag Berlin Heidelberg 2014
Z.M. Solaiman et al. (eds.), *Mycorrhizal Fungi: Use in Sustainable Agriculture and Land Restoration*, Soil Biology 41, DOI 10.1007/978-3-662-45370-4_17

17.2   AM Fungi in Agricultural Systems

The impact of farming practices on AM fungi has been studied extensively (Abbott and Robson 1994; Gavito and Miller 1998; Thompson 1994; Barber et al. 2013). Agricultural practices such as tillage, crop rotation, and use of chemical pesticides and fertilizers (Helgason et al. 2014; Kurle and Pfleger 1994; Ortas et al. 2013) as well as clean fallowing, topsoil removal, fires, and waterlogging (Thompson 1994) have all been shown to influence the abundance of AM fungi, often reducing the level of colonization. Rotation and fertilizer practices are major factors that affect the abundance of AM fungi, especially in Mediterranean agriculture (e.g., Abbott et al. 1995). Difference may not be so marked in soils with high P (e.g., Franke-Snyder et al. 2001). However, geography and other landscape characteristics may override effects of land management on AM fungal communities (Jansa et al. 2014).

Crop rotation is a very important factor in managing nutrient supply. In general, preceding crops can affect the growth and yield of subsequent crops (Karlen et al. 1994; Brito et al. 2012). The choice of crop and crop rotation history can influence the community of AM fungi in agricultural soil because plants differ in their susceptibility to colonization. A study of 27 species of plants demonstrated that AM fungi were present in most legumes whereas Poaceae were poor hosts (Eschen et al. 2013). However, the length of root colonized by AM fungi can be higher in some grasses than in non-grasses grown in the same soil. The Leguminosae can be superior in terms of concentration of fungal hyphae per unit weight or length of root, but in terms of total length of mycorrhizal root available to exploit a given soil volume and in terms of the likely residual population of mycorrhizal propagules, the Gramineae would be superior (Thompson and Wildermuth 1989). Plants with less dense roots (fewer and coarse roots) can have high mycorrhizal colonization (Hetrick 1991) and plants with poorly developed root hairs can be highly dependent on mycorrhizas (Baylis 1970). The growth of highly mycorrhizal-dependent crops like linseed can leave a high level of mycorrhizal inoculum for subsequent crops (Thompson 1994). In addition, other factors that need to be considered when managing crop rotations to obtain maximum benefits from AM fungi include (1) whether AM fungal inoculum in the soil is low after practices such as clean fallowing, (2) whether a nonhost crop has been grown, (3) whether rice has been grown under waterlogged conditions, or (4) whether crops with low mycorrhizal dependency have been grown. If a crop with high dependency is grown for other reasons (e.g., disease control), then a high P fertilizer rate and possibly Zn fertilizer might need to be used to compensate for lower levels of mycorrhizal development if management practices have reduced their infectivity (Thompson 1994).

For soils with naturally high P fertility and high use of P fertilizer, colonization by AM fungi would not be expected to make a contribution to plant growth (Galvez et al. 2001; Kahiluoto et al. 2001). However, this is not always the case as it may depend on the soil type. P applications to field soils may be accompanied by a decrease in the proportion of root length colonized by AM fungi (Abbott and Robson 1984; Clarke and Mosse 1981; Liu et al. 2000) but this is not always the case (Gryndler et al. 1990). The application of phosphate fertilizer to soil can delay in mycorrhiza formation as well as a decrease in the proportion of the root system colonized (Solaiman and Abbott 2008; De Miranda et al. 1989). In contrast, the addition of P fertilizer to soil with extremely low available phosphorus can increase the colonization, possibly through a direct effect on AM fungi (Bolan et al. 1984). Some studies have reported that farms which use alternative (e.g., low input) practices have higher levels of AM colonization than nearby conventional farms because of a lower available soil P associated with reduced applications of soluble P fertilizer (Mäder et al. 2000; Ryan 1999; Ryan and Ash 1999; Kahiluoto et al. 2012).

Nitrogen fertilizer may affect the infectivity of AM fungi but this is less marked than effects of P (Hodge and Storer 2014). Application of high doses of nitrogen fertilizer can reduce colonization by AM fungi (Hayman 1975; Johnson et al. 2003, 2010). Application of ammonium to soil prevented colonization by indigenous AM fungi and nitrate application resulted in a low (6 %) level of root colonization (Ortas and Rowell 2004). AM fungi can also be involved in the decomposition of complex organic material in soil and increase nitrogen capture by plants (Hodge et al. 2001).

Tillage practices can alter AM fungal populations and species composition, reduce root colonization and P uptake (Kurle and Pfleger 1994; McGonigle and Miller 2000; Brito et al. 2012), and disrupt the hyphal network (Jasper et al. 1989; Evans and Miller 1990). The physical disruption of fungal mycelia may change physicochemical properties and influence soil aggregation (Duchicela et al. 2013). Excessive secondary tillage and traffic increased soil bulk density and decreased root growth, mycorrhizal colonization, and top growth of *Phaseolus vulgaris* (Mulligan et al. 1985). On the other hand, reduced tillage intensity can favor higher colonization by AM fungi (Yocum et al. 1985; Mulligan et al. 1985; Brito et al. 2012). Soils in low-input agricultural systems can have higher populations and more propagules of AM fungi than soils under conventional management (Douds et al. 1993, 1995; Galvez et al. 1995; Kahiluoto et al. 2012). An investigation of a 7-year crop rotation and tillage scheme practice showed root length colonized by AM fungi was up to 60 % higher in plants grown in soils from low-input farming systems than in those grown in conventionally fertilized soils (Mäder et al. 2000). Similarly, AM fungal hyphal density was greater in no-till than in reduced tillage systems and lowest in a conventional tillage system (Kabir et al. 1997).

Fallowing land for an extended period without a crop is common practice in some agricultural systems. However, long fallow periods without plant cover may be detrimental to contributions by AM fungi (Douds and Johnson 2003). In some farming systems, weeds are allowed to grow and fallows are grazed by livestock. In other dry-land agricultural systems, fallows are used to accumulate soil water and nitrate and so are kept weed-free. However, longer fallows can result in reduced numbers of spores of AM fungi and lower levels of root colonization (Thompson 1991). Clean fallowing can reduce inoculum levels and colonization by AM fungi in the following crop (Black and Tinker 1979; Thompson 1987). The reduction in abundance of AM fungi in soil during periods of fallow can be substantial, and Harinikumar and Bagyaraj (1988) reported a reduction in AM fungi colonization by 40 % associated with a long fallow period.

The application of pesticides to agricultural soils throughout the production cycle may have a range of effects on AM fungi. Some pesticides may be toxic to AM fungi (Abd-Alla et al. 2000; Jalali and Sharma 1993). Methyl bromide can kill AM propagules deep in the soil profile because it is denser than air (Menge 1982). The use of herbicides can have indirect effects on AM fungi by changing the relative abundance of plant species associated with the length of roots of species that differ in mycorrhizal dependency in the soil.

Grazing livestock can influence AM fungi through influences on root growth, changes in soil structure, and removal and return of nutrients (Harrier and Watson 1997; Davinic et al. 2013). Moderate and intense grazing resulted in increased root colonization and changes in AM fungal species composition of tall grass prairie (Eom et al. 2001). However, grazing can alter root biomass and structure, especially when compounded with other management practices such as N application (Yan et al. 2013) which can further influence communities of AM fungi. However, studies of the effect of grazing (e.g., by domestic animals) on AM fungi in agricultural fields have been inconsistent. In some situations, little effect of grazing on AM fungi has been observed (Torres et al. 2011), but grazing has been shown to have a negative effect on AM fungi in other situations (Saravesi et al. 2013). Furthermore, where domestic animal grazing influences soil structure, there are likely to be associated changes in the abundance and diversity of AM fungi in soil.

17.3   A Conceptual Model of AM Fungi in Soil

A conceptual model of factors influencing the status of AM fungi in agricultural soil is presented in Fig. 17.1. The distribution and abundance of AM fungi in soil can be influenced by a range of factors (e.g., climate, soil properties, management practices, and socioeconomic factors related to the farming enterprise). The dominance of particular influences would be site specific and include soil and geography (Jansa et al. 2014).

Field surveys have shown correlations between the distribution of AM fungi and soil pH. The distribution of some AM fungi can be restricted in either acid or alkaline conditions, while others have been found in both types of soil (Abbott and Robson 1991). For example, in a range of agricultural soils in southwestern Australia, *Acaulospora laevis* spores occurred only in more acid soils (pH in 1/5 0.01 M $CaCl_2$ less than 5.3), and *Glomus monosporum* spores occurred only in soil with pH greater than 4.85 (Abbott and Robson 1977). There was no correlation between the abundance of different spore types and soil pH. The level of root colonization was only slightly affected by pH over a range of soils at pH 4.5–7.5

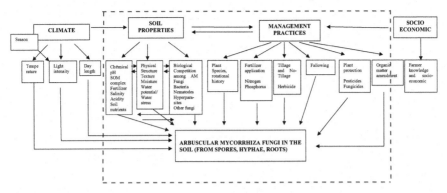

Fig. 17.1 A theoretical model of the factors which could affect the status of AM fungi in agricultural soil

(Wang et al. 1993) or at pH 4.7–7.7 (Porter et al. 1987) but different AM fungi were present at the pH extremes.

Increases in soil salinity in agricultural soils may influence the growth and activity of mycorrhiza fungi. Under saline conditions, AM fungi may have the ability to protect plants from salt stress (Rosendahl and Rosendahl 1991). Salinity can reduce the growth of AM fungi and root colonization in various ecosystems (Juniper and Abbott 1993; McMillen et al. 1998; Carvalho et al. 2003), but this is not always the case (Ruiz-Lozano et al. 1996).

Colonization by AM fungi in pot experiments is commonly reduced by low temperatures (e.g., Baon et al. 1994; Ruotsalainen and Kytöviita 2004) and increased by higher temperatures (e.g., Domisch et al. 2002) when measured as proportion of root length colonized. In the latter case, length of root colonized increased more than did the length of new roots and similar effects could influence the dynamics of mycorrhiza formation under field conditions.

17.4 Conclusion

The infectivity of AM fungi can be influenced by soil factors (chemical, physical, and biological) and agricultural practices, including plant components of agricultural systems. These factors vary across landscapes and geostatistical methods are available for quantifying them. Spatial and temporal variability in infectivity of AM fungi is expected to vary among sites and for different environmental conditions, depending on soil type and soil management. Some soil properties and agricultural practices can enhance the formation of mycorrhizas, but others can apparently be detrimental. Furthermore, as different methods have been used to measure infectivity of AM fungi, this should be considered when interpreting the effects of soil, plant, and environmental factors in these fungi. The conceptual model outlined here for the development of AM could be used to predict the status of AM fungi in agricultural field even though this is not quantitative.

References

Abbott LK, Robson AD (1977) Distribution and abundance of vesicular-arbuscular endophytes in some Western Australian soils. Aust J Bot 25:515–522

Abbott LK, Robson AD (1982) Infectivity of vesicular arbuscular mycorrhizal fungi in agricultural soils. Aust J Agric Res 33:1049–1059

Abbott LK, Robson AD (1984) The effect of mycorrhiza on plant growth. In: Powell CD, Bagyaraj DJ (eds) VA mycorrhiza. CRC, Boca Raton, pp 113–130

Abbott LK, Robson AD (1991) Field management of VA mycorrhizal fungi. In: Kelster DL, Cregan PB (eds) The rhizosphere and plant growth. Kluwer, Norwell, pp 355–362

Abbott LK, Robson AD (1994) The impact of agricultural practices on mycorrhizal fungi. In: Pakhurst CE, Doube BM, Gupta VVSR, Grace PR (eds) Soil biota management in sustainable farming systems. CSIRO, East Melbourne, pp 88–95

Abbott LK, Robson AD, Scheltema MA (1995) Managing soils to enhance mycorrhizal benefits in Mediterranean agriculture. Crit Rev Biotechnol 15(3/4):213–228

Abd-Alla MH, Omar SA, Karanxha S (2000) The impact of pesticides on arbuscular mycorrhizal and nitrogen-fixing symbioses in legumes. Appl Soil Ecol 14:191–200

Baon JB, Smith SE, Altson AM (1994) Phosphorus uptake and growth of barley as affected by soil temperature and mycorrhizal infection. J Plant Nutr 17:479–492

Barber NA, Kiers ET, Theis N, Hazzard RV, Adler LS (2013) Linking agricultural practices, mycorrhizal fungi, and traits mediating plant-insect interactions. Ecol Appl 23:1519–1530

Baylis GTS (1970) Root hairs and phycomycetous mycorrhizas in phosphorus-deficient soil. Plant Soil 33:713–716

Black RLB, Tinker PB (1979) The development of endomycorrhizal root systems. II. Effects of agronomic factors and soil conditions on the development of vesicular-arbuscular mycorrhizal infection in barley and on the endophyte spore density. New Phytol 83:401–413

Boerner REJ, DeMars BG, Leight PN (1996) Spatial patterns of mycorrhizal infectiveness of soils along a successional chronosequence. Mycorrhiza 6:79–90

Bolan NS, Robson AD, Barrow NJ (1984) Increasing phosphorus supply can increase the infection of plant-roots by vesicular arbuscular mycorrhizal fungi. Soil Biol Biochem 16:419–420

Brito I, Goss MJ, de Carvalho M (2012) Effect of tillage and crop on arbuscular mycorrhiza colonization of winter wheat and triticale under Mediterranean conditions. Soil Use Manag 28:202–208

Brundrett M (1991) Mycorrhizas in natural ecosystems. Adv Ecol Res 21:171–313

Carvalho LM, Correia PM, Ryel RJ, Martin-Loucao MA (2003) Spatial variability of arbuscular mycorrhizal fungal spores in two natural plant communities. Plant Soil 251:227–236

Clarke C, Mosse B (1981) Plant growth responses to vesicular-arbuscular mycorrhiza. XII. Field inoculation response of barley at two soil P levels. New Phytol 87:695–705

Crush JR (1978) Changes in effectiveness of soil endomycorrhizal fungal populations during pasture development. N Z J Agric Res 21:683–685

Davinic M, Moore-Kucera J, Acosta-Martinez V, Zak J, Allen V (2013) Soil fungal distribution and functionality as affected by grazing and vegetation components of integrated crop-livestock agroecosystems. Appl Soil Ecol 66:61–70

De Miranda JCC, Harris PJ, Wild A (1989) Effects of soil and plant phosphorus concentrations on vesicular-arbuscular mycorrhiza in sorghum plants. New Phytol 112:405–410

Domisch T, Finer L, Lehto T, Smolander A (2002) Effect of soil temperature on nutrient allocation and mycorrhizas in Scots pine seedlings. Plant Soil 239:173–185

Douds DDJ, Johnson NC (2003) Contributions of arbuscular mycorrhizas to soil biological fertility. In: Abbott LK, Murphy DV (eds) Soil biological fertility-a key to sustainable land use in agriculture. Kluwer Academic, The Netherlands, pp 129–162

Douds DD, Janke RR, Peters SE (1993) VAM fungus spore populations and colonization of roots of maize and soybean under conventional and low-input sustainable agriculture. Agric Ecosyst Environ 43:325–335

Douds DD, Galvez L, Janke R, Wagoner P (1995) Effect of tillage and farming systems upon populations and distribution of vesicular-arbuscular mycorrhiza fungi. Agric Ecosyst Environ 52:111–118

Duchicela J, Sullivan TS, Bontti E, Bever JD (2013) Soil aggregate stability increase is strongly related to fungal community succession along an abandoned agricultural field chronosequence in the Bolivian Altiplano. J Appl Ecol 50:1266–1273

Eom AH, Wilson GWT, Harnett DC (2001) Effects of ungulate grazers on arbuscular mycorrhizal symbiosis and fungal community structure in tall grass prairie. Mycologia 93:233–242

Eschen R, Muller-Scharer H, Schaffner U (2013) Plant interspecific differences in arbuscular mycorrhizal colonization as a result of soil carbon addition. Mycorrhiza 23:61–70

Evans DG, Miller MH (1990) The role of the external mycelial network in the effect of soil disturbance upon vesicular-arbuscular mycorrhizal colonization of maize. New Phytol 114:65–72

Franke-Snyder M, Douds DD, Galvez L, Philips JG, Wagoner P, Drinkwater L, Morton JB (2001) Diversity of communities of arbuscular mycorrhizal (AM) fungi present in conventional versus low-input agricultural sites in eastern Pennsylvania, USA. Appl Soil Ecol 16:35–48

Galvez L, Douds DD, Wagoner P, Longnecker LR, Drinkwater LE, Janke RR (1995) An over-wintering cover crop increases inoculum of VAM fungi in agricultural soil. Am J Altern Agric 10:152–156

Galvez L, Douds DD, Drinkwater LE, Wagoner P (2001) Effect of tillage and farming system upon VAM fungus populations and mycorrhizas and nutrient uptake of maize. Plant Soil 228:299–308

Gavito ME, Miller MH (1998) Changes in mycorrhiza development in maize induced by crop management practices. Plant Soil 198:185–192

Gryndler M, Lestina J, Moravec V, Prikryl Z, Lipavsky J (1990) Colonization of maize roots by VAM fungi under conditions of long-term fertilization of varying intensity. Agric Ecosyst Environ 29:183–186

Harinikumar KM, Bagyaraj DJ (1988) Effect of crop rotation on native vesicular-arbuscular mycorrhizal propagules in soil. Plant Soil 110:77–80

Harley JL, Harley EL (1987) A checklist of mycorrhiza in the British Flora—Addenda, Errata and Index. New Phytol 107:741–749

Harrier LA, Watson CA (1997) The role of Arbuscular mycorrhizal fungi in sustainable cropping systems. In: Sparks DL (ed) Advances in agronomy. Academic, New York, pp 185–225

Hayman DS (1975) The occurrence of mycorrhiza in crops as affected by soil fertility. In: Sanders FE, Mosse B, Tinker PB (eds) Endomycorrhizas. Academic, London, pp 495–510

Helgason T, Daniel TJ, Husband R, Fitter AH, Young JPW (1998) Ploughing up the wood-wide web. Nature 394:431

Helgason T, Feng HY, Sherlock DJ, Young JPW, Fitter AH (2014) Arbuscular mycorrhizal communities associated with maples (*Acer* spp.) in a common garden are influenced by season and host plant. Botanique 92(4). doi:10.1139/cjb-2013-0263

Hetrick BAD (1991) Mycorrhizas and root architecture. Experientia 47:355–362

Hiiesalu I, Paertel M, Davison J, Gerhold P, Metsis M, Moora M, Oepik M, Vasar M, Zobel M, Wilson SD (2014) Species richness of arbuscular mycorrhizal fungi: associations with grass-land plant richness and biomass. New Phytol 203:233–244

Hodge A, Storer K (2014) Arbuscular mycorrhiza and nitrogen: implications for individual plants through to ecosystems. Plant Soil. doi:10.1007/s11104-014-2162-1

Hodge A, Campbell CD, Fitter AH (2001) An arbuscular mycorrhizal fungus accelerates decomposition and acquires nitrogen directly from organic material. Nature 413:297–299

Jacobson KM (1997) Moisture and substrate stability determine VA-mycorrhizal fungal community distribution and structure in an arid grassland. J Arid Environ 35:59–75

Jalali BL, Sharma OP (1993) Biocides and non-target microorganisms: an environmental assessment. Indian J Microbiol 33:83–92

Jansa J, Erb A, Obereholzer HR, Smilauer P, Egli S (2014) Soil and geography are more important determinants of indigenous arbuscular mycorrhizal communities than management practices in Swiss agricultural soils. Mol Ecol 23:2118–2135

Jasper DA, Robson AD, Abbott LK (1989) Soil disturbance reduces the infectivity of external hyphae of vesicular arbuscular mycorrhizal fungi. New Phytol 112:93–99

Johnson NC, Rowland DL, Corkidi L, Egerton-Warburton LM, Allen EB (2003) Nitrogen enrichment alters mycorrhizal allocation at five mesic to semiarid grasslands. Ecology 84:1895–1908

Johnson NC, Wilson GWT, Bowker MA, Wilson JA, Miller RM (2010) Resource limitation is a driver of local adaptation in mycorrhizal symbioses. Proc Natl Acad Sci U S A 107:2093–2098

Juniper S, Abbott LK (1993) The effect of salinity on spore germination and hyphal extension of some VA mycorrhizal fungi. In: Abstracts of the third European symposium on mycorrhizas, 19–23 August 1991, Sheffield

Kabir Z, O'Halloran IP, Hamel C (1997) Seasonal changes of arbuscular mycorrhizal fungi as affected by tillage practices and fertilization: hyphal density and mycorrhiza root colonization. Plant Soil 192:285–293

Kahiluoto H, Ketoja E, Vestberg M, Saarela I (2001) Promotion of AM utilization through reduced P fertilization 2. Field studies. Plant Soil 231:65–79

Kahiluoto H, Ketoja E, Vestberg M (2012) Plant-available P supply is not the main factor determining the benefit from arbuscular mycorrhiza to crop P nutrition and growth in contrasting cropping systems. Plant Soil 350:85–98

Karlen DL, Wollenhaupt NC, Erbach DC, Berry EC, Swan JB, Eash NS, Jordahl JL (1994) Long-term tillage effects on soil quality. Soil Tillage Res 32:313–327

Kurle JE, Pfleger FL (1994) Arbuscular mycorrhizal fungus spore populations respond to conversions between low-input and conventional management practices in a corn-soybean rotation. Agron J 86:467–475

Liu A, Hamel C, Hamilton RI, Smith DL (2000) Mycorrhizae formation and nutrient uptake of new corn (*Zea mays* L.) hybrids with extreme canopy and leaf architecture as influenced by soil N and P levels. Plant Soil 221:157–166

Mäder P, Edenhofer S, Boller T, Wiemken A, Niggli U (2000) Arbuscular mycorrhizae in a long-term field trial comparing low-input (organic, biological) and high input (conventional) farming systems in a crop rotation. Biol Fertil Soils 31:150–156

McGonigle TP, Miller MH (2000) The inconsistent effect of soil disturbance on colonization of roots by arbuscular mycorrhizal fungi: a test of the inoculum density hypothesis. Appl Soil Ecol 14:147–153

McMillen BG, Juniper S, Abbott LK (1998) Inhibition of hyphal growth of a vesicular-arbuscular mycorrhizal containing sodium chloride of infection from fungus in soil limits the spread spores. Soil Biol Biochem 30:1639–1646

Menge JA (1982) Effect of soil fumigants and fungicides on vesicular arbuscular mycorrhizal fungi. Phytopathology 72:1125–1132

Mulligan MF, Smucker AJM, Safir GF (1985) Tillage modifications of dry edible bean root colonization by VAM fungi. Agron J 77:140–144

Newman EI, Reddell P (1987) The distribution of mycorrhizas among families of vascular plants. New Phytol 106:745–751

Ortas I (1996) The influence of use of different rates of mycorrhizal inoculum on root infection, plant growth, and phosphorus uptake. Commun Soil Sci Plant Anal 27:2935–2946

Ortas I, Rowell DL (2004) Effect of ammonium and nitrate on indigenous mycorrhizal infection, rhizosphere pH change and phosphorus uptake by sorghum. Commun Soil Sci Plant Anal 35:1923–1944

Ortas I, Akpinar C, Lal R (2013) Long-term impacts of organic and inorganic fertilizers on carbon sequestration in aggregates of an Entisol in Mediterranean Turkey. Soil Sci 178:12–23

Porter WM, Robson AD, Abbott LK (1987) Field survey of the distribution of vesicular-arbuscular mycorrhizal fungi in relation to soil pH. J Appl Ecol 24:659–662

Rosendahl CN, Rosendahl S (1991) Influence of vesicular-arbuscular mycorrhizal fungi (*Glomus* spp.) on the response of cucumber (*Cucumis sativus* L.) to salt stress. Environ Exp Bot 31:313–318

Ruiz-Lozano JM, Azcon R, Gomez M (1996) Alleviation of salt stress by arbuscular-mycorrhizal *Glomus* species in *Lactuca sativa* plants. Physiol Plant 98:767–772

Ruotsalainen AL, Kytöviita MM (2004) Mycorrhiza does not alter low temperature impact on Gnaphalium norvegicum. Oecologia 140:226–233

Ryan MH (1999) Is an enhanced soil biological community, relative to conventional neighbors, a consistent feature of alternative (organic and biodynamic) agricultural systems? Biol Agric Hortic 17:131–144

Ryan MH, Ash JE (1999) Effects of phosphorus and nitrogen on growth of pasture plants and VAM fungi in SE Australian soils with contrasting fertilizer histories (conventional and biodynamic). Agric Ecosyst Environ 73:51–62

Saravesi K, Ruotsalainen AL, Cahill JF (2013) Contrasting impacts of defoliation on root colonization by arbuscular mycorrhizal and dark septate endophytic fungi of Medicago sativa. Mycorrhiza 24:239–245

Sharma RC, Sarker S, Das D, Banik P (2013) Impact assessment of arbuscular mycorrhiza Azospirillum and chemical fertilizer application on soil health and ecology. Commun Soil Sci Plant Anal 44:1116–1126

Sieverding E (1991) Vesicular-Arbuscular mycorrhizae management in tropical agrosystems. Deutsche Gesellschaft fur Technische Zusammenarbeit (GTZ) GmbH, Eschboran, p 371

Sikes BA, Maherali H, Klironomos JN (2014) Mycorrhizal fungal growth responds to soil characteristics, but not host plant identity, during a primary lacustrine dune succession. Mycorrhiza 24:219–226

Smith SE, Read DJ (2008) Mycorrhizal symbiosis. Academic, London, p 800

Solaiman ZM, Abbott LK (2008) Influence of arbuscular mycorrhizal fungi, inoculum level and phosphorus placement on growth and phosphorus uptake of *Phyllanthus calycinus* in jarrah forest soil. Biol Fertil Soils 44:815–821

Thompson JP (1987) Decline of vesicular-arbuscular mycorrhizas in long fallow disorder of field crops and its expression in phosphorus deficiency in sunflower. Aust J Agric Res 38:847–867

Thompson JP (1991) Improving the mycorrhizal condition of the soil through cultural practices and effect on growth and phosphorus uptake by plants. In: Johansen C, Lee KK, Sahrawat KL (eds) Phosphorus nutrition of grain legumes in the semi-arid tropics. International Crops Research Institute for the Semi-Arid Tropics (ICRISAT), Patancheru, pp 117–137

Thompson JP (1994) What is the potential for management of mycorrhizas in agriculture? In: Robson AD, Abbott LK, Malajcuk N (eds) Management of mycorrhiza in agriculture, horti-culture and forestry. Kluwer Academic, The Netherlands, pp 191–200

Thompson JP, Wildermuth GB (1989) Colonization of crop and pasture species with vesicular arbuscular mycorrhizal fungi and a negative correlation with root infection by Bipolaris-Sorokiniana. Can J Bot 67:687–693

Torres YA, Busso C, Montenegro O, Ithurrart L, Giorgetti H, Rodríguez G, Bentivegna D, Brevedan R, Fernández O, de la Merced MM (2011) Defoliation effects on the arbuscular mycorrhizas of ten perennial grass genotypes in arid Patagonia, Argentina. Appl Soil Ecol 49:208–214

Wang GM, Stribley DP, Tinker PB, Walker C (1993) Effects of pH on arbuscular mycorrhiza. 1. Field observations on the long-term liming experiments at Rothamsted and Woburn. New Phytol 124:465–472

Watts-Williams SJ, Turney TW, Patti AF, Cavagnaro TR (2014) Uptake of zinc and phosphorus by plants is affected by zinc fertiliser material and arbuscular mycorrhizas. Plant Soil 376:165–175

Yan L, Zhou G, Zhang F (2013) Effects of different grazing intensities on grassland production in China: a meta-analysis. PLoS One 8:e81466. doi:10.1371/journal.pone.0081466

Yocum DH, Larsons HJ, Boosalis MG (1985) The effects of tillage treatments and a fallow season on VA mycorrhiza of winter wheat. In: Molina R (ed) Proceedings of the sixth North American conference on mycorrhizae. Forest Research Laboratory, Corvallis, p 297

Chapter 18
Contribution of Arbuscular Mycorrhizal Fungi to Soil Carbon Sequestration

Zakaria M. Solaiman

18.1 Introduction

An arbuscular mycorrhiza (AM) is a mutually beneficial association between species in the fungal phylum Glomeromycota and higher plants roots. The symbiosis is thought to have contributed to plant invasion of dry land ca 450 Ma ago and the vast majority of terrestrial plants currently form this association (Smith and Read 2008). AM fungi perform various ecological functions in exchange for host photosynthetic carbon (C) that almost always contribute to the fitness of hosts from an individual to community level (Willis et al. 2013). Soil contains more C than the atmosphere and vegetation combined (Averill et al. 2014). Understanding the mechanisms controlling the accumulation and stability of soil C is critical to predicting the Earth's future climate change (Averill et al. 2014).

AM symbioses can contribute to C fluxes between the plants and the atmosphere through different pathways (Fellbaum et al. 2012; Zhu and Miller 2003). A commonly known pathway by which AM fungi sequester C in soil is the transfer of photosynthates from the host plants to the AM fungal intraradical hyphae and subsequently to extraradical hyphae before release to the soil matrix (Bago et al. 2002, 2003; Leake et al. 2004; Parniske 2008; Solaiman and Saito 1997). Although the life span of extraradical hyphae attached to the plant roots is difficult to measure, it is believed to be short. The overall contribution of AM fungi to soil C sequestration may be dependent on the volume of hyphal biomass produced, the turnover time of accumulated hyphal biomass and the role played by these fungi in the stabilisation of soil aggregate formation (Zhu and Miller 2003). The turnover of hyphal cell walls, cytoplasm and extracellular polysaccharides represents a

Z.M. Solaiman (✉)
Soil Biology and Molecular Ecology Group, School of Earth and Environment (M087), UWA
Institute of Agriculture, Faculty of Science, The University of Western Australia, Crawley,
Australia
e-mail: zakaria.solaiman@uwa.edu.au

© Springer-Verlag Berlin Heidelberg 2014
Z.M. Solaiman et al. (eds.), *Mycorrhizal Fungi: Use in Sustainable Agriculture and Land Restoration*, Soil Biology 41, DOI 10.1007/978-3-662-45370-4_18

relatively labile organic C pool in soils. For example, the Glomeromycota fungi found in grassland soils represent a significant proportion of the fungal biomass pool, and it has been reported that 20–30 % of microbial biomass C come from AM fungi (Miller et al. 1995; Olsson 1999; Leake et al. 2004; Zhu and Miller 2003). The extraradical hyphae in prairie soil have been assessed to be as high as 28 m/cm^3 soil with an annual hyphal turnover of 26 % (Miller et al. 1995; Miller and Kling 2000). However, a higher hyphal turnover rate for AM fungi has been estimated to be 5–6 days (Staddon et al. 2003). The discrepancy between these studies is most likely associated with differences in sampling and methods of measurement. The former study used a topmost and uneven approach to quantify both a short-lived exploratory hyphae and longer-lived main hyphae, whereas the latter study quantified turnover of exploratory hyphae using ^{14}C isotope as a tracer.

A great proportion of C transferred from plants to AM fungi is incorporated into extraradical hyphal biomass. Fungal hyphae consist of recalcitrant compounds that contribute to a slower turnover of soil organic C (Olsson and Johnson 2005). The cell wall of extraradical hyphae is composed primarily of chitin, a carbohydrate that is recalcitrant to decomposition. Therefore, the rapid turnover of live extraradical hyphae could cause hyphal residues to accumulate within the soil matrix (Staddon et al. 2003). Limited information is available so far on the residence time of chitinous cell wall residues in the soil matrix, although some studies show a residence time of 49 ± 19 years for protein/amino acid/chitin-derived pyrolysis products (Gleixner et al. 2002). The typical dry weight of AM hyphae in a grassland soil has been projected to be between 0.03 and 0.5 mg/g and can characterise a large proportion of soil microbial biomass (Miller et al. 1995; Olsson 1999). At a soil depth of 30 cm with bulk density 1.2 g/cm^3 and 50 % C content of dry hyphae, the amount of soil organic C derived directly from AM fungi ranges from 54 to 900 kg/ha (Zhu and Miller 2003).

Studies of prairies and their restoration provide insights into mechanisms controlling the sequestration of C in soils and the generation of stable soil aggregate structure. They reveal the importance of plant traits in association with AM fungi in the physical protection mechanisms that allow for accumulation of detrital (granular) materials into longer-turnover soil carbon pools. For example, the stable soil aggregate structure can develop under restored prairie vegetation even though soils had been cultivated continuously for nearly 150 years (Miller and Jastrow 1992, 2000; Jastrow et al. 1998).

The soil organic C pool is an important component of terrestrial ecosystems and is a crucial regulator of C fluxes between the biosphere and the atmosphere. Mechanisms influencing soil organic carbon (SOC) storage depend mainly on net primary production and the distribution of photosynthates between above- and below-ground structures. Although primary production is a major determinant in the sequestration of C in soils, it is the size and activity of the microbial biomass of the soil that regulate C accumulation via mineralisation and immobilisation of plant and microbially derived residues in the soil. The exact amount of sequestration appears to depend on land management practices, soil factors, climate change and the amount as well as quality of plant and microbial inputs. The sequestration of C in soils used for agriculture, forestry and land reclamation has been recognised as a potential option to mitigate global change (Batjes 1996; IPCC 1996; Smith et al. 1997; Lal 2003). Recent research suggests that AM fungi might be an important component of the soil organic C pool, in addition to facilitating C sequestration by stabilising soil aggregates along with glomalin formation.

18.2   Role of AM Fungi on Carbon Fluxes Between the Plants and the Atmosphere

The symbiotic association between plant roots and AM fungi is ubiquitous in terrestrial ecosystems (Smith and Read 2008). The role of AM fungi in mediating the ecosystem response to global climate change has been reviewed previously (Zhu and Miller 2003; Rillig et al. 2002; Staddon et al. 2002). In view of the importance of AM fungi in ecosystem processes, these reviews highlight the importance of research addressing the contributions of AM fungi to terrestrial C cycling. Several studies using ^{14}C labelling indicated that photosynthate is transferred from host plants to AM fungi hyphae within hours after labelling (Solaiman and Saito 1997; Johnson et al. 2002). It is also generally accepted that AM fungi receive all their carbohydrate from the host plant and that the association of AM fungi with roots could create a sink (i.e. a demand for carbohydrate) which could result in a 4–20 % drain of carbon from the host plant and could indirectly influence carbon sequestration in soils (Graham 2000). Furthermore, upregulation of photosynthesis by AM fungi is indicated where the amount of fungus in the root system is related directly to net C gain of the host (Miller et al. 2002). Such fungus-mediated effects on plant growth can potentially improve C sequestration by increasing net primary production, especially in nutrient-limited environments (Table 18.1).

18.3   Extraradical Hyphae and Carbon Sequestration in Soils

AM fungi could directly influence soil C sequestration through the growth and turnover of extraradical hyphae in rhizosphere and bulk soil. The decomposition time of extraradical hyphae in soil is relatively short compared to other organic biomass, and it has rarely been estimated. For example, Staddon et al. (2002) used accelerator mass spectrometry microanalysis of ^{14}C to quantify the turnover rate of extraradical hyphae in plants grown in a controlled environment and found that the turnover rate of extraradical hyphae attached to plant roots averaged 5–6 days. The authors indicated that C flow from host plants to AM fungi in soil might quickly be respired back to the atmosphere. More importantly, their findings suggest a rapid pathway for atmospheric C to enter the soil C cycle because AM fungi hyphal cell

Table 18.1 Role of AM fungi in regulating carbon fluxes between the atmosphere, plants and soil

AM fungi functions	Response	Source
Mycorrhizal external mycelium was the dominant pathway through which carbon entered the soil organic matter pool, exceeding the input via leaf litter and fine root turnover	Increase soil carbon sequestration	Godbold et al. (2006)
More carbon sequestration in ectomycorrhizas and ericoid mycorrhizas compared to arbuscular mycorrhizas	The effect of mycorrhizal type on soil carbon	Averill et al. (2014)
Colonisation by AM fungi could increase C transfer from rice to watermelon, while intercropping with watermelon could promote AM fungal colonisation and P uptake by rice	Carbon transfer	Ren et al. (2013), Schulze (2006)
A teleonomic model represents carbon (C), nitrogen (N) and phosphorus (P) substrates with structure in shoot, root and mycorrhiza	Above- and below-ground interactions	Thornley and Parsons (2014)
AM fungal extraradical hyphae form glomalin which is associated with aggregate stability and protection of organic C in soil	Glomalin formation by extraradical hyphae	Wright and Upadhyaya (1996)
1. Carbon acts as an important trigger for fungal N uptake and transport	Carbon availability triggers fungal nitrogen uptake and transport	Fellbaum et al. (2012)
2. The fungus changes its strategy in response to an exogenous supply of carbon		
3. Both plants and fungi reciprocally reward resources to those partners providing more benefit		
AM fungi stimulated under elevated CO_2	Organic carbon decomposition	Cheng et al. (2012)
C and N flow at the soil-root interface is bidirectional with C and N being lost from roots and taken up from the soil simultaneously	Bidirectional transfer of C and N	Jones et al. (2009)
Soil aggregation and C sequestration are tightly correlated with the abundance of AM fungi	Increase C sequestration and stabilisation of soil aggregates	Wilson et al. (2009)

walls are composed of chitin, a carbohydrate that is recalcitrant; the rapid turnover of live extraradical hyphae would still allow for the accumulation of hyphal residues that could remain within the soil matrix for a considerable time. Currently, limited information is available on the residence time of chitinous cell wall residues, particularly in a soil matrix, although recent studies using pyrolysis GC/MS-C-IRMS (gas chromatography-mass spectrometry-combustion interface-isotope ratio mass spectrometry) indicate a residence time of 49 ± 19 years for protein-, amino acid- or chitin-derived pyrolysis products (Gleixner et al. 2002). Hyphal residue accumulation would have been difficult to measure in the short period of time used in this experiment (Staddon et al. 2002). Furthermore, the use of potting medium consisting of sand and attapulgite clay eliminated the involvement of physical protection which is a major mechanism for stabilising hyphal residues exposed to soil microbial activity and hence for increasing residence time in soil. Moreover, the Staddon et al. (2002) study did not determine whether the hyphae (characterised by rapid turnover) were decomposed completely to CO_2 or remained as residues within the potting medium.

The typical dry weight of extraradical hyphae in soil, 0.03–0.5 mg/g, represents a large proportion of soil microbial biomass (Miller et al. 1995; Olsson 1999). For a soil depth of 30 cm with bulk density of 1.2 g/cm^3 and 50 % carbon content of dry hyphae, the amount of SOC derived directly from AM fungi ranges from 54 to 900 kg/ha. This range in extraradical hyphae indicates that despite the rapid turnover of live hyphae, the amount of carbon retained by extraradical hyphae in the soil is measurable, and the maintenance of a stable hyphal network is functionally important for the sequestration of carbon below ground. The rather high turnover values reported by Staddon et al. (2002), in combination with results demonstrating a rather large stock of extraradical hyphae biomass, suggest that more than one pool of extraradical hyphae exists, probably distinguished by hyphal architecture (Staddon et al. 2002; Friese and Allen 1991). One pool is composed of hyphae with relatively fast turnover (days), probably related to the hyphal architectural type known as exploratory or absorptive hyphae. Another pool with relatively slower turnover (weeks) is composed of the thicker-walled extraradical hyphae with arterial architecture. These observations suggest a need for future research to consider extraradical hyphae turnover and the resultant contribution to carbon sequestration owing to hyphal architecture.

The extent to which terrestrial ecosystems can sequester C to mitigate climate change is unknown. The stimulation of AM fungi by elevated atmospheric CO_2 has been assumed to be a major mechanism facilitating soil C sequestration by increasing C inputs to soil and by protecting organic C from decomposition via aggregation. Cheng et al. (2012) presented evidence from four independent microcosm and field experiments demonstrating that CO_2 enhancement of AM fungi results in considerable soil C losses. Their findings challenge the assumption that AM fungi protect against degradation of organic C in soil and raise questions about the current prediction of terrestrial ecosystem C balance under future climate change scenarios.

18.4 Role of AM Fungi on Plant C Rhizodeposition in Soil

Plant roots exudate a significant proportion of C assimilated by photosynthesis to the soil which is within the range of 5–30 % (Philippot et al. 2013). Rhizodeposition of C may be in the form of root exudates, mucilage, dead root cells and C transfer to mycorrhizal fungi (Jones et al. 2009; Badri and Vivanco 2009). This continuous supply of C compounds by plant roots influences soil microbial community composition and activity, which can directly influence plant growth (Philippot et al. 2013). Plants have traded photosynthates with AM hyphae over millions of years of co-evolution (Redecker 2000; Wang et al. 2010).

Most terrestrial plants form symbiosis with one or more kinds of mycorrhizal fungi, of which 80 % of plant species are being associated with AM fungi (Smith and Smith 2011). Several studies have shown that the presence of a mycorrhizal association significantly increases the total C assimilation by plants (Miller et al. 2002; Grimoldi et al. 2006; Calderon et al. 2012) and can induce an extra C flux of 3–8 % of gross photosynthesis into the soil (Grimoldi et al. 2006). AM fungi are constantly provided with recent plant photoassimilates which are used to build up their large extraradical hyphal network. Turnover of this mycorrhizal hyphae network is thought to be a main and quickest process for C input into the soil organic matter pool, possibly even greater than shoot or root litter inputs (Godbold et al. 2006).

The transfer of photosynthetic C from plants to soil occurs on a rapid timescale ranging from hours for grasses to a few days for trees (Kuzyakov and Gavrichkova 2010). Root exudation starts to peak only 3 h after photosynthesis in wheat (Dilkes et al. 2004). The core pathway for transport of recently assimilated C from the shoots to the roots is the plant phloem (Mencuccini and Höltä 2010). However, little is known about rigorous translocation patterns of C along this pathway, and even less is known about the involvement of particular root cells in the process of C being released to the soil matrix (Badri and Vivanco 2009). It is thought that the major proportion of root exudate C is lost passively by the large C concentration gradient between root cytoplasm and the apoplast/soil solution (Farrar et al. 2003; Jones et al. 2009). Thus, not all C exudation into soil occurs directly from roots; some C allocation is associated with the volume of AM fungal hyphal network.

AM fungi colonise roots behind the root hair zone along mature root sections of the roots which have an established phloem and an endodermal layer (Smith and Smith 2011). Carbon is transferred to AM fungi via the mycorrhizal intraradical hyphae in the root cortex (Solaiman and Saito 1997). The intraradical hyphae or arbuscules grow within apoplastic areas of the roots cortex but do not enter the plant cytoplasm. Rather, they form a symbiotic interface made of plasma membranes both of fungus and plant, which is separated by an apoplastic compartment (Smith and Smith 2011). Carbon is supposed to be transferred from the plant as glucose and sucrose to the intraradical hyphae through this symbiotic interface, from where it is transported to the extraradical network which extends into the soil matrix (Bago et al. 2002, 2003; Solaiman and Saito 1997). The pathway along which C is translocated from the phloem cells to the mycorrhizal structures in the cortex remains to be explored.

AM fungi increase plant nutrient uptake especially P in exchange for C, but interactions between these fungi and the soil microbial community have received less attention. Some studies have shown that AM hyphae can exude plant-derived C into the adjacent hyphosphere (Cheng et al. 2012; Johansson et al. 2004; Toljander et al. 2007). In addition, mycorrhizal hyphae are able to transport plant root C and release it beyond the rhizosphere, thereby transferring some of the C to the soil microbial community within the soil matrix that is inaccessible to roots (Herman et al. 2012; Nottingham et al. 2013). A recent study demonstrated that AM fungi may contribute to rhizosphere priming where plants have a role on soil organic matter decomposition (Cheng et al. 2012). Hyphal exudates may be chemically different from root exudates leading to support a different microbial community compared to that of root exudates (Nuccio et al. 2013). The role of C released from hyphae for use by the soil microbial community needs to be elucidated.

18.5   Extraradical Hyphae, Glomalin Exudation and Soil Aggregate Formation

If the turnover values of hyphae reported by Staddon et al. (2002) can be generalised to all extraradical hyphae, it may be necessary to re-evaluate the contributions of AM fungi to soil structure. However, extraradical hyphae may be relatively persistent within soil aggregates, and their influence on soil aggregation might be even more important to the C stock than the influence of the hyphal standing crop alone (Miller and Jastrow 2000). Through their role in soil macroaggregate stabilisation, extraradical hyphae of AM fungi appear to contribute to the formation of aggregates and help to create a mechanism for increasing the residence time of organic biomass within soil macroaggregates. The extraradical hyphae contribute through enmeshment and stabilisation of soil particles within aggregates (Miller and Jastrow 2000; Oades and Waters 1991). The extraradical hyphae are able to ramify through soil pores within macroaggregates.

The contribution of hyphae of AM fungi to carbon cycling occurs in combination with exudates from roots. AM fungi hyphae are responsible for the production of a glycoprotein-like substance, glomalin (Wright and Upadhyaya 1998), which is fairly stable in soils (Steinberg and Rillig 2003). Radiocarbon dating of the operationally defined glomalin extract indicates a residence time in soils of 6–42 years (Rillig et al. 2001), which is longer than the residence time reported for hyphae of AM fungi. In a tropical forest soil, glomalin carbon was shown to represent up to 5 % of total soil carbon, which is much higher than soil microbial biomass carbon (Miller et al. 1995). The close correlation of the amount of glomalin in soil, hyphal length and stability of soil aggregates (Wright and Upadhyaya 1996) is evidence that glomalin could influence soil carbon storage indirectly by stabilising soil aggregates. One of the modes of action of glomalin could be in facilitating the formation of a sticky string bag of hyphae, the primary mode by which AM fungi contribute to soil aggregation (Johnson et al. 2002). However, the relationship between hyphal turnover and glomalin inputs remains largely unknown, and quantification of the relative contribution of glomalin to carbon cycling still needs to be determined.

18.6   Conclusion

AM fungi can contribute C fluxes between the plants and the atmosphere through increased C assimilation in plants. A key AM fungal-mediated process involved in the sequestration of C in soils is the transfer of photosynthate from host plants to AM extraradical hyphae. Although the turnover of extraradical hyphae linked to plant roots is little known, it is understood that the process is rapid. The overall contribution of AM fungi to soil C sequestration could depend significantly on the quantity and quality of hyphae produced, the age and resilience of hyphal residues, the production of glomalin and the role played by AM fungi in the stabilisation of soil aggregates. More detailed investigation is needed to explore the links between C sequestration in soil and nutrient exchange that are associated with AM fungi.

References

Averill C, Turner BL, Finzi AC (2014) Mycorrhiza-mediated competition between plants and decomposers drives soil carbon storage. Nature 505:543–545

Badri DV, Vivanco JM (2009) Regulation and function of root exudates. Plant Cell Environ 32:666–681

Bago B, Pfeffer PE, Zipfel W, Lammers P, Shachar-Hill Y (2002) Tracking metabolism and imaging transport in arbuscular mycorrhizal metabolism and transport in AM fungi. Plant Soil 244:189–197

Bago B, Pfeffer PE, Abubaker J, Jun J, Allen JW, Brouillette J, Douds DD, Lammers PJ, Shachar-Hill Y (2003) Carbon export from arbuscular mycorrhizal roots involves the translocation of carbohydrate as well as lipid. Plant Physiol 131:1496–1507

Batjes NH (1996) The total carbon and nitrogen in the soils of the world. Eur J Soil Sci 47:151–163

Calderon FJ, Schultz DJ, Paul EA (2012) Carbon allocation, belowground transfers and lipid turnover in a plant-microbial association. Soil Sci Soc Am J 76:1614–1623

Cheng L, Booker F, Tu C, Burkey K, Zhou L, Shew H, Rufty TW, Hu S (2012) Arbuscular mycorrhizal fungi increase organic carbon decomposition under elevated CO_2. Science 2:1084–1087

Dilkes NB, Jones DL, Farrar J (2004) Temporal dynamics of carbon partitioning and rhizodeposition in wheat. Plant Physiol 134:706–715

Farrar J, Hawes M, Jones D, Lindow S (2003) How roots control the flux of carbon to the rhizosphere. Ecology 84:827–837

Fellbaum CR, Mensah JA, Pfeffer PE, Kiers ET, Bucking H (2012) The role of carbon in fungal nutrient uptake and transport: implications for resource exchange in the arbuscular mycorrhizal symbiosis. Plant Signal Behav 7:1509–1512

Friese CF, Allen MF (1991) The spread of VA mycorrhizal fungal hyphae in soil: inoculum types and external hyphal architecture. Mycologia 83:409–418

Gleixner G, Poirier N, Bol R, Balesdent J (2002) Molecular dynamics of organic matter in a cultivated soil. Org Geochem 33:357–366

Godbold DL, Hoosbeek MR, Lukac M, Cotrufo MF, Janssens IA, Ceulemans R, Polle A, Velthorst EJ, Scarascia-Mugnozza G, Angelis P, Miglietta F, Peressotti A (2006) Mycorrhizal hyphal turnover as a dominant process for carbon input into soil organic matter. Plant Soil 281:15–24

Graham JH (2000) Assessing the cost of arbuscular mycorrhizal symbiosis in agroecosystems. In: Podila GK, Douds DD (eds) Current advances in mycorrhizal research. The American Phytopathological Society, St. Paul, pp 127–140

Grimoldi AA, Kavanová M, Lattanzi FA, Schäufele R, Schnyder H (2006) Arbuscular mycorrhizal colonization on carbon economy in perennial ryegrass: quantification by $13CO_2/12CO_2$ steady-state labelling and gas exchange. New Phytol 172:544–553

Herman DJ, Firestone MK, Nuccio E, Hodge A (2012) Interactions between an arbuscular mycorrhizal fungus and a soil microbial community mediating litter decomposition. FEMS Microbiol Ecol 80(536):236–247

IPCC (1996) Climate change, impacts, adaptations and mitigation of climate change, scientific-technical analyses. Cambridge University Press, Cambridge

Jastrow JD, Miller RM, Lussenhop J (1998) Contributions of interacting biological mechanisms to soil aggregate stabilization in restored prairie. Soil Biol Biochem 30:905–916

Johansson JF, Paul LR, Finlay RD (2004) Microbial interactions in the mycorrhizosphere and their significance for sustainable agriculture. FEMS Microbiol Ecol 48:1–13

Johnson D, Leake JR, Read DJ (2002) Transfer of recent photosynthate into mycorrhizal mycelium of an upland grassland: short-term respiratory losses and accumulation of ^{14}C. Soil Biol Biochem 34:1521–1524

Jones DL, Nguyen C, Finlay RD (2009) Carbon flow in the rhizosphere: carbon trading at the soil–root interface. Plant Soil 321:5–33

Kuzyakov Y, Gavrichkova O (2010) Time lag between photosynthesis and carbon dioxide efflux from soil: a review of mechanisms and controls. Glob Change Biol 16:3386–3406

Lal R (2003) Global potential of soil carbon sequestration to mitigate the greenhouse effect. Crit Rev Plant Sci 22:151–184

Leake J, Johnson D, Donnelly D, Muckle G, Boddy L, Read D (2004) Networks of power and influence: the role of mycorrhizal mycelium in controlling plant communities and agroecosystem functioning. Can J Bot 82:1016–1045

Mencuccini M, Hölttä T (2010) The significance of phloem transport for the speed with which canopy photosynthesis and belowground respiration are linked. New Phytol 185:189–203

Miller RM, Jastrow JD (1992) The role of mycorrhizal fungi in soil conservation. In: Bethlenfalvay CJ, Linderman RG (eds) Mycorrhizae in sustainable agriculture. Crop Science Society and Soil Science Society of America, Madison, pp 29–44

Miller RM, Jastrow JD (2000) Mycorrhizal fungi influence soil structure. In: Kapulnik Y, Douds D (eds) Arbuscular mycorrhizas: physiology and function. Kluwer Academic, Dordrecht, pp 4–18

Miller RM, Kling M (2000) The importance of integration and scale in the arbuscular mycorrhizal symbiosis. Plant Soil 226:295–309

Miller RM, Reinhardt DR, Jastrow JD (1995) External hyphal production of vesicular arbuscular mycorrhizal fungi in pasture and tallgrass prairie communities. Oecologia 103:17–23

Miller RM, Miller S, Jastrow JD, Rivetta CB (2002) Mycorrhizal mediated feedbacks influence net carbon gain and nutrient uptake in Andropogon gerardii vitman. New Phytol 155:149–162

Nottingham AT, Turner BL, Winter K, Chamberlain PM, Stott A, Tanner EVJ (2013) Root and arbuscular mycorrhizal mycelial interactions with soil microorganisms in lowland tropical forest. FEMS Microbiol Ecol 85:37–50

Nuccio EE, Hodge A, Pett-Ridge J, Herman DJ, Weber PK, Firestone MK (2013) An arbuscular mycorrhizal fungus significantly modifies the soil bacterial community and nitrogen cycling during litter decomposition. Environ Microbiol 15:1870–1881

Oades JM, Waters AG (1991) Aggregate hierarchy in soils. Aust J Soil Res 29:815–828

Olsson PA (1999) Signature fatty acids provide tools for determination of distribution and interactions of mycorrhizal fungi in soil. FEMS Microbiol Ecol 29:303–310

Olsson PA, Johnson NC (2005) Tracking carbon from the atmosphere to the rhizosphere. Ecol Lett 8:1264–1270

Parniske M (2008) Arbuscular mycorrhiza: the mother of plant root endosymbioses. Nat Rev Microbiol 6:763–775

Philippot L, Raaijmakers JM, Lemanceau P, van der Putten WH (2013) Going back to the roots: the microbial ecology of the rhizosphere. Nat Rev Microbiol 11:789–799

Redecker D (2000) Glomalean fungi from the Ordovician. Science 289:1920–1921

Ren LX, Lou YS, Zhang N, Zhu XD, Hao WY, Sun SB, Shen QR, Xu GH (2013) Role of arbuscular mycorrhizal network in carbon and phosphorus transfer between plants. Biol Fertil Soils 49:3–11

Rillig MC, Wright SF, Nichols KA, Schmidt WP, Torn MS (2001) Large contribution of arbuscular mycorrhizal fungi to soil carbon pools in tropical forest soils. Plant Soil 233:167–177

Rillig MC, Treseder KK, Allen MF (2002) Global change and mycorrhizal fungi. In: van der Heijden M, Sanders I et al (eds) Mycorrhizal ecology, vol 157, Ecological studies series. Springer, New York, pp 135–160

Schulze ED (2006) Biological control of the terrestrial carbon sink. Biogeosciences 3:147–166

Smith SE, Read DJ (2008) Mycorrhizal symbiosis. Academic, London, p 800

Smith SE, Smith FA (2011) Roles of arbuscular mycorrhizas in plant nutrition and growth: new paradigms from cellular to ecosystem scales. Annu Rev Plant Biol 62:227–250

Smith P, Powlson DS, Glendining MJ, Smith JU (1997) Potential for carbon sequestration in European soils: preliminary estimates for five scenarios using results from long-term experiments. Glob Change Biol 3:67–79

Solaiman MZ, Saito M (1997) Use of sugars by intraradical hyphae of arbuscular mycorrhizal fungi revealed by radiorespirometry. New Phytol 136:533–538

Staddon PL, Heinemeyer A, Fitter AH (2002) Mycorrhizas and global environmental change: research at different scales. Plant Soil 244:253–261

Staddon PL, Bronk Ramsey C, Ostle N, Ineson P, Fitter AH (2003) Rapid turnover of hyphae of mycorrhizal fungi determined by AMS microanalysis of ^{14}C. Science 300:1138–1140

Steinberg PD, Rillig MC (2003) Differential decomposition of arbuscular mycorrhizal fungal hyphae and glomalin. Soil Biol Biochem 35:191–194

Thornley JHM, Parsons AJ (2014) Allocation of new growth between shoot, root and mycorrhiza in relation to carbon, nitrogen and phosphate supply: teleonomy with maximum growth rate. J Theor Biol 342:1–14

Toljander JF, Lindahl BD, Paul LR, Elfstrand M, Finlay RD (2007) Influence of arbuscular mycorrhizal mycelial exudates on soil bacterial growth and community structure. FEMS Microbiol Ecol 61:295–304

Wang B, Yeun LH, Xue J-Y, Liu Y, Ané J-M, Qiu Y-L (2010) Presence of three mycorrhizal genes in the common ancestor of land plants suggests a key role of mycorrhizas in the colonization of land by plants. New Phytol 186:514–525

Willis A, Rodrigues BF, Harris PJC (2013) The ecology of arbuscular mycorrhizal fungi. Crit Rev Plant Sci 32:1–20

Wilson GWT, Rice CW, Rillig MC, Springer A, Hartnett DC (2009) Soil aggregation and carbon sequestration are tightly correlated with the abundance of arbuscular mycorrhizal fungi: results from long-term field experiments. Ecol Lett 12:452–461

Wright SF, Upadhyaya A (1996) Extraction of an abundant and unusual protein from soil and comparison with hyphal protein of arbuscular mycorrhizal fungi. Soil Sci 161:575–586

Wright SF, Upadhyaya A (1998) A survey of soils for aggregate stability and glomalin, a glycoprotein produced by hyphae of arbuscular mycorrhizal fungi. Plant Soil 198:97–107

Zhu YG, Miller RM (2003) Carbon cycling by arbuscular mycorrhizal fungi in soil-plant systems. Trends Plant Sci 8:407–409

Chapter 19
Biochar as a Habitat for Arbuscular Mycorrhizal Fungi

Noraini M. Jaafar

19.1 Introduction

Biochar, the pyrolised product from pyrolysis of "waste" organic material, has been widely proposed as a soil ameliorant for improving soil properties (Lehmann 2007; Rondon et al. 2007; Lehmann et al. 2011). However, biochar incorporation into soil can have both positive and negative effects on beneficial soil microorganisms, including arbuscular mycorrhizal (AM) fungi (Warnock et al. 2007). Both direct and indirect effects of biochar may be involved (Lehmann et al. 2011).

Direct and indirect mechanisms underlying interactions between AM fungi and biochar include the possibilities that (1) biochar provides a suitable habitat or shelter for soil microorganisms, protecting them from predators; (2) soil conditions and plant growth can be influenced by mycorrhizas after biochar addition through changes in soil physicochemical properties such as soil pH and water; and (3) AM fungi interactions with soil microorganisms may stimulate production of signalling compounds or alleviate production of detrimental compounds (Warnock et al. 2007). Other mechanisms linking biochar to changes in the abundance or functioning of mycorrhizas include potential interference in plant–fungus signalling and detoxification of allelochemicals on biochar (Warnock et al. 2007, 2010). Investigations of how biochar might affect soil microorganisms have mostly focused on microbial attachment, microbial community shift and enzyme activities (Atkinson et al. 2010; Joseph et al. 2010; Sohi et al. 2010; Lehmann et al. 2011).

Two main areas of research on biochar and soil microorganisms require clarification. First, generalisations about responses to biochar application need to be considered in relation to the specific characteristics of the biochar product used. Second, experimental evidence is required to clarify mechanisms by which biochar

N.M. Jaafar (✉)
Department of Land Management, Faculty of Agriculture, Universiti Putra Malaysia (UPM),
Serdang 43400, Malaysia
e-mail: noraini_aj@yahoo.com

© Springer-Verlag Berlin Heidelberg 2014   297
Z.M. Solaiman et al. (eds.), *Mycorrhizal Fungi: Use in Sustainable Agriculture and Land Restoration*, Soil Biology 41, DOI 10.1007/978-3-662-45370-4_19

influences microorganisms in soil. Biochar may also exhibit different interactions over time after its application to soil, but this is not often studied or considered with regard to soil biota (Lehmann et al. 2011). There is a range of factors that could influence the effectiveness of biochar as a soil amendment. Effects of biochar on soil microbial components need to be considered in the context of different biochar and soil backgrounds as well as soil management practices.

As soil microorganisms are sensitive to soil management, knowing the background of soil and biochar is important when managing soils with biochar, especially for determining the amount applied in combination with fertiliser and organic materials for optimal mycorrhizal symbiosis. This review focuses on biochar properties in relation to the factors controlling its variability, function and management leading to how biochar might alter the abundance and activity of soil microorganisms. Mechanisms by which biochar might enhance the contribution of beneficial microorganisms in soil may also depend on other soil management practices. Based on potential similarities between mechanisms underlying interactions between soil microorganisms and biochar, this review focuses on AM fungi (Warnock et al. 2007) as a case study for considering biological influences of biochar on soil microorganisms especially fungi in soil.

19.2   Biochar as a Soil Amendment

There is a general consensus that the incorporation of biochar into soil could be beneficial to soil microorganisms. However, biochars are heterogeneous, with a range in porosity and surface area and pH although they are commonly alkaline. Biological properties of biochar are often overlooked. The beneficial impacts of biochar on soil have been speculated based on observations of the pyrogenic soil containing burned plant and animal materials generally known as Terra Preta soil as well as dark earth soil. AM fungi, used as a biofertiliser, have been considered in combination with biochar for their influence on soil properties such as nutrient retention, availability and uptake by plants (Warnock et al. 2007). However, the value of biochar as a general soil conditioner remains speculative because both positive and negative responses to biochar of soil microbial communities, including AM fungi, have been reported (Atkinson et al. 2010; Blackwell et al. 2010; Joseph et al. 2010; Moskal-del Hoyo et al. 2010; Sohi et al. 2010; Solaiman et al. 2010). Biochars derived from a range of plant and biomass sources have been studied in experiments that include both naturally occurring and inoculated AM fungi. Thus, some of the observed discrepancies in biochar influences on soil biological properties may have resulted from generalisations based on experiments using biochars of different organic origins or for soils with diverse physical and chemical properties (Rillig et al. 2010).

The type and source of biochar is central to estimating the role of biochar as a microbial habitat and its benefit to soil. For example, in Japan, locally available rice husk biochar increased the proportion of AM roots colonised through soil pH modification and absorption of toxic substances and agrochemicals which inhibit root growth and microbial activity (Ishii and Kadoya 1994). In Australia, locally available *Eucalyptus* biochar had a similar positive effect on the percent mycorrhizal colonisation, possibly related to water uptake (Solaiman et al. 2010). In other cases, there was no effect for woody *Eucalyptus* biochar (e.g. Rondon et al. 2007) or woody *Leucaena* biochar (e.g. Habte and Antal 2010).

Most studies have examined the effects of biochar on mycorrhizal colonisation and sporulation (e.g. Ishii and Kadoya 1994; Matsubara et al. 2002; Elmer and Pignatello 2011), while measurements of phosphorus availability in plant and soil are used as indirect indicators of AM fungal effectiveness (Solaiman et al. 2010; Blackwell et al. 2010). Variability in soil characteristics and mycorrhizal inoculation methods (inoculation or indigenous) can influence the responses. These factors need to be considered in relation to identifying mechanisms involved in how biochar affects hyphae of AM fungi, spore germination and sporulation, enzymatic activities and the carbon/phosphorus interchange with plants.

19.3   Factors Influencing Biochar: Soil–Microbe Interactions

Warnock et al. (2007) highlighted the potential mechanisms of biochar–AM fungi interactions, including the potential of biochar as a habitat for soil microorganisms. However, studies of the potential impact of biochar–AM interaction have generally not focused on the mechanisms involved. The nature or physical characteristics of biochar in providing the habitat and protection for AM fungi are emphasised here because it is one of the main mechanisms that may involve direct biochar–mycorrhizal interactions.

The heterogeneous properties of biochar from various materials and pyrolysis processes influence their ameliorative effects on soil microbial colonisation, growth and benefit to plant and soil (Chan et al. 2007, 2008; Kuzyakov et al 2009; Thies and Rillig 2009; Blackwell et al. 2010; Rillig et al. 2010). Below, biochar factors and their effects are discussed in relation to AM fungi in optimising both biochar and AM symbiotic benefit towards improving soil properties and plant growth.

19.3.1   Sources of Biochar

Generalisations about the practical application of biochar have proven to be difficult due to heterogeneity among biochars and interactions with the soil environment into which biochar is applied. The heterogeneous properties of biochar can result from diversity of the original material used in the pyrolysis process (Blackwell et al. 2010). Furthermore, for practical purposes, an appropriate range of biochar particle size, amount and methods of application, especially in the field, need to be considered for different biochar sources (Blackwell et al. 2009; Downie et al. 2009).

The heterogeneity in both physical and chemical properties of biochar is associated with feedstock and pyrolysis parameters (Gundale and DeLuca 2006; Downie et al. 2009). Mycorrhizal interactions with biochar have been compared using different types of biochar in a range of soil environments. Most biochars used have been plant derived and include rice husk, pine and other woody materials (Warnock et al. 2007). Experimental comparisons of the effect of the incorporation of biochars of plant and animal origin into soil on AM fungi are limited (Saito 1990; Warnock et al. 2007). Therefore, the effects of biochar heterogeneity arising from various sources of organic materials, pyrolysis temperature or biochar particle size and application rates on AM fungal growth, symbiosis and functions are not well understood.

Biochar creates a microenvironment in the bulk soil upon its application (Thies and Rillig 2009; Ogawa and Okimori 2010; Lehmann et al. 2011). Within the biochar microenvironment, biochar surfaces and pores can be colonised by bacteria, fungi and soil microfauna (Table 19.1).

Previous studies of microbial colonisation on biochar surfaces included laboratory experiments using biochar retrieved from soil (Ascough et al. 2010a, b; Moskal-del Hoyo et al. 2010), but there has been little characterisation of microbial colonisation of the internal structure of biochar compared to the external surface (Table 19.1). Laboratory studies of fungal colonisation of biochar showed fungal colonisation on surfaces and along cracks (Ascough et al. 2010b). As a consequence, there has been little discussion of experimental conditions and methodologies associated with observations of microorganisms in the biochar microenvironment.

Biochar pores may be structurally stacked and they may be altered by the presence of soil particles. It is expected that microorganisms are preferentially attracted to biochar surfaces rather than to pores (Lehmann et al. 2011). Surface features are important for substrate recognition and attachment by soil microorganisms (Lehmann et al. 2011). Biochar surfaces could provide substrates that are important for biological activity (Thies and Rillig 2009). Furthermore, surface attachment can protect microorganisms and increase the opportunity for synergistic interactions between biochar and soil microorganisms. Biochar pH is usually neutral to alkaline and may contain some phosphorus (Gundale and DeLuca 2006; Yamato et al. 2006) which may be available for microbial uptake.

Fungal hyphae, such as those of AM fungi, have potential to dominate biochar surfaces due to their extensive hyphal networks (Lehmann et al. 2011), and differences in hyphal growth forms have been observed on external compared to internal surfaces of charcoal (Ascough et al. 2010b). Although the surface of biochar has been associated with slow degradation by soil microbial and chemical processes, it can become coated with organic material (Joseph et al. 2010; Lehmann et al. 2011) which contributes to a microbial habitat. Some forms of biochar have been shown to retain moisture and adsorb cations (Liang et al. 2006; Blackwell

Table 19.1 Examples of microscopic observations of biochar as a habitat for soil microorganisms

Experiment	Methodology	Observation	Reference
Comparison of fungal colonisation in biochar feedstocks before and after burning (pyrolysis)	Wood and charcoal fragments were manually broken, observed under reflected light microscope followed by SEM observation on transverse (TS), longitudinal tangential (LTS) and longitudinal radial (LRS) sections	Fungal hyphae observed, some fungal infestation and features of decay were preserved after burning	Moskal-del Hoyo et al. (2010)
Saprophytic white rot fungal colonisation (from laboratory trial on media) on biochar blocks	Blocks were lyophilised, split open, observed using SEM	Distinct fungal growth found on charcoal, hyphal penetration through cracks	Ascough et al. (2010b)
Characterisation of microbial life colonising biochar and biochar-amended soils (fresh corn biochar colonised by microorganisms)	SEM method not available	Fresh corn biochar with microorganisms in pores	Jin (2010); Lehmann et al. (2011)
Fungal hyphae colonisation in fresh biochar pores	Method not available	Fungal hyphae found in fresh biochar	Lehmann and Joseph (2009); Lehmann et al. (2011)
Changes in charcoal particle morphology of 100-year-old char	SEM observation on cross sections, inner and outer parts of biochar, EDX spectroscopy	Filamentous fungi-infiltrated charcoal through larger pores and patches of mineral coating was found	Hockaday et al. (2007)
Ecological study of different ages of wood charcoal from forest humus profiles	SEM observation on transverse and longitudinal plane of biochar	Senescent fungal hyphae in biochar	Zackrisson et al. (1996)

et al. 2010; Solaiman et al. 2010), and this may indirectly influence soil microbial activity on biochar surfaces. A greater number of functional groups and oxidised sites on biochar surfaces could further facilitate microbial oxidation (Hockaday et al. 2007). Higher bacterial growth rates in association with biochar (Pietikainen et al. 2000) indicated that attachment and physical protection may be enhanced by the surface chemistry, including hydrophobicity.

Variation in porosity is expected to alter the suitability of biochar as a habitat for soil microorganisms. Pore size and surface characteristics are likely to influence microbial attachment and presumably the ability of the microorganisms to enter and/or penetrate into the biochar (Lehmann et al. 2011). Biochar includes meso-

(<2 µm), micro- (2–50 µm) and macropore (>50 µm) sizes (Downie et al. 2009) which may create microenvironments. Larger biochar pores may offer a new microhabitat to fungi, but no direct experimental evidence of the extent of pore colonisation by either bacteria or fungi is available. Furthermore, connectivity of pore spaces within biochar particles would influence the availability of important resources for microorganisms such as air and water diffusion through biochar, facilitating colonisation by soil microorganisms. Pores with diameters of 1–4 µm and 2–64 µm would be accessible to soil bacteria and fungal hyphae, respectively (Swift et al. 1979), including hyphae of AM fungi (Saito 1990). However, no studies have qualitatively or quantitatively demonstrated preferential colonisation by fungi and bacteria in biochar pores or on surfaces, and if the connectivity of pores within biochar is restricted, this would limit access by hyphae and bacteria.

Chemical and physical changes in biochar can occur after it is incorporated into soil (Downie et al. 2009). Interactions between soil particles, especially clay, and biochar have been found (Joseph et al. 2010). Quantification of changes in biochar after interaction with soil has not been a focus in investigations of the consequences of microbial colonisation of biochar, but it requires knowledge of the characteristics of biochar pores and surfaces of any biochar applied (Lehmann and Joseph 2009). As soil particles become cemented and the surface area covered, soil may enter biochar pores and alter their porosity and surface area. This could either limit or enhance the habitable spaces of biochar to soil microorganisms depending on the nature of the modification and the soil type. Lehmann et al. (2011) discussed various modes of microbial attachment to biochar, but the role of soil particles in influencing microbial attachment has not been clarified. Biochar surfaces could become cemented by soil and soil could enter biochar pores, but it is not known whether this might have either positive or negative effects on microbial colonisation of biochar.

Among sources of feedstock, woody biochar has potential as a habitat because it has higher porosity compared to the other sources of biochar such as chicken manure (Downie et al. 2009). Pores of 2–80 µm diameter are known to occur in wood-derived biochars and may benefit activity of mycorrhizal fungi (Thies and Rillig 2009). Woody biochar from *Pinus radiata* (Anderson et al. 2011) was able to increase fungal and bacterial abundance and promote P-solubilising bacteria. Fungi, especially saprophytic fungi, may extensively colonise biochar particles due to their association in decomposing fibrous organic matter (Ascough et al. 2010b; Moskal-del Hoyo et al. 2010).

If there is a benefit from provision of habitat, biochar could protect AM fungal hyphae and spores or even stimulate hyphal growth. It has been demonstrated that fungal hyphae penetrate pores of inert material such as vermiculite used for preparation of AM fungal inocula (Douds et al. 2005). Similarly, AM fungi were found sporulating inside the cavities of expanded clay and on the surface of clay material particles (Norris et al. 1992). Saito (1990) stated that the high porosity of charcoal is not an effective substrate for saprophytes, but it can favour AM fungi, although the reason for this is not known. Perhaps hyphae of AM fungi extend into charcoal buried in soil and sporulate preferentially in such particles (Ogawa and

Yamabe 1986; Baltruschat 1987). However, there is little qualitative or quantitative evidence of preferential colonisation by fungi and bacteria in biochar pores or on surfaces compared with soil particles. Furthermore, details of experimental techniques and biochar handling regarding microbial colonisation inside or on biochar surfaces are often lacking.

19.3.2   Method of Biochar Application to Soil

The method of placement of biochar in soil (either as a distinct layer (banded) or mixed through the surface layer) may influence the effects of biochar on soil microorganisms. Biochar banded in soils can increase AM fungal colonisation measured as percentage of roots colonised (Blackwell et al. 2010; Solaiman et al. 2010). Banding biochar into a layer in field soil is normal practice compared to surface application due to the wind problems (Blackwell et al. 2009, 2010). Banding of biochar was effective for both AM fungi and plant growth in a field study at several sites (Blackwell et al. 2010) although this was not compared with any other method of biochar placement.

Banding and surface application of biochar are practical for field conditions, whereas mixing biochar with soils, banding and surface application have been used in pot trials (Blackwell et al. 2009). However, no experimental comparison of these methods is available for AM fungi unlike the pot trial on ectomycorrhizal fungi where responses to different methods of biochar application to soil have been investigated (Makoto et al. 2010). Biochar applied in a layer with ectomycorrhizal inoculum promoted larch plant growth when compared with mixing biochar with soil. This was attributed to the frequency of root contact with biochar enabling effective phosphate utilisation.

Banding biochar in the crop root zone ensures biochar placement in contact with roots at the earliest growth stages (Blackwell et al. 2009). Biochar applied in bands also reduces the potential for biochar and topsoil loss caused by wind erosion and surface disturbance. The improvement in precision of sowing and fertilising machinery provides chances for crops to be sown in, or adjacent to, bands of incorporated biochar. In addition, the appropriate time in applying biochar to soil needs to be considered. Rutto and Mizutani (2006) proposed that biochar is best added once mycorrhizal symbiosis is established. This was based on their conclusion that biochar (or the activated charcoal used in their study) could delay mycorrhizal associations through exudate absorption which adversely affects the fungus signalling process, hence the symbiosis establishment.

19.3.3   Amount of Biochar

The need to apply an appropriate amount of biochar to soil is crucial if it is to restore and maintain soil fertility and to any effects on mycorrhizas, crop growth and nutrition (Ishii and Kadoya 1994; Solaiman et al. 2010). Provision of soil conditions favourable for growth and activities of AM fungi needs to be taken into account when managing mycorrhizas in agricultural soils (Gazey et al. 2004) and this would apply in the presence of biochar. Biochar sourced from vastly different parent materials and pyrolysis conditions may exert different chemical properties including nutrient concentrations, and this needs to be considered when selecting the appropriate level of biochar for soil amendment. The amount may vary among soil types and land use histories which could undermine generalisations about the effects of biochar in soil (Schmidt and Noack 2000).

Several studies have investigated the quantity of biochar applied on soil micro-organisms (e.g. Kolb et al. 2009; Blackwell et al. 2010; Solaiman et al. 2010). The amount of biochar used for agronomic reasons (Blackwell et al. 2010; Solaiman et al. 2010) is likely to change the soil microbial environment, including that of AM fungi (Glaser et al. 2002; Kolb et al. 2009; Cross and Sohi 2011). In terms of soil microbial biomass, Chan et al. (2008) found an increase in soil microbial biomass carbon (MBC) dependent on the type of biochar and N fertiliser addition. MBC at the higher application levels, 25 and 50 t/ha, was significantly greater than that of the unamended control (Chan et al. 2008).

Selection of suitable amounts of biochar for application to soil to enhance colonisation by AM fungi is expected to differ for soil and biochar source (Blackwell et al. 2010; Elmer and Pignatello 2011). Inhibition of growth of AM fungi could result from application of higher than optimum amounts of biochar. The abundance of AM fungi in roots increased when hydrothermal carbonised biochar was added at 20 % w/w, and higher concentrations resulted in reduced mycorrhiza formation. Inoculum dilution at excessive levels of biochar application or an adverse effect on host plants limiting C supply to the AM fungi has been proposed (Rillig et al. 2010). Furthermore, the most appropriate amount of biochar may depend on the fertility of soil and its management, which could include organic matter management and other soil amendments such as fertiliser and lime (Blackwell et al. 2010). Biochar applied in optimal amounts and forms is expected to increase microhabitat availability in topsoils with low clay content (Solaiman et al. 2010). This may deliver mycorrhizal benefits (e.g. improve P acquisition by plants). Degraded soils may require higher amounts of biochar, but this would vary with organic matter or nutrient status (Liang et al. 2006; Chan et al. 2007, 2008; Steiner et al. 2008; Kolb et al. 2009). Most studies involving different amounts of biochar applied to soil show that levels of biochar that are acceptable for one type of soil and plant may not be suitable in another situation (Kolb et al. 2009; Blackwell et al. 2010).

19.3.4 Biochar Particle Size

There have been few studies of the impact of biochar particle size on microbial responses in soil. Different pyrolysis processes and feedstocks (organic origin) create biochar with different chemical, physical and size fractions (Keech et al. 2005; Gundale and DeLuca 2006; Downie et al. 2009; Verheijen et al. 2009).

Some biochars resemble the original cellular structure of the feedstock, in which large fragments correspond with woody plant material (Downie et al. 2009). Biochars also occur as large (>4 mm) through to fine particles (<20 µm) (Glaser et al. 2001, 2002). Commonly, biochar contains a mixture of particle size (Downie et al. 2009) or it is ground after production into smaller fractions (Sohi et al. 2010). Larger particles of biochar may be less practical for agricultural purposes due to their bulky characteristics compared with smaller particle sizes.

The dust portion of biochar has the greatest surface area but may not be the most effective soil amendment due to wind erosion and practicality (Blackwell et al. 2009). Biochar surfaces can gradually oxidise in response to exposure to air, activities of soil microorganisms or roots, and this may increase the cation-exchange capacity (Joseph et al. 2010). Changes to the surface of biochar after exposure to the soil environment may also alter water and nutrient retention properties of the biochar (Joseph et al. 2010). The size of the charcoal pieces amended to soil is not expected to greatly affect nutrient uptake but may alter surface properties which influence microbial attachment (Verheijen et al. 2009). Habte and Antal (2010) found that mycorrhizal colonisation of *Leucaena* roots was reduced when the growth medium was amended with fine (<0.3 mm in diameter) compared to coarse (<2.00 mm) charcoal. Lower levels of colonisation were associated with girdling of stems where the fine charcoal tended to accumulate. The large surface area could also enable greater absorption of toxic compound in soils (Antal and Gronli 2003; Habte and Antal 2010).

The selection of biochar for use in agricultural soils needs to be based on physical characteristics and chemical composition to achieve success in soil amelioration. Theoretically, biochar with higher porosity, a greater density of larger pores or large quantities of smaller particle sizes may benefit soil microorganisms, including mycorrhizal fungi. However, possible subsequent interactions between biochar and soil need to be taken into account. Furthermore, as the amount of biochar applied to soil can influence microbial processes, mixing biochar in the soil may also influence the distribution of microbial microsites in the soil in a different way to banding (Makoto et al. 2010). The application amount and method (mixing or banding) would normally be taken into account when incorporating biochar alone or with other amendments such as fertiliser for optimisation of nutrient capture (Blackwell et al. 2010). This could also change the interactions between biochar and soil microorganisms when organic matter or fertiliser is included.

19.4 Factors Influencing Biochar–Microbe Interactions: Soil Management

It has been suggested that the efficacy of biochar–AM fungal interactions may be reduced in more fertile soils (Lehmann et al. 2011). Incorporating biochar in farming systems that use other soil amendments and practices could be beneficial, but biochar has a longer residence time in soil compared to other sources of carbon. As a carbon-rich material, biochar is affected by soil processes, but the changes in biochar occur slowly in soil and the effect on soil nutrients is not well understood (Lehmann 2007; Lehmann and Joseph 2009; Joseph et al. 2010). Application of fertiliser and labile organic matter has been used with biochar to optimise the benefits of these soil amendments (Blackwell et al. 2010; Graber et al. 2010).

Dual incorporation of biochar with organic matter is normally associated with the goal of improving soil fertility. As a carbon source, biochar when added to soil could contribute to increasing the organic content in soil due to its recalcitrant nature. Addition of biochar as a nutrient source has been suggested (Rajkovich et al. 2012), and biochar may also contain small amount of volatiles, substrate for microbial degradation and activities, and nutrients (Downie et al. 2009). Transformation of organic matter which contains phytotoxic compounds by pyrolysis could be used as a soil amendment to avoid a detrimental effect on plant and soil properties (Ishii and Kadoya 1994), but most of the nutrients may be lost during pyrolysis. The availability of nutrients from soil organic matter may not necessarily be improved by biochar addition (Dempster et al. 2012a, b).

There is a potential role of biochar in improving the microbial status of soil amended with other forms of organic matter. For example, Zackrisson et al. (1996) suggested that microbial activity played a part in reactivating charcoal by decomposing attached materials to the charcoal and providing nutrient sources for microbial activity. Organic materials and minerals can be bound to biochar particles and it is important to note this when managing biochar and other sources of organic matter (Joseph et al. 2010). The structural nature of biochar could facilitate microbial development and indirectly accelerate adsorption and degradation of phenolic compounds (Keech et al. 2005). However, negative implications for soil microorganisms are also possible in certain cases involving organic substances through their interaction with biochar. In a study by Rutto and Mizutani (2006), application of activated charcoal slightly alleviated the negative detrimental effect of root bark extract but reduced the benefits derived from mycorrhizas for plant growth. The large surface area of materials such as activated charcoal enhances its ability to absorb organic compounds for soil detoxification purposes (Uchimiya et al. 2010).

Some biochars may contribute slightly to soil nutrient status through provision of small amounts of nutrients or impurities (Lehmann and Joseph 2009). This has been shown by Graber et al. (2010) whereby tar and labile compounds trapped in pores after pyrolysis provided substrate for microorganisms. Furthermore, biochar application can alter soil phosphorus availability through modification to carbon, nutrient and pH in soil (Glaser et al. 2002; Matsubara et al. 2002). Charcoal may improve the growth and spread of AM fungi in roots by neutralising soil acidity (Ishii and Kadoya 1994). In contrast, addition of carbonised materials to soil can cause a decline in AM fungal colonisation (Gaur and Adholeya 2000).

The ability of biochar to retain nutrients and heavy metals is dependent on the sorption characteristics of biochars which are controlled by the relative carbonised and non-carbonised fractions and their surface and bulk properties (Uchimiya et al. 2010, 2011a, b). However, there are concerns about the ability of AM fungi to develop in biochar-amended soil with high levels of phosphorus. Biochar pH is usually neutral to alkaline and may contain some phosphorus (Gundale and DeLuca 2006; Yamato et al. 2006) which may be available for microbial uptake. AM fungal effectiveness is affected by environmental and biological factors including P availability and mycorrhizal inoculum potential (Smith et al. 1992; Maeder et al. 2002). Thus, AM fungal development and function in soil amended with biochar would depend on biochar characteristics and soil nutrient status.

Wood-based biochar had the capacity to absorb measurable quantities of phosphate ions from a soil-free solution (Verheijen et al. 2009). The sorption of phosphorus to biochar may adversely affect how AM fungal hyphae inhabit the microenvironment of biochar. Mycorrhizal development responded positively to biochar at lower amounts of fertiliser applied to an agricultural soil (Blackwell et al. 2010). In this study, percentage in AM fungal colonisation increased when biochar was applied at 3 t/ha when the low level of phosphorus fertiliser was applied compared to the "full" fertiliser application. In contrast, Yamato et al. (2006) observed that colonisation by AM fungi (measured as proportion of root colonised) was highest for bark charcoal application without phosphorus fertiliser application. A large number of studies on the effect of charcoal application on the enhancement of AM fungal colonisation have been conducted (Ogawa and Yamabe 1986; Saito 1990; Ishii and Kadoya 1994; Ezawa et al. 2002; Ogawa and Okimori 2010) when no fertilisers were incorporated into the soil. However, mycorrhizal–biochar interactions would be expected to depend on the phosphorus status of the soil whether or not phosphate fertiliser was applied (Blackwell et al. 2010; Solaiman et al. 2010).

The absence of fertiliser can be compromised by applying higher amount of biochar. Blackwell et al. (2010) observed that biochar when applied at 3 t/ha resulted in greater root colonisation at the nil or low fertiliser rate. The low-level P fertiliser application in conjunction with biochar seems to have provided better conditions for mycorrhizal colonisation than the unfertilised soil or full fertiliser application. Biochar sorption of labile organic C could serve as a mechanism for decreased soil organic matter decomposition and concurrent P mineralisation and could result in decreased P availability as suggested by Kuzyakov et al. (2009).

19.5 Conclusion

Biochar application to soil involves complex interactions with soil and soil management practices. Biochar is heterogeneous in nature, especially in pore and surface structure associated with pyrolysis processes and feedstock source. These physical features were proposed to be associated with the abundance and development of microorganisms, but quantification of biochar pores and the effect of particle size are inconclusive, making it difficult to support the claim of biochar as a significant habitat for soil microorganisms compared to the soil itself. In summary, soil background characteristics, including pH and P status, may lead to different interactions between soil and biochars. As biochars are normally applied with fertiliser and commonly discussed in terms of a priming effect with labile organic matter addition, further investigations of interactions with soil microorganisms, including AM fungi, are warranted.

Overall, there are significant effects of the type of biochar used, which largely influences the amount of each biochar application to soil that could lead to beneficial effects on AM fungi. Hyphae of AM fungi have mainly been assessed within roots, not in the biochar microenvironment. The optimum amount of biochar application would need to be identified due to potential detrimental effects of higher biochar application levels on soil microorganisms or plant growth. The significance of biochar particle size has rarely been considered in relation to plant benefit or soil changes, but it may influence attachment of soil microorganisms to biochar surfaces. When biochar is applied with organic amendments, the mineralisation of biochar could be enhanced, but a concurrent effect of biochar on organic matter could also be important. On the contrary, although significant interactions between biochar and fertiliser have been shown, the optimal amount of biochar when interacting with fertilisers may vary with biochar type.

References

Anderson CR, Condron LM, Clough TJ, Fiers M, Stewart A, Hill RA, Sherlock RR (2011) Biochar induced soil microbial community change: Implications for biogeochemical cycling of carbon, nitrogen and phosphorus. Pedobiologia 54:309–320

Antal MJ, Gronli M (2003) The art, science, and technology of charcoal production. Ind Eng Chem Res 42:1619–1640

Ascough PL, Bird MI, Scott AC, Collinson ME, Cohen-Ofri I (2010a) Charcoal reflectance measurements: implications for structural characterization and assessment of diagenetic alteration. J Archaeol Sci 37:1590–1599

Ascough PL, Sturrock CJ, Bird MI (2010b) Investigation of growth responses in saprophytic fungi to charred biomass. Isot Environ Health Stud 46:64–77

Atkinson C, Fitzgerald J, Hipps N (2010) Potential mechanisms for achieving agricultural benefits from biochar application to temperate soils: a review. Plant and Soil 337:1–18

Baltruschat H (1987) Evaluation of the suitability of expanded clay as carrier material for VAM spores in field inoculation of maize. Angew Botanik 61:163–169

Blackwell P, Riethmuller G, Collins M (2009) Biochar application to soil. In: Lehmann J, Joseph S (eds) Biochar for environmental management, science and technology. Earthscan, London, pp 207–226

Blackwell P, Krull E, Butler G, Herbert A, Solaiman Z (2010) Effect of banded biochar on dryland wheat production and fertiliser use in south-western Australia: an agronomic and economic perspective. Aust J Soil Res 48:531–545

Chan KY, Van Zwieten L, Meszaros I, Downie A, Joseph S (2007) Agronomic values of greenwaste biochar as a soil amendment. Aust J Soil Res 45:629–634

Chan KY, Van Zwieten L, Meszaros I, Downie A, Joseph S (2008) Using poultry litter biochars as soil amendments. Aust J Soil Res 46:437–444

Cross A, Sohi SP (2011) The priming potential of biochar products in relation to labile carbon contents and soil organic matter status. Soil Biol Biochem 43:2127–2134

Dempster DN, Jones DL, Murphy DV (2012a) Clay and biochar amendments decreased inorganic but not dissolved organic nitrogen leaching in soil. Aust J Soil Res 50:216–221

Dempster DN, Jones DL, Murphy DV (2012b) Organic nitrogen mineralisation in two contrasting agro-ecosystems is unchanged by biochar addition. Soil Biol Biochem 48:47–50

Douds DD Jr, Nagahashi G, Pfeffer PE, Kayser WM, Reider C (2005) On-farm production and utilization of arbuscular mycorrhizal fungus inoculum. Can J Plant Sci 85:15–21

Downie A, Crosky A, Munroe P (2009) Physical properties of biochar. In: Lehmann J, Joseph S (eds) Biochar for environmental management, science and technology. Earthscan, London, pp 13–32

Elmer WH, Pignatello JJ (2011) Effect of biochar amendments on mycorrhizal associations and *Fusarium* crown and root rot of asparagus in replant soils. Plant Dis 95:960–966

Ezawa T, Yamamoto K, Yoshida S (2002) Enhancement of the effectiveness of indigenous arbuscular mycorrhizal fungi by inorganic soil amendments. Soil Sci Plant Nutr 48:897–900

Gaur A, Adholeya A (2000) Effects of the particle size of soil-less substrates upon AM fungus inoculum production. Mycorrhiza 10:43–48

Gazey C, Abbott LK, Robson AD (2004) Indigenous and introduced arbuscular mycorrhizal fungi contribute to plant growth in two agricultural soils from south-western Australia. Mycorrhiza 14:355–362

Glaser B, Haumaier L, Guggenberger G, Zech W (2001) The "Terra Preta" phenomenon: a model for sustainable agriculture in the humid tropics. Naturwissenschaften 88:37–41

Glaser B, Lehmann J, Zech W (2002) Ameliorating physical and chemical properties of highly weathered soils in the tropics with charcoal—a review. Biol Fertil Soils 35:219–230

Graber E, Harel MY, Kolton M, Cytryn E, Silber A, David DR, Tsechansky L, Borenshtein M, Elad Y (2010) Biochar impact on development and productivity of pepper and tomato grown in fertigated soilless media. Plant and Soil 337:481–496

Gundale MJ, DeLuca TH (2006) Temperature and source material influence ecological attributes of ponderosa pine and Douglas-fir charcoal. For Ecol Manag 231:86–93

Habte M, Antal MJ Jr (2010) Reaction of mycorrhizal and nonmycorrhizal *leucaena leucocephala* to charcoal amendment of Mansand and soil. Commun Soil Sci Plant Anal 41:540–552

Hockaday WC, Grannas AM, Kim S, Hatcher PG (2007) The transformation and mobility of charcoal in a fire-impacted watershed. Geochimica Et Cosmochimica Acta 71:3432–3445

Ishii T, Kadoya K (1994) Effects of charcoal as a soil conditioner on citrus growth and vesicular-arbuscular mycorrhizal development. J Jpn Soc Hort Sci 63:529–535

Jin H (2010) Characterization of microbial life colonizing biochar and biochar-amended soils. Ph. D. Dissertation, Cornell University, Ithaca

Joseph SD, Camps-Arbestain M, Lin Y, Munroe P, Chia CH, Hook J, Van Zwieten L, Kimber S, Cowie A, Singh BP, Lehmann J, Foidl N, Smernik RJ, Amonette JE (2010) An investigation into the reactions of biochar in soil. Aust J Soil Res 48:501–515

Keech O, Carcaillet C, Nilsson MC (2005) Adsorption of allelopathic compounds by wood-derived charcoal: the role of wood porosity. Plant and Soil 272:291–300

Kolb SE, Fermanich KJ, Dornbush ME (2009) Effect of charcoal quantity on microbial biomass and activity in temperate soils. Soil Sci Soc Am J 73:1173–1181

Kuzyakov Y, Subbotina I, Chen H, Bogomolova I, Xu X (2009) Black carbon decomposition and incorporation into soil microbial biomass estimated by ^{14}C labeling. Soil Biol Biochem 41:210–219

Lehmann J (2007) Bio-energy in the black. Front Ecol Environ 5:381–387

Lehmann J, Joseph S (eds) (2009) Biochar for environmental management, science and technology. Earthscan, London

Lehmann J, Rillig MC, Thies J, Masiello CA, Hockaday WC, Crowley D (2011) Biochar effects on soil biota: a review. Soil Biol Biochem 43:1812–1836

Liang B, Lehmann J, Solomon D, Kinyangi J, Grossman J, O'Neill B, Skjemstad JO, Thies J, Luizão FJ, Petersen J, Neves EG (2006) Black carbon increases cation exchange capacity in soils. Soil Sci Soc Am J 70:1719–1730

Maeder P, Fliessbach A, Dubois D, Gunst L, Fried P, Niggli U (2002) Soil fertility and biodiversity in organic farming. Science 296:1694–1697

Makoto K, Tamai Y, Kim YS, Koike T (2010) Buried charcoal layer and ectomycorrhizae cooperatively promote the growth of *Larix gmelinii* seedlings. Plant and Soil 327:143–152

Matsubara Y, Hasegawa N, Fukui H (2002) Incidence of *Fusarium* root rot in asparagus seedlings infected with arbuscular mycorrhizal fungus as affected by several soil amendments. J Jpn Soc Hort Sci 71:370–374

Moskal-del Hoyo M, Wachowiak M, Blanchette RA (2010) Preservation of fungi in archaeological charcoal. J Archaeol Sci 37:2106–2116

Norris JR, Reads DJ, Varma AK (eds) (1992) Techniques for the study of mycorrhiza. In: Methods in microbiology, vol. 24. Academic, London

Ogawa M, Okimori Y (2010) Pioneering works in biochar research, Japan. Aust J Soil Res 48:489–500

Ogawa M, Yamabe Y (1986) Effects of charcoal on VA mycorrhizae and nodule formation of soybeans. Bull Green Energy Programme Group II, No. 8 Min Agric For Fish Japan, pp 108–133

Pietikainen J, Kiikkila O, Fritze H (2000) Charcoal as a habitat for microbes and its effect on the microbial community of the underlying humus. Oikos 89:231–242

Rajkovich S, Enders A, Hanley K, Hyland C, Zimmerman A, Lehmann J (2012) Corn growth and nitrogen nutrition after additions of biochars with varying properties to a temperate soil. Biol Fertil Soils 48:271–284

Rillig MC, Wagner M, Salem M, Antunes PM, George C, Ramke H-G, Titirici M-M, Antonietti M (2010) Material derived from hydrothermal carbonization: effects on plant growth and arbuscular mycorrhiza. Appl Soil Ecol 45:238–242

Rondon M, Lehmann J, Ramírez J, Hurtado M (2007) Biological nitrogen fixation by common beans (*Phaseolus vulgaris* L.) increases with bio-char additions. Biol Fertil Soils 43:699–708

Rutto KL, Mizutani F (2006) Effect of Mycorrhizal inoculation and activated charcoal on growth and nutrition in peach *Prunus persica* seedlings treated with peach root-bark extracts. J Jpn Soc Hort Sci 75:463–468

Saito M (1990) Charcoal as a micro-habitat for VA mycorrhizal fungi, and its practical implication. Agr Ecosyst Environ 29:341–344

Schmidt MWI, Noack AG (2000) Black carbon in soils and sediments: analysis, distribution, implications, and current challenges. Global Biogeochem Cycles 14:777–793

Smith SE, Robson AD, Abbott LK (1992) The involvement of mycorrhizas in assessment of genetically dependent efficiency of nutrient uptake and use. Plant and Soil 146:169–179

Sohi SP, Krull E, Lopez-Capel E, Bol R (2010) Chapter 2: A review of biochar and its use and function in soil. In: Donald LS (ed) Advances in agronomy, vol 105. Academic, Burlington, pp 47–82

Solaiman ZM, Blackwell P, Abbott LK, Storer P (2010) Direct and residual effect of biochar application on mycorrhizal root colonisation, growth and nutrition of wheat. Aust J Soil Res 48:546–554

Steiner C, Das KC, Garcia M, Förster B, Zech W (2008) Charcoal and smoke extract stimulate the soil microbial community in a highly weathered xanthic Ferralsol. Pedobiologia 51:359–366

Swift MJ, Heal OW, Anderson JM (1979) Decomposition in terrestrial ecosystems, vol 5. University of California Press, Berkeley, pp 167–219

Thies JE, Rillig MC (2009) Characteristics of biochar: biological properties. In: Lehmann J, Joseph S (eds) Biochar for environmental management, science and technology. Earthscan, London, pp 85–105

Uchimiya M, Lima IM, Klasson KT, Wartelle LH (2010) Contaminant immobilization and nutrient release by biochar soil amendment: Roles of natural organic matter. Chemosphere 80:935–940

Uchimiya M, Chang S, Klasson KT (2011a) Screening biochars for heavy metal retention in soil: role of oxygen functional groups. J Hazard Mater 190:432–441

Uchimiya M, Klasson KT, Wartelle LH, Lima IM (2011b) Influence of soil properties on heavy metal sequestration by biochar amendment: 2. Copper desorption isotherms. Chemosphere 82:1438–1447

Verheijen F, Jeffery S, Bastos AC, Van der Velde M, Diafas I (2009) Biochar application to soils: a critical scientific review of effects on soil properties, processes and functions. Joint Research Centre (JRC) Scientific and Technical Report No EUR 24099 EN, Office for the Official Publications of the European Communities, Luxembourg

Warnock DD, Lehmann J, Kuyper TW, Rillig MC (2007) Mycorrhizal responses to biochar in soil: concepts and mechanisms. Plant and Soil 300:9–20

Warnock DD, Mummey DL, McBride B, Major J, Lehmann J, Rillig MC (2010) Influences of non-herbaceous biochar on arbuscular mycorrhizal fungal abundances in roots and soils: results from growth-chamber and field experiments. Appl Soil Ecol 46:450–456

Yamato M, Okimori Y, Wibowo IF, Anshori S, Ogawa M (2006) Effects of the application of charred bark of Acacia mangium on the yield of maize, cowpea and peanut, and soil chemical properties in South Sumatra, Indonesia. Soil Sci Plant Nutr 52:489–495

Zackrisson O, Nilsson MC, Wardle DA (1996) Key ecological function of charcoal from wildfire in the boreal forest. Oikos 77:10–19

Chapter 20
Application of AM Fungi in Remediation of Saline Soils

Anne Nurbaity

20.1 Introduction

Rehabilitation of saline soil is crucial because large areas of the world have saline soils or are prone to encroaching salinity. Essential to rehabilitation of saline soil is the revegetation of recharge areas (Stirzaker et al. 2002). Management practices are required that improve the quality and productivity of saline soil and the ability of the plant to better withstand salt stress (Al-Karaki 2001; Caravaca et al. 2002a). For sustainable agriculture, solutions to salinity-related problems must acknowledge biological processes as part of rehabilitation. Practices such as organic matter application (Bell and Mann 2004; Caravaca et al. 2002b) and/or microbial inoculation (Aliasgharzadeh et al. 2001) are options for rehabilitation of degraded land. Increased organic matter levels and reactivated microbial activity either as free-living organisms or in association with plant roots are likely to be important for improving soil quality (Caravaca et al. 2002a; Diaz et al. 1994).

Arbuscular mycorrhizal (AM) fungi have been considered as bio-ameliorators of saline soil (Feng et al. 2002) because they may enhance the tolerance of plants to salinity (Al-Karaki 2001; Boyacioglu and Uyanoz 2014; Cantrel and Linderman 2001; Feng et al. 2002; Ruiz-Lozano et al. 1996). The practical use of AM fungi as a form of biological fertiliser and organic matter as low-input technologies for managing soil fertility has been investigated (Gaur and Adholeya 2002; Gryndler et al. 2002). This review examines biological aspects of rehabilitation of saline soil and considers the role of organic matter and AM fungi in saline environments.

A. Nurbaity (✉)
Faculty of Agriculture, Universitas Padjadjaran, Jl. Raya Jatinangor km. 21, Bandung 40600, Indonesia
e-mail: annenurbaity@unpad.ac.id

© Springer-Verlag Berlin Heidelberg 2014   313
Z.M. Solaiman et al. (eds.), *Mycorrhizal Fungi: Use in Sustainable Agriculture and Land Restoration*, Soil Biology 41, DOI 10.1007/978-3-662-45370-4_20

20.2 Effects of Salinity on Plant Growth and Soil Biological Activities

In general, dissolved salts may affect plant growth and soil organisms by direct injury due to specific ion toxicity and/or indirectly via osmotic imbalance effects (Al-Karaki 2001; Ferguson and Grattan 2005). Specific ion toxicities are due to the accumulation of ions such as sodium and/or chloride in the tissue to damaging levels (Al-Karaki 2001; Juniper and Abbott 1993). In plants, these ion accumulations occur as direct foliar accumulation or via root uptake. The damage is visible as foliar chlorosis and necrosis (Ferguson and Grattan 2005). Ionic effects also include interference with essential ions and a lowering of the net rate of photosynthesis (Ruiz-Lozano et al. 1996). Osmotic effects are caused by the total concentration of salt in the soil solution produced by the combination of soil salinity, irrigation water quality and fertilisation (Ferguson and Grattan 2005) and interfere with the ability of the plant to take up water. These effects are associated with the inhibition of cell wall extension and cellular expansion, leading to reduced plant growth (Boughanmi et al. 2003; Ruiz-Lozano et al. 1996).

High salt concentrations can reduce seed germination and initial seedling growth, as well as the growth of established plants (Feng et al. 2000; Esechie et al. 2002; Zedler et al. 2003). Salinity may induce nutrient deficiencies or imbalances in plants due to the competition of Na^+ and Cl^- with nutrients such as K^+, Ca^{2+} and NO_3^- (Hu and Schmidhalter 2005). Most plants have a threshold salt concentration value above which yields decline. High salt concentration can also have an adverse effect on microbiological processes, including mineralisation, soil enzyme activities and soil respiration (Nelson et al. 1996; Ramirez-Fuentes et al. 2002; Wong et al. 2004). Mineralisation of C and N can be negatively correlated with salinity (Nelson et al. 1996; Pathak and Rao 1997). Other examples of an inhibitory effect of salinity on microbial processes include the inhibition of oxidation of NO_2^- (Ramirez-Fuentes et al. 2002) and inhibition of dehydrogenase activity (Batra and Manna 1997).

High salinity levels in soil can have variable impacts on the abundance and activity of AM fungi (Juniper and Abbott 1993). Some investigations have shown that excessive NaCl levels in soil inhibit mycorrhizal formation and restrict the activity of most mycorrhizal fungi (McMillen et al. 1998; Ruiz-Lozano and Azcon 2000). The formation of mycorrhizas will generally be inhibited in plants if the concentration of salt in the soil exceeds 3.3 mg/g (Gupta and Mukerji 2000).

Salts can reduce spore germination, hyphal growth and colonisation and spore production for AM fungi (Juniper and Abbott 2004). In addition, roots of *Parthenium argentatum* treated with NaCl had fewer arbuscules and vesicles (Pfeiffer and Bloss 1988). Consequently, a delay in spore germination phases due to dissolved salts in the soil solution may inhibit or even stop the growth of hyphae and colonisation of plant roots and, hence, the establishment of the symbiosis (Juniper and Abbott 2006).

20.3 Management and Rehabilitation of Saline Lands

Management of saline lands for agricultural use is influenced by many factors but particularly by water availability (FAO 2006; Qadir et al. 2000). As a consequence, the solution to salinity problems is not simple, and techniques or agricultural systems developed will need to be suited to different locations and conditions. Physical, chemical and biological amelioration processes included in rehabilitation programmes need to be considered simultaneously. Physical amelioration can include hydrological processes such as drainage and leaching, land tillage and planting practices, and chemical amelioration practices can include the use of chemical amendments such as gypsum, sulphur and mineral fertilisers (FAO 2006).

20.3.1 Role of Organic Matter in Saline Environments

The application of organic matter during management of saline land has included manure incorporation, mulching and incorporation of crop residues (FAO 2006). Microbial inocula including mycorrhizal fungi have potential in improving the quality of saline soils (Aliasgharzadeh et al. 2001; Caravaca et al. 2002a). Soil organic matter influences a wide range of physical, chemical and biological properties of soil (Caravaca et al. 2002b). The beneficial effects of organic amendments on soil physical characteristics are decreased bulk density and increased aggregate stability, water holding capacity, saturated hydraulic conductivity, water infiltration rate and some measures of biological activity (Caravaca et al. 2002b, 2003; Celik et al. 2004). Furthermore, polysaccharides and other biopolymers from organic matter (including composts) can improve soil aggregate stability, which in turn improves water holding capacity and porosity of the soil (Caravaca et al. 2003; He et al. 1992).

In saline soil, benefits of organic matter on physical, biological and chemical properties as well as on plant growth have been reported. Generally, the utilisation of organic matter as mulch can reduce evaporation losses and thus decrease or prevent soil salinisation (Barrett-Lennard 2003). Mulching can delay the return of salt by lowering surface evaporation after salt is leached downward by rainfall and low evaporation during winter (Badia 2000). Furthermore, addition of organic matter may change the biological properties of the soil through improvement in the condition of soil for the multiplication of microorganisms and hence their activity such as microbial respiration or enzyme function (Badia 2000; Gryndler et al. 2005; Johnson 1998). Microbiological effects of organic matter include changes in total microorganism abundance, relative distribution of different groups of microorganisms (bacteria, fungi, actinomycetes) and activities such as mineralisation, nitrification and denitrification (He et al. 1992).

Finally, the combination of physical and biological ameliorative processes will lead to chemical changes in saline soil. There is evidence that incorporation of organic matter into saline soil (Calcaric Regosols) in a semiarid Mediterranean climate decreased salinity levels (Badia 2000). The production of carbon dioxide from respiration of roots and soil organisms, and the production of organic acids by soil organisms and from organic matter decomposition, can influence dissolution of calcite (Ca^{2+}) (Qadir et al. 2005). Chelation of calcium increased the solubility of $CaCO_3$ and prevented Ca precipitation (Avnimelech et al. 1992), Ca^{2+} exchanges with Na^+ in the clay complex, which can reduce soil salinity (Badia 2000; Qadir et al. 2005). The beneficial effect of organic matter amendment to a great extent depends on the nature, maturity and quantity of organic matter applied (Roldan et al. 1996). For example, non-composted organic residues applied at higher application levels have been shown to be more effective than composted residue in stimulating the microbial activity because they are rich in easily biodegradable compounds (Caravaca et al. 2002b; He et al. 1992).

20.3.2   Role of AM Fungi in Saline Environments

Despite the relatively low mycorrhizal affinity of many halophytic plants, fairly large populations of AM fungi have been reported in some saline soils (Aliasgharzadeh et al. 2001). Experiments that have assessed the status of AM fungi in saline soil have shown a wide range of results, consequently, discussion as to whether biotic or abiotic factors have most influenced the ecology of AM fungi in saline environment is of interest (Carvalho et al. 2001; Mohammad et al. 2003a).

Assessments of spatial and temporal distribution of AM fungi in saline soil show that the abundance of AM fungi is inversely correlated with the level of soil salinity. The number of propagules or the infectivity of fungal isolates decreases with increasing salt (Azcon-Aguilar et al. 2003; Carvalho et al. 2001, 2003; Hildebrandt et al. 2001; Landwehr et al. 2002; Sylvia 1986; Wang et al. 2003). However, spore density had a very weak or no correlation with soil salinity (Carvalho et al. 2003; Mohammad et al. 2003a). In central European salt marshes, a high number of *Glomus* spores was found in saline soils (Hildebrandt et al. 2001), and *Glomus* was the dominant genus in other saline soils (Agwa and Abdel-Fattah 2002; Wang et al. 2003).

Communities of AM fungi have shown significant spatial heterogeneity and non-random associations with different hosts (Husband et al. 2002). For example, the presence of mycorrhizas in a salt marsh was more dependent on host plant species than on environmental stresses (Carvalho et al. 2001). It was also found that differences in patterns of activity of mycorrhizal fungi appeared to be linked to differences in phenology of root growth and not edaphic differences among vegetation zones (Johnson-Green et al. 1995). In contrast, Allen et al. (1995) stated that the diversity of mycorrhizal fungi was not associated with the patterns of plant diversity.

Mycorrhizal symbioses have been shown to improve the ability of some plant species to withstand salt stress (Al-Karaki 2001; Ruiz-Lozano et al. 1996). Some experiments indicated that salt-treated AM plants produced greater shoot and root dry weights than did non-AM controls. For instance, AM fungi promoted growth of *Zea mays* in saline conditions, with the effect increasing as the degree of stress increased (Feng et al. 1998) and elsewhere (Bhoopander and Mukerji 1999; Cantrel and Linderman 2001). Where the growth of both AM and non-AM plants decreased as salinity increased, the decreases were more pronounced in non-mycorrhizal plants (Asghari 2008).

Inoculation with AM fungi under glasshouse conditions increased shoot contents of P and K (e.g. Asghari 2008). However, reports of effects of AM fungi on Na uptake in saline soil have been inconsistent. Sometimes Na content in shoot tissue was higher in mycorrhizal plants (Cantrel and Linderman 2001; Pfeiffer and Bloss 1988). On the contrary, mycorrhizal plants had less Na content in shoots of tomato (Al-Karaki 2001) and barley (Mohammad et al. 2003b) than did non-mycorrhizal plants when grown in a soil with a high level of salinity.

Starch and total carbohydrate concentrations in leaves and roots have been generally shown to be reduced by salt (Ezz and Nawar 1994). Inoculation of plants with AM fungi under saline conditions generally increased the accumulation of leaf and root carbohydrate (Ezz and Nawar 1994), including proline and total free amino acids. In contrast, Aboul-Nasr (1999) found that proline accumulation was considerably less for mycorrhizal plants than for non-mycorrhizal plants (Aboul-Nasr 1999). AM fungi may also influence some plant hormones (Ruiz-Lozano et al. 1996) and improve water uptake (Augé 2001) leading to increased growth and subsequent dilution of toxic ion effects (Al-Karaki 2001; Ruiz-Lozano et al. 1996). Furthermore, AM fungi have been found to affect the activity of some enzymes. Polyphenol oxidase increased with AM fungi inoculation, but peroxidase activity was not affected (Ezz and Nawar 1994; Santos et al. 2001).

20.3.3 Mechanisms of Improved Tolerance of Plants to Salinity

Various mechanisms of salt tolerance by AM fungi have been proposed. These are (a) improved plant mineral nutrition and/or increased leaf sequestration of chlorides (Copeman et al. 1996; Feng et al. 1998; Juniper and Abbott 2003); (b) altered plant water balance such as reduced water stress of the host plants and dilution of toxic ions such as sodium (Na) and chloride (Cl^-) (Gupta and Mukerji 2000; Juniper and Abbott 2003); (c) compartmentation of ions (Jennings and Burke 1990); and (d) osmotic adjustment by production of compatible solutes by the plant (Jennings and Burke 1990; Hampp and Schaeffer 1995).

The mechanisms of salt tolerance by AM fungi include improved plant nutrition, especially P, or enhanced acquisition of low-mobility nutrients (Al-Karaki 2001; Ruiz-Lozano and Azcon 2000). Maintenance of a high K:Na ratio in shoots has been investigated as a mechanism of salt tolerance (Cakmak 2005). The increasing

K concentration in mycorrhizal plants could be an indirect effect associated with better P nutrition in mycorrhizal plants (Poss et al. 1985).

Potassium could alleviate detrimental effects of salt stress because the impairment of K nutrition is a main characteristic of plants under salt stress (Cakmak 2005). At the cellular level, K deficiency might contribute to salt-induced oxidative stress and related cell damage. Accordingly, improving K nutrition of salt-stressed plants could reduce cell damage (Cakmak 2005). Therefore, increased K or decreased Na concentration in mycorrhizal plants are also believed to increase plant salinity tolerance because the internal K:Na ratio increases (Rinaldelli and Mancuso 1996).

Higher water potential of tomato xylem and improved K nutrition associated with AM fungi indicate that mechanisms other than increased P nutrition may be important for mycorrhizal plants grown under saline stress (Poss et al. 1985). Possible mechanisms for the enhancement of salt tolerance by AM fungi are an effect of mycorrhizas on reducing water stress of plants (Al-Karaki 2001; Augé 2001) and dilution of toxic ions such as Na and Cl^- (Al-Karaki 2001; Gupta and Mukerji 2000; Ruiz-Lozano et al. 1996).

Osmotic adjustment is defined as the capacity of the internal concentration of solutes to increase in response to a decrease in external water potential, particularly due to an increase in salinity (Jennings and Burke 1990). This is one of the best known responses of plants to salinity stress, where plants accumulate soluble, low-molecular-mass solutes such as proline and betaine (Ben Khaled et al. 2003; Ruiz-Lozano et al. 1996). Osmotic adjustment by use of compatible solutes assists salinised plants in the maintenance of leaf turgor and other physiological processes such as photosynthesis, transpiration, conductance and water-use efficiency (Augé 2001; Ruiz-Lozano et al. 1996). Correspondingly, one of the mechanisms of adaptation of fungi to high concentrations of salt is osmotic adjustment by the synthesis of compatible solutes such as polyols (glycerol, mannitol, arabitol, sorbitol) and accumulation of proline and betain. These compatible solutes increase in concentration in the cells of many fungi in response to salinity (Jennings and Burke 1990; Hampp and Schaeffer 1995; Naidu 1998).

20.3.4 Interactions Between Organic Matter and AM Fungi

Organic matter is known to have variable effects on AM associations (Soedarjo and Habte 1993). AM fungi may use soil organic C as an energy source (Caravaca et al. 2002b). Apparent preferential associations between AM hyphae and organic-rich microsites have been attributed to the nutrient-rich status of the sites (Gryndler et al. 2005). In contrast, mycorrhizal tissue (particularly hyphae) has been estimated to comprise a significant fraction of soil organic matter or has potential to form a sink or source of C (Treseder and Allen 2000). These examples demonstrate that potential interactions between organic matter and AM fungi are complex.

Interactions between organic matter and AM fungi in nonsaline environments can be either positive (the majority of cases) or negative.

Generally, AM fungal inoculation (with the assumption that these fungi were salt tolerant), in association with organic amendments, increased the abundance of mycorrhizal propagules in soil, leading to potential benefits at later stages of the revegetation process (Palenzuela et al. 2002). Furthermore, investigations of interactions between organic matter and AM fungi may contribute information that is important for the management of plants in soils with low levels of organic matter and nutrients (Gryndler et al. 2002). In the short-term, the use of organic matter and AM fungi together can increase various physical, chemical and biochemical parameters of rhizosphere soil contributing to improved soil quality (Caravaca et al. 2003).

In nonsaline soil, the beneficial effect of the addition of a combination of organic matter and AM fungi is most likely to be due to the reactivation of microbial activity (Caravaca et al. 2002b), including AM fungi. In saline soil, the mechanism of the interaction between organic matter and AM fungi is not simple. As high salinity negatively affects microbial activity (Batra and Manna 1997), an additional step with the aim of lowering the water table and reducing salinity needs to be put into place prior to management of the AM fungi and other soil microorganisms. Incorporation of organic matter (or mulching) could reduce salinity in the top soil by preventing capillary salt rise.

After salt in the soil profile is decreased and because organic matter and AM fungi have combined effects on the improvement of soil physical (increased aggregate stability and porosity) and biochemical (as source of carbon and nutrients or production of enzymes to access nutrient from organic matter) properties, the activity of soil microorganisms is expected to improve. This increased activity of microorganisms may include the reactivation of AM fungi. Consequently, the interaction between organic matter and AM fungi could increase the water balance, thereby enhancing the dilution of toxic ions.

Finally, AM fungi may contribute to nutritional benefits to plant growth in saline soil. The effect of environmental factors such as rainfall and temperature (or season) is an important point that might influence the interaction between organic matter and AM fungi in saline soil. In hot and dry conditions, soil is likely to be saline, but AM fungi may be present even though they are inactive. In wet conditions, the salt concentration is likely to be lower, hence enabling mycorrhizal to function. Consequently, the possible positive interaction between organic matter and AM fungi depends to a great extent on seasonal change and hydrological status of the saline soil.

20.4 Conclusion

Organic matter incorporation can represent a key focus by which to develop longer term ecologically effective strategies to counterbalance degradation processes such as salinisation in agricultural soil. AM fungi have the potential to play an important role in saline soil, but species of these fungi can be affected by salinity to different extents. There are indications that a combined beneficial effect of organic matter and AM fungi in soil would be due to the creation of soil conditions suitable for the growth of hyphae and increased microbial activity overall.

Spatial and seasonal variations of AM fungi infectivity in saline soil are likely to complicate efforts to maximise the benefit of AM fungi in saline soil. Furthermore, there is uncertainty as to whether environmental or other factors such as plant distribution have the most influence on the distribution and infectivity of AM fungi in saline soil. Therefore, assessment of soil salinity and AM fungi across space and time will enable better evaluation of the potential role of AM fungi in rehabilitation of saline soils. It is possible for AM fungi to be present but inactive when the soil is saline (as in summer in a Mediterranean climate) and active after rain has diluted the salt. Thus, inoculation might only be necessary to overcome spatial rather than temporal heterogeneity in the distribution of mycorrhizas in saline soil in this type of environment.

Both organic matter and AM fungi alone can be effective in the amelioration of the effects of salinity on the growth of agricultural plants. However positive synergistic interactions between organic matter and AM fungi mean that AM fungi could be more effective in reducing the effect of salinity in the presence of organic matter.

References

Aboul-Nasr A (1999) Alleviation of salt stress by *Glomus intraradices* on linseed (*Linum usitatissimum* L.) in hydroponic culture. Alex J Agric Res 44:115–127

Agwa HE, Abdel-Fattah GM (2002) Arbuscular mycorrhizal fungi (*Glomales*) in Egypt. II. An ecological view of some saline affected plants in the deltaic Mediterranean coastal land. Acta Bot Hung 4:1–17

Aliasgharzadeh N, Rastin NS, Towfighi H, Alizadeh A (2001) Occurrence of arbuscular mycorrhizal fungi in saline soils in the Tabriz Plain of Iran in relation to some physical and chemical properties of soil. Mycorrhiza 11:119–122

Al-Karaki GN (2001) Salt stress response of salt-sensitive and tolerant durum wheat cultivars inoculated with mycorrhizal fungi. Acta Agron Hung 49:25–34

Allen EB, Allen MF, Helm DJ, Trappe JM, Molina R, Rincon E (1995) Patterns and regulation of mycorrhizal plant and fungal diversity. Plant and Soil 170:47–62

Asghari HR (2008) Vesicular-arbuscular (VA) mycorrhizae improve salinity tolerance in pre-inoculation subterranean clover (Trifolium subterraneum) seedlings. Int J Plant Prod 2:243–256

Augé RM (2001) Water relations, drought and vesicular-arbuscular mycorrhizal symbiosis. Mycorrhiza 11:3–42

Avnimelech Y, Kochva M, Yotal Y, Shkedy D (1992) The use of compost as a soil amendment. Acta Hortic 302:217–236

Azcon-Aguilar C, Palenzuela J, Roldan A, Bautista S, Vallejo R, Barea JM (2003) Analysis of the mycorrhizal potential in the rhizosphere of representative plant species from desertification-threatened mediterranean shrublands. Appl Soil Ecol 22:29–37

Badia D (2000) Straw management effects on organic matter mineralization and salinity in semiarid agricultural soils. Arid Soil Res Rehabil 14:193–203

Barrett-Lennard EG (2003) Saltland pastures in Australia: a practical guide. Land, water and wool sustainable grazing on saline lands sub-program, Australia

Batra L, Manna MC (1997) Dehydrogenase activity and microbial biomass carbon in salt-affected soils of semiarid and arid regions. Arid Soil Res Rehabil 11:295–303

Bell RW, Mann S (2004) Amelioration of salt and waterlogging-affected soils: implications for deep drainage. In: Dogramaci S, Waterhouse A (eds). 1st National salinity engineering conference proceeding, Perth, 9–12 Nov 2004. Institution of Engineers, Australia

Ben Khaled L, Gomez AM, Ouarraqi EM, Oihabi A (2003) Physiological and biochemical responses to salt stress of mycorrhized and/or nodulated clover seeding (Trifolium alexandrinum L.). Agronomie 23:571–580

Bhoopander G, Mukerji KG (1999) Improved growth and productivity of Sesbania grandiflora (Pers) under salinity stress through mycorrhizal technology. J Phytol Res 12:35–38

Boughanmi N, Michonneau P, Verdus M, Piton F, Ferjani E, Bizid E, Fleurat-Lessard P (2003) Structural changes induced by NaCl in companion and transfer cells of Medicago sativa blades. Protoplasma 220:179–187

Boyacioglu TU, Uyanoz R (2014) Effects of mycorrhizal fungi on tolerance capability of corn grown under salt stress condition. J Plant Nutr 37:107–122

Cakmak I (2005) The role of potassium in alleviating detrimental effects of abiotic stresses in plants. J Plant Nutr Soil Sci 168:521–530

Cantrel IC, Linderman RG (2001) Preinoculation of lettuce and onion with VA mycorrhizal fungi reduces deleterious effects of soil salinity. Plant and Soil 233:269–281

Caravaca F, Garcia C, Hernandez MT, Roldan A (2002a) Aggregate stability changes after organic amendment and mycorrhizal inoculation in the afforestation of a semiarid site with Pinus halepensis. Appl Soil Ecol 19:199–208

Caravaca F, Barea JM, Figueroa D, Roldan A (2002b) Assessing the effectiveness of mycorrhizal inoculation and soil compost addition for enhancing reafforestation with Olea europaea subsp. sylvestris through changes in soil biological and physical parameters. Appl Soil Ecol 20:107–118

Caravaca F, Figueroa D, Roldan A (2003) Alteration in rhizosphere soil properties of afforested Rhamnus lycioides seedlings in short-term response to mycorrhizal inoculation with Glomus intraradices and organic amendment. Environ Manage 31:412–420

Carvalho LM, Cacador I, Martins-Loucao MA (2001) Temporal and spatial variation of arbuscular mycorrhizas in salt marsh plants of the Tagus estuary (Portugal). Mycorrhiza 11:303–309

Carvalho LM, Correia PM, Ryel RJ, Martins-Loucao MA (2003) Spatial variability of arbuscular mycorrhizal fungal spores in two natural plant communities. Plant and Soil 251:227–236

Celik I, Ortas I, Kilic S (2004) Effects of compost, mycorrhiza, manure and fertilizer on some physical properties of a Chromoxerert soil. Soil Tillage Res 78:59–67

Copeman RH, Martin CA, Stutz JC (1996) Tomato growth in response to salinity and mycorrhizal fungi from saline or non saline soils. HortSci 31:341–344

Diaz E, Roldan A, Lax A, Albaladejo J (1994) Formation of stable aggregates in degraded soil by amendment with urban refuse and peat. Geoderma 63:277–288

Esechie H, Al-Saidi A, Al-Khanjari S (2002) Effect of sodium chloride salinity on seedling emergence in chickpea. J Agron Crop Sci 188:155–160

Ezz T, Nawar A (1994) Salinity and mycorrhizal association in relation to carbohydrate status, leaf chlorophyll and activity of peroxidase and polyphenol oxidase enzymes in sour orange seedlings. Alex J Agric Res 39:263–280

FAO (2006) Global Network on Integrated soil management of sustainable use of salt affected soils (SPUSH): extent and causes of salt-affected soils in participating countries. http://www.fao.org/ag/agl/agll/spush/default.htm

Feng G, Yang M, Bai D (1998) Influence of VAM fungi on mineral elements concentration and composition in *Bromus inermis* under salinity stress. Acta Pratacult Sin 7:21–28

Feng G, Li X, Zhang F, Li S (2000) Effect of phosphorus and arbuscular mycorrhizal fungus on response of maize plant to saline environment. J Plant Resourc Environ 9:22–26

Feng G, Zhang FS, Li XL, Tian CY, Tang C, Rengel Z (2002) Improved tolerance of maize plants to salt stress by arbuscular mycorrhiza is related to higher accumulation of soluble sugars in roots. Mycorrhiza 12:185–190

Ferguson L, Grattan SR (2005) How salinity damages citrus: osmotic effects and specific ion toxicities. HortTechnol 15:95–99

Gaur A, Adholeya A (2002) Arbuscular-mycorrhizal inoculation of five tropical fodder crops and inoculum production in marginal soil amended with organic matter. Biol Fertil Soils 35:214–218

Gryndler M, Vosatka M, Hrselova H, Cvatalova I, Jansa J (2002) Interaction between arbuscular mycorrhizal fungi and cellulose in growth substrate. Appl Soil Ecol 19:279–288

Gryndler M, Hrselova H, Sudova R, Gryndlerova H, Rezacova V, Merhautova V (2005) Hyphal growth and mycorrhiza formation by the arbuscular mycorrhizal fungus *Glomus claroideum* BEG 23 is stimulated by humic substances. Mycorrhiza 15:483–488

Gupta R, Mukerji KG (2000) The growth of VAM fungi under stress conditions. In: Mukerji KG et al (eds) Mycorrhizal biology. Kluwer, New York, pp 57–62

Hampp R, Schaeffer C (1995) Mycorrhiza-carbohydrate and energy metabolism. In: Varma A, Hock B (eds) Mycorrhiza. Springer, Berlin, pp 267–296

He XT, Traina SJ, Logan TJ (1992) Chemical properties of municipal solid waste composts. J Environ Qual 21:318–329

Hildebrandt U, Janetta K, Ouziad F, Renne B, Nawrath K, Bothe H (2001) Arbuscular mycorrhizal colonization of halophytes in Central European salt marshes. Mycorrhiza 10:175–183

Hu Y, Schmidhalter U (2005) Drought and salinity: a comparison of their effects on mineral nutrition of plants. J Plant Nutr Soil Sci 168:541–549

Husband R, Herre EA, Turner SL, Gallery R, Young JPW (2002) Molecular diversity of arbuscular mycorrhizal fungi and patterns of host association over time and space in a tropical forest. Mol Ecol 11:2669–2678

Jennings DH, Burke RM (1990) Compatible solutes—the mycological dimension and their role as physiological buffering agents. New Phytol 116:277–283

Johnson NC (1998) Responses of *Salsola kali* and *Panicum virgatum* to mycorrhizal fungi, phosphorus and soil organic matter: implications for reclamation. J Appl Ecol 35:86–94

Johnson-Green PC, Kenkel NC, Booth T (1995) The distribution and phenology of arbuscular mycorrhizae along an inland salinity gradient. Can J Bot 73:1318–1327

Juniper S, Abbott LK (1993) Vesicular-arbuscular mycorrhizas and soil salinity. Mycorrhiza 4:45–57

Juniper S, Abbott LK (2003) Arbuscular mycorrhizas and soil salinity—the continuing story. The fourth international conference on mycorrhizae ICOM4

Juniper S, Abbott LK (2004) A change in the concentration of NaCl in soil alters the rate of hyphal extension of some arbuscular mycorrhizal fungi. Can J Bot 82:1235–1242

Juniper S, Abbott LK (2006) Soil salinity delays germination and limits growth of hyphae from propagules of arbuscular mycorrhizal fungi. Mycorrhiza 16:371–397

Landwehr M, Hildebrandt U, Wilde P, Nawrath K, Toth T, Biro B, Bothe H (2002) The arbuscular mycorrhizal fungus *Glomus geosporum* in European saline, sodic and gypsum soils. Mycorrhiza 12:199–211

McMillen BG, Juniper S, Abbott LK (1998) Inhibition of hyphal growth of a vesicular-arbuscular mycorrhizal fungus in soil containing sodium chloride limits the spread of infection from spores. Soil Biol Biochem 30:1639–1646

Mohammad MJ, Hamad SR, Malkawi HI (2003a) Population of arbuscular mycorrhizal fungi in semi-arid environment of Jordan as influenced by biotic and abiotic factors. J Arid Environ 53:409–417

Mohammad M, Malkawi H, Shibli R (2003b) Effects of arbuscular mycorrhizal fungi and phosphorus fertilization on growth and nutrient uptake on barley grown on soils with different levels of salts. J Plant Nutr 26:125–137

Naidu BP (1998) Separation of sugars, polyols, proline analogues, and betaines in stressed plant extracts by high performance liquid chromatography and quantification by ultra violet detection. Aust J Plant Physiol 25:793–800

Nelson PN, Ladd JN, Oades JM (1996) Decomposition of ^{14}C-labelled plant material in a salt-affected soil. Soil Biol Biochem 28:433–441

Palenzuela J, Azcon-Aguilar C, Figueroa D, Caravaca F, Roldan A, Barea JM (2002) Effects of mycorrhizal inoculation of shrubs from Mediterranean ecosystems and composted residue application on transplant performance and mycorrhizal developments in a desertified soil. Biol Fertil Soils 36:170–175

Pathak H, Rao DLN (1997) Carbon and nitrogen mineralization from added organic matter saline and alkali soils. Soil Biol Biochem 30:695–702

Pfeiffer CM, Bloss HE (1988) Growth and nutrition of guayule (*Parthenium argentatum*) in a saline soil as influenced by vesicular-arbuscular mycorrhiza and phosphorus fertilization. New Phytol 108:315–321

Poss JA, Pond E, Menge JA, Jarrell WM (1985) Effect of salinity on mycorrhizal onion and tomato in soil with and without additional phosphate. Plant and Soil 88:307–319

Qadir M, Ghafoor A, Murtaza G (2000) Amelioration strategies for saline soils: a review. Land Degrad Dev 11:501–521

Qadir M, Noble AD, Schobert S, Gafoor A (2005) Phytoremediation of sodic and sodic-saline soils. International Salinity Forum, Riverside, CA, pp 383–386

Ramirez-Fuentes E, Luna-Guido ML, Ponce-Mendoza A, Van den Broeck E, Dendooven L (2002) Incorporation of glucose ^{14}C and NH_4^+ in microbial biomass of alkaline saline soil. Biol Fertil Soils 36:269–275

Rinaldelli E, Mancuso S (1996) Response of young mycorrhizal and non-mycorrhizal plants of olive tree (*Olea europaea* L.) to saline conditions. I. short-term electrophysiological and long-term vegetative salt effects. Adv Hortic Sci 10:126–134

Roldan A, Albaladejo J, Thornes JB (1996) Aggregate stability changes in a semiarid soil after treatment with different organic amendments. Arid Soil Res Rehabil 10:139–148

Ruiz-Lozano J, Azcon R (2000) Symbiotic efficiency and infectivity of an autochthonous arbuscular mycorrhizal *Glomus* sp from saline soils and *Glomus deserticola* under salinity. Mycorrhiza 10:137–143

Ruiz-Lozano JM, Azcon R, Gomez M (1996) Alleviation of salt stress by arbuscular-mycorrhizal *Glomus* species in *Lactuca sativa* plants. Physiol Plant 98:767–772

Santos BA, Maia LC, Cavalcante UMT, Correia MTS, Coelo LCBB (2001) Effect of arbuscular mycorrhizal fungi and soil phosphorus level on expression of protein and activity of peroxidase on passion fruit roots. Braz J Biol 61:693–700

Soedarjo M, Habte M (1993) Vesicular-arbuscular mycorrhizal effectiveness in an acid soil amended with fresh organic matter. Plant and Soil 149:197–203

Stirzaker R, Vertessy R, Sarre A (2002) Trees, water and salt: an Australian guide to using trees for healthy catchments and productive farms. Joint Venture Agroforestry Program

Sylvia DM (1986) Spatial and temporal distribution of vesicular-arbuscular mycorrhizal fungi associated with *Uniola paniculata* in Florida foredunes. Mycologia 78:728–734

Treseder KK, Allen MF (2000) Mycorrhizal fungi have a potential role in soil carbon storage under elevated CO_2 and nitrogen deposition. New Phytol 147:189–200

Wang FY, Liu RJ, Lin XG, Zhou JM (2003) Arbuscular mycorrhizal status of wild plants in saline-alkaline soils of the Yellow River Delta. Mycorrhiza 13:123–127

Wong VNL, Greene RSB, Murphy B, Dalal R (2004) The effects of salinity and sodicity on soil carbon turnover. Proceeding of 3rd Australian New Zealand soils conference, 5–9 Dec 2004. University of Sydney, Australia

Zedler JB, Morzaria-Luna H, Ward K (2003) The challenge of restoring vegetation on tidal, hypersaline substrates. Plant and Soil 253:259–273

Chapter 21
Use of Mycorrhizal Fungi for Forest Plantations and Minesite Rehabilitation

Ying Long Chen, Run Jin Liu, Yin Li Bi, and Gu Feng

21.1 Introduction

The integral role of mycorrhizal symbioses in natural and managed ecosystems has been widely recognised. In the recent decades, more attention has been paid to establishing efficient mycorrhizal fungi on plants at the nursery or seedling stage for forest plantations and minesite rehabilitation. The number of people depending on forests for their livelihoods reaches 1.6 billion, including some 300 million living in them (FAO 2012). Forestry plantations have become an increasingly important supply for wood during the era of rapid deforestation of primary habitats. Due to the increasing demand for consumption of plant products, including timber, fuel wood, leaves, twigs, fruits and other non-wood products, forest plantations have been established largely in recent decades. Exotic trees are preferred over native ones because of their shorter rotation, well-studied biology and paucity of pests in new

Y.L. Chen (✉)
State Key Laboratory of Soil Erosion and Dryland Farming on the Loess Plateau, Research Center of Soil and Water Conservation and Eco-environment, Chinese Academy of Sciences and Ministry of Education, Yangling 712100, China

School of Earth and Environment, The University of Western Australia, Crawley, WA 6009, Australia
e-mail: yinglongchen@hotmail.com

R.J. Liu
Institute of Mycorrhizal Biotechnology, Qingdao Agricultural University, Qingdao 266109, China

Y.L. Bi
School of Geoscience and Surveying Engineering, China University of Mining and Technology, Beijing 100083, China

G. Feng
College of Resource and Environmental Science, China Agricultural University, Beijing 100094, China

© Springer-Verlag Berlin Heidelberg 2014
Z.M. Solaiman et al. (eds.), *Mycorrhizal Fungi: Use in Sustainable Agriculture and Land Restoration*, Soil Biology 41, DOI 10.1007/978-3-662-45370-4_21

habitats. Among the hundreds of commercial trees, species of *Eucalyptus*, *Pinus* and *Acacia* dominate in forestry plantations worldwide (West 2006). On the other hand, exploitation of mineral resources around the world leaves many closed minesites facing difficulties and challenges in land rehabilitation. It has been recognised that mycorrhizal technology can profitably be applied in forest planta- tions and land restoration as well as agricultural and horticultural crops for better nutrient utilisation and more effective land use. Introduction of appropriate mycor- rhizal symbioses to improve the soil and crop productivity permits a satisfactory reduction of chemical fertilisers and pesticide inputs, thus offsetting ecological and environmental concerns. Appropriate symbiotic fungal partners are critical for some plants to become established, grow normally and withstand better in adverse climatic and soil conditions such as high temperatures and infertile, salinity or polluted soils (Behie and Bidochka 2013). Recent studies suggest that mycorrhizal fungi may exhibit some degree of heavy metal tolerance and, as a result, confer heavy metal tolerance in host plants (e.g. Zaefarian et al. 2013). The conceptual background including the biology and ecology of mycorrhizal symbioses is well addressed in the literature and in other chapters of this volume. This chapter discusses developments and insights regarding the potential of mycorrhizal tech- nologies in forest plantations and minesite rehabilitation with particular reference to the ectomycorrhizal (ECM) fungi.

21.2 Constraints in Plantation Establishment

New plantations may need to be established in cleared forest lands, denuded uplands, contaminated areas or minesites either for the production of timber, wood chips and non-wood products or for revegetation and rehabilitation. Climatic characters and soil physical and chemical properties should be taken into account to integrate the plantation and vegetation programmes. The utilisation of local species for minesite revegetation takes advantage of the inherent attributes of fitness conferred on those species by natural selection, and thus the species are assumed to adapt to local climatic, edaphic and ecological processes (Corbett 1999). The mined environment, however, may be hostile to some local species, and the importance of selecting species suited to the 'new' local conditions must be recognised. Impediments limiting the establishment of new plantations include abiotic factors, such as high temperature, infertile soil, alkaline or salinity soil and heavy metal-polluted soil, and biotic constrains, such as risk of soilborne pathogens and lack of plant beneficial microbes (e.g. mycorrhizal fungi, nitrogen- fixing bacteria). Among these, saline soil is one of the most common types of devastated land, with some 932 million hectares in arid and semiarid regions affected by severe salt accumulation (Summer et al. 1998). The diversity of mycorrhizal fungi in alkaline–saline soil is known to be low (Ishida et al. 2009) and is also affected by various abiotic and biotic environmental factors (Chai et al. 2013).

Low mycorrhizal diversity is common in many new commercial plantation sites (Chen et al. 2007). Fast-growing exotic species are preferred for forest plantations. Establishment and nutrient acquisition strategies of those species are likely to be highly dependent on ECM fungi. Several reports claim that plantations of *Pinus* species fail in the absence of co-introduced symbiotic fungi in exotic habitats, and subsequent success in re-establishing plantations with mycorrhizal seedlings suggests that the failure was due to lack of compatible ECM fungi in the soil (Kohout et al. 2011; Liu and Chen 2007). The diversity of ECM fungi is found to be low in exotic plantations, such as in eucalypt plantations in south China (Chen et al. 2007; Dell et al. 2002). Inoculation with appropriate ECM fungi promotes tree survival and growth of eucalypt plantations in exotic lands (e.g. Chen et al. 2000a; Grove and Le Tacon 1993). The introduction of compatible mycorrhizal fungi is recommended to the new plantation sites along with the plant species when mycorrhizal status in the soil is poor.

Large-scale surface mining represents severe ecological disruption at the landscape level. Soil disturbance, associated with stripping and respreading of topsoil during mining, is known to reduce the diversity and propagule levels of ECM fungi (Malajczuk et al. 1994), arbuscular mycorrhizal (AM) fungi (Jasper et al. 1989), ericoid (Hutton et al. 1997) and orchid mycorrhizal fungi (Collins et al. 2007). Plant species in natural woodland communities surrounding the Ranger lease area of northern Australia are dominated by ECM fungi (Corbett 1999; Reddell and Milnes 1992). With a large-scale population survey of glomalean fungi in disturbed and natural habitats in tropical Australia, Brundrett and Ashwath (2013) concluded that the diversity of AM fungi was substantially lower in disturbed sites than in natural habitats. This reduced diversity in young sites may have resulted from limitations in their dispersal mechanisms resulting in delays in fungal introductions, the absence of appropriate host plants or the inability of fungi to adapt to site conditions resulting in establishment failure (Malajczuk et al. 1994). Establishment of symbiotic microorganisms is often recognised as one of critical issues for the success of minesite rehabilitation (Corbett 1999).

21.3   Applications of ECM Fungi

A number of factors associated with the use of mycorrhizal fungi need to be considered in inoculation programmes for the establishment and production of forest plantations, land revegetation and minesite rehabilitation. The criteria for selecting the optimal fungus, the compatibility of the fungus and host, the suitability of the fungus to the site and the ease of inoculum production must be achieved before large-scale application in the field (Rincón et al. 2001). Once compatibility of the plant and fungus is established, the development of suitable methods for inoculum production and application is necessary. This section therefore analyses advances of the use of various kinds of inoculants, mycorrhization and evaluation, and inoculation effectiveness.

21.3.1 Sources of ECM Inoculants

Various types of ECM fungal inoculants have been developed and tested for applications in plantation nurseries and field trials. Sources of ECM inoculants can be categorised generally as (1) natural inoculants in the form of either airborne spores or colonised soil, (2) mycorrhizal seedlings, (3) vegetative inoculants of ECM fungal mycelium and (4) spores. Because the merits may vary among each kind of inoculum, selection of appropriate types of inoculants for particular application case is recommended. Here the advantages and limitations of each kind of inoculants are compared along with particular references for their use in forest nurseries and the field.

The use of natural inoculants is a simple and effective practice for introducing ECM fungi into new plantation sites. The advantages and disadvantages of the use of natural inoculants have been reviewed by Kendrick and Berch (1985). The practice of natural airborne spore inoculum, however, relies largely on the season in which ECM fungi produce fruiting bodies, and therefore this is unsuited to forestry applications when considering restriction of the availability and the low level of inoculum. Obvious disadvantages of the use of soil inoculants include the risk of introducing diseases and unsuitability for large-scale applications due to the difficulties in transportation of heavy bulky soil to new sites (Kendrick and Berch 1985). For example, hemlock (*Tsuga canadensis*) seedlings propagated in hemlock forest soils had good ECM colonisation, and growth increment of outplanted mycorrhizal plants was observed when comparing seedlings raised in sterilised field soil (O'Brien et al. 2011). Using *E. urophylla* seedlings as bait in a bioassay experiment, Chen et al. (2007) determined inoculation potential of ECM fungi in field soils from various locations in south China where eucalypt plantations are being established. Four morphotypes of ECM were identified (Fig. 21.1) including an indigenous *Laccaria* species which also produced basidiomes in one soil. However the poor colonisation suggested low level of ECM fungal inoculants in the field soils.

Transplanting of mycorrhizal seedlings colonised by a known ECM fungus is another approach to introduce compatible fungal partner to its desired host species. This method has been used especially in the establishment of new mushroom orchards. Attempts to grow edible mycorrhizal mushrooms (EMMs) commenced on the Périgord black truffle (*Tuber melanosporum*) in France and in Italy by transplanting truffle-colonised seedlings raised in forest nurseries (Chevalier and Frochot 2000). This technique has been extensively used in establishment of truffière in European countries and also in Asian and Oceanic countries (Chap. 23). Mycorrhizal seedlings of *Pinus yunnanensis* and *P. armandii* colonised by matsutake (*Tricholoma matsutake*) are outplanted into the matsutake-producing forests in southwest China aiming to the increased mushroom production (Chen 2004). Quality control of mycorrhizal seedlings from nurseries is critical to ensure the extent of colonisation of the proposed fungus and to reduce the risk of introduction of pests and pathogens along with the mycorrhizal seedlings.

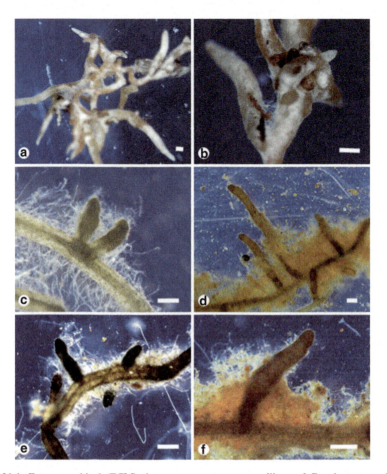

Fig. 21.1 Ectomycorrhizal (ECM) tips on nursery-grown seedlings of *Eucalyptus urophylla* viewed under light microscope. (**a**, **b**) Typical ECMs formed by *Scleroderma*; (**c**) *white* ECM tips formed by *Laccaria* (ECM I); (**d**) *yellowish Pisolithus*-like ECMs (ECM II); (**e**) *jet-black* ECMs formed by an unknown fungus (ECM III); (**f**) *brown* to *dark brown* ECMs formed by an unknown contaminant (ECM IV) (*bar* = 1.0 mm). This figure is extracted from Ying Long Chen's PhD Thesis (2006, Murdoch University, Australia)

Vegetative inoculants of fungal mycelium produced in axenic culture on either solid or liquid medium have been frequently used in plantation nurseries (Brundrett et al. 1996; Kendrick and Berch 1985). The merits of using mycelial inoculants in forests include the facts of known fungal species or isolates involved, the absence of pests and pathogens and the year-round availability. However, cultivation of ECM fungi is expensive because some fungi are difficult to isolate or grow slowly in pure culture. Molina and Palmer (1982) tested a wide range of ECM fungi and found that members of 18 genera, such as *Amanita*, *Boletus*, *Cortinarius*, *Hebeloma*, *Laccaria*, *Paxillus*, *Pisolithus*, *Rhizopogon*, *Scleroderma*, *Suillus* and *Tricholoma*, are fairly easy to isolate for axenic cultivation. Attempts to produce mycelium inoculants from a few other ECM fungi (e.g. *Gomphidius*) were unsuccessful, while many are yet to be attempted. Kendrick and Berch (1985) commented that most ECM fungi could be cultivated as long as the nutritional and other growth conditions meet the requirements by the fungus for its development.

Supriyanto (1999) tested the effectiveness of some ECM fungi in alginate beads in promoting the growth of several dipterocarp seedlings. Since ECM fungi are obligately biotrophic in the natural habitat, vegetative inoculants are grown very slowly through the soil before colonising a suitable host root (Kendrick and Berch 1985). The survival, competition with indigenous organisms and persistence in the soil of the introduced vegetative inoculants require further examination. Another drawback is that the storage of mycelial inoculants usually adversely influences its effectiveness, while large quantities of viable inoculants are needed for application on an operational scale. These impediments shadow the use of cultivated ECM fungal mycelium in a wider forestry practice.

Fungal spore inoculants including spores or sclerotia specifically collected for the purpose are commonly used because of the ease of application in plantation nurseries and the availability of large quantities of spores from a few sporocarps (Chen et al. 2007; Dell et al. 2002; Marx and Cordell 1990). It is estimated that the top 5-cm soil in a Douglas fir (*Pseudotsuga menziesii*) stand contained 2,785 kg ha^{-1} dry weight of *Cenococcum geophilum* sclerotia (Fogel and Hunt 1979). Hunt and Trappe (1987) found that in a western Oregon Douglas fir stand, the production of sporocarps of *Hysterangium setchellii* was up to 3,770 ha^{-1}, i.e. 842 g dry weight ha^{-1}. The main disadvantages of spore inoculants are genetic variability, the lack of reliable laboratory methods to determine spore viability and the delay in mycorrhization compared with vegetative inoculants (Brundrett et al. 1996; Chen et al. 2006a). Spores of ECM fungi are generally harder to germinate than are those of saprophytes. Germination is often promoted by the presence of other microbes, growing hyphae of same species or activated charcoal. Coating spores individually on to roots of young seedlings of *P. radiata* enhances germination success to about 30 % in both *Suillus* and *Rhizopogon* (Theodorou and Bowen 1987). Fries (1988) found abietic acid, a diterpene resin acid, in pine roots induced germination of *Suillus* spores. *Descomyces* spores are known to have slow growth, mostly aerial with some submerged hyphae. In contrast, *Pisolithus* and *Scleroderma* spores have moderate to rapid growth, primarily aerial, readily culturable.

Basidiocarps of ECM fungi do not necessarily mature when inoculation programmes are required in the nursery. Therefore, it is often necessary to collect and store spores for a considerable time. Spores of several ECM fungi can tolerate long storage periods. Dry spores of *Pisolithus* could be stored at 5 °C for up to 34 months without significant loss of spore viability (Marx 1976). Castellano and Molina (1989) found that *Rhizopogon* spores could be stored for up to 3 years. *Scleroderma* spores kept at 4 °C for 5 years germinated and formed mycorrhizas on six important plantation tree species and showed effective in producing mycorrhizas as freshly collected spores (Chen et al. 2006a). Torres and Honrubia (1994) examined the viability of basidiospores of 14 species in the genera *Cortinarius*, *Hebeloma*, *Inocybe*, *Laccaria*, *Rhizopogon*, *Russula*, *Suillus* and *Tricholoma* using nuclear and fluorescein diacetate staining methods. They observed nearly total loss of viability of all fungi stored in refrigerated or frozen slurries after 6 months. However, many successful mycorrhizal experiments have been performed using spore suspensions previously stored at either room temperature or low temperature for up to 10 months (e.g. Chen et al. 2006c; Duñabeitia et al. 2004). Cold treatment enhances spore germination and mycorrhization of *Tuber* (Chen 2002; Chevalier and Frochot 2000).

Spore preparations of *Pisolithus tinctorius* are being produced for commercial use as pellets, sprays or encapsulated on seed (Martin et al. 2003; Marx et al. 1989). Plantation nurseries commonly prefer to handle spore inoculants due to the ease of inoculum production and delivery. Spores of *Pisolithus* and *Scleroderma* collected from the field in south China are being used to effectively inoculate clonal eucalypts in commercial nurseries. Inoculation with a spore rate as low as 10^4 spores seedling^{-1} is appropriate for eucalypt seedlings to form mycorrhizas in containerised nurseries (Chen et al. 2006b). The greatest increases in ECM formation occurred at low to medium spore densities, and there was no inhibition of mycorrhizal development at the highest spore density used (10^8 spores seedling^{-1}). By contrast, Marx (1976) inoculated *P. taeda* seedlings with *Pisolithus tinctorius* basidiospores and found a density of 5.5×10^7 spores per 800 cm^3 soil produced significantly more ECMs than other spore densities tested. Similarly, Torres and Honrubia (1994) observed threshold spore densities when inoculating *P. halepensis* seedlings with basidiospores of *Pisolithus*, *Rhizopogon* or *Suillus*, beyond which high spore densities reduced ECM formation and seedling growth. The recommended dose used for nursery inoculation is usually 10^4–10^8 spores seedling^{-1}. This recommendation has been applied for a range of fungus–host partners: *Abies* (*Scleroderma*) (Parladé et al. 1997), *Afzelia* (*Lactarius*, *Pisolithus*, *Russula*, *Suillus*) (Munyanziza and Kuyper 1995), *Eucalyptus* (*Cortinarius*, *Hydnangium*, *Laccaria*, *Pisolithus*, *Scleroderma*) (Chen et al. 2000a, b; Lu et al. 1998), *Pinus* (*Lactarius*, *Melanogaster*, *Pisolithus*, *Rhizopogon*, *Scleroderma*, *Suillus*) (Duñabeitia et al. 2004; Marx et al. 1989; Ortega et al. 2004; Rincón et al. 2001), *Pseudotsuga* (*Melanogaster*, *Rhizopogon*, *Tuber*) (Parladé et al. 1997) and *Quercus* (*Pisolithus*) (Martin et al. 2003; Marx et al. 1997; Parladé et al. 1997).

Spore inoculants of *Scleroderma* were effective in forming mycorrhizas on eucalypts when a suitable spore density was applied (Chen et al. 2006b). It is not known how many spores germinated in the glasshouse and resulted in mycorrhizal formation with seedlings of commercial plantation species such as *Acacia*, *Eucalyptus* and *Pinus*. Generally, it would be desirable to measure the percentage of spores able to germinate in inoculants being tested. Spores of several *Scleroderma* spp. were taken from sporocarps prior to spore release to test germination in vitro on agar in the presence of eucalypt roots. Where spores germinated, the viability varied from 0.1 to 0.8 % in *Scleroderma* (Chen, unpublished data).

21.3.2   ECM Formation and Evaluation

The development and formation of mycorrhizal associations involves physical, molecular and physiological interactions between the host plant root and the fungal hyphae. Experimental work of Horan and Chilvers (1990) using compatible and incompatible isolates of *Pisolithus tinctorius* and *Paxillus involutus* indicated that specific root exudates may be involved in the ECM formation. It is known that root-released chemical compounds, such as cytokinins and other hormones, promote hyphal branching and growth (Gogala 1991). The presence of fluorescent pseudomonads, so-called mycorrhizal helper bacteria (MHBs), enhances the rate and extent of ECM formation (Garbaye 1994). Some symbiosis-related proteins (i.e. ectomycorrhizins) are found in *E. globulus–P. tinctorius* association (Hilbert and Martin 1988). Hydrophobins have also been strongly implicated in hyphal recognition (Talbot et al. 1993). Smith and Read (2008) concluded that molecular probing with hydrophobin genes or genes for specific membrane transport proteins or host defence responses would be likely to prove valuable, especially using plant–fungus associations with different levels of compatibility. Establishment of functional links between gene expression and the key events of recognition and mycorrhiza synthesis remains challenging.

Horan et al. (1988) developed the paper–sandwich method to investigate synchronous colonisation of lateral roots over a period of days. The cellophane-over-agar method of Malajczuk et al. (1990) enables the same sequence of events to be completed on the primary root within hours. Mycorrhizas synthesised in this way have permitted investigations of fungal and host physiology and interactions, such as rapidity and extent of mycorrhization, host specificity, nutrient uptake and responses to biotic and abiotic stresses.

Hartig net formation is considered as a good indicator of fungus–host compatibility and is correlated with growth responses. A typically compatible fungus–host pair has mantle on the thick-branched roots viewed with eye, well-developed mantle and Hartig net, with elongated epidermal cells or hyphae extending into the cortex (Chen et al. 2006c). Thin roots may or may not have a mantle, often with a wound reaction and no Hartig net indicating incompatible. Partially compatible pair produces branched roots (thicker or not) with or without mantle, in which Hartig net may or may not develop, and limited hyphal penetration between epidermal cells. Superficial ECM has thin roots and thin mantle; although Hartig net is present, generally it is thin due to the lack of expansion of epidermal cells (Chen et al. 2006c). Different fungus–host combinations may have different structure of ECM. Eucalypts tend to have a typical morphology of a well-developed epidermal Hartig net and thick mantle. *Cortinarius* and *Hysterangium* species often form thin mantles and superficial sheathing mycorrhizas on eucalypts indicating host incompatibility (Malajczuk et al. 1987).

Numerous parameters are adopted for quantifying the degree of mycorrhization. Parameters include root length, ECM root tip density (numbers of ECM root tips in a unit soil volume), ramification indices (numbers of ECM root tips per cm root), frequencies (percentage of ECM root tips over the total root tips), specific root length (cm root g^{-1} root fresh mass), root length density (cm root ml^{-1} soil) and mycorrhizal dependence (dry biomass of mycorrhizal plant over non-mycorrhizal plant).

21.3.3 ECM Fungus–Host Specificity

It is generally accepted that ECM fungi are often host specific in spite of contradictory observations concerning host specificity in some fungal genera (Table 21.1; Cairney and Chambers 1999; Jairus et al. 2011; Zhou and Hyde 2001). Molina and Trappe (1994) recorded three general responses among 20 species of *Rhizopogon* on *Pseudotsuga menziesii*, *Pinus contorta* and *Tsuga heterophylla*: strong specificity to *Pseudotsuga menziesii*, specificity or strongest development on *Pinus contorta* and an intermediate response where ECM formed on two or three of the hosts. The unsuccessful pure culture syntheses make *Eucalyptus–Rhizopogon* association seem unlikely (Chilvers 1973). With few exceptions, field observations and pure culture syntheses confirm the specificity of *Rhizopogon* species for Pinaceae (Molina and Trappe 1994). A few species of *Hysterangium* are widely distributed, but some often display high levels of endemism and discrete host ranges, such as *Quercus*-specific species (Párladé et al. 1997; Rincón et al. 2001). Malajczuk et al. (1987) described the superficial ECM of *Hysterangium inflatum* with *E. diversicolor* as having abundant calcium oxalate crystals on its hyphae although the fungal mantle was only one to five cells deep.

We examined the compatibility of 15 *Scleroderma* collections to form mycorrhizas with seedlings of six plantation trees (*Acacia mangium*, *A. mearnsii*, *E. globulus*, *E. urophylla*, *Pinus elliottii* and *P. radiata*) in a nursery potting mix (Chen et al. 2006c). Observations on mycorrhizal structure confirmed that most collections were able to aggressively colonise eucalypts and pines, while roots of acacias were poorly colonised. The findings demonstrated that the Australian collections were more effective in colonising short roots on eucalypts than the Chinese collections.

Plantation tree species may also perform differently in establishing symbiosis with ECM fungi. Studies showed that *Pinus* spp. had difficulty to establish ECM associations with native fungi in tropical habitats (Walbert et al. 2010). Experimental work also confirms that some *Eucalyptus* species perform better with their co-introduced fungi than with locally available mycobionts, emphasising the importance of a long-term coevolution and enhancement of histological and functional compatibility (Chen et al. 2007; Malajczuk et al. 1984). Compatibility in ECM associations with potential host range deduced from laboratory experiments may not reflect that under field conditions. Thus, the concept of ecological specificity (Molina et al. 1992), embracing all environmental abiotic and biotic factors that affect the ability of plants to form functional ECM with particular fungi, may

Table 21.1 Host range of some key ECM fungal genera

Fungal genus	Host genus
Amanita	Abies; Allocasuarina; Betula; Carpinus; Castanea; Castanopsis; Casuarina; Eucalyptus; Fagus; Larix; Monotropa; Nothofagus; Picea; Pinus; Platanus; Polygonum; Pseudotsuga; Quercus; Salix; Tilia; Tsuga
Cantharellus	Abies; Betula; Carpinus; Castanea; Corylus; Eucalyptus; Fagus; Picea; Picea; Pinus; Populus; Pseudotsuga; Quercus; Shorea; Tsuga
Cenococcum	Abies; Acer; Eucalyptus; Juniperus; Larix; Pinus; Pseudotsuga
Hebeloma	Alnus; Arbutus; Arctostaphylos; Betula; Castanea; Cistus; Dryas; Larix; Picea; Pinus; Pseudotsuga; Quercus; Tsuga
Hysterangium	Arbutus; Arctostaphylos; Larix; Pinus; Pseudotsuga; Tsuga
Laccaria	Abies; Betula; Betula; Dipterocarpus; Eucalyptus; Fagus; Larix; Leptospermum; Nothofagus; Picea; Pinus; Pseudotsuga; Quercus; Salix; Tilia; Tsuga
Lactarius	Alnus; Arbutus; Arctostaphylos; Betula; Eucalyptus; Fagus; Picea; Pinus; Pseudotsuga; Quercus; Salix; Tsuga
Paxillus	Allocasuarina; Alnus; Betula; Castanea; Dryas; Eucalyptus; Fagus; Picea; Pinus; Populus; Pseudotsuga; Quercus; Salix
Pisolithus	Abies; Acacia; Afzelia; Allocasuarina; Alnus; Arbutus; Arctostaphylos; Betula; Carya; Castanea; Castanopsis; Casuarina; Eucalyptus; Hopea; Larix; Pinus; Populus; Pseudotsuga; Quercus; Tsuga
Rhizopogon	Adenostoma; Pinus; Pseudotsuga; Tsuga
Scleroderma	Abies; Acacia; Afzelia; Alnus; Betula; Brachystegia; Carya; Casuarina; Eucalyptus; Hopea; Isoberlinia; Larix; Picea; Pinus; Populus; Pseudotsuga; Quercus; Shorea; Tsuga; Uapaca
Suillus	Arbutus; Arctostaphylos; Larix; Larix; Monotropa; Picea; Pinus; Pseudotsuga; Pseudotsuga; Quercus
Thelephora	Abies; Acacia; Allocasuarina; Alnus; Arbutus; Arctostaphylos; Betula; Castanea; Castanopsis; Casuarina; Eucalyptus; Fagus; Hudsonia; Larix; Lithocarpus; Picea; Pinus; Pinus; Populus; Pseudotsuga; Quercus; Salix; Tsuga
Tricholoma	Abies; Castanopsis; Cedrus; Lithocarpus; Picea; Pinus; Pseudotsuga; Quercus; Tsuga
Tuber	Abies; Alnus; Carpinus; Carya; Castanea; Cedrus; Cistus; Corylus; Fagus; Fumana; Helianthemum; Ostrya; Picea; Pinus; Populus; Pseudotsuga; Quercus; Salix; Tilia

explain why *Suillus* species may exhibit a broader host range under experimental conditions than observed in nature (Dahlberg and Finlay 1999).

Host plant specificity is considered one of the most important factors influencing fungal diversity, particularly for ECM fungi. Nelson (1979) suggested that evolutionary and ecological processes that determine specificity act differently on hosts and their fungal symbionts. Thus, the level of host specificity among mycorrhizal fungi is dynamic and depends not only on symbiotic partners but also on ecological opportunities (Zhou and Hyde 2001). A concern on the invasion of exotic fungi due to host shift has been raised. An Australian fungus *Laccaria fraterna* can colonise European Cistaceae in natural conditions indicating occurrences of host shifts and invasion of exotic fungal species (Jairus et al. 2011). Host shifting among the Australian and particularly native African ECM fungi in mixed eucalypt plantations in Zambia, south-central Africa, has been identified by analysing rDNA and plastid intron sequences. *Amanita muscaria*, a mycobiont of Pinaceae and Fagaceae, has only recently become invasive in Australia and New Zealand despite two centuries of known introduction history (Orlovich and Cairney 2004). Jairus et al. (2011) recommended that exotic forestry plantations could ideally be established by use of seeds of seedlings pre-inoculated with native ECM fungi, preferably edible mycorrhizal mushrooms (see Chap. 23) as a case in Zambia, to reduce the potential for microbial invasion and encourage utilisation of forestry 'by-products'. Thus, availability and compatibility of native fungal resources must be examined to optimise production of exotic tree plantations.

21.3.4 Effectiveness of ECM Inoculation

Numerous measures can be taken to evaluate effectiveness of ECM inoculation. The positive effect of mycorrhizas on plant growth through increased phosphorus availability is well documented (e.g. Smith and Read 2008). Increased tolerance of saline conditions, uptake of zinc, protection against pathogens and enhanced water uptake are some of the other potential benefits conferred by mycorrhizas. Here we discuss inoculation effectiveness in three general aspects: (1) host response, (2) response to abiotic stresses and (3) response to other biotic organisms.

Commercial plantations have a privilege in the use of fast-growing exotic species such as trees of the genera *Acacia*, *Eucalyptus* and *Pinus* (West 2006). For example, more than 100 species of *Eucalyptus* native to Australia and the surrounding islands are used in plantation trails in tropical to temperate regions around the world (Jairus et al. 2011). Inoculating *Acacia*, *Eucalyptus* and *Pinus* with compatible ECM fungi has been shown to be beneficial in many parts of the world (Chen et al. 2006c; Duponnois et al. 2005; Duñabeitia et al. 2004).

ECM fungi can help improve the establishment and productivity of eucalypt plantations in China (Chen et al. 2000b; Dell et al. 2002). A number of ECM fungi, collected from under *Eucalyptus* in Australia, have been introduced in research trials into eucalypt plantations in south China since the 1990s (Chen et al. 2000b; Dell and Malajczuk 1995). The *Scleroderma* genus is favoured for introduction because it readily colonises eucalypt roots in disturbed habitats. It is easy to collect spores of *Scleroderma* from species that form large epigeous basidiocarps and then to produce spore inoculants for nursery inoculation programmes. This fungal genus has potential for application in commercial plantation forests in the region where mycorrhizal status is poor (Chen et al. 2007). Beneficial *Scleroderma* isolates can vigorously compete with other ECM fungi in the field (Dell et al. 2002; Hall et al. 1994; Martin et al. 2003). In plantations of exotic acacias, eucalypts and pines, these fungi are desirable as inoculants if they are compatible with the host tree and are effective in promoting survival and production in the field (Chen et al. 2007). Eucalypt seedlings inoculated with cold-stored spores of several *Scleroderma* species were taller than those inoculated with fresh spores (Chen et al. 2006b). Plant growth-promoting bacteria (PGPRs) have been isolated from Western Australian sporocarps of some ECM fungi (B. Dell, pers. comm.). The presence of these bacteria in the cold-stored spore slurry may account for the extra growth stimulation of the host.

Benefits from inoculation with *Scleroderma* fungi were also observed on other *Eucalyptus* species, such as *E. diversicolor*, *E. grandis*, *E. pellita*, *E. tereticornis* and *E. urophylla* (e.g. Reddell and Milnes 1992; Chen et al. 2000a). Stimulations of *Scleroderma* inoculation on the growth of other tree genera were obtained, particularly on *Acacia mangium* and *A. holosericea* (Founoune et al. 2002); *Castanopsis hystrix* (Chen et al. 2001); *Pinus caribaea* in axenic culture (Rangarajan et al. 1990); *P. contorta* in mine spoil sites (Fay et al. 1997); *P. kesiya* in forest and degraded soils (Rao et al. 1996); *Hopea odorata*, *Vatica sumatrana*, *Shorea stenoptera*, *Sh. compressa* and *Sh. pinanga* (Santoso 1991); and *Sh. leprosula* cuttings (Omon 1996). These results, however, are controversial to some other reports where no significant stimulation or even depression of some *Scleroderma* fungi on the growths of several hosts was observed (e.g. Seva et al. 1996; Lu et al. 1998). As soil characteristics, particularly soil nutrient levels, may affect spore germination and mycorrhization, there are challenges in practical applications of ECM fungi in nurseries and the field.

Numerous studies suggest that there is a connection between growth enhancement of host plants following inoculation practices and increased phosphorus accumulations in the host (e.g. Jansa et al. 2011). These studies also claim that enhanced P accumulation appears to relate to the level of ECM colonisation and the surface area of the extrametrical mycelial phase. The pioneering studies of Melin and Nilsson (1950) provided the first experimental evidence of P and N translocation through ECM mycelia. Finlay et al. (1988) examined mycelial uptake, translocation and assimilation of nitrogen from ^{15}N-labelled ammonium by *Pinus sylvestris* plants colonised by four different ECM fungi, *Paxillus involutus*, *Pisolithus tinctorius*, *Rhizopogon roseolus* and *Suillus bovinus*. Absorbed NH_4^+ by the fungi appears to be rapidly incorporated into amino acid precursors within the extrametrical mycelium and is translocated to the host. The extensive rhizomorph network of ECM fungi is largely responsible for enhanced nutrient uptake and seedling growth. Environmental factors often have impacts on mycorrhization and consequently alter the inoculation effectiveness on the host plants. The carbon location from the host plants to the fungi is a cost; therefore, under limiting resources the plant–fungi interaction could change from benefit to cost resulting in a continuum of behaviours from mutualistic to antagonistic (Johnson et al. 1997). However, recent studies of Saner et al. (2011) suggest that a light-constrained environment may not influence seedling growth due to ECM colonisation.

The external mycelium of ECM fungi transports water to the host plant (Duddridge et al. 1980). Radiate pine inoculated with *Rhizopogon roseolus* and *Scleroderma citrinum* performed better particularly in a dry site when compared with non-mycorrhizal plants (Ortega et al. 2004). Analyses of the internal tran-scribed spacer (ITS) and large subunit (LSU) regions of the ribosomal DNA revealed that pezizalean ECM fungi associated with pinyon pine (*Pinus edulis*) respond positively to dry conditions in drought-stressed woodlands (Gordon and Gehring 2011). A field experiment of Dixon et al. (1983) showed that black oak (*Quercus velutina*) inoculated with *Pisolithus tinctorius* had higher water potentials and higher soil-to-plant conductance than non-mycorrhizal ones consistently over the growing season. The mycorrhizal roots of *Nothofagus dombeyi* accumulated considerably more N and P during drought and had greater activities of glutamate synthase, glutamine synthetase, glutamate dehydrogenase, nitrate reductase and acid phosphomonoesterase than the non-mycorrhizal ones (Alvarez et al. 2009). Read and Boyd (1986) pointed out that rhizomorphs in *Rhizopogon* play an important role in water uptake and movement in ECM systems. Possible mecha-nisms of mycorrhizal effects on plant water uptake and drought resistance are discussed in Lehto and Zwiazek (2011).

There is an increased interest in the role of ECM fungi in interactions with soil toxicity such as from heavy metals. ECM symbionts display differential effective-ness in providing resistance to metal toxicity. Field studies have shown that sporocarps of ECM fungi are able to accumulate heavy metals in high concentra-tions when present on metal-polluted sites (Zaefarian et al. 2013). Barcan et al. (1998) reported the Ni concentrations in *Suillus luteus* growing near the Severonickel plant on the Kola Peninsula were up to 40 times the background level. This fungus was also presented in Zn-contaminated soils and was able to grow at concentrations of 1,000 μg g^{-1} (Colpaert and van Assche 1987). Jones and Hutchinson (1986) suggested that the morphology of *Scleroderma flavidum* mycor-rhizas was important in providing Ni tolerance to their host plants. Research on metal tolerance of *Suillus* species indicated that mycobionts with an extensive extramatrical mycelium, a thick mantle and massive carpophores may be more suitable to accomplish a filter function than fungi without these features (Colpaert et al. 1992).

There is evidence that many organisms adapt to high levels for one metal indicating a rather specific biochemical mechanism for metal tolerance. Possible mechanisms for passive binding or metabolic detoxification by the mycobiont, which can lead to metal tolerance, are discussed by various researchers (e.g. Zaefarian et al. 2013). Jourand et al. (2010) reported that Ni-tolerant *Pisolithus albus* isolated from nickel mines in New Caledonia strongly enhanced the growth of the host plant *Eucalyptus globulus* at toxic nickel concentrations. These studies suggest that the use of metal tolerant mycobionts for practical inoculation of nursery plants could be helpful for revegetation of heavy metal-polluted sites.

The study of rehabilitation of the bauxite-mined areas in south-western Western Australia involves a successional process, and the re-establishment of vegetation cover and species composition, soil microbial population size and diversity and soil development have been investigated (e.g. Grant and Loneragan 2001; Ward 2000). ECM fungi are likely to follow successional pathways in rehabilitated bauxite mines (Gardner and Malajczuk 1988). A similar pattern was observed in orchid mycorrhizal (OM) fungi in rehabilitated bauxite mines when a correlation of OM fungi detection rates with litter measurements and other environmental factors that increase with time in the post-bauxite mining landscape (Collins et al. 2007). They conclude that Jarrah forest OM fungi are expected to re-establish at the rehabilitated sites provided there is continued vegetation development. Meharg (2003) claimed that the exudation of organic acids to alter pollutant availability in the rhizosphere could be the only direct evidence of mycorrhizal adaptation to metal cation pollutants. There may be other mechanism of adaptation, but conclusive evidence of adaptive mechanisms of tolerance needs further exploration.

Most ECM fungi studied exhibit optimal growth at a pH of 5 or 6, and high salinity is less toxic to most ECM fungi than others (Bois et al. 2006; Kernaghan et al. 2002). Some pH-tolerant ECM fungi are identified by in vitro cultivating under alkaline and/or saline conditions (Bois et al. 2006; Kernaghan et al. 2002). Ishida et al. (2009) characterised ECM fungal community in alkaline–saline soil (pH 7.8–9.2) in north-eastern China and identified 11 T-RFLP types from 57 ECM root tips suggesting poor fungal diversity. An uncommon ECM fungus, *Geopora* spp., was dominant in this extreme environment. With respect to low-pH environments, acid-tolerant ECM species have also been observed in the tropics. For example, Kasuya et al. (1990) reported the impact of aluminium on ECM fungi and mycorrhizal formation on *Pinus caribaea* seedlings. Marx and Altman (1979) observed an enhanced survival and growth of pine seedlings inoculated with *Pisolithus tinctorius* on acid coal spoils.

The higher pH of Ranger mine spoil of northern Australia due to high concentrations of magnesium sulphate ($MgSO_4$) is considered a significant problem for rehabilitation (Ashwath et al. 1993). The natural dispersal and re-establishment of ECM fungi on Ranger waste rock dump occur at a very slow rate which may significantly impact on the rate of development and the resilience of the plant community in the area (Malajczuk et al. 1994). The importance of mycorrhizas in the establishment and growth of native vegetation has been recognised. Hinz (1997) believes the growth of the dominant woody species *Eucalyptus tetrodonta* is dependent on an effective association with mycorrhizal fungi. Reddell et al. (1993) found that fungal root colonisation increased with age of rehabilitation and that ECM and fungal fruiting bodies were most indicative of the development of rehabilitated areas. Therefore, future research emphasis should be placed on identifying the factors affecting the establishment of viable mycorrhizal populations on mines with extreme pH.

A number of studies have demonstrated the promoting effect of MHBs on mycorrhizal formation (e.g. Dunstan et al. 1998; Garbaye 1994; Mogge et al. 2000). Frey-Klett et al. (2007) revisited the concept of MHB and discussed three critical functions of practical significance. A range of bacteria associated with ECM and their role in improving the host plants is summarised in Reddy and Satyanarayana (2006). Detrimental effects of rhizosphere microbes on mycelial growth and ECM formation have also been investigated. Bending et al. (2002) found that two bacterial isolates, *Burkholderia* and *Serratia*, from *Pinus sylvestris–Suillus luteus* mycorrhizosphere, inhibited ECM formation.

The presence of ECM fungi on the roots of trees has repeatedly been shown to confer some protection against the effects of several important root pathogenic fungi. *Boletus bovinus* helped to protect *Picea abies* from *Fomes annosus* (Stack and Sinclair 1975), and *Pisolithus tinctorius* increased the survival rate of *Pinus taeda* seedlings exposed to *Rhizoctonia solani* (Viljoen et al. 1992). Lei et al. (2005) examined antagonistic interactions between a wide range of ECM fungi and root pathogenic fungi in culture experiment. Strong antagonistic interactions between *Suillus grevillei* and *Boletus* sp. and pathogenic fungi *Fusarium solani* and *Rhizoctonia solani* were confirmed by plate-culture experiments and nursery inoculation experiments (Li et al. 2005). Reddy and Satyanarayana (2006) addressed mycorrhizas may also affect herbivores through alteration of plant growth or foliar chemistry or influencing anti-herbivore defences and/or herbivory tolerance (Gange et al. 2005).

21.3.5 Mycosilviculture

Based on the effects on human health, the fruiting bodies of ECM fungi are edible, medicinal or poisonous including suspected poisonous (Chang 2008). Species known to have toxic fruit bodies should not be introduced to new areas as mycorrhizal partners. For example, the Australian Government refused to allow the importation of cultures of *Amanita phalloides*, which is a good mycorrhizal partner, but its basidiomata contains dangerous levels of ibotenic acid (Trappe 1977). In contrast, EMMs with ecological and economic importance are introduced to exotic mushroom orchards. *Tuber melanosporum, T. magnatum, Tricholoma matsutake, Boletus edulis* and *Cantharellus cibarius* are the most expensive and sought-after edible mushrooms, and their biological and ecological characters have been well studied, and attempts for commercial cultivation are in progress (see Chap. 23). The development of a science-based production of EMMs becomes a new industrial crop referred as mycosilviculture although attempts for the majority of EMMs remain a challenge (Savoie and Largeteau 2011). Understanding the ecology of EMMs and the adapted forest management practices appears to be the means to improve natural mushroom production and introduced new species in forest plantations using mycorrhizal seedlings from nurseries. Application of appropriate mycorrhizal technology enables production of valuable forest mushrooms for human consumption; on the other hand, it also promotes the healthy growth of host plants and other products such as timber, hazelnuts, etc. from mushroom orchards.

21.4   Applications of AM Fungi

21.4.1   Source of AM Inoculum

Unlike ECM fungi, AM fungi cannot be grown in axenic culture, and therefore the sources of AM inoculants are restricted to colonised roots, spores or colonised soil mixed with mycorrhizal root segments, spores and hyphae. These forms of inoculants can be derived from naturally colonised soil or from propagation in a dual culture system with host plant. The root-based hyphal network in soils is the primary inoculum for seedlings that become established on natural grasslands. However, the inoculants of natural soil or colonised roots have some profound disadvantages since they may contain more than one mycorrhizal fungus and may also contain pathogenic organisms, as discussed for ECM above. Spores are perhaps the best inoculants for laboratory experiments because the features diagnostic of individual species are present only in the spores developed primarily on extrametrical hyphae. Natural soil of agricultural crops and forests may contain varying numbers of spores of different AM fungal species. It is estimated that the upper 10 cm of soil in an undisturbed *Acer*-dominated hardwood forest in Michigan contained nearly seven million sporocarps ha^{-1} of AM spores (Kessler and Blank 1972). The dual culture using sterile soil with some kind of quality control is believed a practical approach to produce high level of inoculants for commercial applications. A pot culture of *Glomus versiforme* on Sudan grass (*Sorghum vulgare*) can produce up to 1.8×10^{7} spores per month over an extended period (Daniels et al. 1981). Spores from colonised soil near the colonised roots collected from field or pot cultures can be extracted using the traditional wet-sieve method. This approach and the later modified techniques are widely used in extracting spores from soils with modifications.

21.4.2   Evaluation and Selection of AM Fungi

Several properties inherent in all symbiotic systems are also required for evaluation in AM associations, including mycorrhizal dependency, compatibility and specificity. These properties in AM are determined by mycotrophy (plant acquisition of nutrients via a fungus), fungal dependency and mycorrhizal dependency of a plant. Considering that over 200 species of AM fungi form associations with most vascular plant species, the combined response diversity of the fungus–plant symbiosis is likely high. It is generally accepted that AM fungi have no or limited host specificity as they can associate with a wide range of host plants. However, AM fungi are believed to have a certain type of specificity termed 'functional compatibility' (Gianinazzi-Pearson 1984) or 'ecological specificity' (McGonigle and Fitter 1990) since the extent of colonisation on plant roots may vary among different fungus–host partners.

The extent of AM fungi colonising roots together with propagules produced in soil can be detected using appropriate approaches. The forms of AM fungi occur in the roots are as hyphae, arbuscules and/or vesicles (except for *Gigaspora* and *Scutellospora*) and in the soil as spores, sporocarps and hyphae. Propagules of AM fungi include colonised roots, spores or sporocarps, dead root fragments, other colonised organic materials and networks of hyphae in soil, which are sources of AM inoculum. Techniques have been developed for assessing the level of root colonisation, quantifying spores and determining inoculum potential in the soil (Abbott and Robson 1977). Colonisation characteristics can be assessed using the magnified intersection method (McGonigle et al. 1990). The incidences of some microscopic features of AM root at each intersection between the root and the crosshair can be noted to calculate the percentage incidence of each structure over total colonised intersections. Total proportion of root length that was colonised was based on the presence of any mycorrhizal structure. AM features which can be measured using the McGonigle's method include intraradical hyphae, arbuscules, intraradical spores (thick-walled structures, often occluded by a septum or plug, typical of those found in *G. intraradices*), hyphal coils, vesicles (thin-walled sac-like structures lacking occlusion, typical of fungi in the genus *Acaulospora*), entry points and external hyphae (Tibbett et al. 2008).

Bioassays using bait plants grown in intact soil cores provide a better estimate of mycorrhizal inoculum potential than assays using mixed soil or methods for counting propagules such as spores, root fragments, other colonised organic materials and networks of hyphae in soil (see Djuuna et al. 2009). Using a bioassay with clover (bait for AM fungi) and *Eucalyptus globulus* (for ECM), Chen et al. (1999) assessed inoculum potential of both types of mycobionts in established eucalypt plantations of varying ages in Western Australia. Brundrett and Ashwath (2013) compared the results of bioassay, spore survey and culturing experiments using the same soils collected from both natural and disturbed habitats and found differences in the propagule strategies of AM fungi for survival and spread within tropical Australian soils.

The growth and branching of AM fungal hyphae are induced by root factors exuded by host plants and are followed by the formation of an appressorium leading to the hyphal penetration in the root system (Ramos et al. 2008). These root signalling factors seem to be specifically synthesised by host plants, as exudates from non-host plants are not able to promote either hyphal differentiation or appressorium formation (Giovannetti et al. 1996).

To assist evaluation and selecting AM fungi, fungal effect can be measured in several ways. Dry weight production and mycorrhizal dependence are the two most widely used expressions for evaluating AM effect on host plants (e.g. Kendrick and Berch 1985). Fungal influences on plant physiology such as mineral nutrition particularly phosphorus, plant performance and plant protection are important components in assessing fungal efficiency.

21.4.3 Effectiveness of AM Inoculation

The use of AM inoculants in forest nurseries is far less than the use in the agriculture and is also not as often as the use of the ECM fungi in the plantation practices. This may be due to our understanding of relatively less predominance of AM symbioses in commercial plantation species and variable effects on plant growth (Smith and Smith 2011). Liu and Luo (1988) inoculated *Prunus pseudocerasus* with *G. mosseae* and *G. versiforme* and demonstrated substantial increase of the acquisition on the growth, mineral nutrition and water of host plant. Application of AM inoculants in China for some woody plants is reviewed in Zhang (1995) including species of *Abrus*, *Calamus*, *Casuarina*, *Citrus*, *Dimocarpus* and *Malus*. Inoculation of *Acacia mangium* with AM fungi was less convincing probably due to the presence of native efficient strains in the soil (De la Cruz and Yantasath 1993).

Occurrence of mycorrhizal symbionts is widespread in acid soils in the tropics indicating that mycorrhizal functions and selection of acid-tolerant fungal strains may be important for both trees and crops (Haselwandter and Bowen 1996). Some AM fungi have the capacity to reduce the absorption of toxic metals by plants (e.g. Amir et al. 2008; Bi et al. 2005). Gildon and Tinker (1983) found a heavy metal tolerant strain of the AM fungus *G. mosseae*, collected on a heavily zinc- and cadmium-contaminated site. Mycorrhizal seedlings of *Betula* performed better when exposed to the toxic metals Cu and Ni compared to non-mycorrhizal ones. Furthermore, heavy metal-induced genes encoding glutathione S-transferases in *G. intraradices* are identified (Waschke et al. 2006). However, incidences of AM fungi conferring to toxic metals in plantation species are less well addressed.

There are few reports of resistance of AM to pathogens in woody plants. Tang and Chen (1995) found that *G. mosseae* helped to protect *Populus* seedlings from a canker fungus (*Dothiorella gregaria*) by promoting acquisition of water and P and inducing peroxidise and polyphenoloxidase activities in host. Induction of defence responses in pre-inoculated plants with *G. mosseae* was much higher and quicker than that in non-mycorrhizal plants upon colonisation of *Rhizoctonia solani* (Song et al. 2011). This indicates that induction of accumulation of DIMBOA, an important phytoalexin in corn, and systemic defence responses by AM fungus, plays a vital role in enhanced disease resistance of mycorrhizal plants against sheath blight. However, the effectiveness of AM fungi in biocontrol is dependent on the AM fungus involved, as well as the substrate and host plant.

AM fungi may also have interactions with plant growth-promoting rhizosphere (PGPR) organisms. The concept and role of PGPR plant growth and protection is well documented (e.g. Whipps 2001). The presence of a biocontrol PGPR *Trichoderma harzianum* suppressed hyphal length of *G. intraradices* but no effect on hyphal biomass (Green et al. 1999). Another biocontrol agent *Gliocladium virens* had no detrimental effect on *G. etunicatum* and *G. mosseae* (Paulitz and Linderman 1991).

The effect of AM inoculation may vary since many factors can influence the occurrence of AM (Abbott and Robson 1991). The interplay between environmental factors (phosphorus, pH, nitrogen, water and temperature) and the host–fungus relationship is discussed in Smith and Smith (2011). Abbott and Robson (1982) stressed the importance of knowing the response curve of mycorrhizal and non-mycorrhizal plants to P application when evaluations of the impact of AM fungi on the growth.

21.5 Dual Inoculation

21.5.1 Dual AM and ECM

A few tree genera are ecologically interesting because they can form dual associations with both AM and ECM fungi (Lodge 2000). Plants reported to have dual AM/ECM associations belong to the genera *Casuarina*, *Allocasuarina* (Casuarinaceae), *Eucalyptus*, *Melaleuca* (Myrtaceae) and *Acacia* and *Leucaena* (Mimosaceae) from Australia (Brundrett et al. 1996; Chen et al. 2000a; Saravesi et al. 2011) and *Alnus*, *Populus*, *Salix* and *Uapaca* from Northern Hemisphere (Lodge and Wentworth 1990; Moyersoen and Fitter 1999; Saravesi et al. 2011; Zhao 1995). These genera include some major species used in commercial plantation forestry. Despite the ecological importance of the tripartite associations involving plant, AM and ECM fungi, only a few studies explored the relative benefits from each fungal type to the host plant and interactions between ECM and AM fungi colonising the same root systems (Chen et al. 1998, 2000a; Kariman et al. 2012; Lodge and Wentworth 1990).

The existence of dual association in the same root systems of *Eucalyptus* species has been confirmed both in plantation soils and under controlled conditions (Chen et al. 1998, 2000a; Lodge 2000; Oliveira et al. 1997). Jones et al. (1998) compared the growth response, phosphorus uptake efficiency and external hyphal production of AM and ECM fungi in *Eucalyptus coccifera*. Seedlings of *Eucalyptus urophylla* colonised by both AM and ECM fungi enhanced plant growth, root activity and acquisition of nutrients, amino acids and polysaccharides in root exudates when compared with non-mycorrhizal plants or plants colonised by one type of fungus (Chen et al. 1998). Chen et al. (2000a) established an experimental model to study dual colonisation in *Eucalyptus* and investigated the relative benefits of each type of fungi provided to two tree species and demonstrated several different mechanisms involved in successional replacement. Succession within a root system from predominantly AM to dominance by ECM has previously been reported for *Eucalyptus*, *Populus* and *Salix* in both field observations (Gardner and Malajczuk 1988; Lodge and Wentworth 1990) and glasshouse experiments (Chen et al. 1998; Dos Santos et al. 2001).

Proposed mechanisms to explain successional replacement in tripartite associations include mechanical barriers posed by the ECM sheath, chemicals of fungal or host origin, competition for root carbohydrates and effects on rhizosphere communities (Chen et al. 2000a; Lodge and Wentworth 1990). ECM fungi may have a greater impact on colonisation by AM fungi by causing their host to reduce production of fine roots, thereby limiting the availability of new roots to the fungus. These studies indicate that ECM associations are usually more important that AM associations for *Eucalyptus* species. However, there is evidence that AM associations can provide benefits to eucalypts, especially during seedling establishment despite some ambiguous reports on AM efficiency. There are also cases where the benefits provided by AM and ECM together can exceed those provided by either one alone (Chen et al. 1998, 2000a). The importance of AM associations of eucalypts is likely to be greater in disturbed habitats or exotic locations where there are few eucalypt compatible ECM fungi (Dell et al. 2002), since AM fungi generally exhibit little host specificity (see Sect. ECM section above 21.3.3). Variation of inoculation efficiency between tree species suggests that careful matching of host and fungal species (and genotype) is needed to obtain the best results.

Osonubi et al. (1991) examined effects of ECM and AM fungi on drought tolerance of four leguminous species (*Acacia auriculiformis*, *Albizia lebbeck*, *Gliricidia sepium* and *Leucaena leucocephala*). Under well-watered conditions, there were significant differences between species in development of both ECM and AM associations. They found that imposition of drought stress after colonisation had become established, showing significant reduction of ECM colonisation in *Gliricidia* only. Growth simulation and drought tolerance were observed for all tree species inoculated with ECM and/or AM fungi.

Gange et al. (2005) examined AM and ECM fungi and the interactions between them, on foliar-feeding insect attack of *Eucalyptus urophylla*. Both fungal types affected levels of damage by insect herbivores. Most importantly, herbivory by the pest insects *Anomala cupripes* (Coleoptera) and *Strepsicrates* spp. (Lepidoptera) was decreased by ECM. It is suggested that mycorrhizal effects on eucalypt insects may be determined by carbon allocation within the plant (Gange et al. 2005). This study that has enhanced our understanding of how these different fungi affect insect performance may help in unravelling the complex and little understood phenomenon of dual mycorrhizal plants. A study by Gehring and Whitham (2002) reported that AM colonisation of hybrid cottonwood trees (*Populus angustifolia* × *P. fremontii*) reduced populations of a specialist aphid, *Chaitophorus populicola*, whereas ECM colonisation enhanced aphid numbers. Future studies of mycorrhizal effects on plant growth should include a consideration of the insect herbivores present. These fungi clearly have the potential to influence insect herbivore attack rates, and experiments need to be performed in which fungal species and soil conditions are varied, to determine which, if any, mycorrhizal combinations could be used to reduce potential pest insect levels.

21.5.2   Dual Mycorrhizal Fungi and Nodule-Forming Organisms

Some species in Mimosaceae and Casuarinaceae forming dual AM/ECM associations also have nitrogen-fixing root nodule symbioses. There are two different types of nitrogen-fixing symbioses: the legume–rhizobia symbioses that form between *c.* 80 % of all legumes and rhizobia of the genera *Rhizobium* and *Bradyrhizobium* and actinorhizal symbioses that form between actinorhizal plants and *Frankia* (Katharina et al. 2011). Mycorrhizal and rhizobial/*Frankia* symbioses often act synergistically on colonisation rate, nitrogen-fixing efficiency, mineral nutrition and plant growth (Amora-Lazcano et al. 1998). The mycorrhizal fungi associated with legumes or actinorhizal plants are an essential link for adequate phosphorus nutrition, leading to enhanced nitrogenase activity that in turn promotes root and mycorrhizal growth (Reddy and Satyanarayana 2006).

Dual inoculation with AM fungi and rhizobia enhanced survival and growth of *Centrolobium tomentosum* plantations in the field, and AM fungi seemed to favour the nodule occupation by rhizobia strains as compared to the non-mycorrhizal plants (Marques et al. 2001). Cao et al. (2005) stimulated the growth of *Acacia* and *Leucaena* by dual inoculation with two *Glomus* species and *Rhizobium*. Nodulation, mycorrhizal colonisation, dry weight and nitrogen and phosphorus content of *Leucaena leucocephala* seedlings were improved by dual inoculation with *G. fasciculatum* and *Rhizobium* compared to single inoculation with either organism (Manjunath et al. 1984). This study also showed that inoculation with *Glomus* only improved nodulation by native rhizobia grown in a phosphorus-deficient unsterile soil, and the *Rhizobium*-only treatment improved colonisation of roots by native AM fungi. Diem and Gauthier (1982) demonstrated that mycorrhization of *Casuarina equisetifolia* saplings with *G. mosseae* improved plant growth, *Frankia* nodulation and nitrogen fixation. *Acacia* species are spontaneously associated to the three symbionts in their native soils (Warcup 1980), while in exotic area, local colonisation of mycorrhizal fungi seemed inefficient due to the lack of compatible mycobionts. A pot experiment showed *Medicago sativa* plants co-inoculated with *G. mosseae* and *Rhizobium* greatly increased the survival rate and nutrient uptake in coal mine substrates (Wu et al. 2009). There are fewer studies reported the effect of ECM than AM on nitrogen-fixing plants. Duponnois et al. (2002) observed the positive effect of the controlled dual ECM and rhizobial symbioses on the growth of *Acacia mangium* provenances, the indigenous symbiotic microflora and the structure of plant parasitic nematode communities. However inoculation with *Hebeloma crustuliniforme* alone or in combination with *Frankia* had no effect on the growth and root nodulation of *Alnus crispa* due to the failure of mycorrhization which may suggest incompatibility of the fungus (Quoreshi et al. 2007). The three types of symbioses have been shown to coexist on the same root system, but their functional relevance remains unclear.

21.6 Conclusions

Mycorrhizal colonisation can help establishment of plantations, particularly in eroded, degraded or heavy metal-contaminated areas. In novel habitats, mycorrhizal fungi may transform soil carbon cycling (Chapela et al. 2001), affect mineral nutrient dynamics (Phillips and Fahey 2006) and alter surrounding vegetation (Richardson and Rejmánek 2004). Appropriate mycorrhizal fungi incorporated in forest nurseries for raising mycorrhizal seedlings and transfer of seedlings to the field is a practical inoculation technique currently suitable in plantation crops and trees. Experience of the use of inoculated seedlings has indicated that responses to mycorrhizal inoculation are often greatest under the most extreme conditions, particularly those involving exposure to infertile soils, drought, metal contamination or pathogens (Smith and Read 2008).

Field surveys for the mycorrhizal community associated with the given tree species, combined with estimations of the extent of mycorrhizal colonisation on the roots and propagules such as basidiomata of agarics and spores of AM fungi, can assist in defining the range of fungal symbionts available for the tree. However, this should not rule out the possibility of introducing new and efficient fungal partners to the area. The low diversity of fungi currently being used in Australasian eucalypt plantations may give minimal benefit to tree production because the fungi may not necessarily be well suited to the local site characteristics (climate, soil type, host plants, etc.). In the long term, maintenance of soil structure, fertility and general ecosystem stability in the face of environmental changes and disturbances may be enhanced by the presence of a broader diversity of fungi.

Inoculation of seedlings with mycorrhizal fungi should aim to ensure that seedlings have extensive colonisation at the time of transplanting from the nursery to the field. There are still relatively few examples in the use of AM inoculants under forest field conditions. Thus, more precise experimental work with thoughtful design should be carried out to overcome potential constraints limiting the development and function of introduced symbionts. A mixed inoculum containing fungi with differing ecological strategies might give more consistent and permanent results in promoting plant growth. The additive beneficial effects from insuring simultaneous colonisation by multiple types of symbionts could be useful for the establishment of commercial timber species in adverse sites. Cultural practices may have to be modified to produce conditions which are optimal for the development of symbioses in the nursery. As nutrient supply and composition can influence hyphal development in the nursery, application of fertilisers at appropriate regimes is essential for optimising the potential benefits of inoculation programmes. The practical application of mycorrhizal fungi may be integrated in the disease management by producing mycorrhizal seedlings, so as to prevent primary and secondary colonisation by pathogenic fungi, herbivore insects and other harmful organisms. Further research on optimising mycorrhizal inoculants and seedlings in forest nurseries is required to maximise efficiency and productivity of fungal inoculation. There is also a need for long-term field studies to monitor the performance and persistence of introduced fungi in the plantation and revegetation sites and their impacts on native microflora.

Acknowledgements Y.L.C. appreciates Professor Bernard Dell, Professor Lyn Abbott, Dr. Nick Malajczuk, Dr. Mark Brundrett, Dr. Chris Walker and Dr. François Le Tacon for their support and long-term collaboration. This study was financially supported by National Natural Science Foundation of China (31471946), International Foundation for Science (D-2894/1 & D-2894/2), Australian Centre for International Agricultural Research (F/9425) and International Tropical Timber Organization (PD 38/98 Rev.2(F)). The Australian Department of Education, Employment and Workplace Relations, Murdoch University and Chinese Academy of Forestry are acknowledged for granting scholarships and travel funds for a part of this study.

References

Abbott LK, Robson AD (1977) The distribution and abundance of vesicular arbuscular endophytes in some Western Australian soils. Aust J Bot 25:515–522

Abbott LK, Robson AD (1982) The role of vesicular-arbuscular mycorrhizal fungi and selection of fungi for inoculation. Aust J Agric Res 33:389–408

Abbott LK, Robson AD (1991) Factors influencing the occurrence of vesicular-arbuscular mycorrhizas. Agric Ecosyst Environ 35:121–150

Alvarez M, Huygens D, Olivares E, Saavedra I, Alberdi M, Valenzuela E (2009) Ectomycorrhizal fungi enhance nitrogen and phosphorus nutrition of *Nothofagus dombeyi* under drought conditions by regulating assimilative enzyme activities. Physiol Plant 136:426–436

Amir H, Jasper DA, Abbott LK (2008) Tolerance and induction of tolerance to Ni of arbuscular mycorrhizal fungi from New Caledonian ultramafic soils. Mycorrhiza 19:1–6

Amora-Lazcano E, Vazquez MM, Azcon R (1998) Response of nitrogen-transforming microorganisms to arbuscular mycorrhizal fungi. Biol Fert Soils 27:65–70

Ashwath N, Cusbert PC, Bayliss B, McLaughlin M, Hunt C (1993) Chemical properties of mine spoils and selected natural soils of the Alligator Rivers Region – implications for establishing native plants on mine spoils. In: Proceedings of the waste rock dump symposium, Darwin N.T., AGPS, Canberra

Barcan VSH, Kovnatsky EF, Smetannikova MS (1998) Absorption of heavy metals in wild berries and edible mushrooms in an area affected by smelter emissions. Water Air Soil Pollut 103:173–195

Behie SW, Bidochka MJ (2013) Potential agricultural benefits through biotechnological manipulation of plant fungal associations. Bioessays. doi:10.1002/bies.201200147

Bending GD, Poole EJ, Whipps JM, Read DJ (2002) Characterisation of bacteria from *Pinus sylvestris-Suillus luteus* mycorrhizas and their effects on root–fungus interactions and plant growth. FEMS Microbiol Ecol 39:219–227

Bi YL, Wu FU, Wu YK (2005) Application of arbuscular mycorrhizas in ecological restoration of areas affected by coal mining in China. Acta Ecol Sin 25:2068–2073

Bois G, Bertrand A, Piche Y, Fung M, Khasa DP (2006) Growth, compatible solute and salt accumulation of five mycorrhizal fungal species grown over a range of NaCl concentrations. Mycorrhiza 16:99–109

Brundrett MC, Ashwath N (2013) Glomeromycotan mycorrhizal fungi from tropical Australia III. Measuring diversity in natural and disturbed habitats. Plant Soil. doi:10.1007/s11104-013-1613-4

Brundrett MC, Ashwath N, Jasper DA (1996) Mycorrhizas in the Kakadu region of tropical Australia. Plant Soil 184:173–184

Cairney JWG, Chambers SM (1999) Ectomycorrhizal fungi: key genera in profile. Springer, Berlin

Cao J, Tang Y, Qin S, Hou Y (2005) Dual inoculation of Mimosaceae seedlings with vesicular arbuscular mycorrhizal fungi and Rhizobium. In: Brundrett M, Dell B, Malajczuk N, Gong M (eds) Mycorrhizal research for forestry in Asia. ACIAR proceedings no. 62. ACIAR, Canberra, pp 119–121

Castellano MA, Molina R (1989) Mycorrhiza. In: Landis TD, Tinus RW, McDonald SE, Barnett JP (eds) The biological component: nursery pests and mycorrhizae. Vol. 5. The container tree nursery manual. Agric. handbook 674. USDA Forest Service, Washington, DC, pp 101–167

Chai DD, Guo SJ, Sun XB, Qin TT (2013) The major factors affecting ectomycorrhizal fungi diversity in the forest ecosystem. Adv J Food and Technol 5(7):879–890

Chang ST (2008) Overview of mushroom cultivation and utilization as functional foods. In: Cheung PCK (ed) Mushrooms as functional foods. Wiley, Hoboken, NJ, pp 1–33

Chapela IH, Osher LJ, Horton TR, Henn MR (2001) Ectomycorrhizal fungi introduced with exotic pine plantations induce soil carbon depletion. Soil Biol Biochem 33:1733–1740

Chen YL (2002) Mycorrhizal synthesis of *Tuber melanosporum* on *Castanopsis*. Edible Fungi China 21:15–17

Chen YL (2004) Song rong (*Tricholoma matsutake*), a valuable mushroom from China: consumption, development and sustainability. In: Kusters K, Belcher B (eds) Forest products, livelihoods and conservation: case studies of NTFP systems, vol 1. CIFOR, Bogor, pp 56–71

Chen YL, Gong MQ, Wang FZ, Chen Y (1998) Study on mycorrhizal physiology of *Eucalyptus urophylla* inoculated with dual ECM and VAM fungi. For Res 11:237–242

Chen YL, Brundrett M, Dell B, Gong MQ, Malajczuk N (1999) Bioassay measurements of mycorrhizal inoculum in soils from eucalypt plantations of varying ages in Western Australia. For Stud China 1(2):26–32

Chen YL, Brundrett MC, Dell B (2000a) Effects of ectomycorrhizas and vesicular–arbuscular mycorrhizas, alone or in competition, on root colonization and growth of *Eucalyptus globulus* and *E. urophylla*. New Phytol 146:545–556

Chen YL, Gong MQ, Xu DP, Zhong CL, Wang FZ, Chen Y (2000b) Screening and inoculant efficacy of Australian ectomycorrhizal fungi on *Eucalyptus urophylla* in field. For Res 13:569–576

Chen YL, Gong MQ, Wang FZ, Chen Y (2001) Effects of inoculation with 11 ectomycorrhizal fungal isolates on growth and photosynthesis of *Castanopsis hystrix* saplings. For Res 14:515–522

Chen YL, Dell B, Malajczuk N (2006a) Effect of *Scleroderma* spore density and age on mycorrhiza formation and growth of containerized *Eucalyptus globulus* and *E. urophylla* seedlings. New Forests 31:453–467

Chen YL, Kang LH, Dell B (2006b) Inoculation of *Eucalyptus urophylla* with spores of *Scleroderma* in a nursery in south China: comparison of field soil and potting mix. For Ecol Manage 222:439–449

Chen YL, Kang LH, Malajczuk N, Dell B (2006c) Selecting ectomycorrhizal fungi for inoculating plantations in south China: effect of *Scleroderma* on colonization and growth of exotic *Eucalyptus globulus*, *E. urophylla*, *Pinus elliottii* and *P. radiata*. Mycorrhiza 16:251–259

Chen YL, Liu S, Dell B (2007) Mycorrhizal status of *Eucalyptus* plantations in south China and implications for management. Mycorrhiza 17:527–535

Chevalier G, Frochot H (2000) La Truffe De Bourgogne (*Tuber unicinatum* Chatin). INRA, Cedex, France

Chilvers GA (1973) Host range of some eucalypt mycorrhizal fungi. Aust J Bot 21:103–111

Collins M, Brundrett M, Koch J, Sivasithamparam K (2007) Colonisation of jarrah forest bauxite-mine rehabilitation areas by orchid mycorrhizal fungi. Aust J Bot 55:653–664

Colpaert JV, van Assche JA (1987) Heavy metal tolerance in some ectomycorrhizal fungi. Funct Ecol 1:415–421

Colpaert JV, van Assche JA, Luijtens K, Van Assche JA (1992) The growth of the extramatrical mycelium of ectomycorrhizal fungi and the growth response of *Pinus sylvestris* L. New Phytol 120:127–135

Corbett MH (1999) Revegetation of mined land in the wet–dry tropics of northern Australia: a review. Supervising Scientist Report 150, Supervising Scientist, Canberra

Dahlberg A, Finlay RD (1999) *Suillus*. In: Cairney JWG, Chambers SM (eds) Ectomycorrhizal fungi: key genera in profile. Springer, Berlin, pp 34–64

Daniels BA, McCool PM, Menge JA (1981) Evaluation of the commercial potential of the vesicular-arbuscular mycorrhizal fungus. New Phytol 87:345–354

De la Cruz RE, Yantasath K (1993) Symbiotic associations. In: Kamis A, Taylor D (eds) *Acacia mangium*, growing and utilization. Winrock International and The Food and Agriculture Organization of the United Nations, Bangkok, pp 101–111

Dell B, Malajczuk N (1995) Fertiliser requirements for ectomycorrhizal eucalypts in forest nurseries and field plantings in Southern China. In: Brundrett MC, Dell D, Malajczuk N, Gong M (eds) Mycorrhizas for plantation forestry in Asia, ACIAR monograph 62. ACIAR, Canberra, pp 96–100

Dell B, Malajczuk N, Dunstan WA (2002) Persistence of some Australian *Pisolithus* species introduced into eucalypt plantations in China. For Ecol Manage 169:271–281

Diem HG, Gauthier D (1982) Effet de l'infection endomycorhizienne (Glomus mosseae) sur la nodulation et la croissance de Casuarina equisetifolia. CR Acad Sci (Paris) 294:215–218

Dixon RK, Pallardy SK, Garrett HE, Cox GS, Sander IL (1983) Comparative water relations of container-grown and bare-root ectomycorrhizal and nonmyocorrhizal *Quercus velutina* seedlings. Can J Bot 61:1559–1565

Djuuna IAF, Abbott LK, Solaiman MZ (2009) Use of mycorrhiza bioassays in ecological studies. In: Varma A, Kharkwal AC (eds) Symbiotic fungi, Soil biology 18. Springer, Berlin

Dos Santos VL, Muchovej RM, Borges AC, Neves JCL, Kasuya MCM (2001) Vesicular-arbuscular-/ecto-mycorrhiza succession in seedlings of Eucalyptus spp. Braz J Microbiol 32:81–86

Duddridge JA, Malibari A, Read DJ (1980) Structure and function of mycorrhizal rhizomorphs with special reference to their role in water transport. Nature 287:834–836

Duñabeitia MK, Hormilla S, Garcia-Plazaola JI, Txarterina K, Arteche U, Becerril JM (2004) Differential responses of three fungal species to environmental factors and their role in the mycorrhization of *Pinus radiata* D. Don. Mycorrhiza 14:11–18

Dunstan WA, Malajczuk N, Dell B (1998) Effects of bacteria on mycorrhizal development and growth of container grown *Eucalyptus diversicolor* F. Muell seedlings. Plant Soil 201:243–251

Duponnois R, Founoune H, Lesueur D (2002) Influence of the controlled dual ectomycorrhizal and rhizobal symbiosis on the growth of Acacia mangium provenances, the indigenous symbiotic microflora and the structure of plant parasitic nematode communities. Geoderma 109:85–102

Duponnois R, Founoune H, Masse D, Pontainer R (2005) Inoculation of *Acacia holosericea* with ectomycorrhizal fungi on a semiarid site in Senegal: growth response and influences on the mycorrhizal soil infectivity after 2 years plantation. For Ecol Manage 207:351–362

FAO (2012) FAO statistical yearbook 2012. Food and Agriculture Organization of the United Nations (FAO), Rome

Fay DA, Mitchell DT, Parkes MA (1997) A preliminary study of the mycorrhizal associations of tree seedlings growing on mine spoil at Avoca, Co. Wicklow. Special issue. In: The ecology of old mine sites. Proceedings of a workshop organised by the Mining Heritage Society of Ireland at the Geological Survey of Ireland, 18–19 Oct 1997

Finlay RD, Odham G, Söderström B (1988) examined mycelial uptake, translocation and assimilation of nitrogen from ^{15}N-labelled ammonium by *Pinus sylvestris* plants infected with four different ECM fungi. New Phytol 110:59–66

Fogel E, Hunt G (1979) Fungal and arboreal biomass in a western Oregon Douglas fir ecosystem: distribution patterns and turnover. Can J For Res 9:245–256

Founoune H, Duponnois R, Bâ AM (2002) Ectomycorrhization of *Acacia mangium* Willd. and *Acacia holosericea* A. Cunn. ex G. Don in Senegal: impact on plant growth, populations of indigenous symbiotic microorganisms and plant parasitic nematodes. J Arid Environ 50:325–332

Frey-Klett P, Garbaye J, Tarkka M (2007) The mycorrhiza helper bacteria revisited. New Phytol 176:22–36

Fries N (1988) specific effects of diterpene resin acids on spore germination of ectomycorrhizal basidiomycetes. Experientia 44:1027–1030

Gange AC, Gane DRJ, Chen YL, Gong MQ (2005) Dual colonization of *Eucalyptus urophylla* S. T. Blake by arbuscular and ectomycorrhizal fungi affects levels of insect herbivore attack. Agric For Entomol 7:253–263

Garbaye J (1994) Helper bacteria: a new dimension to the mycorrhizal symbiosis. New Phytol 128:197–210

Gardner JH, Malajczuk N (1988) Recolonisation of rehabilitated bauxite mine sites in Western Australia by mycorrhizal fungi. For Ecol Manage 24:27–42

Gehring CA, Whitham TG (2002) Mycorrhizae-herbivore interactions: population and community consequences. In: van der Heijden MGA, Sanders IR (eds) Mycorrhizal ecology. Springer, Heidelberg, pp 295–320

Gianinazzi-Pearson V (1984) Host-fungus specificity in mycorrhizae. In: Verma DPS, Hohn TH (eds) Genes involved in plant–microbe interactions. Springer, Vienna, pp 225–253

Gildon A, Tinker PB (1983) Interactions of vesicular-arbuscular mycorrhizal infection and heavy metal in plants.1: The effect of heavy metals on the development of vesicular–arbuscular mycorrhizas. New Phytol 95:247–261

Giovannetti M, Sbrana C, Citernesi AS, Avio L (1996) Analysis of factors involved in fungal recognition responses to host derived signals by arbuscular mycorrhizal fungi. New Phytol 133:65–71

Gogala N (1991) Regulation of mycorrhizal infection by hormonal factors produced by hosts and fungi. Experientia 47:331–340

Gordon GJ, Gehring CA (2011) Molecular characterization of pezizalean ectomycorrhizas associated with pinyon pine during drought. Mycorrhiza 21:431–441

Grant CD, Loneragan WA (2001) The effect of burning on the understorey composition and vegetation succession of 11–13 year-old rehabilitated bauxite mines in Western Australia: community changes and vegetation succession. For Ecol Manage 145:255–279

Green H, Larsen J, Olsson PA, Funck Jensen D, Jakobsen I (1999) Suppression of the biocontrol agent *Trichoderma harzianum* by mycelium of the arbuscular mycorrhizal fungus *Glomus intraradices* in root-free soil. Appl Environ Microbiol 65:1428–1434

Grove TS, Le Tacon F (1993) Mycorrhiza in plantation forestry. In: Tommerup IC (ed) Mycorrhiza synthesis, vol 9, Advances in plant pathology. Academic, New York, pp 191–227

Hall I, Brown G, Byars J (1994) The black truffle, 2nd edn. New Zealand Institute for Crop and Food Research, Christchurch

Haselwandter K, Bowen GD (1996) Mycorrhizal relations in trees and agroforestry and land rehabilitation. For Ecol Manage 81:1–17

Hilbert JL, Martin F (1988) Regulation of gene expression in ectomycorrhizas. I. Protein changes and he presence of ectomycorrhiza specific polypeptides in the *Pisolithus-Eucalyptus* symbiosis. New Phytol 110:339–346

Hinz DA (1997) The return of the Gadayka tree after bauxite mining. Minerals Council of Australia, November 1997

Horan DP, Chilvers GA (1990) Chemotropism, the key to ectomycorrhizal formation. New Phytol 116:297–302

Horan DP, Chilvers GA, Lapeyrie FE (1988) Time sequence of the infection process in eucalypt ectomycorrhiza. New Phytol 109:451–458

Hunt GA, Trappe JM (1987) Seasonal hypogeous sporocarp production in a western Oregon Douglas-fir stand. Can J Bot 65:438–445

Hutton BJ, Dixon KW, Sivasithamparam K, Pate JS (1997) Effect of habitat disturbance on inoculum potential of ericoid endophytes of Western Australian heaths (Epacridaceae). New Phytol 135:739–744

Ishida TA, Nara K, Ma S, Takano T, Liu S (2009) Ectomycorrhizal fungal community in alkaline–saline soil in northeastern China. Mycorrhiza 19:329–335

Jairus T, Mpumba R, Chinoya S, Tedersoo L (2011) Invasion potential and host shifts of Australian and African ectomycorrhizal fungi in mixed eucalypt plantations. New Phytol 192:179–187

Jansa J, Finlay R, Wallander H, Smith FA, Smith SE (2011) Role of mycorrhizal symbioses in phosphorus cycling. In: Bünemann EK (ed) Phosphorus in action. Springer, Berlin, pp 137–168

Jasper DA, Abbott LK, Robson AD (1989) The loss of VA mycorrhizal infectivity during bauxite mining may limit the growth of *Acacia pulchella* R.Br. Aust J Bot 37:33–42

Johnson NC, Graham JH, Smith FA (1997) Functioning of mycorrhizal associations along the mutualism–parasitism continuum. New Phytol 135:575–585

Jones MD, Hutchinson TC (1986) The effect of mycorrhizal infection on the response of *Betula papyrifera* to nickel and copper. New Phytol 102:429–442

Jones MD, Durall DM, Tinker PB (1998) A comparison of arbuscular and ectomycorrhizal *Eucalyptus coccifera*: growth response, phosphorus uptake efficiency and external hyphal production. New Phytol 140:125–134

Jourand P, Ducousso M, Reid R, Majorel C, Richert C, Riss J, Lebrun M, Epron D (2010) Nickel-tolerant ectomycorrhizal *Pisolithus albus* ultramafic ecotype isolated from nickel mines in New Caledonia strongly enhance growth of the host plant *Eucalyptus globulus* at toxic nickel concentrations. Tree Physiol 30:1311–1319

Kariman K, Barker SJ, Finnegan PM, Tibbett M (2012) Dual mycorrhizal associations of jarrah (*Eucalyptus marginata*) in a nurse-pot system. Aust J Bot 60(8):661–668

Kasuya MCM, Muchovej RMC, Muchovej JJ (1990) Influence of aluminium on *in vitro* formation of *Pinus caribaea* mycorrhiza. Plant Soil 124:73–79

Katharina P, Didier B, Ana R, Berry AM (2011) Progress on research on actinorhizal plants. Funct Plant Biol 38:633–638

Kendrick B, Berch S (1985) Mycorrhizae: applications in agriculture and forestry. In: Robinson CW (ed) Comprehensive biotechnology, vol 4. Pergamon, Oxford, pp 109–152

Kernaghan G, Hambling B, Fung M, Khasa D (2002) In vitro selection of boreal ectomycorrhizal fungi for use in reclamation of saline–alkaline habitats. Restor Ecol 10:43–51

Kessler KJ Jr, Blank RW (1972) Endogone sporocarps associated with sugar maple. Mycologia 64:634–638

Kohout P, Sýkorová Z, Bahram M, Hadincová V, Albrechtová J, Tedersoo L, Vohník M (2011) Ericaceous dwarf shrubs affect ectomycorrhizal fungal community of the invasive *Pinus strobus* and native *Pinus sylvestris* in a pot experiment. Mycorrhiza 21:403–412

Lehto T, Zwiazek JJ (2011) Ectomycorrhizas and water relations of trees: a review. Mycorrhiza 21:71–90

Lei Z, Jin J, Wang C (2005) Antagonism between ectomycorrhizal fungi and plant pathogens. In: Brundrett M, Dell B, Malajczuk N, Gong M (eds) Mycorrhizal research for forestry in Asia. ACIAR proceedings no. 62. ACIAR, Canberra, pp 34–40

Li X, Fu B, Yu J (2005) Inoculation of Pinus massoniana with ectomycorrhizal fungi: growth responses and suppression of pathogenic fungi. In: Brundrett M, Dell B, Malajczuk N, Gong M (eds) Mycorrhizal research for forestry in Asia. ACIAR proceedings no. 62. ACIAR, Canberra, pp 34–40

Liu RJ, Chen YL (2007) Mycorrhizology. Science Press, Beijing

Liu RJ, Luo XS (1988) Effects of VAM on the growth, mineral nutrition and water relations of cherry. J Laiyang Agric Coll 5:6–13

Lodge DJ (2000) Ecto- or arbuscular mycorrhizas, which are best? New Phytol 146:353–354

Lodge DJ, Wentworth TR (1990) Negative associations among VA-mycorrhizal fungi and some ectomycorrhizal fungi inhabiting the same root system. Oikos 57:347–356

Lu XH, Malajczuk N, Dell B (1998) Mycorrhiza formation and growth of *Eucalyptus globulus* seedlings inoculated with spores of various ectomycorrhizal fungi. Mycorrhiza 8:81–86

Malajczuk N, Molina R, Trappe JM (1984) Ectomycorrhiza formation in Eucalyptus. II. The ultrastructure of compatible and incompatible mycorrhizal fungi and associated roots. New Phytol 96:43–53

Malajczuk N, Dell B, Bougher NL (1987) Ectomycorrhiza formation in Eucalyptus. III. Superficial ectomycorrhizas initiated by *Hysterangium* and *Cortinarius* species. New Phytol 105:421–428

Malajczuk N, Lapeyrie F, Garbaye J (1990) Infectivity of pine and eucalypt isolates of *Pisolithus tinctorius* on the roots of *Eucalyptus urophylla in vitro*. New Phytol 114:627–631

Malajczuk N, Reddell P, Brundrett M (1994) Role of ectomycorrhizal fungi in minesite reclamation. In: Pfleger FL, Linderman RG (eds) Mycorrhizae and plant health. American Phytopathology Society symposium series, St Paul, MN

Manjunath A, Bagyaraj DJ, Gopala Gowda HS (1984) Dual inoculation with VA mycorrhiza and *Rhizobium* is beneficial to *Leucaena*. Plant Soil 78:445–448

Marques MS, Pagano M, Scotti MRMML (2001) Dual inoculation of a woody legume (*Centrolobium tomentosum*) with rhizobia and mycorrhizal fungi in south-eastern Brazil. Agroforest Syst 50:107–117

Martin TP, Harris JR, Eaton GK, Miller OK (2003) The efficacy of ectomycorrhizal colonization of pin and scarlet oak in nursery production. J Environ Horticult 21:45–50

Marx DH (1976) Synthesis of ectomycorrhizae on loblolly pine seedling with basidiospores of *Pisolithus tinctorius*. For Sci 22:13–20

Marx DH, Altman JD (1979) Pisolithus tinctorius ectomycorrhiza improve survival and growth of pine seedlings on acid coal spoil in Kentucky and Virginia. Reclam Rev 2:23–37

Marx DH, Cordell CE (1990) Development of *Pisolithus tinctorius* ectomycorrhizae on loblolly pine from spores sprayed at different times and rates. USDA Forest Service. Research Note SE-356

Marx DH, Cordell CE, Maul SB, Ruehle JL (1989) Ectomycorrhizal development on pine by *Pisolithus tinctorius* in bare-root and container seedling nurseries. II. Efficacy of various vegetative and spore inocula. New Forests 3:57–66

Marx DH, Murphy M, Parrish T, Marx S, Haigler D, Eckard D (1997) Root response of mature live oaks in coastal South Carolina to root zone inoculations with ectomycorrhizal fungal inoculants. J Arboricult 23:257–263

McGonigle TP, Fitter AH (1990) Ecological specificity of vesicular-arbuscular mycorrhizal associations. Mycol Res 94:120–122

McGonigle TP, Miller MH, Evans DG, Fairchild GL, Swan JA (1990) A new method which gives an objective measure of colonization of roots by vesicular-arbuscular mycorrhizal fungi. New Phytol 115:495–501

Meharg AA (2003) The mechanistic basis of interactions between mycorrhizal associations and toxic metal cations. Mycol Res 107:1253–1265

Melin E, Nilsson H (1950) Transfer of radioactive phosphorus to pine seedlings by means of mycorrhizal hyphae. Physiol Plant 3:88–92

Mogge B, Loferer C, Agerer R, Hutzler P, Hartmann A (2000) Bacterial community structure and colonization patterns of *Fagus sylvatica* L. ectomycorrhizospheres as determined by florescence in situ hybridization and confocal laser scanning microscopy. Mycorrhiza 9:271–278

Molina R, Palmer JG (1982) Isolation, maintenance, and pure culture manipulation of ectomycorrhizal fungi. In: Schenck NC (ed) Methods and principles of mycorrhizal research. American Phytopathological Society, St Paul, MN, pp 115–129

Molina R, Trappe JM (1994) Biology of the ectomycorrhizal genus Rhizopogon. I. Host associations, host-specificity and pure culture syntheses. New Phytol 125:653–675

Molina R, Massicotte H, Trappe JM (1992) Specificity phenomena in mycorrhizal symbioises: community-ecological consequences and practical implications. In: Allen MF (ed) Mycorrhizal functioning. Chapman and Hall, New York, pp 357–423

Moyersoen B, Fitter AH (1999) Presence of arbuscular mycorrhizas in typical ectomycorrhizal host species from Cameroon and New Zealand. Mycorrhiza 8:247–253

Munyanziza E, Kuyper TW (1995) Ectomycorrhizal synthesis on seedlings of *Afzelia quanzensis* Welw. using various types of inoculum. Mycorrhiza 5:283–287

Nelson RR (1979) Some thoughts on coevolution of plant pathogenic fungi and their hosts. In: Nickel BB (ed) Host–parasite interfaces. Academic, New York, pp 17–25

O'Brien MJ, Gomola CE, Horton TR (2011) The effect of forest soil and community composition on ectomycorrhizal colonization and seedling growth. Plant Soil 341:321–331

Oliveira VL, Schmidt VDB, Bellei MM (1997) Patterns of arbuscular- and ecto-mycorrhizal colonization of *Eucalyptus dunnii* in southern Brazil. Ann For Sci 54:473–481

Omon RM (1996) The effect of some mycorrhizal fungi and media on the growth of *Shorea leprosula* Miq. cuttings. Bull Penelitian Hutan 603:27–36

Orlovich DA, Cairney JWG (2004) Ectomycorrhizal fungi in New Zealand: current perspectives and future directions. N Z J Bot 42:721–738

Ortega U, Dunabeitia M, Menendez S, González-Murua C, Majada J (2004) Effectiveness of mycorrhizal inoculation in the nursery on growth and water relations of Pinus radiata in different water regimes. Tree Physiol 24:65–73

Osonubi O, Mulongoy K, Awotoye OO, Atayese MO, Okali DUU (1991) Effects of ectomycorrhizal and vesicular–arbuscular mycorrhizal fungi on drought tolerance of four leguminous woody seedlings. Plant Soil 136:131–143

Parladé J, Pera J, Alvarez IF (1997) La mycorhization contrôlée du douglas dans le nord de l'espagne: premiers rèsultats en plantation. In: Le Tacon F (ed) Champignons et mycorhizes en foret. revue forestière française, Numéro spécial 1997, pp 163–173

Paulitz TC, Linderman RG (1991) Lack of antagonism between the biocontrol agent *Gliocladium virens* and vesicular-arbuscular mycorrhizal fungi. New Phytol 117:303–308

Phillips RP, Fahey TJ (2006) Tree species and mycorrhizal associations influence the magnitude of rhizosphere effects. Ecology 87:1302–1313

Quoreshi AM, Roy S, Greer CW, Beaudin J, McCurdy D, Khasa DP (2007) Inoculation of green alder (*Alnus crispa*) with Frankia-ectomycorrhizal fungal inoculant under commercial nursery production conditions. Native Plants J 8:271–281

Ramos AC, Façanha AR, Feijó JA (2008) Proton (H⁺) flux signature for the presymbiotic development of the arbuscular mycorrhizal fungi. New Phytol 178:177–188

Rangarajan M, Narayanan R, Kandasamy D, Oblisami G (1990) Studies on the growth of certain ectomycorrhizal fungi in culture media and in the host under axenic conditions. In: Jalali BL, Chand H (ed) Trends in mycorrhizal research. Proceedings of the national conference on mycorrhiza, Hisar, India, pp 126–127

Rao CS, Sharma GD, Shukla AK (1996) Ectomycorrhizal efficiency of various mycobionts with *Pinus kesiya* seedlings in forest and degraded soils. Proc Indian Natl Sci Acad B Biol Sci 62:427–434

Read DJ, Boyd R (1986) Water relations of mycorrhizal fungi and their host plats. In: Ayres PG, Boddy L (eds) Water, fungi and plants. Cambridge University Press, Cambridge, pp 287–303

Reddell P, Milnes AR (1992) Mycorrhizas and other specialized nutrient-acquisition strategies: Their occurrence in woodland plants from Kakadu and their role in rehabilitation of waste rock dumps at a local uranium mine. Aust J Bot 40:223–242

Reddell P, Spain AV, Milnes AR, Hopkins M, Hignett CT, Joyce S, Playfair LA (1993) Indicators of ecosystem recovery in rehabilitated areas of the open strip bauxite mine, Gove, NT. Final report for Nabalco Pty Ltd, CSIRO Minesite Rehabilitation research Program, Adelaide

Reddy MS, Satyanarayana T (2006) Interactions between ectomycorrhizal fungi and rhizospheric microbes. In: Mukerji KG, Manoharachary JS (eds) Microbial activity in the rhizosphere, vol 7, Soil biology. Springer, Berlin, pp 245–263

Richardson DM, Rejmánek M (2004) Invasive conifers: a global survey and predictive framework. Divers Distrib 10:321–331

Rincón A, Alvarez IF, Pera J (2001) Inoculation of containerized *Pinus pinea* L. seedlings with seven ectomycorrhizal fungi. Mycorrhiza 11:265–271

Saner P, Philipson C, Ong RC, Majalap N, Egli S, Hector A (2011) Positive effects of ectomycorrhizal colonization on growth of seedlings of a tropical tree across a range of forest floor light conditions. Plant Soil 338:411–421

Santoso E (1991) Effect of mycorrhizal fungi on nutrient uptake of five dipterocarp seedlings. Bull Penelitian Hutan 532:11–18

Saravesi K, Markkola A, Rautio P, Tuomi J (2011) Simulated mammal browsing and host gender effects on dual mycorrhizal *Salix repens*. Botany 89(1):35–42

Savoie JM, Largeteau ML (2011) Production of edible mushrooms in forests: trends in development of a mycosilviculture. Appl Microbiol Biotechnol 89:971–979

Seva JP, Vilagrosa A, Valdecantos A, Cortina J, Vallejo VR, Bellot J (1996) Mycorrhization and application of urban compost for the improvement of survival and growth of *Quercus ilex* subsp. *ballota* seedlings in an arid zone. Cahiers Options Mediterraneennes 20:105–121

Smith S, Read D (2008) Mycorrhizal symbiosis, 3rd edn. Elsevier, Amsterdam

Smith FA, Smith SE (2011) What is the significance of the arbuscular mycorrhizal colonisation of many economically important crop plants? Plant Soil 348:63–79

Song YY, Cao M, Xie LJ, Liang XT, Zeng RS, Su YJ, Huang JH, Wang RL, Luo SM (2011) Induction of DIMBOA accumulation and systemic defense responses as a mechanism of enhanced resistance of mycorrhizal corn (*Zea mays* L.) to sheath blight. Mycorrhiza 21:721–731

Stack RW, Sinclair WA (1975) Protection of Douglas-fir seedlings against *Fusarium* root rot by a mycorrhizal fungus in the absence of mycorrhiza formation. Phytopathology 65:468–472

Summer ME, Rengasamy P, Naidu R (1998) Sodic soils: a reappraisal. In: Summer ME, Naidu R (eds) Sodic soils. Oxford University Press, New York, pp 3–17

Supriyanto (1999) The effectiveness of some ectomycorrhizal fungi in alginate beads in promoting the growth of several dipterocarp seedlings. Biotropia 12:59–77

Talbot NJ, Ebbole DJ, Hamer JE (1993) Identification and characterization of MPG1, a gene involved in pathogenicity from the rice blast fungus *Magnaporthe grisea*. Plant Cell 5:1575–1590

Tang M, Chen H (1995) The effect of vesicular–arbuscular mycorrhizas on the resistance of poplar to a canker fungus (*Dothiorella gregaria*). In: Brundrett M, Dell B, Malajczuk N, Gong M (eds) Mycorrhizal research for forestry in Asia. ACIAR proceedings no. 62. ACIAR, Canberra, pp 34–40

Theodorou C, Bowen GD (1987) Germination of basidiospores of mycorrhizal fungi in the rhizosphere of *Pinus radiata* D. Don. New Phytol 106:217–223

Tibbett M, Ryan M, Barker S, Chen YL, Denton M, Edmonds-Tibbett T, Walker C (2008) The diversity of arbuscular mycorrhizas of selected Australian Fabaceae. Plant Biosyst 142:420–427

Torres P, Honrubia M (1994) Basidiospore viability in stored slurries. Mycol Res 98:527–530

Trappe JM (1977) Selection of fungi for ectomycorrhizal inoculation in nurseries. Annu Rev Phytopathol 15:203–222

Viljoen A, Wingfield MJ, Crous PW (1992) Fungal pathogens in *Pinus* and *Eucalyptus* seedling nurseries in South Africa: a review. S Afr Forest J 161(1):45–51

Walbert K, Ramsfield TD, Ridgway HJ, Jones EE (2010) Ectomycorrhizal species associated with *Pinus radiata* in New Zealand including novel association determined by molecular analysis. Mycorrhiza 20:209–215

Warcup JH (1980) Ectomycorrhizal associations of Australian indigenous plants. New Phytol 85:531–535

Ward SC (2000) Soil development on rehabilitated bauxite mines in southwest Australia. Aust J Soil Res 38:453–464

Waschke A, Sieh D, Tamasloukht M, Fischer K, Mann P, Franken P (2006) Identification of heavy metal-induced genes encoding glutathione S-transferases in the arbuscular mycorrhizal fungus *Glomus intraradices*. Mycorrhiza 17:1–10

West P (2006) Growing plantation forests. Springer, Berlin

Whipps JM (2001) Microbial interactions and biocontrol in the rhizosphere. J Exp Bot 52:487–511

Wu FY, Bi YL, Wong MH (2009) Dual inoculation with an arbuscular mycorrhizal fungus and Rhizobium to facilitate the growth of alfalfa on coal mine substrates. J Plant Nutr 32:755–771

Zaefarian F, Rezvani M, Ardakani MR, Rejali F, Miransari M (2013) Impact of mycorrhizae formation on the phosphorus and heavy-metal uptake of Alfalfa. Commun Soil Sci Plant Anal 44(8):1340–1352

Zhang MQ (1995) Use of vesicular–arbuscular mycorrhizal fungi to promote tree growth in China. In: Brundrett M, Dell B, Malajczuk N, Gong M (eds) Mycorrhizal research for forestry in Asia. ACIAR proceedings no. 62. ACIAR, Canberra, pp 110–113

Zhao Z (1995) Research on mycorrhizal associations of poplar. In: Brundrett MC, Dell B, Malajczuk N, Gong M (eds) Mycorrhizas for plantation forestry in Asia. ACIAR proceedings no. 62. ACIAR, Canberra, pp 62–66

Zhou D, Hyde KD (2001) Host-specificity, host-exclusivity, and host-recurrence in saprobic fungi. Mycol Res 105:1449–1457

Chapter 22
Use of Arbuscular Mycorrhizal Fungi for Reforestation of Degraded Tropical Forests

Keitaro Tawaraya and Maman Turjaman

22.1 Introduction

Tropical forests are important for their diverse bioresources as well as the significance of the carbon pool. Tropical forests are disappearing at the rate of 13.5 million hectares (ha) each year, largely due to logging, burning and clearing for agricultural land, and shifting cultivation (Kobayashi 2004). Timber harvesting has resulted in the transformation of more than five million ha of tropical forest annually into over-logged, poorly managed, and degraded forests. Degraded tropical forests require wide-scale rehabilitation and it is not easy to rehabilitate degraded tropical forests because a major obstacle in the rehabilitation of tropical forests is slow tree growth and high mortality of seedlings in the nursery. It is also necessary to understand the physical, chemical, and biological factors of forest soils, in order to remediate degraded tropical forests. Among these properties, biological properties are least well known. Arbuscular mycorrhizal (AM) fungi affect the maintenance of vegetation in various ecosystems and may play an important role in tropical forests. Most tropical tree species form arbuscular mycorrhizas.

The diversity of AM fungi and the breadth of their associations with plant species in natural environments are crucial to understanding the ecological role of AM fungi in plant coexistence. AM fungal community structures differ significantly between host species and have been reported to increase the growth and survival rate of some tropical tree seedlings (Wubet et al. 2009). Phosphorus (P) limits the productivity of trees in many forests and plantations especially in

K. Tawaraya (✉)
Faculty of Agriculture, Yamagata University, Tsuruoka 997-8555, Japan
e-mail: tawaraya@tds1.tr.yamagata-u.ac.jp

M. Turjaman
Forest Microbiology Laboratory, Forest and Nature Conservation Research and Development Centre (FNCRDC), Ministry of Forestry, Bogor 16610, Indonesia

© Springer-Verlag Berlin Heidelberg 2014
Z.M. Solaiman et al. (eds.), *Mycorrhizal Fungi: Use in Sustainable Agriculture and Land Restoration*, Soil Biology 41, DOI 10.1007/978-3-662-45370-4_22

highly weathered, acidic, or calcareous profiles in the world. Most trees form mycorrhizal associations which are prevalent in the organic and mineral soil horizons. Mycorrhizal tree roots have a greater capacity to take up phosphate (Pi) from the soil solution than non-mycorrhizal roots (Plassard and Dell 2010). Rehabilitation of degraded tropical forests following inoculation of AM fungi has potential to restore important ecosystem functions. The purpose of this chapter is to review the effect of inoculation of AM fungi on growth of native tree species from tropical forests.

22.2   Degraded Tropical Forest and Reforestation

The total world's forests cover nearly 3.9 billion ha or nearly 30 % of the world's land area (FAO 2001; Fenning and Gershenzon 2002). The number of tropical forests has been declining owing to illegal logging, fire, conversion into agricultural lands, rubber tree and palm oil plantation, and use of the forest plantation estate as pulp trees. Degraded forests are considered to be low-value resources because they are characterized by the vegetation such as ferns, sedges, and scrub. However, it is not easy to rehabilitate this ecosystem in a short term, because it is necessary to select and produce high-quality tree seedling species that have high survival rates during the rehabilitation process.

Tropical forests contribute considerably in sustaining global biodiversity (Laurence 1999). They are homes to indigenous people, pharmacopeias of natural products, and providers of vital ecosystem services, such as flood amelioration and soil conservation. At regional and global scales, tropical forests also have a major influence on climate and carbon storage. Tropical forestlands have been disappearing at the rate of 13.5 million ha each year. Furthermore, timber harvesting has resulted in the transformation of more than five million ha of tropical forest annually into logged-over, poorly managed, and degraded forests.

One of the most serious world problems affecting tropical rain forest is desertification. This is a complex and dynamic process which is claiming several 100 million ha annually. Tropical forests are particularly affected, resulting in a rapid reduction in area. Human activities can cause or accelerate desertification and the loss of most plant species as well as their associated symbioses. The reduction and degradation caused by anthropological activities affect not only the sustainable production of timber but also the global environment. Accurate scientific information will enable managers to devise silvicultural systems to enhance soil properties and forest resources important for sustainable production and for minimizing deleterious impacts of harvesting and short-rotation plantation. Degraded tropical forested lands require wide-scale rehabilitation and it is necessary to improve the biological diversity of tropical forestlands and to enhance the commercial value of timber.

The rapid production of forest planting stock seedlings of high quality in nurseries is important for replenishing degraded tropical forestlands. Moreover, many soils of tropical forests are nutrient poor (Hattenschwiler et al. 2011). Soil nutrient availability is one of the limiting factors for the early growth of transplanted seedlings in degraded tropical forestlands. Degraded tropical forest-lands are recognized as low-value forest resources without successful natural regeneration that are dominated by grasslands including fern, sedges, or scrub. Nowadays, reforestation programs have to prepare millions of seedling stocks annually. The use of vigorous seedlings in reforestation programs is important. However, seedling stocks of tropical forest species are usually weak, often N and P deficient, and have high mortality rates after transplanting in the field. Phosphorus was the most limiting nutrient for plant growth of four woody legume species (Moreira et al. 2010). Ultimately, rehabilitation can increase the area of forest as well as contribute to conservation of the remaining primary forests and environ-mental quality.

22.3   Ecology of Arbuscular Mycorrhizal Fungi in Tropical Forests

Tropical rain forest soils often have high P adsorption because of their strong affinity to P to form iron and aluminum oxides and hydroxides, whereas in neutral and alkaline soils, P is adsorbed on the surface of Ca and Mg carbonates (Holford 1997; Whitmore 1989). Soil P concentration of tropical soil is very low (Table 22.1). In most experiment with tropical rain forest plant species, the influ-ence of AM fungi on P nutrition has been evaluated by measuring the growth response of inoculated and non-inoculated plants cultivated in soils with controlled levels of P (Janos 1980). Moyersoen et al. (1998) reported that AM colonization of the tropical tree *Oubanguia alata* (Scytopetalaceae) was positively correlated with increased P uptake despite low P availability in Korup National Park rain forest, Cameroon.

Early studies focused primarily on mycorrhizas of the temperate forests, but attention turned toward mycorrhizas of the tropical rain forests (Torti et al. 1997). In contrast to the temperate zone, where mycorrhizal associations of trees tend to be formed by ectomycorrhizal fungi, the majority of tropical tree species surveyed thus far are formed by AM fungi (Janos 1980). Notable exceptions of tropical trees forming ectomycorrhizas occur in the families Myrtaceae, Caesalpiniaceae, Euphorbiaceae, Fagaceae, and Dipterocarpaceae (Munyanziza et al. 1997). The highest number of species and spores of AM fungi was observed during the dry season, with a marked decrease during the rainy season in a tropical rain forest in Veracruz, Mexico (Guadarrama and Alvarez-Sanchez 1999). Moyersoen et al. (2001) reported that AM colonization was about 40 % in tree species in heath forests and mixed Dipterocarpaceae forest in Brunei. Tawaraya et al. (2003) showed that 17 of 22 tree species in a tropical peat swamp forest in Kalimantan, Indonesia, had mycorrhizas formed by AM fungi. Of the 142 species of trees and

Table 22.1 Arbuscular mycorrhizal colonization, mycorrhizal dependency (MD) of different tree species grown in tropical forests, and soil phosphorus concentration

Family	Species	Growth period (d)	Fungal species	Colonization (%)	MD (%)	Soil P (mg P/kg)	References
Anacardiaceae	*Lithraea molleoides*	90	*Glomus etunicatum*	27	97	1 (Mehlich I)	Siqueira and Saggin-Junior (2001)
Anacardiaceae	*Anacardium occidentale*	90	*Glomus aggregatum*	71	23	6.6 (Bray-1)	Bá et al. (2000)
Anacardiaceae	*Schinus terebinthifolius*	120	Gigaspora margarita, Glomus etunicatum	20	92	2 (Olsen)	Siqueira et al. (1998)
Anacardiaceae	*Sclerocarya birrea*	90	*Glomus aggregatum*	75	17	6.6 (Bray-1)	Bá et al. (2000)
					57		
Apocynaceae	*Aspidosperma parvifolium*	180	*Glomus etunicatum*	–	68	1 (Mehlich I)	Siqueira and Saggin-Junior (2001)
Apocynaceae	*Dyera polyphylla*	202	*Glomus clarum*	39	61	4.8	Turjaman et al. (2006)
Apocynaceae	*Dyera polyphylla*	202	*Glomus decipiens*	22	62	4.8	Turjaman et al. (2006)
Apocynaceae	*Stemmadenia donnell-smithii*	180	Mixture* 1	10	55	N.D.	Guadarrama et al. (2004)
					62		
Araucariaceae	*Araucaria angustifolia*	686	*Glomus clarum*	81	71	39	Zandavalli et al. (2004)
Bignoniaceae	*Jacaranda mimosaefolia*	120	Gigaspora margarita, Glomus etunicatum	77	95	2 (Olsen)	Siqueira et al. (1998)
Bignoniaceae	*Stenolobium stans*	120	Gigaspora margarita, Glomus etunicatum	72	85	2 (Olsen)	Siqueira et al. (1998)
Bignoniaceae	*Tabebuia serratifolia*	129	*Glomus etunicatum*	72	89	1 (Mehlich I)	Siqueira and Saggin-Junior (2001)

Family	Species		AMF				Reference
Bignoniaceae	*Tabebuia impetiginosa*	84	*Glomus etunicatum*	41	58 / 82	1 (Mehlich I)	Siqueira and Saggin-Junior (2001)
Boraginaceae	*Cordia trichotoma*	90	*Glomus etunicatum*	–	98	1 (Mehlich I)	Siqueira and Saggin-Junior (2001)
Caesalpiniaceae	*Caesalpinia ferrea*	97	*Glomus etunicatum*	30	76	1 (Mehlich I)	Siqueira and Saggin-Junior (2001)
Caesalpiniaceae	*Copaifera langsdorffii*	262	*Glomus etunicatum*	–	50	1 (Mehlich I)	Siqueira and Saggin-Junior (2001)
Caesalpiniaceae	*Senna macranthera*	120	*Glomus etunicatum*	20	87	1 (Mehlich I)	Siqueira and Saggin-Junior (2001)
Caesalpiniaceae	*Senna spectabilis*	90	*Glomus etunicatum*	63	95	1 (Mehlich I)	Siqueira and Saggin-Junior (2001)
Caesalpiniaceae	*Dicorynia guianensis*	350	Indigenous	62	52	N.D.	Bereau et al. (2000)
Caesalpiniaceae	*Dicorynia guianensis*	281	Indigenous	60–95	71	21	de Grandcourt et al. (2004)
Caesalpiniaceae	*Eperua falcata*	281	Indigenous	45–75	22	21	de Grandcourt et al. (2004)
Caesalpinoideae	*Bauhinia sp.*	120	*Gigaspora margarita, Glomus etunicatum*	9	6	2 (Olsen)	Siqueira et al. (1998)
Caesalpinoideae	*Caesalpinia ferrea*	120	*Gigaspora margarita, Glomus etunicatum*	30	75	2 (Olsen)	Siqueira et al. (1998)
Caesalpinoideae	*Caesalpinia peltophoroides*	120	*Gigaspora margarita, Glomus etunicatum*	33	48	2 (Olsen)	Siqueira et al. (1998)
Caesalpinoideae	*Cassia grandis*	120	*Gigaspora margarita, Glomus etunicatum*	63	71	2 (Olsen)	Siqueira et al. (1998)
Caesalpinoideae	*Copaifera langsdorffii*	120	*Gigaspora margarita, Glomus etunicatum*	–	–4	2 (Olsen)	Siqueira et al. (1998)

(continued)

Table 22.1 (continued)

Family	Species	Growth period (d)	Fungal species	Colonization (%)	MD (%)	Soil P (mg P/kg)	References
Caesalpinoideae	Peltophorum dubium	120	Gigaspora margarita, Glomus etunicatum	18	61	2 (Olsen)	Siqueira et al. (1998)
Caesalpinoideae	Schizolobium parahyba	120	Gigaspora margarita, Glomus etunicatum	8	−25	2 (Olsen)	Siqueira et al. (1998)
Caesalpinoideae	Senna macranthera	120	Gigaspora margarita, Glomus etunicatum	81	84	2 (Olsen)	Siqueira et al. (1998)
Caesalpinoideae	Senna multijuga	120	Gigaspora margarita, Glomus etunicatum	–	38	2 (Olsen)	Siqueira et al. (1998)
Caesalpinoideae	Senna spectabilis	120	Gigaspora margarita, Glomus etunicatum	17	85	2 (Olsen)	Siqueira et al. (1998)
					52		
Casuarinaceae	Casuarina equisetifolia	144	Glomus geosporum	45	55	0.34 (Olsen)	Muthukumar and Udaiyan (2010)
Cecropiaceae	Cecropia pachystachya	98	Glomus etunicatum	62	100	1 (Mehlich I)	Siqueira and Saggin-Junior (2001)
Clusiaceae	Clusia minor	420	Scutellospora fulgida	100	99	3.05	Cáceres and Cuenca (2006)
Clusiaceae	Clusia minor	420	Scutellospora fulgida	75	−42	39	Cáceres and Cuenca (2006)
Clusiaceae	Clusia multiflora	180	Scutellospora fulgida	98	71	3.05	Cáceres and Cuenca (2006)
Clusiaceae	Clusia multiflora	180	Scutellospora fulgida	92	−21	39	Cáceres and Cuenca (2006)
					27		
Euphorbiaceae	Croton floribundus	101	Glomus etunicatum	48	92	1 (Mehlich I)	Siqueira and Saggin-Junior (2001)

Family	Species		Inoculum					Reference
Euphorbiaceae	Macaranga denticulata	120	Indigenous	70		42	4.10 (Bray II)	Youpensuk et al. (2004)
Fabaceae	Dalbergia sissoo	111	G. albida, G. intraradices, A. scrobiculata	70	67	61	14.9 (Olsen)	Bisht et al. (2009)
Fabaceae	Dalbergia sissoo	111	G. albida, G. intraradices, A. scrobiculata	30		59	6.3 (Olsen)	Bisht et al. (2009)
Fabaceae	Dialium guineensis	90	Glomus aggregatum	50		45	6.6 (Bray-1)	Bá et al. (2000)
Fabaceae	Leucaena diversifolia	45	G. aggregatum	73		73	0.02 mg/L	Manjunath and Habte (1991)
Fabaceae	Leucaena leucocephala	45	G. aggregatum	76		79	0.02 mg/L	Manjunath and Habte (1991)
Fabaceae	Leucaena leucocephala	56	Mixture	98		77	4.22 (Bray II)	Saif (1987)
Fabaceae	Leucaena retusa	45	G. aggregatum	56		35	0.02 mg/L	Manjunath and Habte (1991)
Fabaceae	Leucaena trichodes	45	G. aggregatum	58		70	0.02 mg/L	Manjunath and Habte (1991)
Fabaceae	Sesbania pubescens	120	G. aggregatum	83		−69	4.8 (Olsen)	Duponnois et al. (2001)
Fabaceae	Sesbania formosa	45	G. aggregatum	75		24	0.02 mg/L	Manjunath and Habte (1991)
Fabaceae	Sesbania grandiflora	120	G. aggregatum	90		−4	4.8 (Olsen)	Duponnois et al. (2001)
Fabaceae	Sesbania grandiflora	45	G. aggregatum	80		35	0.02 mg/L	Manjunath and Habte (1991)
Fabaceae	Sesbanica nubica	120	G. aggregatum	97		−32	4.8 (Olsen)	Duponnois et al. (2001)
Fabaceae	Sesbania pachycarpa	45	G. aggregatum	54		17	0.02 mg/L	Manjunath and Habte (1991)
Fabaceae	Sesbania paludosa	120	G. aggregatum	79		59	4.8 (Olsen)	Duponnois et al. (2001)

(continued)

Table 22.1 (continued)

Family	Species	Growth period (d)	Fungal species	Colonization (%)	MD (%)	Soil P (mg P/kg)	References
Fabaceae	Sesbania sesban	45	G. aggregatum	70	14	0.02 mg/L	Manjunath and Habte (1991)
Fabaceae	Tamarindus indica	90	Glomus aggregatum	88	53	6.6 (Bray-1)	Bá et al. (2000)
					35		
Faboideae	Platycyamus regnellii	120	Gigaspora margarita, Glomus etunicatum	–	20	2 (Olsen)	Siqueira et al. (1998)
Faboideae	Tipuana tipu	120	Gigaspora margarita, Glomus etunicatum	75	74	2 (Olsen)	Siqueira et al. (1998)
					47		
Guttiferae	Calophyllum hosei	270	Glomus aggregatum	18	60	0.17	Turjaman et al. (2008)
Guttiferae	Calophyllum hosei	270	Glomus clarum	19	57	0.17	Turjaman et al. (2008)
Guttiferae	Ploiarium alternifolium	270	Glomus aggregatum	32	51	0.17	Turjaman et al. (2008)
Guttiferae	Ploiarium alternifolium	270	Glomus clarum	27	56	0.17	Turjaman et al. (2008)
					56		
Malvaceae	Adansonia digitata	90	Glomus aggregatum	63	8	6.6 (Bray-1)	Bá et al. (2000)
Melastomaceae	Tibouchina granulosa	144	Glomus etunicatum	20	100	1 (Mehlich I)	Siqueira and Saggin-Junior (2001)
Meliaceae	Azadirachta indica	120	Glomus geosporum	25	8	N.D.	Muthukumar et al. (2001)
Meliaceae	Azadirachta indica	120	Glomus intraradices	28	20	N.D.	Muthukumar et al. (2001)
Meliaceae	Azadirachta indica	120	G. geosporum, G. intraradices	44	26	N.D.	Muthukumar et al. (2001)

Family	Species		AMF			P	Reference
Meliaceae	Cedrela fissilis	96	Glomus etunicatum	70	95	1 (Mehlich I)	Siqueira and Saggin-Junior (2001)
Meliaceae	Cedrela fissilis	120	Gigaspora margarita, Glomus etunicatum	34	41	2 (Olsen)	Siqueira et al. (1998)
					38		
Mimosoideae	Albizia lebbeck	120	Gigaspora margarita, Glomus etunicatum	55	62	2 (Olsen)	Siqueira et al. (1998)
Mimosoideae	Anadenanthera falcata	120	Gigaspora margarita, Glomus etunicatum	11	81	2 (Olsen)	Siqueira et al. (1998)
Mimosoideae	Leucaena leucocephala	136	Glomus etunicatum	17	92	1 (Mehlich I)	Siqueira and Saggin-Junior (2001)
Mimosoideae	Parkia biglobosa	90	Glomus aggregatum	68	32	6.6 (Bray-1)	Bá et al. (2000)
					67		
Myrsinaceae	Myrsine umbellata	111	Glomus etunicatum	52	98	1 (Mehlich I)	Siqueira and Saggin-Junior (2001)
Myrtaceae	Syzygium jambolanum	120	Gigaspora margarita, Glomus etunicatum	50	91	2 (Olsen)	Siqueira et al. (1998)
Rhamnaceae	Hovenia dulcis	120	Gigaspora margarita, Glomus etunicatum	47	63	2 (Olsen)	Siqueira et al. (1998)
Rhamnaceae	Zizyphus mauritiana	90	Glomus aggregatum	98	78	6.6 (Bray-1)	Bá et al. (2000)
					71		
Rubiaceae	Coffea arabica	150	Glomus clarum	12	−15	3 (Olsen)	Vaast et al. (1996)
Rubiaceae	Coffea arabica	150	Glomus clarum	20	61	13	Vaast et al. (1996)
Rubiaceae	Coffea arabica	150	Glomus clarum	22	46	27	Vaast et al. (1996)
Rubiaceae	Coffea arabica	150	Glomus clarum	26	19	42	Vaast et al. (1996)
Rubiaceae	Coffea arabica	150	Acaulospora mellea	26	−31	3	Vaast et al. (1996)
Rubiaceae	Coffea arabica	150	Acaulospora mellea	50	55	13	Vaast et al. (1996)

(continued)

Table 22.1 (continued)

Family	Species	Growth period (d)	Fungal species	Colonization (%)	MD (%)	Soil P (mg P/kg)	References
Rubiaceae	Coffea arabica	150	Acaulospora mellea	32	16	27	Vaast et al. (1996)
Rubiaceae	Coffea arabica	150	Acaulospora mellea	22	−18	42	Vaast et al. (1996)
					17		
Sapindaceae	Sapindus saponaria	116	Glomus etunicatum	24	36	1 (Mehlich I)	Siqueira and Saggin-Junior (2001)
Sapindaceae	Aphania senegalensis	90	Glomus aggregatum	40	21	6.6 (Bray-1)	Bá et al. (2000)
Sapindaceae	Sapindus saponaria	120	Gigaspora margarita, Glomus etunicatum	41	66	2 (Olsen)	Siqueira et al. (1998)
					41		
Thymelaeaceae	Aquilaria filaria	202	Glomus clarum	93	53	4.8	Turjaman et al. (2006)
Thymelaeaceae	Aquilaria filaria	202	Glomus decipiens	87	48	4.8	Turjaman et al. (2006)
					51		
Tiliaceae	Heliocarpus appendiculatus	180	Mixture*1	83	−47	N.D.	Guadarrama et al. (2004)
Tiliaceae	Luehea grandiflora	120	Gigaspora margarita, Glomus etunicatum	60	93	2 (Olsen)	Siqueira et al. (1998)
					23		
Ulmaceae	Luehea grandiflora	93	Glomus etunicatum	21	98	1 (Mehlich I)	Siqueira and Saggin-Junior (2001)
Ulmaceae	Trema micrantha	70	Glomus etunicatum	32	98	1 (Mehlich I)	Siqueira and Saggin-Junior (2001)
Ulmaceae	Trema micrantha	120	Gigaspora margarita, Glomus etunicatum	64	92	2 (Olsen)	Siqueira et al. (1998)

Zygophyllaceae	Balanites aegyptiaca	90	Glomus aggregatum	52	96	0	6.6 (Bray-1)	Bá et al. (2000)
Average (all species)						50		

Average of MD of each family was also shown

Mixture*1: Sclerocystis, Acaulospora and Gigaspora

liana surveyed in Guyana, 137 were exclusively formed by AM fungi (McGuire et al. 2008). A light microscopy investigation showed arbuscular mycorrhizas in 112 tree species from 53 families on mineral as well as organic soils in Ecuador (Kottke et al. 2004). In a related study, a segment of fungal 18S rDNA was sequenced from the mycorrhizas of *Cedrela montana*, *Heliocarpus americanus*, *Juglans neotropica*, and *Tabebuia chrysantha* in reforestation plots from degraded pastures in Ecuador and observed distinct species-rich AM communities (Haug et al. 2010). Dual ectomycorrhizal and AM colonization was observed in 4 of 14 ectomycorrhizal tree species belonging to Caesalpiniaceae and Uapacaceae from rain forest in Cameroon (Moyersoen and Fitter 1999). In total, 193 glomeromycotan sequences were analyzed, 130 of them previously unpublished.

Spores of AM fungi have been isolated from soils of tropical forests and their population and richness were affected by environmental conditions. Spore density and richness based on soil cores were higher in the dry season than in the rainy season in a tropical sclerophyllous shrubland in the Venezuelan Guayana (Cuenca and Lovera 2010). Spore numbers of AM fungi were higher in young secondary forest and pastures and lower in pristine forest in the Amazon region (Sturmer et al. 2009), and AM fungal diversity was high in dry tropical Afromontane forests of Ethiopia (Wubet et al. 2009). AM fungal spores in soil decreased from an early plant succession to mature tropical forest in a Brazilian study (Zangaro et al. 2008). AM fungal types that were dominant in the newly germinated seedlings were almost entirely replaced by previously rare types in the surviving seedlings the following years (Husband et al. 2002a). As the seedlings matured in a tropical forest in the Republic of Panama, the fungal diversity decreased and there was a significant shift (Husband et al. 2002b). Based on spore morphology, 29 species of AM fungi were found in the rhizosphere of *Macaranga denticulata* (Youpensuk et al. 2004).

22.4 Inoculation of Tropical Tree Species with AM Fungi

AM fungi have been reported to increase growth of some tropical trees (Table 22.1). AM fungi increased seedling growth of 23 of 28 species from a lowland tropical rain forest in Costa Rica under nursery conditions (Janos 1980). AM colonization of the tropical tree *Oubanguia alata* (Scytopetalaceae) was positively correlated with increased P uptake despite low P availability in a study in Cameroon (Moyersoen et al. 1998). AM fungi improved growth of the Brazilian pine *Araucaria angustifolia* (Araucariaceae) (Zandavalli et al. 2004). There are also reports of improved growth of non-timber forest product tree species following AM fungal inoculation in tropical forests. For example, Muthukumar et al. (2001) reported that inoculation of *Azadirachta indica* (Meliaceae) with AM fungi improved seedling growth. Furthermore, the inoculation of AM fungi with phosphate-solubilizing and nitrogen-fixing bacteria increased the growth of *A. indica*. Conversely, *A. excelsa* inoculated with AM fungi (without fertilizer) grew more slowly than did the uninoculated plants (Huat et al. 2002). Kashyap et al. (2004) showed that inoculation of *Morus alba* (Moraceae) with both AM fungi and *Azotobacter* increased the survival percentage of saplings.

Clusia minor and *Clusia multiflora* inoculated with *Scutellospora fulgida* in acidic soil had greater shoot and root biomass, leaf area, and height in comparison to the biomass of P-fertilized plants and non-mycorrhizal plants (Cáceres and Cuenca 2006). Inoculation with the AM fungus *Glomus geosporum* improved the growth, nutrient acquisition, and seedling quality of *Casuarina equisetifolia* seedlings under nursery conditions (Muthukumar and Udaiyan 2010). Seedlings of *Araucaria angustifolia* inoculated with *Glomus clarum* had higher shoot biomass; leaf concentrations of P, K, Na, and Cu; and lower concentrations of Ca, Mg, Fe, Mn, and B than controls (Zandavalli et al. 2004). Inoculation with soil-containing AM fungi increased shoot growth nutrient contents when P was limiting but N was applied (Youpensuk et al. 2004). Inoculation with AM fungi *Glomus clarum* and *Gigaspora decipiens* increased shoot N and P uptake of non-timber forest product species *Dyera polyphylla* and *Aquilaria filaria* under greenhouse conditions, indicating that AM fungi can reduce the application of chemical fertilizer (Turjaman et al. 2006). Other studies have used mycorrhizal roots from individual tree species or from a mixture of the four trap species with resulting improvement in growth of 6-month-old *Cedrela montana* and *Heliocarpus americanus* (Urgiles et al. 2009). This latter technique is much easier to handle and has lower costs than spore production for tropical countries with limited facilities for storage of inoculum.

AM fungi increased the growth of *Acacia nilotica* and *Leucaena leucocephala* (Leguminosae) 12 weeks after transplantation under greenhouse conditions (Michelsen and Rosendahl 1990), and similar observations were made for three multipurpose fruit-tree species: *Parkia biglobosa*, *Tamarindus indica*, and *Ziziphus mauritiana* 2 months after inoculation (Guissou et al. 1998). The AM fungus *Glomus aggregatum* stimulated plant growth of 17 leguminous plants (Duponnois et al. 2001), and *Glomus macrocarpum* increased the growth of two species: *Sesbania aegyptiaca* and *S. grandiflora* (Giri et al. 2004). Some studies have successfully used mixed inocula of AM fungi including two (Bá et al. 2000), three (Adjoud et al. 1996), and nine species (Rajan et al. 2000).

Mycorrhizal dependency was calculated to compare the degree of plant growth change associated with AM colonization of 76 species, 25 families (Table 22.1). The average mycorrhizal dependency value of all the plants was 50 % (−69 Min. and 100 Max.). Mycorrhizal dependency was also different among families. It was higher in Ulmaceae and Bignoniaceae. Guissou et al. (1998) reported that mycorrhizal dependency of *Parkia biglobosa* and *Tamarindus indica* was similar, reaching no more than 36 %, while *Ziziphus mauritiana* showed higher mycorrhizal dependency values, reaching up to 78 %. A similar effectiveness of AM fungi for different plant species was also reported by Adjoud et al. (1996). Mycorrhizal dependency is frequently related to the morphological properties of the root of different plant species, and also root systems with only a few, short root hairs are indicative of a high mycorrhizal dependency of the plant species concerned (Baylis

1970). Responses of 12 native woody species to the inoculation of AM fungi were related to root morphological plasticity of the plant (Zangaro et al. 2007).

The survival rate of seedling stocks in the field is vital to reforestation. In one study, the survival rates of AM-inoculated cuttings of *Ploiarium alternifolium* and *Calophyllum hosei* were 100 % after 6 months (Turjaman et al. 2008). These values were higher than the survival rates of two tropical tree species from Panama inoculated with AM fungi, which were *Ochroma pyramidale* (97 %) and *Luehea seemannii* (52 %), respectively (Kiers et al. 2000). Inoculation with AM fungi can reduce the cost of seedling production for reforesting vast areas of disturbed tropical forests. Despite extensive studies of inoculation of tree species under controlled conditions, there are few reports about the effect of AM fungal inoculation on growth of tropical tree species under field conditions. Recently, Graham et al. (2013) showed that inoculation of *Glomus clarum* and *Gigaspora decipiens* increased N and P content of *Dyera polyphylla* under tropical peat swamp forest in Central Kalimantan, Indonesia.

22.5 Conclusion

Colonization of roots by AM fungi can improve growth of many tree species that occur in tropical forests. Survival rate of seedlings is a key measure of success in reforestation and afforestation. Survival rates of inoculated seedlings can be higher than those of non-inoculated seedlings. Inoculation with AM fungi at the nursery stage is a useful technique to include in large-scale reforestation programs. However, mycorrhizal dependency differs among plant species and with species of AM fungi. Therefore, selection of appropriate combination of plant species and fungal species is also important for reforestation programs.

References

Adjoud D, Plenchette C, Hallihargas R, Lapeyrie F (1996) Response of 11 eucalyptus species to inoculation with three arbuscular mycorrhizal fungi. Mycorrhiza 6:129–135
Bá AM, Plenchette C, Danthu P, Duponnois R, Guissou T (2000) Functional compatibility of two arbuscular mycorrhizae with thirteen fruit trees in Senegal. Agrofor Syst 50:95–105
Baylis GTS (1970) Root hairs and phycomycetous mycorrhizas in phosphorus-deficient soil. Plant and Soil 33:713–716
Bereau M, Barigah TS, Louisanna E, Garbaye J (2000) Effects of endomycorrhizal development and light regimes on the growth of Dicorynia guianensis Amshoff seedlings. Ann For Sci 57:725–733
Bisht R, Chaturvedi S, Srivastava R, Sharma AK, Johri BN (2009) Effect of arbuscular mycorrhizal fungi, Pseudomonas fluorescens and Rhizobium leguminosarum on the growth and nutrient status of Dalbergia sissoo Roxb. Trop Ecol 50:231–242

Cáceres A, Cuenca G (2006) Contrasting response of seedlings of two tropical species Clusia minor and Clusia multiflora to mycorrhizal inoculation in two soils with different pH. Trees 20:593–600

Cuenca G, Lovera M (2010) Seasonal variation and distribution at different soil depths of arbuscular mycorrhizal fungi spores in a tropical sclerophyllous shrubland. Botany 88:54–64

de Grandcourt A, Epron D, Montpied P, Louisanna E, Bereau M, Garbaye J, Guehl JM (2004) Contrasting responses to mycorrhizal inoculation and phosphorus availability in seedlings of two tropical rainforest tree species. New Phytol 161:865–875

Duponnois R, Plenchette C, Ba AM (2001) Growth stimulation of seventeen fallow leguminous plants inoculated with Glomus aggregatum in Senegal. Eur J Soil Biol 37:181–186

FAO (2001) The global forest resources assessment 2000. http://www.fao.org/docrep/meeting/003/X9835e/X9835e00.htm#P469_24024

Fenning TM, Gershenzon J (2002) Where will the wood come from? Plantation forests and the role of biotechnology. Trends Biotechnol 1–6

Giri B, Kapoor R, Agarwal L, Mukerji KG (2004) Preinoculation with arbuscular mycorrhizae helps Acacia auriculiformis grow in degraded Indian wasteland soil. Commun Soil Sci Plant Anal 35:193–204

Graham LL, Turjaman M, Page SE (2013) Shorea balangeran and Dyera polyphylla (syn. Dyera lowii) as tropical peat swamp forest restoration transplant species: effects of mycorrhizae and level of disturbance. Wetl Ecol Manag 21(5):307–321

Guadarrama P, Alvarez-Sanchez FJ (1999) Abundance of arbuscular mycorrhizal fungi spores in different environments in a tropical rain forest, Veracruz, Mexico. Mycorrhiza 8:267–270

Guadarrama P, Alvarez-Sanchez FJ, Briones O (2004) Seedling growth of two pioneer tree species in competition: the role of arbuscular mycorrhizae. Euphytica 138:113–121

Guissou T, Ba AM, Ouadba JM, Guinko S, Duponnois R (1998) Responses of Parkia biglobosa (Jacq.) Benth, Tamarindus indica L. and Zizyphus mauritiana Lam. to arbuscular mycorrhizal fungi in a phosphorus-deficient sandy soil. Biol Fertil Soils 26:194–198

Hattenschwiler S, Coq S, Barantal S, Handa IT (2011) Leaf traits and decomposition in tropical rainforests: revisiting some commonly held views and towards a new hypothesis. New Phytol 189:950–965

Haug I, Wubet T, Weiss M, Aguirre N, Weber M, Gunter S, Kottke I (2010) Species-rich but distinct arbuscular mycorrhizal communities in reforestation plots on degraded pastures and in neighboring pristine tropical mountain rain forest. Trop Ecol 51:125–148

Holford ICR (1997) Soil phosphorus: its measurement, and its uptake by plants. Aust J Soil Res 35:227–239

Huat OK, Awang K, Hashim A, Majid NM (2002) Effects of fertilizers and vesicular-arbuscular mycorrhizas on the growth and photosynthesis of Azadirachta excelsa (Jack) Jacobs seedlings. For Ecol Manag 158:51–58

Husband R, Herre EA, Turner SL, Gallery R, Young JPW (2002a) Molecular diversity of arbuscular mycorrhizal fungi and patterns of host association over time and space in a tropical forest. Mol Ecol 11:2669–2678

Husband R, Herre EA, Young JPW (2002b) Temporal variation in the arbuscular mycorrhizal communities colonising seedlings in a tropical forest. FEMS Microbiol Ecol 42:131–136

Janos DP (1980) Vesicular-arbuscular mycorrhizae affect lowland tropical rain forest plant growth. Ecology 61:151–162

Kashyap S, Sharma S, Vasudevan P (2004) Role of bioinoculants in development of salt-resistant saplings of Morus alba (var. sujanpuri) in vivo. Sci Hortic 100:291–307

Kiers ET, Lovelock CE, Krueger EL, Herre EA (2000) Differential effects of tropical arbuscular mycorrhizal fungal inocula on root colonization and tree seedling growth: implications for tropical forest diversity. Ecol Lett 3:106–113

Kobayashi S (2004) Landscape rehabilitation of degraded tropical forest ecosystems. Case study of the CIFOR/Japan project in Indonesia and Peru. For Ecol Manage 201:13–22

Kottke I, Beck A, Oberwinkler F, Homeier J, Neill D (2004) Arbuscular endomycorrhizas are dominant in the organic soil of a neotropical montane cloud forest. J Trop Ecol 20:125–129

Laurance WF (1999) Reflection on the tropical deforestation crisis. Biol Conserv 91:109–117

Manjunath A, Habte M (1991) Root morphological characteristics of host species having distinct mycorrhizal dependency. Can J Bot 69:671–676

McGuire KL, Henkel TW, de la Cerda IG, Villa G, Edmund F, Andrew C (2008) Dual mycorrhizal colonization of forest-dominating tropical trees and the mycorrhizal status of non-dominant tree and liana species. Mycorrhiza 18:217–222

Michelsen A, Rosendahl S (1990) The effect of VA mycorrhizal fungi, phosphorus and drought stress on the growth of Acacia nilotica and Leucaena leucocephala seedlings. Plant and Soil 124:7–13

Moreira FMD, de Carvalho TS, Siqueira JO (2010) Effect of fertilizers, lime, and inoculation with rhizobia and mycorrhizal fungi on the growth of four leguminous tree species in a low-fertility soil. Biol Fertil Soils 46:771–779

Moyersoen B, Fitter AH (1999) Presence of arbuscular mycorrhizas in typically ectomycorrhizal host species from Cameroon and New Zealand. Mycorrhiza 8:247–253

Moyersoen B, Alexander IJ, Fitter AH (1998) Phosphorus nutrition of ectomycorrhizal and arbuscular mycorrhizal tree seedlings from a lowland tropical rain forest in Korup National Park, Cameroon. J Trop Ecol 14:47–61

Moyersoen B, Becker P, Alexander IJ (2001) Are ectomycorrhizas more abundant than arbuscular mycorrhizas in tropical heath forests? New Phytol 150:591–599

Munyanziza E, Kehri HK, Bagyaraj DJ (1997) Agricultural intensification, soil biodiversity and agro-ecosystem function in the tropics: the role of mycorrhiza in crops and trees. Appl Soil Ecol 6:77–85

Muthukumar T, Udaiyan K (2010) Growth response and nutrient utilization of Casuarina equisetifolia seedlings inoculated with bioinoculants under tropical nursery conditions. New For 40:101–118

Muthukumar T, Udaiyan K, Rajeshkannan V (2001) Response of neem (Azadirachta indica A. Juss) to indigenous arbuscular mycorrhizal fungi, phosphate-solubilizing and asymbiotic nitrogen-fixing bacteria under tropical nursery conditions. Biol Fertil Soils 34:417–426

Plassard C, Dell B (2010) Phosphorus nutrition of mycorrhizal trees. Tree Physiol 30:1129–1139

Rajan SK, Reddy BJD, Bagyaraj DJ (2000) Screening of arbuscular mycorrhizal fungi for their symbiotic efficiency with Tectana grandis. For Ecol Manage 126:91–95

Saif SR (1987) Growth-responses of tropical forage plant-species to vesicular-arbuscular mycorrhizae. I. Growth, mineral uptake and mycorrhizal dependency. Plant and Soil 97:25–35

Siqueira JO, Saggin-Junior OJ (2001) Dependency on arbuscular mycorrhizal fungi and responsiveness of some Brazilian native woody species. Mycorrhiza 11:245–255

Siqueira JO, Saggin-Junior OJ, Flores-Aylas WW, Guimaraes PTG (1998) Arbuscular mycorrhizal inoculation and superphosphate application influence plant development and yield of coffee in Brazil. Mycorrhiza 7:293–300

Sturmer SL, Leal PL, Siqueira JO (2009) Occurrence and diversity of arbuscular mycorrhizal fungi in trap cultures from soils under different land use systems in the Amazon, Brazil. Braz J Microbiol 40:111–121

Tawaraya K, Takaya Y, Turjaman M, Tuah SJ, Limin SH, Tamai Y, Cha JY, Wagatsuma T, Osaki M (2003) Arbuscular mycorrhizal colonization of tree species grown in peat swamp forests of Central Kalimantan, Indonesia. For Ecol Manage 182:381–386

Torti SD, Coley PD, Janos DP (1997) Vesicular-arbuscular mycorrhizae in two tropical monodominant trees. J Trop Ecol 13:623–629

Turjaman M, Tamai Y, Santoso E, Osaki M, Tawaraya K (2006) Arbuscular mycorrhizal fungi increased early growth of two nontimber forest product species Dyera polyphylla and Aquilaria filaria under greenhouse conditions. Mycorrhiza 16:459–464

Turjaman M, Tamai Y, Sitepu IR, Santoso E, Osaki M, Tawaraya K (2008) Improvement of early growth of two tropical peat-swamp forest tree species Ploiarium alternifolium and

Calophyllum hosei by two arbuscular mycorrhizal fungi under greenhouse conditions. New For 36:1–12

Urgiles N, Lojan P, Aguirre N, Blaschke H, Gunter S, Stimm B, Kottke I (2009) Application of mycorrhizal roots improves growth of tropical tree seedlings in the nursery: a step towards reforestation with native species in the Andes of Ecuador. New For 38:229–239

Vaast P, Zasoski RJ, Bledsoe CS (1996) Effects of vesicular-arbuscular mycorrhizal inoculation at different soil P availabilities on growth and nutrient uptake of in vitro propagated coffee (Coffee arabica L) plant. Mycorrhiza 6:493–497

Whitmore TC (1989) Southeast Asian tropical forests. In: Leith H, Werger MJA (eds) Biogeographical and ecological studies. Ecosystems of the world 14B: tropical rain forest ecosystems. Elsevier, Amsterdam, pp 195–218

Wubet T, Kottke I, Teketay D, Oberwinkler F (2009) Arbuscular mycorrhizal fungal community structures differ between co-occurring tree species of dry Afromontane tropical forest, and their seedlings exhibit potential to trap isolates suited for reforestation. Mycol Prog 8:317–328

Youpensuk S, Lumyong S, Dell B, Rerkasem B (2004) Arbuscular mycorrhizal fungi in the rhizosphere of Macaranga denticulata Muell. Arg., and their effect on the host plant. Agrofor Syst 60:239–246

Zandavalli RB, Dillenburg LR, de Souza PVD (2004) Growth responses of *Araucaria angustifolia* (Araucariaceae) to inoculation with the mycorrhizal fungus *Glomus clarum*. Appl Soil Ecol 25:245–255

Zangaro W, Nishidate FR, Vandresen J, Andrade G, Nogueira MA (2007) Root mycorrhizal colonization and plant responsiveness are related to root plasticity, soil fertility and successional status of native woody species in southern Brazil. J Trop Ecol 23:53–62

Zangaro W, de Assis RL, Rostirola LV, Souza P, Goncalves MC, Andrade G, Nogueira MA (2008) Changes in arbuscular mycorrhizal associations and fine root traits in sites under different plant successional phases in southern Brazil. Mycorrhiza 19:37–45

Chapter 23
Recent Advances in Cultivation of Edible Mycorrhizal Mushrooms

Yun Wang and Ying Long Chen

23.1 Introduction

Edible mushrooms are becoming more popular and important food on our table. There are hundreds of edible mushrooms available in the markets, of which most are saprophytic fungi, such as button mushroom (*Agaricus bisporus*), shiitake mushroom (*Lentinus edodes*) and oyster mushroom (*Pleurotus ostreatus*). However, the most expensive and sought-after edible mushrooms belong to the mycorrhizal group, including *Tuber melanosporum* Vitt., *T. magnatum* Pico & Vitt., *T. aestivum* Vitt. (*T. uncinatum*), the *T. indicum* complex, *Tricholoma matsutake* (S. Ito et Imai) Sing., *Boletus edulis* Bull: Fr. sensu lato, *Cantharellus cibarius* Fr., *Amanita caesarea* (Scop.: Fr.) Pers: Schw., *Lyophyllum shimeji* (Kawam.) Hongo, *Lactarius sanguifluus* (Paul) Fr. and *L. deliciosus* (L. Fr.) Gray (Hall et al. 1998a; Wang and Hall 2004; Hall and Zambonelli 2012). Edible mycorrhizal mushrooms (EMMs) comprise a specific group of fungal species belonging to either the Basidiomycetes or Ascomycetes, which form symbiotic associations with their host plants (Smith and Read 2008; Hall and Zambonelli 2012). EMMs are not only gourmet food but they are also a source of livelihood in many countries (Boa 2001; Molina 1998; Román and Boa 2006; Wang and Hall 2004). *Tricholoma matsutake* in Japan and truffles such as *Tuber melanosporum* and *T. magnatum* in

Y. Wang (✉)
Yunnan Institute for Tropical Crop Research, Jinghong, Sipsongpanna 666100, China
e-mail: wangy@crop.cri.nz

Y.L. Chen
State Key Laboratory of Soil Erosion and Dryland Farming on the Loess Plateau, Research Center of Soil and Water Conservation and Eco-environment, Chinese Academy of Sciences and Ministry of Education, Yangling 712100, China

School of Earth and Environment, The University of Western Australia, Crawley, WA 6009, Australia

© Springer-Verlag Berlin Heidelberg 2014   375
Z.M. Solaiman et al. (eds.), *Mycorrhizal Fungi: Use in Sustainable Agriculture and Land Restoration*, Soil Biology 41, DOI 10.1007/978-3-662-45370-4_23

Fig. 23.1 Cultivation of Basidiomycete edible ectomycorrhizal mushrooms: Tricholoma, Lactarius, and Rhizopogon. (**a**) A productive plantation of *Tuber melanosporum* at Charente, France, 2011; (**b**) A plantation of *Tuber melanosporum* at West Australia (Photo provided by Nick Malajczuk); (**c**) A truffle plantation at Guizhou Province, China with production of *Tuber melanosporum* and *T. indicum* since 2008; (**d**) A big ascocarp produced from a plantation of *Tuber borchii* at Cantherbury, New Zealand, 2010; (**e**) A small trial of cultivation of *Rhizopogon roseollus* at Cantherbury, New Zealand, 2011; (**f**) Mushrooms produced from an experimental plantation of 4-year old *Pinus radiate* with *Lactarius deliciosus* at Cantherbury, New Zealand, 2011

France and Italy are also an important part of the culture (Chen 2004a; Hall et al. 2007; Ogawa 1978; Renowden 2005; Sourzat 2009; Trappe 1990).

Unlike saprophytic edible mushrooms, the market of EMMs is supplied almost solely from what can be harvested from natural forests. Unfortunately, harvests of many wild forest mushrooms have declined over the past century, due to worldwide environmental changes caused by various natural and social factors (Wang and Hall 2004; Wang and Liu 2011). The falls in the availability of EMMs and increased demand have encouraged scientific research into developing technologies for the cultivation of EMMs and for sustainable mushroom production in forests. A few species of truffles have been produced in commercial quantities, although methods have been developed for many years. Despite numerous scientific publications and the establishment of thousands of hectares of plantations, the downward trend in EMM production continues (Bencivenga et al. 2009; Hall et al. 2003; Hall and Zambonelli 2012; Hosford et al. 1997; Pilz et al. 2003; Reyna et al. 2002; Sourzat 2009; Wang and Liu 2011). Many of the most expensive mycorrhizal mushrooms, including *Tuber magnatum* and *Tricholoma matsutake*, have defied cultivation.

In recent decades, cultivation of EMMs has made good progress. More new EMM species can be cultivated and more plantations of EMMs have been established in different countries (Fig. 23.1) (Savoie and Largeteau 2011; Hall and Zambonelli 2012). In particular, research on genetics and sexuality of truffles has made good progress (Rubini et al. 2010). The genome of *Tuber melanosporum* has recently been sequenced which provides better understanding of truffle fructification (Martin et al. 2010). Furthermore, advanced molecular tools have been developed and are being used to identify truffle species and their mycorrhizal symbioses and microorganism compositions in truffières. Surely, this new achievement would make cultivation of truffles and other EMMs more successful.

23.2 Cultivation Progress

23.2.1 Hypogeous EMMs

23.2.1.1 Truffles

Truffles are the fungi in the genus *Tuber* (Ascomycetes), which form below-ground ascocarps. Some truffle species are highly prized culinary commodities. Eight *Tuber* species, namely, *T. melanosporum*, *T. magnatum*, *T. aestivum* (=*T. uncinatum*), *T. indicum* species complex, *T. macrosporum*, *T. mesentericum*, *T. borchii* and *T. brumale*, are sold on international, mainly European markets and *T. gibbosum* and *T. oregonense* on North American markets. Desert truffles, species in the genera *Terfezia* and *Tirmania*, are popular delicacies in Arabic countries. These commercial truffle mushrooms are mainly harvested from natural forests. Only *Tuber melanosporum* and *T. aestivum* (=*T. uncinatum*) have been cultivated on large commercial scales together with a small proportion of production of cultivated species of *Tuber borchii, T. brumale*, the *T. indicum* complex and the desert truffles (Bencivenga et al. 2009; Chevalier 2009; Hall and Zambonelli 2012; Sourzat 2009; Wang and Hall 2004; Wang and Liu 2011).

Attempts to grow truffles began in the eighteenth century in France and in Italy (Bencivenga et al. 2009; Chevalier 1998; Pierre 2009). The seedlings were raised around black truffle (*T. melanosporum*) trees and transplanted to new areas. However, it was not until the early 1970s that methods of producing truffle mycorrhised seedlings of suitable host plants with truffle spores were developed and in 1978 that truffières yielded first truffles (Chevalier 1998; Bencivenga et al. 2009; Olivier 2000; Sourzat 2009). Since then the technique has been extensively used and hundreds of truffières have been established in European countries. The largest black truffle plantation established in Spain covers 600 ha. The production of truffles from truffières varies: on average, 2–50 kg ha^{-1} without irrigation and up to 150 kg ha^{-1} with irrigation (Chevalier 1998; Reyna et al. 2002; Wang and Hall 2002; Zambonelli, pers. comm.). However, in France, a truffière is considered successful if 10 years after planting, 50 % of the trees produce truffles with yields reaching 15–20 kg ha^{-1} (Chevalier 1998). Similar technology has been used to successfully grow other truffles species, e.g. *T. aestivum* (Chevalier and Frochot 1997; Chevalier and Frochot 1990; Wehrlen et al. 2009) in France, Italy (Bencivenga et al. 2009) and Sweden (Wedén 2004), *T. borchii* (Zambonelli et al. 2002) in Italy and New Zealand (Guerin-Laguette et al. 2009), and desert truffles, such as *Terfezia claveryi*, in Spain (Honrubia, pers. comm.).

Cultivation of truffles has also been successful in several countries outside Europe. In the 1980s, a few Périgord black truffle truffières were established in the USA using inoculated truffle seedlings imported from France or produced by the US company, Agri-Truffle (Picart 1980). In 1991, Northern California and North Carolina welcomed their first harvest of Périgord black truffles (Garland 1999). Many plantations have been established across the USA since then, but less than ten of those plantations have began production with annual production of up to 40 kg. Other species including *T. aestivum* and *T. borchii* were introduced into the USA for cultivation, but there is no report on their production. Attempts on cultivation of several native edible species, including *T. gibbosum*, *T. oregonense, T. lyonii* and *Leucangium carthusianum*, have been unsuccessful (Bruhn et al. 2009; Lefevre 2008).

In 1993, the first harvest of Périgord black truffles in the southern hemisphere came from a truffière established in New Zealand in 1987. More than 100 truffières have been established in both the North and South islands of New Zealand since 1987. So far more than ten truffières have produced truffles. The latest one to become productive was established in 2002. All successful truffières are in the warmer parts of the country (Guerin-Laguette et al. 2009).

In Australia, the first black truffle truffière was established in Tasmania island in 1993 with the first harvest of truffles in 1999. Twenty-eight truffières have been established in the island, of which six truffières have produced truffles in a small amount since 2001. A total of 600 ha of truffle plantations have been established in Australia, including areas in New South Wales, Victoria and Western Australia

(Carol 2003; Cooper 2001; Malajczuk and Amaranthus 2007). According to Graham Duell (pers. comm.), the production of truffles of *T. melanosporum* in Australia is 4,900 kg. It is noted that a considerable proportion of truffles in truffières in both New Zealand and Australia have become rotten before they got matured. Unsuitable weather conditions could be the possible reasons for mushroom decay.

Culture techniques for *T. melanosporum* were introduced in Chile at around 2003 (Pérez et al. 2007; Ramírez et al. 2003; Santelices and Palfner 2010), and a few black truffle plantations were established on hazelnut (*Corylus avellana* L.) and produced a small quantity of truffles.

Israel has been reported to produce a small quantity of truffles in 2000 from one of the truffières established in 1993 (Kagan-Zur et al. 2002; Pinkas et al. 2000). There are no further reports on the productivity in Israel. Experimental black truffle plantations were established in Morocco in which truffles have been produced (Kahbar et al. 2008). In southern British Columbia, Canada, three experimental plantations of *T. melanosporum* and *T. aestivum* have been established with two sites limed to raise pH to 7.8 (Berch et al. 2009). In 2013, truffles were produced from the plantations (Berch S, pers. comm.)

In China, *Tuber indicum* and *T. aestivum* (*T. uncinatum*) have been collected and traded for centuries (Wang and Liu 2011), but it was not until the late 1980s that research on cultivation of truffles was initiated. The first successful cultivation of a *Tuber* species in China was *Tuber formosanum* which is closely related to *T. indicum* in a truffière established in 1989, producing truffles in 1996 (Huang et al. 2009). Since then, a few plantations have been established in Guizhou, Hunan, Sichuan and Yunnan provinces. They have no production yet except one report of harvest of *T. indicum* and *T. melanosporum* in Guizhou in 2008. Research on the cultivation of *T. melanosporum* and *T. aestivum* in China is in progress (Chen 2002; Wang and Liu 2011).

Experimental plots of *T. uncinatum* were established in Sweden and produced truffles in 2005 (Wedén and Danell 2008). In Finland, the first plantation of *T. aestivum* (=*T. uncinatum*) was established in 2002 with oak, hazel and *Tilia cordata*. The seedlings were protected during the cold winter and the mycorrhizas survived under −7 °C in 2007 (Shamekb and Leisola 2008). A few small plantations of *T. uncinatum* were also established in New Zealand and one of them produced truffles in 2007 (Guerin-Laguette et al. 2009). Research on cultivation of *T. uncinatum* is carried out in the USA, Germany, Austria, the UK, Slovenia, Slovakia, Serbia and Switzerland (Wehrlen et al. 2009).

Since the successful cultivation of *T. borchii* (bianchetto) in Italy in 1994, a few truffières have also been established with *T. borchii* mycorrhised trees of *Quercus robur*, *Corylus avellana*, *Pinus radiata*, *P. pinea* and *P. pinaster* in modified acid soils in New Zealand. The first bianchetto was produced from the demo plantation of the Plant and Food Research in New Zealand in 2006 and five *T. borchii* plantations are producing bianchetto truffles. The bianchetto truffles are well accepted by the New Zealand markets.

23.2.1.2 Desert Truffles

Desert truffles are species from the genus of *Terfezia*, *Tirmania* and *Leucangium* which are used as food in Africa and the Middle East (Trappe 1990). They can form ectomycorrhizas, endomycorrhizas and ectendomycorrhizas with *Helianthemum* and other members of Cistaceae under various conditions (Awameh 1981; Fortas and Chevalier 1992; Kagan-Zur 1998; Honrubia et al. 2002). Mycorrhizas of desert truffles have been produced in semi-axenic culture and in vitro (Awameh 1981; Fortas and Chevalier 1992; Kagan-Zur 1998; Honrubia et al. 2002; Morte and Honrubia 1995). Plantations were established in Spain in 2000, and desert truffles were harvested 2 years later. Yields of desert truffles from natural bushes typically range from 50 to 170 kg ha^{-1} annually in Spain. However, irrigated truffières (e.g. 90 L m^{-2}) produced truffles as high as 300 kg ha^{-1} indicating a potentially high profitable industry in semi-desert areas of warm countries (Honrubia, pers. comm.). Unfortunately, little progress in cultivation of desert truffles has been made since 2002.

23.2.1.3 Shoro

The shoro (*Rhizopogon roseolus* Corda) as delicacy has been harvested and eaten for many years in Japan and recorded in ancient fungal books (Wang et al. 2008). Shoro is also collected and traded in China. Shoro was the fourth most commonly consumable mushroom in Japan 200 years ago (Okumura 1989). However, production of shoro has declined since the nineteenth century and hence attempts for cultivation of shoro commenced in the 1980s. In the Shimane and Kyoto Prefecture, fruiting bodies were produced from infected seedlings in 1988 and 1991, respectively (Iwase, pers. comm.) A few plantations have been established in New Zealand using pine seedlings mycorrhised with spores from a shoro species that was accidentally introduced to this country with European settlers. Since 1999 all plantations have produced mushrooms (Wang and Hall 2002). Recently, a group of New Zealand scientists from Plant and Food Research used multiplex PCR to analysis phylogeographic variation among collections in the *Rhizopogon* subgenus *Roseoli* and showed that the shoro species which was reported as *Rhizopogon rubescens* and commonly found in New Zealand is different from the Japanese shoro species, *R. roseolus* (Visnovsky et al. 2010). New Zealand shoro species is more closely related to the American collections in the subgenus. However, data were insufficient to determine whether the genetic differences observed between the two types of shoro were of significance at the species or subspecies levels. The Japanese isolates of shoro have since been introduced into New Zealand for producing mycorrhizal seedlings. A small experimental trial established in 2007 has produced fruiting bodies since 2009 (Visnovsky et al. 2010).

23.2.2 Epigeous EMMs

Compared to truffles, cultivation of epigeous EMMs is more difficult and has been much less successful. So far, only saffron milk cap (*Lactarius deliciosus*) in France (Poitou et al. 1984) and then in New Zealand (Wang and Hall 2002), *Lactarius hatsutake* in China (Tan et al. 2008) and honshimeji (*Lyophyllum shimeji*) in Japan (Yamanaka 2008) have been successfully cultivated (Wang et al. 2012).

23.2.2.1 Saffron Milk Cap

Poitou et al. (1984) pioneered the cultivation of *Lactarius deliciosus*. They produced fruiting bodies in the field from outplanted mycorrhizal seedlings of *Pinus pinaster*. After a silent period for the development of commercial cultivation of saffron milk cap, a New Zealand *Pinus radiata* plantation produced the first saffron milk cap fruiting body in 2002, 18 months after outplanting (Wang and Hall 2002). Presently hundreds of hectares of saffron milk cap plantations have been established in New Zealand and nearly all of them are producing fruiting bodies every year. Recently, New Zealand mycologists have found out that the production of saffron milk cap is significantly related to initial mycorrhizal level and to plantation management including irrigation and mulching (Wang et al. 2011; Wang et al. 2012). Seedlings of *Pinus massoniana* mycorrhised with *L. hatsutake* in a nursery in Hunan, China, have produced fruiting bodies in plantations since 2001 (3–4 years after inoculation) with an average yearly production of 670 kg ha^{-1} (Tan et al. 2008). The fruiting bodies of several *Lactarius* species were also obtained in open pot containers under growth chamber conditions, i.e. *L. deliciosus* on *P. sylvestris* seedlings (Guerin-Laguette et al. 2000a) and *L. akahatsu* on *P. densiflora* seedlings (Yamada et al. 2001). *Lactarius sanguifluus* mycorrhised plants have been produced in vitro (González-Ochoa et al. 2003), but infections failed to develop after outplanting.

23.2.2.2 Honshimeji

Honshimeji (*Lyophyllum shimeji*) is a delicacy in Japan and China, which is equally famous as matsutake in some regions in Japan. In Nara and Kyoto Prefecture, Japan, fruiting bodies of shimeji were produced from artificial mycorrhised seedlings in 1998 and 1996, respectively (Iwase, pers. comm.). Fruiting bodies of *Lyophyllum shimeji* have been also produced from inoculated seedlings growing in open pots in a greenhouse and pure cultures (Kawai 1997; Ohta 1994, 1998; Yoshida and Fujimoto 1994). Research on *L. shimeji* in New Zealand has made good progress. Seedlings of *Pinus radiata*, *P. densiflora* and *Picea alba* formed mycorrhizas with shimeji isolates from Japan and China.

23.2.2.3 Chanterelle

The chanterelle (*Cantharellus cibarius*) and related species are one of the most popular EMMs. Their fruiting bodies are found to have bacteria living inside which makes pure culture of their mycelia very difficult. Danell (2002) successfully obtained pure cultures from fruiting bodies of *C. cibarius* and produced mycorrhizal seedlings. The growth and establishment of mycorrhizal formation by *C. cibarius* depended on the co-bacteria (Danell 1994). Fruiting bodies of *C. cibarius* have formed on inoculated seedlings growing in open pots in a greenhouse (Danell 2002; Danell and Camacho 1997). However, experimental plots of *Cantharellus cibarius* in Sweden failed to produce fruiting bodies even though the mycorrhizal formation has spread onto new roots (Danell 2002).

23.2.2.4 Matsutake

Matsutake, pine mushroom (*Tricholoma matsutake*) and related species are the most appreciated EMM in Japan. There has been a high demand for this mushroom due to production decline since WWII. Cultivation of matsutake has become a hot spot in research in Japan and Korea, but no successful attempt is reported so far although *T. matsutake* mycorrhizal plants have been produced in vitro and mycorrhizal formation in situ (Hu 1994; González-Ochoa et al. 2003; Guerin-Laguette et al. 2000a; Wang et al. 1997; Yamada et al. 1999, 2006; Wang et al. 2012). Recently, South Korean mycologists transplanted 150 matsutake mycorrhised seedlings of *Pinus densiflora* into a non-matsutake-produced *P. densiflora* forest at Hongcheon, Korea. The matsutake seedlings were produced by planting the pine seedlings in mesh pots near the front shiro. The matsutake mycelia started to grow into neighbouring soils 1 year after planting (Ka, pers. comm. 2008). The Mountain Environmental Research Institute, Gyeongsangbuk, South Korea, patented their methods of producing matsutake mycorrhised pine seedlings in 2007 (Park et al. 2007, patent No.: US7269923). Under sterile conditions 15-day-old seedlings of *Pinus densiflora* were inoculated with matsutake mycelia and incubated at a clean room. The 2-month-old mycorrhised seedlings were transplanted into a greenhouse for 4 years before outplanting to exiting *P. densiflora* forests. More than 40,000 seedlings have been outplanted but there is no report on its progress.

With respect to the growth of matsutake in liquid medium, Kawagishi et al. (2004) found that the addition of D-isoleucine to the culture medium of matsutake significantly enhanced mycelia growth. Kim et al. (2010) investigated the optimal medium composition of liquid culture with the goal of shortening the culture period and to maximise polysaccharide production and mycelial growth of matsutake. The experimental results showed that the optimal medium contained 40 g L^{-1} glucose, 30 g L^{-1} yeast extract, 1.5 g L^{-1} KH_2PO_4 and 1 g L^{-1} $MgSO_4 \cdot 7H_2O$.

23.2.2.5 Porcini

Boletus edulis Bull.: Fr. sensu stricto, *B. aereus* Bull.: Fr, *B. aestivalis* Fr., *B. pinophilus* Pilát et Dermek and *B. reticulatus* Boud. are a group of allied porcini species that are often grouped together as *B. edulis* sensu lato. *B. edulis* sensu lato is one of the highly prized edible mushrooms, sold freshly or dried worldwide with approximately 20,000–100,000 tons consumption annually (Hall et al. 1998b). The commercialisation of *B. edulis* depends on the collection of fruiting bodies from natural forests.

Modifications of the techniques described by Molina and Palmer (1982) for other ectomycorrhizal fungi have been successful in producing laboratory-scale numbers of plants mycorrhised with *B. edulis* and other Boletaceae (Olivier et al. 1997; Wang et al. 1998; Zuccherelli 1988). However, mycorrhizas often collapse once infected plants are transferred into unsterile media. Although the germination of *B. edulis* spores can be enhanced with abietic acid (Fries et al. 1987), mycorrhisation failed to form even with the application rate of 10^7 spores per seedling (Guerin-Laguette et al. 2011). The New Zealand Plant and Food Research mycologists successfully produced porcini mycorrhised pine seedlings with Talon's method, which has been used to produce *T. melanosporum* mycorrhised seedlings. Several small experimental plantations have been established with mycorrhizal seedlings since 2007 (Wang and Guerin-Laguette, unpublished data). Future research is required for efficient production of large numbers of mycorrhised seedlings. Spanish mycologists produced porcini mycorrhised seedlings with *Cistus* species in vitro (Agueda et al. 2008) and produced fruiting bodies in *Cistus* plantation as early as 3 years after outplanting (Oria de Rueda et al. 2008). This discovery and achievement might be the new hope for cultivation of porcini.

23.2.2.6 Caesar's Mushroom

Amanita caesarea (Scop.: Fr.) personally known as Caesar's mushroom is another delicious EMM (Wang and Hall 2004). Daza et al. (2006) studied the effect of carbon and nitrogen sources, pH and temperature on in vitro culture of several isolates of *Amanita caesarea* in association with *Quercus suber* and *Castanea sativa* in southwest of Spain. The growth condition was optimised to maximise production of mycelia (24–28 °C, pH 6–7, mannitol and glucose as carbon sources and ammonium as a source of nitrogen). The knowledge of in vitro growth requirements of *A. caesarea* is a first step towards inoculum production for nursery and field applications to increase the sporocarp production and ecological benefits to trees.

23.3   Advance in Cultivation Technology

23.3.1   Production of Mycorrhizal Seedlings

The major challenge to cultivate EMMs is producing stable mycorrhizal seedlings. The method used to produce truffle mycorrhizal seedlings pioneered by Joseph Talon and Francolini in the early 1800s (Bencivenga et al. 2009; Sourzat 2009) is still widely used presently. This method has been used to produce matsutake-infected pine seedlings in Japan (Iwase 1997), Korea (Lee 1981) and China (Wang et al. 1997) and porcini trees in New Zealand (Wang and Guerin-Laguette, unpublished data). However, spore inoculation remains the most popular method and has been used successfully to produce trees mycorrhised by many common truffle species with the exception of *T. magnatum*, *T gibbosum* and *T. oregonense*. This method has also been used to produce trees infected with *Rhizopogon* and desert truffle species (Honrubia et al. 2002; Visnovsky et al. 2010). Generally the starting point is the preparation of spore suspension, mycorrhizal-free seedlings and suitable substrate (Bencivenga et al. 2009; Hall et al. 2007; Sourzat 2009). The use of mycelium inoculum to produce mycorrhizal seedlings is a successful practice in forestry (Grove and Malajczuk 1994). The use of mycelium inoculum has been reported to be successful for mycorrhisation of pine seedlings with *Lactarius deliciosus* (Diáz et al. 2009; Guerin-Laguette et al. 2000a; Wang and Hall 2002), *L. hatsutake* (Tan et al. 2008) and *Suillus bovinus* (Chen et al. 2004). However, mycelium inoculum has been shown ineffective in producing mycorrhizal seedlings with other EMMs.

Rossi et al. (2007) pointed out that the fungal inocula preparation is the crucial point in the production of mycorrhizal seedlings. Diáz et al. (2009) studied on the production of saffron milk cap mycorrhised *Pinus halepensis* seedlings under nursery conditions. They concluded that (1) mycelial slurry at a dose of 10 mL plant^{-1} was efficient when compared with mycelia entrapped in alginated beads and solid inoculum, (2) mycorrhizal formation performed better in sphagnum peat substrate than the mixture of sphagnum peat and vermiculate and (3) addition of moderate N (35 mg plant^{-1}) and P (27 mg plant^{-1}) produced better mycorrhisation. Knowledge on mating type genes opens up the possibility of using mycelial inoculation technology to produce truffle mycorrhizal seedlings with compatible mating strains (Rubini et al. 2010). In addition, this technology provides the possibility to select the best genetic strains for improving the productivity or environmental adaptability of EMMs. Mycorrhiza helper bacteria (MHB) could be essential for EMM mycorrhisation as shown in a case of Danell and Camacho (1997). Kataoka et al. (2009) also confirmed that MHB strains of *Ralstonia basilensis* and *Bacillus subtilis* are fungal selective when tested with four ECM fungi including EMM fungi *Rhizopogon* sp. and *Suillus granulatus*. Savoie and Largeteau (2011) addressed that the MHB may help *Tuber magnatum* and other uncultivated EMM species to produce mycorrhizal trees. Appropriate selection of suitable host plant species is essential for successful mycorrhisation.

For the cultivation of Périgord black truffle (*T. melanosporum*), *Quercus pubescens* and *Q. ilex* are considered better candidates than hazels (*Corylus* spp.) in European counties. However, hazels performed better in EMM production than oaks in Australia and New Zealand. Furthermore, among the species of hazels, *Corylus colurna* was better than *C. avellana*. *T. uncinatum* (*T. aestivum*) has a wider range of host plants than the Périgord black truffle, including broad-leaf tree species such as oak, hazel and conifers such as *Pinus nigra*, *P. armandii* and *Cedrus atlantica*. Wang and Guerin-Laguette (unpublished data) have observed that *T. borchii* associated with pine species showed a better production of mycorrhizal seedlings with less contaminants than with broadleaved trees (hazels, oaks) in New Zealand environment. More research on the relative advantages of different host species or superior strains is needed. Is host plant cloning improving EMM mycorrhizal quality and sustainability? This respect has been less studied and needed further exploration. A recent report of Ortega-Martínez et al. (2010) demonstrates that tree age influences the speed of sporocarp growth of *Boletus edulis* and *Lactarius deliciosus* in a *Pinus sylvestris* stand.

The Asian black truffle *T. indicum* is considered to be a broad host spectrum, an ecological trait that may be important to its invasion ecology. *T. indicum* was found fruiting in a forest in Oregon, USA, and was invaded into a nearby truffle orchard in which only *T. melanosporum* was introduced via molecular authentication (Bonito et al. 2010). Asian *T. indicum* was also observed on the roots of several North American endemic trees, such as loblolly pine (*Pinus taeda*) and pecan (*Carya illinoinensis*).

The quality control of EMM seedlings is an important issue but hard to do both technically and legally. Reyna et al. (2002) discussed methods for root sampling and the measurement levels of infection suggesting 250–500 infected root tips equivalent to 10–25 % of a root system were acceptable. Bencivenga et al. (1995) believed that the infection rate of about 33 % was acceptable and contaminations should never be higher than 25 %. However, performance on sampling seedlings and root subsamples is problematic. Morphological identification of mycorrhizal roots of different *Tuber* species is sometimes difficult or impossible (Mabru et al. 2001). Molecular methods may provide more accurate for quality control in a large scale. This practice and the associated cost require further study.

23.3.2   Establishing and Managing Plantations

The term "mycosilviculture" has been recently used to refer to the development of a science-based production of EMMs as a new industrial crop (see Sect. 21.3.5). Soil properties and climatic characters are the most important factors for truffières establishing. The climatic condition is more decisive than the soil because soil conditions are more manageable than climatic conditions (Chevalier 1998; Olivier 2000; Hall et al. 2007; Sourzat 2009). For instance, adding lime to acidic soils to raise the pH can make it suitable for the cultivation of the Périgord black truffle

(Chevalier 1998; Hall et al. 2007; Olivier 2000). Sourzat (2009) generated the details of soil requirements for growing truffles. The physical soil characteristics are perhaps even more important because the Périgord black truffle requires good drainage to avoid the soil conditions favouring competitive ectomycorrhizal fungi (Chevalier 1998; Hall et al. 2007; Olivier 2000). Spanish mycologists recently showed that the content of active Ca in soil played an important role for black truffle mycorrhisation and production (Garcia-Montero et al. 2008). In general, warm and moist spring and fall, hot summer with rainstorms and mild winter without heavy frost are good for growing black truffles (Sourzat 2009). Too much rain or snow in late fall and early winter and less sunny hours in summer are not suitable for cultivation of black truffles.

Methods of management of truffières are variable, from intensive procedures termed the Pallier technique to minimal management—the Tanguy method (Bencivenga et al. 2009; Chevalier 1998; Sourzat 2009). The Pallier method, adapted from orchard management, includes soil tillage, irrigation, weed control and tree pruning. It is expensive, but sometimes gives good production in return. Sourzat (2009) described a new method, "the Tanguy method," as an alternative cultivation method for black truffles. This method is based on natural truffle grounds and was created by Marcel Tanguy in Le Périgord, France. The method manages trees growing slower and produces truffles later with high production of 1 kg per trees on average.

Irrigation has been proved to be important for the Périgord black truffle (Hall et al. 2007), burgundy truffles (Wehrlen et al. 2009), the desert truffle (Honrubia et al. 2002) and saffron milk cap (Wang and Hall 2002; Wang et al. 2012). Irrigation is, in particular, necessary in the first 2 years after establishment and during truffle formation period. But heavy irrigation encourages *T. brumale* development that will lead to the replacement of *T. melanosporum* by *T. brumale* (Bencivenga et al. 2009; Sourzat 2009). The soil physiology and biology (microflora, micro- and mesofauna) are probably particularly important, but our knowledge about them is very limited. Recent research revealed that irrigation is not only important but also closely related to mushroom production of saffron milk cap (Wang et al. 2012).

23.3.3 Molecular Technology

Recent advances in genetics and molecular biological techniques in EMMs have provided better understanding of biology for an enhanced production of EMM. Most significantly, the genome of *Tuber melanosporum* has recently been sequenced which provided information to identify genes involved in the reproductive processes of this truffle. It is known that *T. melanosporum* is heterothallic and requires two master genes for mating. Mating type-specific primer pairs were developed to screen asci and gleba to provide definitive evidence of the presence of the mating genes (Rubini et al. 2010). The comparison of genomes of *Tuber melanosporum* and *Laccaria bicolor* showed that their genetic predispositions for symbiosis evolved in different ways between Ascomycetes and Basidiomycetes (Martin et al. 2010). It might be a good idea to apply the technology of growing truffles to cultivate EMMs of Basidiomycetes (Savoie and Largeteau 2011).

Molecular methods, such as PCR with specific primers and multiplex PCR and sequencing, have potential in verifying inoculum and mycorrhizal trees. Molecular technologies also provide a new approach for understanding mycorrhizal associations and therefore have implications for cultivation and management of EMM plantations (Amicucci et al. 1998; Franken and Requena 2001; Martin 2001; Nehls et al. 2001; Visnovsky et al. 2010). Murata et al. (2008) found four genotypes and some sub-genotypes in Asian matsutake based on retroelement-based DNA markers. Using single-nucleotide polymorphic (SNP) DNA markers, Amend et al. (2009a) investigated isolation by distance patterns on eight populations of matsutake mushrooms within and between watersheds suggesting an important determinant of air-dispersed ectomycorrhizal species population structure in heterogeneous landscape. The nuclear-encoded large subunit ribosomal DNA of isolates in the genus *Amanita* including Caesar's mushroom was sequenced to explore phylogenetic relationships among collections (Drehmel et al. 1999). Visnovsky et al. (2010) investigated the phylogenetic relationships of *Rhizopogon roseolus* and other closely related fungi belonging to *Rhizopogon* subgenus *Roseoli* by sequencing the ITS1-5.8S-ITS2 region of rRNA. The designed multiplex PCR approach is used to track the establishment of ectomycorrhizal symbioses on *Pinus radiata* seedlings inoculated with commercially valuable *R. roseolus*. This diagnostic demonstrated the first fruiting of Japanese shoro cultivated on *P. radiata* in the southern hemisphere. The haplotype networking method was employed to assess intraspecific ITS rDNA diversity among Asian and North American *T. indicum* group B isolates (Bonito et al. 2010).

Great progress in molecular studies has been made in particular for the *Périgord* truffle in recent decades. Phylogenetic information such as single-based polymorphisms has been used to design species-specific primers for white (Mello et al. 1999) and black truffles (Mabru et al. 2001; Paolocci et al. 1999; Rubini et al. 1998). PCR-RFLP using a SNP on the mitochondrial LSU-rDNA is an easy method to differentiate *Tuber melanosporum* from other truffle species like *T. aestivum*, *T. brumale* or *T. indicum* (Mabru et al. 2004). The genetic differentiation among *T. melanosporum* populations, highlighted by Murat et al. (2004) and Riccioni et al. (2008), suggested that the characterisation of molecular markers to identify the regional origin of ascocarps is within reach (Murat and Martin 2008). In the fungal genome, several thousands of simple sequence repeat (SSR) motifs can be identified (Lim et al. 2004) given a large set of polymorphic markers. By analysing SSRs, Murat et al. (2011) claimed that *T. melanosporum* genomes is rich and highly polymorphic in SSRs. Out of the 139 isolates, 132 different multilocus genotypes were identified indicating high genotypic diversity (0.999). Furthermore, Tisserant et al. (2011) applied high-throughput Illumina RNA sequencing (RNA-Seq) to the transcriptome of *T. melanosporum* at different major developmental stages and identified a substantial number of novel transcripts, antisense transcripts, new exons, untranslated regions (UTRs), alternative upstream initiation codons and upstream open reading frames. These researches provide new molecular markers to analyse the natural populations of this truffle and to authenticate mycorrhised seedlings.

23.4 Management and Conservation of EMM Resources

Since most EMMs, including *Tuber magnatum* and *Tricholoma matsutake*, are not able to be cultivated, their products are all gathered from the natural forests. The last decade's environmental deterioration has become a worldwide problem due to varied reasons, causing a general decline of EMM production. And thus protection and management of EMM nature resources has become an urgent matter particularly in developing countries.

Most EEMs are found growing in remote mountainous regions where EEMs are important sources of food and revenue to local people. Under the pressure of hunger and cold, deforestation and overharvesting are common. Environmental deterioration has been the cause for EEM species to become endangered or to disappear from some areas. For example, commercial harvest of matsutake began in the late 1970s in northeast China and the late 1980s in southwest China. Production of matsutake in both regions has been dramatically declined since the time commercial harvest started. Matsutake production in northeast China dropped from several hundreds to less than 100 tonnes per year. The southwest China matsutake production followed the similar pattern of depletion. For example, the production of matsutake in Chuxiong Prefecture of Yunnan Province, southwest China, reduced from more than 1,000 tonnes in the 1980s to around 250 tonnes in 2005. Matsutake production in Diqing Prefecture, another productive area of Yunnan, dropped from 865 tonnes in 1999 to 469 tonnes in 2005 (Yang et al. 2008, 2009). Nevertheless, production of the Chinese black truffle (the *Tuber indicum* complex) in southwest China is decreasing due to environmental deterioration caused by large-scale commercial harvesting. The environmental conditions have been damaged so much by unrestricted plundering of these natural resources during large-scale commercial harvesting since the 1990s that the black truffle species disappeared or became endangered in many counties of Sichuan and Yunnan Province (Chen 2004b; Wang and Liu 2009). Similar situations of deforestation and overharvesting EMMs occurred in most developing countries, where the range of EMM species present and their market value are poorly understood (Wang and Hall 2004). Compared with developing countries, Canada, South Korea and the USA show better management of their matsutake forests, and a relatively sustainable production of matsutake is maintained. The decline of matsutake production in South Korea by 7 % annually since the middle of the 1980s is regarded as forest ageing rather than commercial harvesting. Tree ageing caused the similar problem in Japan (Wang and Hall 2004). Therefore, improved forest management and conservation of existing EEM environments is an urgent matter particularly in developing countries.

Savoie and Largeteau (2011) termed the management of forest stands for mushroom production as "mycosilviculture" which is becoming increasingly important in Europe (Honrubia et al. 2011; Rondet 2011). Egli and Ayer (2009) found that thinning old-growth forest induced a significant increase of fungal diversity and their production, especially EMMs. Pilz et al. (2006) found that the number and weight of *Cantharellus formosus* were significantly decreased by thinning in the first year after logging, but the differences disappeared within the following 6 years. In Japan, management techniques for maximising *T. matsutake* production in existing forests are developed. Practical measurements include reducing the thickness of the litter layer; tree thinning removing competing ectomycorrhizal fungal fruiting bodies; protecting *T. matsutake* forests from diseases, insects, birds and other animals; and inoculating the soil by spraying *T. matsutake* spores and/or retaining some of the mature fruiting bodies on the forest floor. Similar management technologies were also employed in South Korea and China (Wang and Hall 2004). Commercial harvesting of EMMs has removed the opportunity for new trees or roots to become mycorrhised by spores and damaged growth environment for fungal hyphae in the topsoils (Chen 2004b). In order to correct this problem, the method for inoculating existing mature trees was developed (Guerin-Laguette et al. 2000b; Wang and Hall 2004). The results showed that black truffle mycorrhizas were successfully established on the new roots. Spores to reinoculate existing trees have also been used in attempts to replace *Tuber brumale* with *T. uncinatum* in France (Frochot et al. 1999). Similar methods have been used successfully to inoculate pine trees with saffron milk cap spores in Spain (Marcos, pers. comm.). Liu and co-workers used spores of *Lactarius volemus* to inoculate existing mature pine trees in Yunnan, China, and observed increased production of the mushroom from the inoculated trees (Liu et al. 2007).

Basic research on fungal succession as vegetation successions and growth in natural forests and plantations has provided better understanding of the dynamics of diversity and productivity of EMMs (Savoie and Largeteau 2011; Wang and He 2004; Wang and Hall 2004). Molecular analysis of EMM populations revealed that the matsutake population structure is related to heterogeneous landscape and its dispersal strategy (Amend et al. 2009a, b). Ecological studies on EMMs, including mycorrhizal communities and relationship of fructification to climatic and soil conditions, have provided the scientific basis for management of EMM plantations (Savoie and Largeteau 2011; Wang and Hall 2004).

23.5   Conclusion

EMMs are not only a gourmet food but also significant sources of livelihood. The most expensive and sought-after edible mushrooms belong to this group, for example, *Tuber melanosporum*, *T. magnatum*, *Tricholoma matsutake*, *Boletus edulis*, *Cantharellus cibarius*, *Amanita caesarea*, *Lyophyllum shimeji* and *Lactarius deliciosus*. Over the past 100 years, natural production of many mycorrhizal mushrooms has declined dramatically. This has prompted interest and the need for developing appropriate methods for their cultivation. A few species of truffles, mostly in the genus *Tuber*, have been cultivated commercially. Techniques have been extended and developed for the cultivation of epigeous species, but few species, e.g. *Lactarius deliciosus* and *Lyophyllum shimeji*, are successful at a commercial scale. *Tuber magnatum* and *Tricholoma matsutake* and many valuable mycorrhizal mushrooms have defied cultivation. The last decade's environmental deterioration has become a worldwide problem due to varied reasons. Protection and management of EMM resources has become urgent matter, particularly in developing countries. Some new technologies have been developed for the management of EMM plantations, in order to maximise their production. Modem technologies involving the use of molecular approaches for truffle genome studies have provided better understanding of the biology and plant-fungus symbioses of EMMs. Cultivation and management of EMMs is in progress although cultivation on some mycorrhizal mushrooms remains challenging.

Acknowledgments The authors are grateful to colleagues around the world for providing data used in this chapter. Special thanks go to Mario Honrubia, Hung-Tao Hu, Koji Iwase, Chang-Duck Koo, Pei Gui Liu, Akira Ohta, Marcos Morcillo, Dave Pilz, Santiago Reyna, Zhuming Tan, Akiyoshi Yamada, Nick Malckzuck and Alessandra Zambonelli. We appreciate Alexis Gerin-Laguette for his special contribution to the text.

References

Agueda B, Parladé J, Femandez-Toiran L, Cisneros O, de Miguel AM, Modrego MP, Martinez-Pena F, Pera J (2008) Mycorrhizal synthesis between *Boletus edulis* species complex and rockroses (*Cistus* sp.). Mycorrhiza 18:443–449

Amend A, Garbelotto M, Fang Z, Keeley S (2009a) Isolation by landscape in populations of a prized edible mushroom *Tricholoma matsutake*. Conserv Genet 11:795–802

Amend A, Keeley S, Garbelotto M (2009b) Forest age correlates with fine-scale spatial structure of Matsutake mycorrhizas. Mycol Res 113:541–551

Amicucci A, Zambonelli A, Giomaro G, Potenza L, Stocchi V (1998) Identification of ectomycorrhizal fungi of the genus *Tuber* by species-specific ITS primers. Mol Ecol 7:273–277

Awameh MS (1981) The response of *Helianthemum salicifolium* and *H. ledifolium* to infection by the desert truffle *Terfezia boudieri*. In: Mushroom science XI. Proceedings of the 11th international scientific congress on the cultivation of edible fungi, Australia

Bencivenga M, Govi G, Granetii B, Palenzona M, Pacioni G, Tocci A, Zambonelli A (1995) Presentazione del metodo di valutazione delle piante micorrhizate con fungi del gen. Tuber bassato caractterzzazione morfologica delle miccorrize

Bencivenga M, Massimo DG, Domizia D (2009) The cultivation of truffles in Italy. Acta Bot Yunnanica (suppl XVI):21–28

Berch SM, Gamiet S, Haddow W, Wyne Q, Lestock-Kay D (2009) Towards producing Perodord black truffles (*Tuber melanosporum*) in southern British Columbia, Canada. Acta Bot Yunnanica (suppl XVI):37–38

Boa E (2001) How do local people make use of wild edible fungi? Personal narratives from Malawi. In: Hall IR, Wang Y, Danell E, Zambonelli A (eds) Edible mycorrhizal mushrooms and their cultivation. Proceedings of the second international conference on edible mycorrhizal mushrooms, Christchurch, New Zealand, 3–6 July 2002. New Zealand Institute for Crop and Food Research Limited, Christchurch

Bonito G, Trappe JM, Donovan S, Vilgalys R (2010) The Asian black truffle *Tuber indicum* can form ectomycorrhizas with North American host plants and complete its life cycle in non-native soils. Fungal Ecol 4:83–93

Bruhn JN, Pruett GE, Mihail JD (2009) Progress toward truffle cultivation in the central USA. Acta Bot Yunnanica (suppl XVI):52–54

Carol A (2003) It's no truffle to Pickles to find this treasure. The Australia, July 9 2000, p 5

Chen YL (2002) Mycorrhizal formation of *Tuber melanosporum*. Edible Fungi China 21(5):15–17

Chen YL (2004a) Song rong mushroom: good fortune and fertility from a fungus. In: López C, Shanley P (eds) Riches from the forest: food, spices, crafts and resins of Asia. CIFOR Publication, Bogor, pp 25–28

Chen YL (2004b) Song rong, a valuable mushroom from China: consumption, development and sustainability. In: Kusters K, Belcher B (eds) Forest products, livelihoods and conservation: case studies of NTFP systems, vol 1, Asia. CIFOR Publication, Bogor, pp 56–71

Chen YL, Dell B, Kang LH (2004) Cultivation of *Suillus bovinus* (Boletaceae) on *Pinus elliottii* in south China. In: Abstracts of IV Asian-Pacific mycological congress, Chiang Mai, Thailand, 14–19 Nov 2004

Chevalier G (1998) The truffle cultivation in France: assessment of the situation after 25 years of intensive use of mycorrhizal seedlings. In: Proceedings of the first international meeting on ecology, physiology and cultivation of edible mycorrhizal mushrooms, Uppsala, Sweden, July 1998. http://www.mykopat.slu.se/mycorrhiza/edible/proceed/chevalier.html

Chevalier G (2009) The truffle of Europe (*Tuber aestivum* Vitt.): ecology and possibility of cultivation. In: Abstract of the first conference on the European truffle *Tuber aestivum/uncinatum*, 6–8 Nov 2009. University of Vienna, Vienna

Chevalier G, Frochot H (1990) Ecologie et possibilités de culture en Europe de la truffe de Bourgogne (*Tuber uncinatum* Ch.). In: Proceedings of the second international congress on truffle, Spoleto, Italy, 24–27 Nov 1988. Communità Montana dei Monti Martani e del Serrano, Spoleto, pp 323–330

Chevalier G, Frochot H (1997) La truffe de Bourgogne, Tuber uncinatum Chatin. Pètrarque, Levallois-Perret

Cooper M (2001) Festive fungi. New Scientists 22/29 December, pp 65–67

Danell E (1994) Formation and growth of the ectomycorrhiza of Cantharellus cibarius. Mycorrhiza 5:89–97

Danell E (2002) Current research on chanterelle cultivation in Sweden. In: Hall IR, Wang Y, Danell E, Zambonelli A (eds) Edible mycorrhizal mushrooms and their cultivation. Proceedings of the second international conference on edible mycorrhizal mushrooms, Christchurch, New Zealand, 3–6 July 2002. New Zealand Institute for Crop and Food Research Limited, Christchurch

Danell E, Camacho FJ (1997) Successful cultivation of the golden chanterelle. Nature 385:303

Daza A, Manjón JL, Camacho M, Romero de la Osa L, Aguilar A, Santamaría C (2006) Effect of carbon and nitrogen sources, pH and temperature on *in vitro* culture of several isolates of *Amanita caesarea* (Scop.:Fr.) Pers. Mycorrhiza 16:133–136

Diáz G, Carrillo C, Honrubia M (2009) Production of *Pinus halepensis* inoculated with the edible fungus *Lactarius deliciosus* under nursery condition. New For 38:215–227

Drehmel D, Moncalvo JM, Vilgalys R (1999) Molecular phylogeny of *Amanita* based on large-subunit ribosomal DNA sequences: implications for taxonomy and character evolution. Mycologia 91:610–618

Egli S, Ayer F (2009) Thinning an old-growth forest increased diversity and productivity of mushrooms. Acta Bot Yunnanica (suppl XVI):62–68

Fortas Z, Chevalier G (1992) Effect des conditions de culture sur la mycorrhization de l'*Helianthemum guttatum* par tros espéces de ferfez des genres *Terfezia* et *Tirmania* d'Algérie. Can J Bot 70:2453–2460

Franken P, Requena N (2001) Analysis of gene expression in arbuscular mycorrhizas: new approach and challenges. New Phytol 150:517–523

Fries N, Serck-Hanssen K, Dimberg LH, Theander O (1987) Abietic acid, an activator of basidiospore germination in ectomycorrhizal species of the genus *Suillus* (Boletaceae). Exp Mycol 11:360–363

Frochot H, Chevalier G, Barbotin P, Beaucamp F, Grillot JJ, Menu CP (1999) Avances sur la culture de la truffe de bougogne. 5 Congres International Science et la Culture de la truffe. Aix en Provence, France

Garcia-Montero LG, Diaz P, Martin-Fernandez S, Casermeiro MA (2008) Soil factors that favour the production of *Tuber melanosporum* carpophores over other truffle species: a multivariate statistical approach. Acta Agric Scand B Soil Plant Sci 58:322–329

Garland F (1999) Growing *Tuber melanosporum* under adverse acid soil conditions in the Unite States of America. Science et Culture de la Truffe, Actes du Ve Congres International, 4 au 6 mars 1999. Aix-en-Provence, France

González-Ochoa AI, Heras J, Toras H, Sánchez-Gómez E (2003) Mycorrhization of *Pinus halepensis* Mill. and *Pinus pinaster* Aiton seedlings in two commercial nurseries. Ann For Sci 6:43–48

Grove TS, Malajczuk N (1994) The potential for management of ectomycorrhiza in forestry. In: Robson AD, Abbott LK, Malajczuk N (eds) Management of mycorrhizas in agriculture, horticulture and forestry. Kluwer Academic, Dordrect, pp 201–210

Guerin-Laguette A, Plassard C, Mousa C (2000a) Effects of experimental conditions on mycorrhizal relationships between *Pinus sylvestris* and *Lactarius deliciosus* and unpredicted fruitbody formation of the saffron milk cap under control soil-less conditions. Can J Microbiol 46:790–799

Guerin-Laguette A, Vaario LM, Gill WM, Lapeyrie F, Matsushita N, Suzuki K (2000b) Rapid *in vitro* ectomycorrhizal infection on *Pinus densiflora* roots by *Tricholoma matsutake*. Mycoscience 41:389–393

Guerin-Laguette A, Heson-Williams N, Parmener G, Strong G (2009) Field research and cultivation of truffles in New Zealand: an update. Acta Bot Yunnanica (suppl XVI):90–93

Guerin-Laguette A, Cummings N, Wang Y (2011) Overview of research on edible mycorrhizal fungi in New Zealand. In: Abstract of the sixth international workshop on edible mycorrhizal mushrooms, Rbat, Morocco, 6–10 April 2011

Hall IR, Zambonelli A (2012) Laying the foundation. In: Zambonelli A, Bonito GM (eds) Edible ectomycorrhizal mushrooms. Springer, Berlin, pp 3–16

Hall IR, Buchanan P, Wang Y, Cole T (1998a) Edible and poisonous mushrooms of the world: an introduction. New Zealand Institute for Crop and Food Research Limited, Christchurch

Hall IR, Lyon AJE, Wang Y, Sinclair L (1998b) Ectomycorrhizal fungi with edible fruiting bodies. 2. *Boletus edulis*. Econ Bot 51:44–56

Hall IR, Amcucci A, Wang Y (2003) Cultivation of edible ectomycorrhizal mushrooms. Trends Biotechnol 21:433–438

Hall IR, Brown GT, Zambonelli A (2007) Taming the truffle. Timber Press, Portland

Honrubia M, Guterrez A, Morte A (2002) Desert truffle plantation from south-east Spain. In: Hall IR, Wang Y, Danell E, Zambonelli A (eds) Edible mycorrhizal mushrooms and their cultivation. Proceedings of the second international conference on edible mycorrhizal mushrooms, Christchurch, New Zealand, 3–6 July 2002. New Zealand Institute for Crop and Food Research Limited, Christchurch

Honrubia M, Figueroa V, Morte A (2011) Micodes: applied mycology for environment conservation and rural development. A first step towards a balanced forest management. In: Abstract of the sixth international workshop on edible mycorrhizal mushrooms, Rbat, Morocco, 6–10 April 2011

Hosford D, Pilz D, Molina R, Amanranthus R (1997) Ecology and management of the commercially harvested American matsutake mushroom. USDA, Forest Service, PNW Research Station, Portland

Hu HT (1994) Study on the relationship between *Pinus taiwanensis* and *Tricholoma matsutake* (I) Semi-aseptic mycorrhizal synthesis of *Tricholoma matsutake* and *Pinus taiwanensis*. Q J Exp For Natl Taiwan Univ 8:47–54

Huang JY, Hu HT, Shen WC (2009) Phylogenetic study of two truffles, *Tuber formosanum* and *Tuber furfuraceum*, identified from Taiwan. FEMS Microbiol Lett 294:157–171

Iwase K (1997) Cultivation of mycorrhizal mushrooms. Food Rev Int 13(3):431–442

Kagan-Zur V (1998) Terfezias–a family of mycorrhizal edible mushrooms for arid zones. In: Proceedings of the first international meeting on ecology, physiology and cultivation of edible mycorrhizal mushrooms, Uppsala, Sweden, July 1998

Kagan-Zur V, Wenkart S, Mills D, Freeman S (2002) *Tuber melanosporum* research in Israel. In: Hall IR, Wang Y, Danell E, Zambonelli A (eds) Edible mycorrhizal mushrooms and their cultivation. Proceedings of the second international conference on edible mycorrhizal mushrooms, Christchurch, New Zealand, 3–6 July 2002. New Zealand Institute for Crop and Food Research Limited, Christchurch

Kahbar L, Vevalier G, Laqbaqbi A (2008) Truffe et trufficulture au Maroc:etat actuel des recherché et perspectives. In: Chevalier G (ed) La culture de La truffe dans Le monde. Actes du colloque, Brive-La-Gaillarde, France, 2 Février 2007. Le Causse Corrézien, pp 135–150

Kataoka R, Taniguchi T, Futai K (2009) Fungal selectivity of two mycorrhiza helper bacteria on five mycorrhizal fungi associated with *Pinus thunbergii*. World J Microbiol Biotechnol 25:1815–1819

Kawagishi H, Hamajima K, Takanami R, Nakamura T, Sato Y, Akiyama Y, Sano M, Tanaka O (2004) Growth promotion of mycelia of the matsutake mushroom *Tricholoma mats*utake by D-isoleucine. Biosci Biotechnol Biochem 68:2405–2407

Kawai M (1997) Artificial ectomycorrhizal formation on roots of air-layered *Pinus densiflora* saplings by inoculation with *Lyophyllum shimeji*. Mycologia 89:228–232

Kim SS, Lee JS, Cho JY, Kim YE, Honh EK (2010) Effects of C/N ratio and trace elements on mycelial growth and exo-polysaccharide production of *Tricholoma matsutake*. Biotechnol Bioproc Eng 15:293–298

Lee TS (1981) Ecological environments and yield production of matsutake in Korea. In: Report of symposium on matsutake cultivation method. Forestry Experimental Farm, Korea Forest Administration, Seoul, Korea, pp 36–44

Lefevre C (2008) Truffle research and cultivation in North America. In: Third Congresso internazionale di spoleto sul tartufo, spoleto, 25–28 Nov 2009, p 7

Lim S, Notley-McRobb L, Lim M, Carter DA (2004) A comparison of the nature and abundance of microsatellites in 14 fungal genomes. Fungal Genet Biol 41:1025–1036

Liu PG, Yu FQ, Wang XH, Chen J, Zheng HD, Tian XF (2007) Mycorrhizal edible fungi stimulate solutions and their preparation. China Patents, Patent No. CN1958784A

Mabru D, Dupre C, Duet JP, Leroy P, Ravel C, Richard JM, Medina B, Castroviejo M, Chevalier G (2001) Rapid molecular typing method for detection of Asiatic black truffle (*Tuber indicum*) in commercial products: fruiting bodies and mycorrhizal seedlings. Mycorrhiza 11:89–94

Mabru D, Douet JP, Mouton A, Dupré C, Ricard JM, Médina B, Castroviejo M, Chevalier G (2004) PCR-RFLP using a SNP on the mitochondrial LSU-rDNA as an easy method to differentiate *Tuber melanosporum* (*Périgord* truffle) and other truffle species in cans. Int J Food Microbiol 94:33–42

Malajczuk N, Amaranthus M (2007) Cultivation of *Tuber* species in Australia. In: Chevalier G (ed) La culture de La truffe dans Le monde. Actes du colloque, Brive-La-Gaillarde, France, 2 Février 2007. Le Causse Corrézien, pp 9–18

Martin F (2001) Frontiers in molecular mycorrhizal research—genes, loci, dots and spins. New Phytol 150:499–504

Martin F, Kohler A, Murat C, Balestrini R, Coutinho PM, Jaillon O, Montanini B, Morin E, Noel B, Percudani R, Porcel B, Rubini A, Amicucci A, Amselem J, Anthouard V, Arcioni S, Artiguenave F, Aury JM, Ballario P, Bolchi A, Brenna A, Brun A, Buée M, Cantarel B, Chevalier G, Couloux A, Da Silva C, Denoeud F, Duplessis S, Ghignone S, Hilselberger B,

Iotti M, Marçais B, Mello A, Miranda M, Pacioni G, Quesneville H, Riccioni C, Ruotolo R, Splivallo R, Stocchi V, Tisserant E, Viscomi AR, Zambonelli A, Zampieri E, Henrissat B, Lebrun MH, Paolocci F, Bonfante P, Ottonello S, Wincker P (2010) Périgord black truffle genome uncovers evolutionary origins and mechanisms of symbiosis. Nature 464:1033–1038

Mello A, Garnero L, Bonfante P (1999) Specific PCR-primers as a reliable tool for the detection of white truffles in mycorrhizal roots. New Phytol 141:511–516

Molina R (1998) Concluding remarks. In: Proceedings of the first international meeting on ecology, physiology and cultivation of edible mycorrhizal mushrooms, Uppsala, Sweden, July 1998

Molina R, Palmer JG (1982) Isolation, maintenance and pure culture manipulation of ectomycorrhizal fungi. In: Schenck NC (ed) Methods and principles of mycorrhizal research. American Phytopathological Society, Minnesota, pp 115–129

Morte MA, Honrubia M (1995) Improvement of mycorrhizal synthesis between micropropagated *Helianthemum almeriense* plantlets with *Terfezia claveryi* (desert truffle). In: Elliot TJ (ed) Science and cultivation of edible fungi. Balkema, Rotterdam, p 863

Murat C, Martin F (2008) Sex and truffles: first evidence of *Périgord* black truffle outcrosses. New Phytol 180:260–263

Murat C, Díez J, Luis P, Delaruelle C, Dupré C, Chevalier G, Bonfante P, Martin F (2004) Polymorphism at the ribosomal DNA ITS and its relation to postglacial re-colonization routes of the *Périgord* truffle *Tuber melanosporum*. New Phytol 164:401–411

Murat C, Riccioni C, Belfiori B, Cichocki N, Labbé J, Morin E, Tisserant E, Paolocci F, Rubini A, Martin F (2011) Distribution and localization of microsatellites in the *Périgord* black truffle genome and identification of new molecular markers. Fungal Genet Biol 48:592–601

Murata H, Babasaki K, Saegusa T, Takemoto K, Yamada A, Ohta A (2008) Traceability of Asian matsutake, specialty mushrooms produced by the ectomycorrhizal Basidiomycete *Tricholoma matsutake*, on the basis of retroelement-based DNA markers. Appl Environ Microbiol 74:2023–2031

Nehls U, Mikolajewski S, Magel E, Hamp R (2001) Carbohydrate metabolism in ectomycorrhizas: gene expression, monosaccharide transport and metabolic control. New Phytol 150:533–541

Ogawa M (1978) Biology of Matsutake mushroom. Tsukiji Shokan, Tokyo (In Japanese)

Ohta A (1994) Production of fruiting-bodies of a mycorrhizal fungus, *Lyophyllum shimeji*, in pure culture. Mycoscience 35:147–151

Ohta A (1998) Culture conditions for commercial production of *Lyophyllum shimeji*. Mycoscience 39:13–20

Okumura A (1989) Healthy mushroom dishes. Hoikusha, Osaka (In Japanese)

Olivier JM (2000) Progress in the cultivation of truffles. In: Griensven VLJLD (ed) Science and cultivation of edible fungi. Proceedings of the 15th international congress on the science and cultivation of edible fungi, Maastricht, The Netherlands, pp 937–942

Olivier JM, Guinberteau J, Rondet J, Mamoun M (1997) Vers l'inoculation contrôlée des cèpes et bolets comestibles? Rev For Fr XLIX:222–234

Oria de Rueda JA, Martin P, Olaizola J (2008) Bolete productivity of cistaceous scrublands in Northwestern Spain. Econ Bot 62:323–330

Ortega-Martínez P, Águeda B, Fernández-Toirán LM, Martínez-Peña F (2010) Tree age influences on the development of edible ectomycorrhizal fungi sporocarps in *Pinus sylvestris* stands. Mycorrhiza 21:65–70

Paolocci F, Capucci R, Arcioni S, Damiani F (1999) Birdsfoot trefoil, a model for studying the synthesis of condensed tannins. In: Groos GG, Heminghway RW, Yoshida T (eds) Plant polyphenols. 2. Chemistry, biology, pharmacology, ecology. Kluwer/Plenum, New York, pp 343–356

Park MC, Sim SJ, Cheon WJ (2007) Methods of preparing *Tricholoma matsutake*-infected young pine by culturing aseptic pine seedlings and *T. matsutake*. United States Patents, Patent No. 7.269.923 B2

Pérez F, Palfner G, Brunel N, Santelices R (2007) Synthesis and establishment of *Tuber melanosporum* Vitt. ectomycorrhizae on two *Nothofagus* species in Chile. Mycorrhiza 17:627–632

Picart F (1980) Truffle: the black diamond. Agri-Truffle, Santa Rosa

Pierre S (2009) The truffle and its cultivation in France. Plant Divers Resour 31:72–80

Pilz D, Norvell Denall E, Molina R (2003) Ecology and management of commercially harvested *Chanterelle* mushrooms. USDA Pacific Northwest Research Station, Portland

Pilz D, Molina R, Mayo J (2006) Effect of thinning young forests on *Chanterelle* mushroom production. J For 104(1):9–14

Pinkas Y, Maimon M, Shabi E, Elisha S, Shmulewic Y, Freeman S (2000) Inoculation, isolation and identification of *Tuber melanosporum* from old and new oak hosts in Israel. Mycol Res 104:472–477

Poitou N, Mamoun M, Ducamp M, Delmas M (1984) Aprés le bolet granuleux, le Lactaire délicieux obtenu en fructification au champ a partir de plants mycorhozés. PHM Rev Hortic 244:65–68

Ramírez R, Pérez F, Reyna S, Santelices R (2003) Introduction and cultivation of *Tuber melanosporum* Vitt. in Chile. In: Abstracts of third international workshop on edible mycorrhizal mushrooms, Victoria, 16–22 Aug 2003. University of Victoria, British Columbia, Canada

Renowden G (2005) The Truffles. Limestone Hills Publishing, Amberly

Reyna S, Rodriguez Barreal JA, Folch L, Pérez-Badía R, Domínguez A, Saiz-De-Omecana JA, Zazo (2002) Techniques for inoculating mature trees with *Tuber melanosporum* Vitt. in New Zealand. In: Hall IR, Wang Y, Danell E, Zambonelli A (eds) Edible mycorrhizal mushrooms and their cultivation. Proceedings of the second international conference on edible mycorrhizal mushrooms, Christchurch, New Zealand, 3–6 July 2002. New Zealand Institute for Crop and Food Research Limited, Christchurch

Riccioni C, Belfiori B, Rubini A, Passeri V, Arcioni S, Paolocci F (2008) *Tuber melanosporum* outcrosses: analysis of the genetic diversity within and among its natural populations under this new scenario. New Phytol 180:466–478

Román M, Boa E (2006) The marketing of *Lactarius deliciosus* in northern Spain. Econ Bot 60:284–290

Rondet J (2011) Methods et outils de sensibilisation a l'importance des champignons dens les ecosystemes forestiers. In: Abstract of the sixth international workshop on edible mycorrhizal mushrooms, Rbat, Morocco, 6–10 April 2001

Rossi MJ, Furigo A, Oliveira VL (2007) Inoculant production of ectomycorrhizal fungi by solid and submerged fermentations. Food Technol Biotechnol 45:277–286

Rubini A, Paolocci F, Granetti B, Arcioni S (1998) Single step molecular characterization of morphologically similar black truffle species. FEMS Microbiol Lett 164:7–12

Rubini A, Belfiori B, Riccioni C, Tisserant E, Arcioni S, Martin F, Paolocci F (2010) Isolation and characterization of MAT genes in the symbiotic ascomycete *Tuber melanosporum*. New Phytol 189:710–722

Santelices R, Palfner G (2010) Controlled rhizogenesis and mycorrhization of hazelnut (*Corylus avellana* L) cuttings with Black Truffle (*Tuber melanosporum* Vitt.). Chil J Agric Res 70:204–212

Savoie JM, Largeteau ML (2011) Production of edible mushrooms in forests: trends in development of a mycosilviculture. Appl Microbiol Biotechnol 89:971–997

Shamekb S, Leisola M (2008) Trufflle orchards in Finland. In: Tuber 2008 third Congresso internazionale di Spolrto sul tratufo, Chiostro Di San Nicoló, Spoleto, 25–28 Nov 2008

Smith S, Read D (2008) Mycorrhizal symbiosis, 3rd edn. Elsevier, New York

Sourzat P (2009) The truffle and its cultivation in France. Acta Bot Yunnanica (suppl XVI):72–80

Tan ZM, Danell E, Shen AR, Fu SC (2008) Successful cultivation of *Lactarius hatsutake*—an evaluation with molecular methods. Acta Edulis Fungi 15(3):85–88 (In Chinese)

Tisserant E, Silva CD, Kohler A, Morin E, Wincker P, Martin F (2011) Deep RNA sequencing improved the structural annotation of the *Tuber melanosporum* transcriptome. New Phytol 189:883–891

Trappe JM (1990) Use of truffles and false-truffles around the world. In: Bencivenga M, Granetti B (eds) Atti del secondo congresso internazionale sul tartufo, Spoleto, Italy 1988. Com Mont Dei Martini, Italy, pp 19–30

Visnovsky SB, Guerin-Laguette A, Wang Y, Pitman AR (2010) Traceability of marketable Japanese shoro in New Zealand: using multiplex PCR to exploit phylogeographic variation among taxa in the *Rhizopogon* subgenus Roseoli. Appl Environ Mircobiol 76:294–302

Wang Y, Hall RI (2002) The cultivation of *Lactarius deliciosus* and *Rhizopogon rubescens* (shoro) in New Zealand. In: Hall IR, Wang Y, Danell E, Zambonelli A (eds) Edible mycorrhizal mushrooms and their cultivation. Proceedings of the second international conference on edible mycorrhizal mushrooms, Christchurch, New Zealand, 3–6 July 2002. New Zealand Institute for Crop and Food Research Limited, Christchurch

Wang Y, Hall IR (2004) Edible mycorrhizal mushrooms: challenges and achievements. Can J Bot 82:1063–1073

Wang Y, He XY (2004) Matsutake. A collection of the forum on development of matsutake industry. China Yunnan Shagri-La Matsutake Festival, 28–30 July 2005, pp 132–139

Wang Y, Liu PG (2009) Achievements and challenges of research on Chinese truffles. Acta Bot Yunnanica (suppl XIV):1−9

Wang Y, Liu PG (2011) Truffle research and cultivation in China. In: Third Congresso Internazionale Di Spoleto Sul Tartufo, Spoleto, 25–28 Nov 2009, pp 258–263

Wang Y, Evans LA, Hall IR (1997) Ectomycorrhizal fungi with edible fruiting bodies. 1. *Tricholoma matsutake* and related species. Econ Bot 51:311–327

Wang Y, Zanbonelli A, Hall IR (1998) *Boletus edulis* in New Zealand. In: Ahonen-Jonnarth U, Danell E, Fransson P, Karen O, Lindahl B, Rangel I, Finlay R (eds) Second international conference on mycorrhizae, 5–10 July 1998, Uppsala, Sweden, Program and Abstracts, SLU, 182

Wang Y, Liu PG, Chen J, Hu HD (2008) China—a newly emerging truffle-producing nation. In: Chevalier G (ed) La culture de La truffe dans Le monde. Actes du colloque, Brive-La-Gaillarde, France, 2 Février 2007. Le Causse Corrézien, pp 35–44

Wang Y, Cummings C, Guerin-Laguette A (2011) Research on cultivation of saffron milk cap (*Lactarius deliciosus*) in New Zealand. In: Abstract of the sixth international workshop on edible mycorrhizal mushrooms, Rbat, Morocco, 6–10 April 2001

Wang Y, Commings N, Guerin-Laguette A (2012) Cultivation of Basidiomycete edible ectomycorrhizal mushrooms: *Tricholoma*, *Lactarius* and *Rhizopogon*. In: Zambonelli A, Bonito GM (eds) Edible ectomycorrhizal mushrooms. Springer, Berlin, pp 281–304

Wedén C (2004) Black truffles of Sweden. Systematics, population studies, ecology and cultivation of *T. aestivum* syn. *T. uncinatum*. PhD thesis, Uppsala University, Sweden

Wedén CH, Danell E (2008) Truffle cultivation in Sweden. In: Chevalier G (ed) La culture de La truffe dans Le monde. Actes du colloque, Brive-La-Gaillarde, France, 2 Février 2007. Le Causse Corrézien, pp 193–207

Wehrlen L, Chevalier G, Besancon G, Frochot H (2009) Truffle cultivation-forestry: a new strategy of produce the Burgundy truffle (*Tuber uncinatum* Chatin). Acta Bot Yunnanica (suppl XVI):97−99

Yamada A, Maeda K, Ohmasa M (1999) Ectomycorrhiza formation of *Tricholoma matsutake* isolates on seedlings of *Pinus densiflora in vitro*. Mycoscience 40:455–463

Yamada A, Ogura T, Ohmasa M (2001) Cultivation of mushrooms of edible ectomycorrhizal fungi associated with *Pinus densiflora* by *in vitro* mycorrhizal synthesis. II. Morphology of mycorrhizas in open-pot soil. Mycorrhiza 11:67–81

Yamada A, Maeda K, Kobayahsi H, Murata H (2006) Ectomycorrhizal symbiosis *in vitro* between *Tricholoma matsutake* and *Pinus densiflora* seedlings that resembles naturally occurring "shiro". Mycorrhiza 16:111–116

Yamanaka K (2008) Commercial cultivation of *Lyophyllum shimeji*. In: Lelley JI, Buswell JA (eds) Mushroom biology and mushroom products. Proceedings of the sixth international conference on mushroom biology and mushroom products, Bonn, Germany, 29 September–3 October 2008, pp 197–202

Yang XF, He J, Li C, Ma JZ, Yang YP, Xu JC (2008) Matsutake trade in Yunnan Province, China: an overview. Econ Bot 62:269–277

Yang XF, Wilkes A, Yang YP, Xu JC, Geslani CS, Yang XQ, Gao F, Yang JK, Robinson B (2009) Common and privatized conditions for wise management of matsutake mushrooms in northwest Yunnan Province, China. Ecol Soc 14(2):30

Yoshida H, Fujimoto S (1994) A trial cultivation of *Lyophyllum shimeji* on solid media. Trans Mycol Soc Jpn 35:192

Zambonelli A, Lotti M, Giomaro G, Hall I, Stocchi V (2002) *T. borchii* cultivation: an interesting perspective. In: Hall IR, Wang Y, Danell E, Zambonelli A (eds) Edible mycorrhizal mushrooms and their cultivation. Proceedings of the second international conference on edible mycorrhizal mushrooms, Christchurch, New Zealand, 3–6 July 2002. New Zealand Institute for Crop and Food Research Limited, Christchurch

Zuccherelli G (1988) Prime esperienze sulla produzione di piante forestali micorrizate con *Boletus edulis*. Monti e Boschi 3:11–14

Index

A

Abelmoschus moschatu, 178
Abies, 9, 334
Abrus, 342
Acacia, 326, 331, 334, 335, 343
    A. mangium, 333, 342, 345
    A. mearnsii, 333
    A. nilotica, 369
Acanthaceae, 177
Acaulospora, 3, 155, 157, 173, 242, 268
    A. appendicula, 175
    A. appendicula (HR0201), 55
    A. bireticulata, 175
    A. cavernata, 174
    A. colossica, 154
    A. delicata, 56
    A. delicate, 176
    A. denticulata, 174, 176
    A. denticulata (HR0406), 56
    A. denticulata (RA2106), 55
    A. excavate, 175
    A. foveata, 52
    A. foveata (HR0602), 55
    A. mellea, 365, 366
    A. nicolsonii, 175
    A. paniculata, 180
    A. rehmi, 175
    A. rehmii, 174
    A. rugosa, 56
    A. scrobiculata, 174, 175, 363
    A. spinosa, 104, 174, 176
    A. tuberculata, 125, 174
Acaulosporaceae, 155
Acer, 334
Acer buergerianum, 50

Adansonia digitata, 364
Adenostoma, 334
Aenic, 73
Aeroponic, 47, 72
Afzelia, 334
Agrobacterium
    A. rhizogenes, 136
Agroecosystems, 191
Agrostis, 269
    A. castellana, 269
    A. delicatula (bentgrass), 262, 269
    A. scabra, 207
Agrostis castellana (bentgrass or dryland
    browntop), 262
Albizia lebbeck, 365
Alfalfa, 54
Allium
    A. cepa, 82, 83
    A. fistulosum, 82–85
    A. fistulosum, 85
    A. macrostemon, 175
    A. porrum, 54, 82, 84, 192
    A. sativum, 82, 84, 85
    A. tuberosum, 82
Allocasuarina, 334
Alnus, 334
Aloe vera, 175
Alpinia galanga, 176
AM. *See* Arbuscular mycorrhizal (AM)
Amanita, 9, 244, 329, 334
    A. caesarea, 375, 383
    A. muscaria, 335
    A. phalloides, 339
Amaranthaceae, 46, 177
Amaranthus spinosus, 174, 177

© Springer-Verlag Berlin Heidelberg 2014
Z.M. Solaiman et al. (eds.), *Mycorrhizal Fungi: Use in Sustainable Agriculture and
Land Restoration*, Soil Biology 41, DOI 10.1007/978-3-662-45370-4

CPSIA information can be obtained at www.ICGtesting.com
Printed in the USA
LVOW02*1438040815

448796LV00002B/19/P

9 783662 453698